Graduate Texts in Mathematics 212

T0202830

Springer
New York
Berlin
Heidelberg
Barcelona
Hong Kong
London
Milan
Paris
Singapore
Tokyo

Graduate Texts in Mathematics

(continued after index)

Jiří Matoušek

Lectures on
Discrete Geometry

With 206 Illustrations

 Springer

Jiří Matoušek
Department of Applied Mathematics
Charles University
Malostranské nám. 25
118 00 Praha 1
Czech Republic
matousek@kam.mff.cuni.cz

Mathematics Subject Classification (2000): 52-01

Library of Congress Cataloging-in-Publication Data
Matoušek, Jiří.
    Lectures on discrete geometry / Jiří Matoušek.
        p. cm. — (Graduate texts in mathematics ; 212)
    Includes bibliographical references and index.
    ISBN 0-387-95373-6 (alk. paper) — ISBN 0-387-95374-4 (softcover : alk. paper)
        1. Convex geometry.   2. Combinatorial geometry.   I. Title.   II. Series.
QA639.5 .M37 2002
516—dc21   2001054915

Printed on acid-free paper.

Production managed by Michael Koy; manufacturing supervised by Jacqui Ashri.
Typesetting: Pages created by author using Springer TeX macro package.
Printed and bound by Sheridan Books, Inc., Ann Arbor, MI.
Printed in the United States of America.

9 8 7 6 5 4 3 2 1

ISBN 0-387-95373-6   SPIN 10854370   (hardcover)
ISBN 0-387-95374-4   SPIN 10854388   (softcover)

Springer-Verlag   New York Berlin Heidelberg
A member of BertelsmannSpringer Science+Business Media GmbH

Preface

The next several pages describe the goals and the main topics of this book.

Questions in discrete geometry typically involve finite sets of points, lines, circles, planes, or other simple geometric objects. For example, one can ask, what is the largest number of regions into which n lines can partition the plane, or what is the minimum possible number of distinct distances occurring among n points in the plane? (The former question is easy, the latter one is hard.) More complicated objects are investigated, too, such as convex polytopes or finite families of convex sets. The emphasis is on "combinatorial" properties: Which of the given objects intersect, or how many points are needed to intersect all of them, and so on.

Many questions in discrete geometry are very natural and worth studying for their own sake. Some of them, such as the structure of 3-dimensional convex polytopes, go back to the antiquity, and many of them are motivated by other areas of mathematics. To a working mathematician or computer scientist, contemporary discrete geometry offers results and techniques of great diversity, a useful enhancement of the "bag of tricks" for attacking problems in her or his field. My experience in this respect comes mainly from combinatorics and the design of efficient algorithms, where, as time progresses, more and more of the first-rate results are proved by methods drawn from seemingly distant areas of mathematics and where geometric methods are among the most prominent.

The development of *computational geometry* and of geometric methods in *combinatorial optimization* in the last 20–30 years has stimulated research in discrete geometry a great deal and contributed new problems and motivation. Parts of discrete geometry are indispensable as a foundation for any serious study of these fields. I personally became involved in discrete geometry while working on geometric algorithms, and the present book gradually grew out of lecture notes initially focused on computational geometry. (In the meantime, several books on computational geometry have appeared, and so I decided to concentrate on the nonalgorithmic part.)

In order to explain the path chosen in this book for exploring its subject, let me compare discrete geometry to an Alpine mountain range. Mountains can be explored by bus tours, by walking, by serious climbing, by playing in the local casino, and in many other ways. The book should provide safe trails to a few peaks and lookout points (key results from various subfields of discrete geometry). To some of them, convenient paths have been marked in the literature, but for others, where only climbers' routes exist in research papers, I tried to add some handrails, steps, and ropes at the critical places, in the form of intuitive explanations, pictures, and concrete and elementary proofs.[1] However, I do not know how to build cable cars in this landscape: Reaching the higher peaks, the results traditionally considered difficult, still needs substantial effort. I wish everyone a clear view of the beautiful ideas in the area, and I hope that the trails of this book will help some readers climb yet unconquered summits by their own research. (Here the shortcomings of the Alpine analogy become clear: The range of discrete geometry is infinite and no doubt, many discoveries lie ahead, while the Alps are a small spot on the all too finite Earth.)

This book is primarily an *introductory textbook*. It does not require any special background besides the usual undergraduate mathematics (linear algebra, calculus, and a little of combinatorics, graph theory, and probability). It should be accessible to early graduate students, although mastering the more advanced proofs probably needs some mathematical maturity. The first and main part of each section is intended for teaching in class. I have actually taught most of the material, mainly in an advanced course in Prague whose contents varied over the years, and a large part has also been presented by students, based on my writing, in lectures at special seminars (Spring Schools of Combinatorics). A *short summary* at the end of the book can be useful for reviewing the covered material.

The book can also serve as a collection of *surveys* in several narrower subfields of discrete geometry, where, as far as I know, no adequate recent treatment is available. The sections are accompanied by remarks and bibliographic notes. For well-established material, such as convex polytopes, these parts usually refer to the original sources, point to modern treatments and surveys, and present a sample of key results in the area. For the less well covered topics, I have aimed at surveying most of the important recent results. For some of them, proof outlines are provided, which should convey the main ideas and make it easy to fill in the details from the original source.

Topics. The material in the book can be divided into several groups:

- *Foundations* (Sections 1.1–1.3, 2.1, 5.1–5.4, 5.7, 6.1). Here truly basic things are covered, suitable for any introductory course: linear and affine subspaces, fundamentals of convex sets, Minkowski's theorem on lattice points in convex bodies, duality, and the first steps in convex polytopes, Voronoi diagrams, and hyperplane arrangements. The remaining sections of Chapters 1, 2, and 5 go a little further in these topics.

---

[1] I also wanted to invent fitting names for the important theorems, in order to make them easier to remember. Only few of these names are in standard usage.

- *Combinatorial complexity of geometric configurations* (Chapters 4, 6, 7, and 11). The problems studied here include line–point incidences, complexity of arrangements and lower envelopes, Davenport–Schinzel sequences, and the k-set problem. Powerful methods, mainly probabilistic, developed in this area are explained step by step on concrete nontrivial examples. Many of the questions were motivated by the analysis of algorithms in computational geometry.

- *Intersection patterns and transversals of convex sets.* Chapters 8–10 contain, among others, a proof of the celebrated (p,q)-theorem of Alon and Kleitman, including all the tools used in it. This theorem gives a sufficient condition guaranteeing that all sets in a given family of convex sets can be intersected by a bounded (small) number of points. Such results can be seen as far-reaching generalizations of the well-known Helly's theorem. Some of the finest pieces of the weaponry of contemporary discrete and computational geometry, such as the theory of the VC-dimension or the regularity lemma, appear in these chapters.

- *Geometric Ramsey theory* (Chapters 3 and 9). Ramsey-type theorems guarantee the existence of a certain "regular" subconfiguration in every sufficiently large configuration; in our case we deal with geometric objects. One of the historically first results here is the theorem of Erdős and Szekeres on convex independent subsets in every sufficiently large point set.

- *Polyhedral combinatorics and high-dimensional convexity* (Chapters 12–14). Two famous results are proved as a sample of polyhedral combinatorics, one in graph theory (the weak perfect graph conjecture) and one in theoretical computer science (on sorting with partial information). Then the behavior of convex bodies in high dimensions is explored; the highlights include a theorem on the volume of an N-vertex convex polytope in the unit ball (related to algorithmic hardness of volume approximation), measure concentration on the sphere, and Dvoretzky's theorem on almost-spherical sections of convex bodies.

- *Representing finite metric spaces by coordinates* (Chapter 15). Given an n-point metric space, we would like to visualize it or at least make it computationally more tractable by placing the points in a Euclidean space, in such a way that the Euclidean distances approximate the given distances in the finite metric space. We investigate the necessary error of such approximation. Such results are of great interest in several areas; for example, recently they have been used in approximation algorithms in combinatorial optimization (multicommodity flows, VLSI layout, and others).

These topics surely do not cover all of discrete geometry, which is a rather vague term anyway. The selection is (necessarily) subjective, and naturally I preferred areas that I knew better and/or had been working in. (Unfortunately, I have had no access to supernatural opinions on proofs as a more reliable guide.) Many interesting topics are neglected completely, such as the wide area of packing and covering, where very accessible treatments exist, or the celebrated negative solution by Kahn and Kalai of the Borsuk conjecture, which I consider sufficiently popularized by now. Many more chapters analogous to the fifteen of this book could be added, and each of the fifteen chapters could be expanded into a thick volume. But the extent of the book, as well as the time for its writing, are limited.

Exercises. The sections are complemented by exercises. The little framed numbers indicate their difficulty: ① is routine, ⑤ may need quite a bright idea. Some of the exercises used to be a part of homework assignments in my courses and the classification is based on some experience, but for others it is just an unreliable subjective guess. Some of the exercises, especially those conveying important results, are accompanied by hints given at the end of the book.

Additional results that did not fit into the main text are often included as exercises, which saves much space. However, this greatly enlarges the danger of making false claims, so the reader who wants to use such information may want to check it carefully.

Sources and further reading. A great inspiration for this book project and the source of much material was the book *Combinatorial Geometry* of Pach and Agarwal [PA95]. Too late did I become aware of the lecture notes by Ball [Bal97] on modern convex geometry; had I known these earlier I would probably have hesitated to write Chapters 13 and 14 on high-dimensional convexity, as I would not dare to compete with this masterpiece of mathematical exposition. Ziegler's book [Zie94] can be recommended for studying convex polytopes. Many other sources are mentioned in the notes in each chapter. For looking up information in discrete geometry, a good starting point can be one of the several handbooks pertaining to the area: *Handbook of Convex Geometry* [GW93], *Handbook of Discrete and Computational Geometry* [GO97], *Handbook of Computational Geometry* [SU00], and (to some extent) *Handbook of Combinatorics* [GGL95], with numerous valuable surveys. Many of the important new results in the field keep appearing in the journal *Discrete and Computational Geometry*.

Acknowledgments. For invaluable advice and/or very helpful comments on preliminary versions of this book I would like to thank Micha Sharir, Günter M. Ziegler, Yuri Rabinovich, Pankaj K. Agarwal, Pavel Valtr, Martin Klazar, Nati Linial, Günter Rote, János Pach, Keith Ball, Uli Wagner, Imre Bárány, Eli Goodman, György Elekes, Johannes Blömer, Eva Matoušková, Gil Kalai, Joram Lindenstrauss, Emo Welzl, Komei Fukuda, Rephael Wenger, Piotr Indyk, Sariel Har-Peled, Vojtěch Rödl, Géza Tóth, Károly Böröczky Jr., Radoš Radoičić, Helena Nyklová, Vojtěch Franěk, Jakub Šimek, Avner Magen, Gregor Baudis, and Andreas Marwinski (I apologize if I forgot someone; my notes are not perfect, not to speak of my memory). Their remarks and suggestions allowed me to improve the manuscript considerably and to eliminate many of the embarrassing mistakes. I thank David Kramer for a careful copy-editing and finding many more mistakes (as well as offering me a glimpse into the exotic realm of English punctuation). I also wish to thank everyone who participated in creating the friendly and supportive environments in which I have been working on the book.

Errors. If you find errors in the book, especially serious ones, I would appreciate it if you would let me know (email: `matousek@kam.mff.cuni.cz`). I plan to post a list of errors at `http://www.ms.mff.cuni.cz/~matousek`.

Prague, July 2001 *Jiří Matoušek*

Contents

Notation and Terminology

This section summarizes rather standard things, and it is mainly for reference. More special notions are introduced gradually throughout the book. In order to facilitate independent reading of various parts, some of the definitions are even repeated several times.

If X is a set, $|X|$ denotes the number of elements (cardinality) of X. If X is a *multiset*, in which some elements may be repeated, then $|X|$ counts each element with its multiplicity.

The very slowly growing function $\log^* x$ is defined by $\log^* x = 0$ for $x \le 1$ and $\log^* x = 1 + \log^*(\log_2 x)$ for $x > 1$.

For a real number x, $\lfloor x \rfloor$ denotes the largest integer less than or equal to x, and $\lceil x \rceil$ means the smallest integer greater than or equal to x. The boldface letters $\mathbf{R}$ and $\mathbf{Z}$ stand for the real numbers and for the integers, respectively, while $\mathbf{R}^d$ denotes the d-dimensional Euclidean space. For a point $x = (x_1, x_2, \ldots, x_d) \in \mathbf{R}^d$, $\|x\| = \sqrt{x_1^2 + x_2^2 + \cdots + x_d^2}$ is the Euclidean norm of x, and for $x, y \in \mathbf{R}^d$, $\langle x, y \rangle = x_1 y_1 + x_2 y_2 + \cdots + x_d y_d$ is the scalar product. Points of $\mathbf{R}^d$ are usually considered as column vectors.

The symbol $B(x, r)$ denotes the closed ball of radius r centered at x in some metric space (usually in $\mathbf{R}^d$ with the Euclidean distance), i.e., the set of all points with distance at most r from x. We write B^n for the unit ball $B(0, 1)$ in $\mathbf{R}^n$. The symbol ∂A denotes the boundary of a set $A \subseteq \mathbf{R}^d$, that is, the set of points at zero distance from both A and its complement.

For a measurable set $A \subseteq \mathbf{R}^d$, $\mathrm{vol}(A)$ is the d-dimensional Lebesgue measure of A (in most cases the usual volume).

Let f and g be real functions (of one or several variables). The notation $f = O(g)$ means that there exists a number C such that $|f| \le C|g|$ for all values of the variables. Normally, C should be an absolute constant, but if f and g depend on some parameter(s) that we explicitly declare to be fixed (such as the space dimension d), then C may depend on these parameters as well. The notation $f = \Omega(g)$ is equivalent to $g = O(f)$, $f(n) = o(g(n))$ to $\lim_{n \to \infty} (f(n)/g(n)) = 0$, and $f = \Theta(g)$ means that both $f = O(g)$ and $f = \Omega(g)$.

For a random variable X, the symbol $\mathbf{E}[X]$ denotes the expectation of X, and $\mathrm{Prob}[A]$ stands for the probability of an event A.

Graphs are considered simple and undirected in this book unless stated otherwise, so a graph G is a pair (V, E), where V is a set (the *vertex set*) and $E \subseteq \binom{V}{2}$ is the *edge set*. Here $\binom{V}{k}$ denotes the set of all k-element subsets of V. For a *multigraph*, the edges form a multiset, so two vertices can be connected by several edges. For a given (multi)graph G, we write $V(G)$ for the vertex set and $E(G)$ for the edge set. A *complete graph* has all possible edges; that is, it is of the form $\left(V, \binom{V}{2}\right)$. A complete graph on n vertices is denoted by K_n. A graph G is *bipartite* if the vertex set can be partitioned into two subsets V_1 and V_2, the *(color) classes*, in such a way that each edge connects a vertex of V_1 to a vertex of V_2. A graph $G' = (V', E')$ is a *subgraph* of a graph $G = (V, E)$ if $V' \subseteq V$ and $E' \subseteq E$. We also say that G *contains a copy* of H if there is a subgraph G' of G isomorphic to H, where G' and H are *isomorphic* if there is a bijective map $\varphi: V(G') \to V(H)$ such that $\{u, v\} \in E(G')$ if and only if $\{\varphi(u), \varphi(v)\} \in E(H)$ for all $u, v \in V(G')$. The *degree* of a vertex v in a graph G is the number of edges of G containing v. An *r-regular graph* has all degrees equal to r. Paths and cycles are graphs as in the following picture,

paths   cycles

and a path or cycle in G is a subgraph isomorphic to a path or cycle, respectively. A graph G is *connected* if every two vertices can be connected by a path in G.

We recall that a set $X \subseteq \mathbf{R}^d$ is *compact* if and only if it is closed and bounded, and that a continuous function $f: X \to \mathbf{R}$ defined on a compact X attains its minimum (there exists $x_0 \in X$ with $f(x_0) \leq f(x)$ for all $x \in X$).

The *Cauchy–Schwarz inequality* is perhaps best remembered in the form $\langle x, y \rangle \leq \|x\| \cdot \|y\|$ for all $x, y \in \mathbf{R}^n$.

A real function f defined on an interval $A \subseteq \mathbf{R}$ (or, more generally, on a convex set $A \subseteq \mathbf{R}^d$) is *convex* if $f(tx + (1-t)y) \leq tf(x) + (1-t)f(y)$ for all $x, y \in A$ and $t \in [0, 1]$. Geometrically, the graph of f on $[x, y]$ lies below the segment connecting the points $(x, f(x))$ and $(y, f(y))$. If the second derivative satisfies $f''(x) \geq 0$ for all x in an (open) interval $A \subseteq \mathbf{R}$, then f is convex on A. *Jensen's inequality* is a straightforward generalization of the definition of convexity: $f(t_1 x_1 + t_2 x_2 + \cdots + t_n x_n) \leq t_1 f(x_1) + t_2 f(x_2) + \cdots + t_n f(x_n)$ for all choices of nonnegative t_i summing to 1 and all $x_1, \ldots, x_n \in A$. Or in integral form, if μ is a probability measure on A and f is convex on A, we have $f\left(\int_A x \, d\mu(x)\right) \leq \int_A f(x) \, d\mu(x)$. In the language of probability theory, if X is a real random variable and $f: \mathbf{R} \to \mathbf{R}$ is convex, then $f(\mathbf{E}[X]) \leq \mathbf{E}[f(X)]$; for example, $(\mathbf{E}[X])^2 \leq \mathbf{E}[X^2]$.

1

Convexity

We begin with a review of basic geometric notions such as hyperplanes and affine subspaces in $\mathbf{R}^d$, and we spend some time by discussing the notion of general position. Then we consider fundamental properties of convex sets in $\mathbf{R}^d$, such as a theorem about the separation of disjoint convex sets by a hyperplane and Helly's theorem.

1.1 Linear and Affine Subspaces, General Position

Linear subspaces. Let $\mathbf{R}^d$ denote the d-dimensional Euclidean space. The points are d-tuples of real numbers, $x = (x_1, x_2, \ldots, x_d)$.

The space $\mathbf{R}^d$ is a vector space, and so we may speak of linear subspaces, linear dependence of points, linear span of a set, and so on. A linear subspace of $\mathbf{R}^d$ is a subset closed under addition of vectors and under multiplication by real numbers. What is the geometric meaning? For instance, the linear subspaces of $\mathbf{R}^2$ are the origin itself, all lines passing through the origin, and the whole of $\mathbf{R}^2$. In $\mathbf{R}^3$, we have the origin, all lines and planes passing through the origin, and $\mathbf{R}^3$.

Affine notions. An arbitrary line in $\mathbf{R}^2$, say, is *not* a linear subspace unless it passes through 0. General lines are what are called *affine subspaces*. An affine subspace of $\mathbf{R}^d$ has the form $x + L$, where $x \in \mathbf{R}^d$ is some vector and L is a linear subspace of $\mathbf{R}^d$. Having defined affine subspaces, the other "affine" notions can be constructed by imitating the "linear" notions.

What is the *affine hull* of a set $X \subseteq \mathbf{R}^d$? It is the intersection of all affine subspaces of $\mathbf{R}^d$ containing X. As is well known, the linear span of a set X can be described as the set of all linear combinations of points of X. What is an *affine combination* of points $a_1, a_2, \ldots, a_n \in \mathbf{R}^d$ that would play an analogous role? To see this, we translate the whole set by $-a_n$, so that a_n becomes the origin, we make a linear combination, and we translate back by

$+a_n$. This yields an expression of the form $\beta_1(a_1 - a_n) + \beta_2(a_2 - a_n) + \cdots + \beta_n(a_n - a_n) + a_n = \beta_1 a_1 + \beta_2 a_2 + \cdots + \beta_{n-1} a_{n-1} + (1 - \beta_1 - \beta_2 - \cdots - \beta_{n-1})a_n$, where $\beta_1, \ldots, \beta_n$ are arbitrary real numbers. Thus, an affine combination of points $a_1, \ldots, a_n \in \mathbf{R}^d$ is an expression of the form

$$\alpha_1 a_1 + \cdots + \alpha_n a_n, \text{ where } \alpha_1, \ldots, \alpha_n \in \mathbf{R} \text{ and } \alpha_1 + \cdots + \alpha_n = 1.$$

Then indeed, it is not hard to check that the affine hull of X is the set of all affine combinations of points of X.

The *affine dependence* of points $a_1, \ldots, a_n$ means that one of them can be written as an affine combination of the others. This is the same as the existence of real numbers $\alpha_1, \alpha_2, \ldots \alpha_n$, at least one of them nonzero, such that both

$$\alpha_1 a_1 + \alpha_2 a_2 + \cdots + \alpha_n a_n = 0 \text{ and } \alpha_1 + \alpha_2 + \cdots + \alpha_n = 0.$$

(Note the difference: In an affine *combination*, the α_i sum to 1, while in an affine *dependence*, they sum to 0.)

Affine dependence of $a_1, \ldots, a_n$ is equivalent to linear dependence of the $n-1$ vectors $a_1 - a_n, a_2 - a_n, \ldots, a_{n-1} - a_n$. Therefore, the maximum possible number of affinely independent points in $\mathbf{R}^d$ is $d+1$.

Another way of expressing affine dependence uses "lifting" one dimension higher. Let $b_i = (a_i, 1)$ be the vector in $\mathbf{R}^{d+1}$ obtained by appending a new coordinate equal to 1 to a_i; then $a_1, \ldots, a_n$ are affinely dependent if and only if $b_1, \ldots, b_n$ are linearly dependent. This correspondence of affine notions in $\mathbf{R}^d$ with linear notions in $\mathbf{R}^{d+1}$ is quite general. For example, if we identify $\mathbf{R}^2$ with the plane $x_3 = 1$ in $\mathbf{R}^3$ as in the picture,

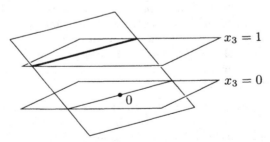

then we obtain a bijective correspondence of the k-dimensional linear subspaces of $\mathbf{R}^3$ that do not lie in the plane $x_3 = 0$ with $(k-1)$-dimensional affine subspaces of $\mathbf{R}^2$. The drawing shows a 2-dimensional linear subspace of $\mathbf{R}^3$ and the corresponding line in the plane $x_3 = 1$. (The same works for affine subspaces of $\mathbf{R}^d$ and linear subspaces of $\mathbf{R}^{d+1}$ not contained in the subspace $x_{d+1} = 0$.)

This correspondence also leads directly to extending the affine plane $\mathbf{R}^2$ into the *projective plane*: To the points of $\mathbf{R}^2$ corresponding to nonhorizontal lines through 0 in $\mathbf{R}^3$ we add points "at infinity," that correspond to horizontal lines through 0 in $\mathbf{R}^3$. But in this book we remain in the affine space most of the time, and we do not use the projective notions.

Let $a_1, a_2, \ldots, a_{d+1}$ be points in $\mathbf{R}^d$, and let A be the $d \times d$ matrix with $a_i - a_{d+1}$ as the ith column, $i = 1, 2, \ldots, d$. Then $a_1, \ldots, a_{d+1}$ are affinely independent if and only if A has d linearly independent columns, and this is equivalent to $\det(A) \neq 0$. We have a useful criterion of affine independence using a determinant.

Affine subspaces of $\mathbf{R}^d$ of certain dimensions have special names. A $(d-1)$-dimensional affine subspace of $\mathbf{R}^d$ is called a *hyperplane* (while the word *plane* usually means a 2-dimensional subspace of $\mathbf{R}^d$ for any d). One-dimensional subspaces are lines, and a k-dimensional affine subspace is often called a k-*flat*.

A hyperplane is usually specified by a single linear equation of the form $a_1 x_1 + a_2 x_2 + \cdots + a_d x_d = b$. We usually write the left-hand side as the scalar product $\langle a, x \rangle$. So a hyperplane can be expressed as the set $\{x \in \mathbf{R}^d : \langle a, x \rangle = b\}$ where $a \in \mathbf{R}^d \setminus \{0\}$ and $b \in \mathbf{R}$. A (closed) *half-space* in $\mathbf{R}^d$ is a set of the form $\{x \in \mathbf{R}^d : \langle a, x \rangle \geq b\}$ for some $a \in \mathbf{R}^d \setminus \{0\}$; the hyperplane $\{x \in \mathbf{R}^d : \langle a, x \rangle = b\}$ is its boundary.

General k-flats can be given either as intersections of hyperplanes or as affine images of $\mathbf{R}^k$ (parametric expression). In the first case, an intersection of k hyperplanes can also be viewed as a solution to a system $Ax = b$ of linear equations, where $x \in \mathbf{R}^d$ is regarded as a column vector, A is a $k \times d$ matrix, and $b \in \mathbf{R}^k$. (As a rule, in formulas involving matrices, we interpret points of $\mathbf{R}^d$ as *column* vectors.)

An *affine mapping* $f \colon \mathbf{R}^k \to \mathbf{R}^d$ has the form $f \colon y \mapsto By + c$ for some $d \times k$ matrix B and some $c \in \mathbf{R}^d$, so it is a composition of a linear map with a translation. The image of f is a k'-flat for some $k' \leq \min(k, d)$. This k' equals the rank of the matrix B.

General position. *"We assume that the points (lines, hyperplanes, . . .) are in general position."* This magical phrase appears in many proofs. Intuitively, general position means that no "unlikely coincidences" happen in the considered configuration. For example, if 3 points are chosen in the plane without any special intention, "randomly," they are unlikely to lie on a common line. For a planar point set in general position, we always require that no three of its points be collinear. For points in $\mathbf{R}^d$ in general position, we assume similarly that no unnecessary affine dependencies exist: No $k \leq d+1$ points lie in a common $(k-2)$-flat. For lines in the plane in general position, we postulate that no 3 lines have a common point and no 2 are parallel.

The precise meaning of general position is not fully standard: It may depend on the particular context, and to the usual conditions mentioned above we sometimes add others where convenient. For example, for a planar point set in general position we can also suppose that no two points have the same x-coordinate.

What conditions are suitable for including into a "general position" assumption? In other words, what can be considered as an unlikely coincidence? For example, let X be an n-point set in the plane, and let the coordinates of the ith point be (x_i, y_i). Then the vector $v(X) = (x_1, x_2, \ldots, x_n, y_1, y_2, \ldots, y_n)$ can be regarded as a point of $\mathbf{R}^{2n}$. For a configuration X in which $x_1 = x_2$, i.e., the first and second points have the same x-coordinate, the point $v(X)$ lies on the hyperplane $\{x_1 = x_2\}$ in $\mathbf{R}^{2n}$. The configurations X where *some* two points share the x-coordinate thus correspond to the union of $\binom{n}{2}$ hyperplanes in $\mathbf{R}^{2n}$. Since a hyperplane in $\mathbf{R}^{2n}$ has ($2n$-dimensional) measure zero, almost all points of $\mathbf{R}^{2n}$ correspond to planar configurations X with all the points having distinct x-coordinates. In particular, if X is any n-point planar configuration and $\varepsilon > 0$ is any given real number, then there is a configuration X', obtained from X by moving each point by distance at most ε, such that all points of X' have distinct x-coordinates. Not only that: Almost all small movements (*perturbations*) of X result in X' with this property.

This is the key property of general position: Configurations in general position lie arbitrarily close to any given configuration (and they abound in any small neighborhood of any given configuration). Here is a fairly general type of condition with this property. Suppose that a configuration X is specified by a vector $t = (t_1, t_2, \ldots, t_m)$ of m real numbers (coordinates). The objects of X can be points in $\mathbf{R}^d$, in which case $m = dn$ and the t_j are the coordinates of the points, but they can also be circles in the plane, with $m = 3n$ and the t_j expressing the center and the radius of each circle, and so on. The general position condition we can put on the configuration X is $p(t) = p(t_1, t_2, \ldots, t_m) \neq 0$, where p is some nonzero polynomial in m variables. Here we use the following well-known fact (a consequence of Sard's theorem; see, e.g., Bredon [Bre93], Appendix C): *For any nonzero m-variate polynomial $p(t_1, \ldots, t_m)$, the zero set $\{t \in \mathbf{R}^m : p(t) = 0\}$ has measure 0 in $\mathbf{R}^m$.*

Therefore, almost all configurations X satisfy $p(t) \neq 0$. So any condition that can be expressed as $p(t) \neq 0$ for a certain polynomial p in m real variables, or, more generally, as $p_1(t) \neq 0$ or $p_2(t) \neq 0$ or $\ldots$, for finitely or countably many polynomials $p_1, p_2, \ldots$, can be included in a general position assumption.

For example, let X be an n-point set in $\mathbf{R}^d$, and let us consider the condition "no $d+1$ points of X lie in a common hyperplane." In other words, no $d+1$ points should be affinely dependent. As we know, the affine dependence of $d+1$ points means that a suitable $d \times d$ determinant equals 0. This determinant is a polynomial (of degree d) in the coordinates of these $d+1$ points. Introducing one polynomial for every $(d+1)$-tuple of the points, we obtain $\binom{n}{d+1}$ polynomials such that at least one of them is 0 for any configuration X with $d+1$ points in a common hyperplane. Other usual conditions for general position can be expressed similarly.

In many proofs, assuming general position simplifies matters considerably. But what do we do with configurations X_0 that are not in general position? We have to argue, somehow, that if the statement being proved is valid for configurations X arbitrarily close to our X_0, then it must be valid for X_0 itself, too. Such proofs, usually called *perturbation arguments*, are often rather simple, and almost always somewhat boring. But sometimes they can be tricky, and one should not underestimate them, no matter how tempting this may be. A nontrivial example will be demonstrated in Section 5.5 (Lemma 5.5.4).

Exercises

1. Verify that the affine hull of a set $X \subseteq \mathbf{R}^d$ equals the set of all affine combinations of points of X. ②
2. Let A be a 2×3 matrix and let $b \in \mathbf{R}^2$. Interpret the solution of the system $Ax = b$ geometrically (in most cases, as an intersection of two planes) and discuss the possible cases in algebraic and geometric terms. ②
3. (a) What are the possible intersections of two (2-dimensional) planes in $\mathbf{R}^4$? What is the "typical" case (general position)? What about two hyperplanes in $\mathbf{R}^4$? ③
   (b) Objects in $\mathbf{R}^4$ can sometimes be "visualized" as objects in $\mathbf{R}^3$ moving in time (so time is interpreted as the fourth coordinate). Try to visualize the intersection of two planes in $\mathbf{R}^4$ discussed (a) in this way.

1.2 Convex Sets, Convex Combinations, Separation

Intuitively, a set is convex if its surface has no "dips":

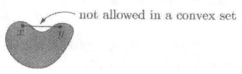

1.2.1 Definition (Convex set). *A set $C \subseteq \mathbf{R}^d$ is* convex *if for every two points $x, y \in C$ the whole segment xy is also contained in C. In other words, for every $t \in [0, 1]$, the point $tx + (1 - t)y$ belongs to C.*

The intersection of an arbitrary family of convex sets is obviously convex. So we can define the *convex hull* of a set $X \subseteq \mathbf{R}^d$, denoted by $\mathrm{conv}(X)$, as the intersection of all convex sets in $\mathbf{R}^d$ containing X. Here is a planar example with a finite X:

An alternative description of the convex hull can be given using convex combinations.

1.2.2 Claim. *A point x belongs to* $\mathrm{conv}(X)$ *if and only if there exist points $x_1, x_2, \ldots x_n \in X$ and nonnegative real numbers $t_1, t_2, \ldots, t_n$ with $\sum_{i=1}^n t_i = 1$ such that $x = \sum_{i=1}^n t_i x_i$.*

The expression $\sum_{i=1}^n t_i x_i$ as in the claim is called a *convex combination* of the points $x_1, x_2, \ldots, x_n$. (Compare this with the definitions of linear and affine combinations.)

Sketch of proof. Each convex combination of points of X must lie in $\mathrm{conv}(X)$: For $n = 2$ this is by definition, and for larger n by induction. Conversely, the set of all convex combinations obviously contains X, and it is convex. □

In $\mathbf{R}^d$, it is sufficient to consider convex combinations involving at most $d+1$ points:

1.2.3 Theorem (Carathéodory's theorem). *Let $X \subseteq \mathbf{R}^d$. Then each point of $\mathrm{conv}(X)$ is a convex combination of at most $d+1$ points of X.*

For example, in the plane, $\mathrm{conv}(X)$ is the union of all triangles with vertices at points of X. The proof of the theorem is left as an exercise to the subsequent section.

A basic result about convex sets is the separability of disjoint convex sets by a hyperplane.

1.2.4 Theorem (Separation theorem). *Let $C, D \subseteq \mathbf{R}^d$ be convex sets with $C \cap D = \emptyset$. Then there exists a hyperplane h such that C lies in one of the closed half-spaces determined by h, and D lies in the opposite closed half-space. In other words, there exist a unit vector $a \in \mathbf{R}^d$ and a number $b \in \mathbf{R}$ such that for all $x \in C$ we have $\langle a, x \rangle \geq b$, and for all $x \in D$ we have $\langle a, x \rangle \leq b$.*

If C and D are closed and at least one of them is bounded, they can be separated strictly; in such a way that $C \cap h = D \cap h = \emptyset$.

In particular, a closed convex set can be strictly separated from a point. This implies that the convex hull of a closed set X equals the intersection of all closed half-spaces containing X.

Sketch of proof. First assume that C and D are compact (i.e., closed and bounded). Then the Cartesian product $C \times D$ is a compact space, too, and the distance function $(x, y) \mapsto \|x - y\|$ attains its minimum on $C \times D$. That is, there exist points $p \in C$ and $q \in D$ such that the distance of C and D equals the distance of p and q.

The desired separating hyperplane h can be taken as the one perpendicular to the segment pq and passing through its midpoint:

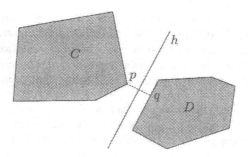

It is easy to check that h indeed avoids both C and D.

If D is compact and C closed, we can intersect C with a large ball and get a compact set C'. If the ball is sufficiently large, then C and C' have the same distance to D. So the distance of C and D is attained at some $p \in C'$ and $q \in D$, and we can use the previous argument.

For arbitrary disjoint convex sets C and D, we choose a sequence $C_1 \subseteq C_2 \subseteq C_3 \subseteq \cdots$ of compact convex subsets of C with $\bigcup_{n=1}^{\infty} C_n = C$. For example, assuming that $0 \in C$, we can let C_n be the intersection of the closure of $(1 - \frac{1}{n})C$ with the ball of radius n centered at 0. A similar sequence $D_1 \subseteq D_2 \subseteq \cdots$ is chosen for D, and we let $h_n = \{x \in \mathbf{R}^d : \langle a_n, x \rangle = b_n\}$ be a hyperplane separating C_n from D_n, where a_n is a unit vector and $b_n \in \mathbf{R}$. The sequence $(b_n)_{n=1}^{\infty}$ is bounded, and by compactness, the sequence of $(d+1)$-component vectors $(a_n, b_n) \in \mathbf{R}^{d+1}$ has a cluster point (a, b). One can verify, by contradiction, that the hyperplane $h = \{x \in \mathbf{R}^d : \langle a, x \rangle = b\}$ separates C and D (nonstrictly). $\qquad\square$

The importance of the separation theorem is documented by its presence in several branches of mathematics in various disguises. Its home territory is probably functional analysis, where it is formulated and proved for infinite-dimensional spaces; essentially it is the so-called Hahn–Banach theorem. The usual functional-analytic proof is different from the one we gave, and in a way it is more elegant and conceptual. The proof sketched above uses more special properties of $\mathbf{R}^d$, but it is quite short and intuitive in the case of compact C and D.

Connection to linear programming. A basic result in the theory of linear programming is the Farkas lemma. It is a special case of the duality of linear programming (discussed in Section 10.1) as well as the key step in its proof.

1.2.5 Lemma (Farkas lemma, one of many versions). *For every $d \times n$ real matrix A, exactly one of the following cases occurs:*

(i) *The system of linear equations $Ax = 0$ has a nontrivial nonnegative solution $x \in \mathbf{R}^n$ (all components of x are nonnegative and at least one of them is strictly positive).*

(ii) *There exists a $y \in \mathbf{R}^d$ such that $y^T A$ is a vector with all entries strictly
negative. Thus, if we multiply the jth equation in the system $Ax = 0$ by
y_j and add these equations together, we obtain an equation that obviously
has no nontrivial nonnegative solution, since all the coefficients on the
left-hand sides are strictly negative, while the right-hand side is 0.*

Proof. Let us see why this is yet another version of the separation theorem.
Let $V \subset \mathbf{R}^d$ be the set of n points given by the column vectors of the
matrix A. We distinguish two cases: Either $0 \in \mathrm{conv}(V)$ or $0 \notin \mathrm{conv}(V)$.

In the former case, we know that 0 is a convex combination of the points
of V, and the coefficients of this convex combination determine a nontrivial
nonnegative solution to $Ax = 0$.

In the latter case, there exists a hyperplane strictly separating V from 0,
i.e., a unit vector $y \in \mathbf{R}^d$ such that $\langle y, v \rangle < \langle y, 0 \rangle = 0$ for each $v \in V$. This is
just the y from the second alternative in the Farkas lemma.   □

Bibliography and remarks. Most of the material in this chapter is
quite old and can be found in many surveys and textbooks. Providing
historical accounts of such well-covered areas is not among the goals
of this book, and so we mention only a few references for the specific
results discussed in the text and add some remarks concerning related
results.

The concept of convexity and the rudiments of convex geometry
have been around since antiquity. The initial chapter of the *Handbook
of Convex Geometry* [GW93] succinctly describes the history, and the
handbook can be recommended as the basic source on questions related to convexity, although knowledge has progressed significantly
since its publication.

For an introduction to functional analysis, including the Hahn–
Banach theorem, see Rudin [Rud91], for example. The Farkas lemma
originated in [Far94] (nineteenth century!). More on the history of the
duality of linear programming can be found, e.g., in Schrijver's book
[Sch86].

As for the origins, generalizations, and applications of Carathéodory's theorem, as well as of Radon's lemma and Helly's theorem discussed in the subsequent sections, a recommendable survey is Eckhoff
[Eck93], and an older well-known source is Danzer, Grünbaum, and
Klee [DGK63].

Carathéodory's theorem comes from the paper [Car07], concerning
power series and harmonic analysis. A somewhat similar theorem, due
to Steinitz [Ste16], asserts that if x lies in the interior of $\mathrm{conv}(X)$
for an $X \subseteq \mathbf{R}^d$, then it also lies in the interior of $\mathrm{conv}(Y)$ for some
$Y \subseteq X$ with $|Y| \leq 2d$. Bonnice and Klee [BK63] proved a common
generalization of both these theorems: Any k-interior point of X is
a k-interior point of Y for some $Y \subseteq X$ with at most $\max(2k, d+1)$

points, where x is called a **k-interior point** of X if it lies in the relative interior of the convex hull of some $k+1$ affinely independent points of X.

Exercises

1. Give a detailed proof of Claim 1.2.2. ▢2
2. Write down a detailed proof of the separation theorem. ▢3
3. Find an example of two disjoint closed convex sets in the plane that are not strictly separable. ▢1
4. Let $f: \mathbf{R}^d \to \mathbf{R}^k$ be an affine map.
   (a) Prove that if $C \subseteq \mathbf{R}^d$ is convex, then $f(C)$ is convex as well. Is the preimage of a convex set always convex? ▢2
   (b) For $X \subseteq \mathbf{R}^d$ arbitrary, prove that $\mathrm{conv}(f(X)) = \mathrm{conv}(f(X))$. ▢1
5. Let $X \subseteq \mathbf{R}^d$. Prove that $\mathrm{diam}(\mathrm{conv}(X)) = \mathrm{diam}(X)$, where the diameter $\mathrm{diam}(Y)$ of a set Y is $\sup\{\|x - y\|: x, y \in Y\}$. ▢3
6. A set $C \subseteq \mathbf{R}^d$ is a *convex cone* if it is convex and for each $x \in C$, the ray $\overrightarrow{0x}$ is fully contained in C.
   (a) Analogously to the convex and affine hulls, define the appropriate "conic hull" and the corresponding notion of "combination" (analogous to the convex and affine combinations). ▢3
   (b) Let C be a convex cone in $\mathbf{R}^d$ and $b \notin C$ a point. Prove that there exists a vector a with $\langle a, x \rangle \geq 0$ for all $x \in C$ and $\langle a, b \rangle < 0$. ▢2
7. (Variations on the Farkas lemma) Let A be a $d \times n$ matrix and let $b \in \mathbf{R}^d$.
   (a) Prove that the system $Ax = b$ has a nonnegative solution $x \in \mathbf{R}^n$ if and only if every $y \in \mathbf{R}^d$ satisfying $y^T A \geq 0$ also satisfies $y^T b \geq 0$. ▢3
   (b) Prove that the system of inequalities $Ax \leq b$ has a nonnegative solution x if and only if every nonnegative $y \in \mathbf{R}^d$ with $y^T A \geq 0$ also satisfies $y^T b \geq 0$. ▢3
8. (a) Let $C \subset \mathbf{R}^d$ be a compact convex set with a nonempty interior, and let $p \in C$ be an interior point. Show that there exists a line ℓ passing through p such that the segment $\ell \cap C$ is at least as long as any segment parallel to ℓ and contained in C. ▢4
   (b) Show that (a) may fail for C compact but not convex. ▢1

1.3 Radon's Lemma and Helly's Theorem

Carathéodory's theorem from the previous section, together with Radon's lemma and Helly's theorem presented here, are three basic properties of convexity in $\mathbf{R}^d$ involving the dimension. We begin with Radon's lemma.

1.3.1 Theorem (Radon's lemma). *Let A be a set of $d+2$ points in $\mathbf{R}^d$. Then there exist two disjoint subsets $A_1, A_2 \subset A$ such that*

$$\mathrm{conv}(A_1) \cap \mathrm{conv}(A_2) \neq \emptyset.$$

A point $x \in \text{conv}(A_1) \cap \text{conv}(A_2)$, where A_1 and A_2 are as in the theorem, is called a *Radon point* of A, and the pair (A_1, A_2) is called a *Radon partition* of A (it is easily seen that we can require $A_1 \cup A_2 = A$).

Here are two possible cases in the plane:

Proof. Let $A = \{a_1, a_2, \ldots, a_{d+2}\}$. These $d+2$ points are necessarily affinely dependent. That is, there exist real numbers $\alpha_1, \ldots, \alpha_{d+2}$, not all of them 0, such that $\sum_{i=1}^{d+2} \alpha_i = 0$ and $\sum_{i=1}^{d+2} \alpha_i a_i = 0$.

Set $P = \{i \colon \alpha_i > 0\}$ and $N = \{i \colon \alpha_i < 0\}$. Both P and N are nonempty. We claim that P and N determine the desired subsets. Let us put $A_1 = \{a_i \colon i \in P\}$ and $A_2 = \{a_i \colon i \in N\}$. We are going to exhibit a point x that is contained in the convex hulls of both these sets.

Put $S = \sum_{i \in P} \alpha_i$; we also have $S = -\sum_{i \in N} \alpha_i$. Then we define

$$x = \sum_{i \in P} \frac{\alpha_i}{S} a_i . \tag{1.1}$$

Since $\sum_{i=1}^{d+2} \alpha_i a_i = 0 = \sum_{i \in P} \alpha_i a_i + \sum_{i \in N} \alpha_i a_i$, we also have

$$x = \sum_{i \in N} \frac{-\alpha_i}{S} a_i . \tag{1.2}$$

The coefficients of the a_i in (1.1) are nonnegative and sum to 1, so x is a convex combination of points of A_1. Similarly, (1.2) expresses x as a convex combination of points of A_2. □

Helly's theorem is one of the most famous results of a combinatorial nature about convex sets.

1.3.2 Theorem (Helly's theorem). *Let $C_1, C_2, \ldots, C_n$ be convex sets in $\mathbf{R}^d$, $n \geq d+1$. Suppose that the intersection of every $d+1$ of these sets is nonempty. Then the intersection of all the C_i is nonempty.*

The first nontrivial case states that if every 3 among 4 convex sets in the plane intersect, then there is a point common to all 4 sets. This can be proved by an elementary geometric argument, perhaps distinguishing a few cases, and the reader may want to try to find a proof before reading further.

In a contrapositive form, Helly's theorem guarantees that whenever $C_1, C_2, \ldots, C_n$ are convex sets with $\bigcap_{i=1}^n C_i = \emptyset$, then this is witnessed by some at most $d+1$ sets with empty intersection among the C_i. In this way, many proofs are greatly simplified, since in planar problems, say, one can deal with 3 convex sets instead of an arbitrary number, as is amply illustrated in the exercises below.

It is very tempting and quite usual to formulate Helly's theorem as follows: "If every $d+1$ among n convex sets in $\mathbf{R}^d$ intersect, then all the sets intersect." But, strictly speaking, this is false, for a trivial reason: For $d \geq 2$, the assumption as stated here is met by $n = 2$ disjoint convex sets.

Proof of Helly's theorem. (Using Radon's lemma.) For a fixed d, we proceed by induction on n. The case $n = d+1$ is clear, so we suppose that $n \geq d+2$ and that the statement of Helly's theorem holds for smaller n. Actually, $n = d+2$ is the crucial case; the result for larger n follows at once by a simple induction.

Consider sets $C_1, C_2, \ldots, C_n$ satisfying the assumptions. If we leave out any one of these sets, the remaining sets have a nonempty intersection by the inductive assumption. Let us fix a point $a_i \in \bigcap_{j \neq i} C_j$ and consider the points $a_1, a_2, \ldots, a_{d+2}$. By Radon's lemma, there exist disjoint index sets $I_1, I_2 \subset \{1, 2, \ldots, d+2\}$ such that

$$\operatorname{conv}(\{a_i \colon i \in I_1\}) \cap \operatorname{conv}(\{a_i \colon i \in I_2\}) \neq \emptyset.$$

We pick a point x in this intersection. The following picture illustrates the case $d = 2$ and $n = 4$:

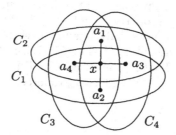

We claim that x lies in the intersection of all the C_i. Consider some $i \in \{1, 2, \ldots, n\}$; then $i \notin I_1$ or $i \notin I_2$. In the former case, each a_j with $j \in I_1$ lies in C_i, and so $x \in \operatorname{conv}(\{a_j \colon j \in I_1\}) \subseteq C_i$. For $i \notin I_2$ we similarly conclude that $x \in \operatorname{conv}(\{a_j \colon j \in I_2\}) \subseteq C_i$. Therefore, $x \in \bigcap_{i=1}^n C_i$. $\square$

An infinite version of Helly's theorem. If we have an infinite collection of convex sets in $\mathbf{R}^d$ such that any $d+1$ of them have a common point, the entire collection still need not have a common point. Two examples in $\mathbf{R}^1$ are the families of intervals $\{(0, 1/n) \colon n = 1, 2, \ldots\}$ and $\{[n, \infty) \colon n = 1, 2, \ldots\}$. The sets in the first example are not closed, and the second example uses unbounded sets. For *compact* (i.e., closed and bounded) sets, the theorem holds:

1.3.3 Theorem (Infinite version of Helly's theorem). *Let $\mathcal{C}$ be an arbitrary infinite family of compact convex sets in $\mathbf{R}^d$ such that any $d+1$ of the sets have a nonempty intersection. Then all the sets of $\mathcal{C}$ have a nonempty intersection.*

Proof. By Helly's theorem, any finite subfamily of $\mathcal{C}$ has a nonempty intersection. By a basic property of compactness, if we have an arbitrary family of compact sets such that each of its finite subfamilies has a nonempty intersection, then the entire family has a nonempty intersection.   □

Several nice applications of Helly's theorem are indicated in the exercises below, and we will meet a few more later in this book.

Bibliography and remarks.   Helly proved Theorem 1.3.2 in 1913 and communicated it to Radon, who published a proof in [Rad21]. This proof uses Radon's lemma, although the statement wasn't explicitly formulated in Radon's paper. References to many other proofs and generalizations can be found in the already mentioned surveys [Eck93] and [DGK63].

Helly's theorem inspired a whole industry of Helly-type theorems. A family $\mathcal{B}$ of sets is said to have *Helly number h* if the following holds: Whenever a finite subfamily $\mathcal{F} \subseteq \mathcal{B}$ is such that every h or fewer sets of $\mathcal{F}$ have a common point, then $\bigcap \mathcal{F} \neq \emptyset$. So Helly's theorem says that the family of all convex sets in $\mathbf{R}^d$ has Helly number $d{+}1$. More generally, let P be some property of families of sets that is hereditary, meaning that if $\mathcal{F}$ has property P and $\mathcal{F}' \subseteq \mathcal{F}$, then $\mathcal{F}'$ has P as well. A family $\mathcal{B}$ is said to have Helly number h with respect to P if for every finite $\mathcal{F} \subseteq \mathcal{B}$, all subfamilies of $\mathcal{F}$ of size at most h having P implies $\mathcal{F}$ having P. That is, the absence of P is always witnessed by some at most h sets, so it is a "local" property.

Exercises

1. Prove Carathéodory's theorem (you may use Radon's lemma). [4]
2. Let $K \subset \mathbf{R}^d$ be a convex set and let $C_1, \ldots, C_n \subseteq \mathbf{R}^d$, $n \geq d{+}1$, be convex sets such that the intersection of every $d{+}1$ of them contains a translated copy of K. Prove that then the intersection of all the sets C_i also contains a translated copy of K. [2]
   This result was noted by Vincensini [Vin39] and by Klee [Kle53].
3. Find an example of 4 convex sets in the plane such that the intersection of each 3 of them contains a segment of length 1, but the intersection of all 4 contains no segment of length 1. [1]
4. *A strip of width w* is a part of the plane bounded by two parallel lines at distance w. The *width* of a set $X \subseteq \mathbf{R}^2$ is the smallest width of a strip containing X.
   (a) Prove that a compact convex set of width 1 contains a segment of length 1 of every direction. [3]
   (b) Let $\{C_1, C_2, \ldots, C_n\}$ be closed convex sets in the plane, $n \geq 3$, such that the intersection of every 3 of them has width at least 1. Prove that $\bigcap_{i=1}^n C_i$ has width at least 1. [2]

The result as in (b), for arbitrary dimension d, was proved by Sallee [Sal75], and a simple argument using Helly's theorem was noted by Buchman and Valentine [BV82].

5. Statement: Each set $X \subset \mathbf{R}^2$ of diameter at most 1 (i.e., any 2 points have distance at most 1) is contained in some disc of radius $1/\sqrt{3}$.
   (a) Prove the statement for 3-element sets X. ②
   (b) Prove the statement for all finite sets X. ②
   (c) Generalize the statement to $\mathbf{R}^d$: determine the smallest $r = r(d)$ such that every set of diameter 1 in $\mathbf{R}^d$ is contained in a ball of radius r (prove your claim). ④
   The result as in (c) is due to Jung; see [DGK63].

6. Let $C \subset \mathbf{R}^d$ be a compact convex set. Prove that the mirror image of C can be covered by a suitable translate of C blown up by the factor of d; that is, there is an $x \in \mathbf{R}^d$ with $-C \subseteq x + dC$. ④

7. (a) Prove that if the intersection of each 4 or fewer among convex sets $C_1, \ldots, C_n \subseteq \mathbf{R}^2$ contains a ray then $\bigcap_{i=1}^n C_i$ also contains a ray. ④
   (b) Show that the number 4 in (a) cannot be replaced by 3. ②
   This result, and an analogous one in $\mathbf{R}^d$ with the Helly number $2d$, are due to Katchalski [Kat78].

8. For a set $X \subseteq \mathbf{R}^2$ and a point $x \in X$, let us denote by $V(x)$ the set of all points $y \in X$ that can "see" x, i.e., points such that the segment xy is contained in X. The *kernel* of X is defined as the set of all points $x \in X$ such that $V(x) = X$. A set with a nonempty kernel is called *star-shaped*.
   (a) Prove that the kernel of any set is convex. ①
   (b) Prove that if $V(x) \cap V(y) \cap V(z) \neq \emptyset$ for every $x, y, z \in X$ and X is compact, then X is star-shaped. That is, if every 3 paintings in a (planar) art gallery can be seen at the same time from some location (possibly different for different triples of paintings), then all paintings can be seen simultaneously from somewhere. If it helps, assume that X is a polygon. ⑤
   (c) Construct a nonempty set $X \subseteq \mathbf{R}^2$ such that each of its finite subsets can be seen from some point of X but X is not star-shaped. ②
   The result in (b), as well as the d-dimensional generalization (with every $d+1$ regions $V(x)$ intersecting), is called Krasnosel'skiĭ's theorem; see [Eck93] for references and related results.

9. In the situation of Radon's lemma (A is a $(d+2)$-point set in $\mathbf{R}^d$), call a point $x \in \mathbf{R}^d$ a *Radon point* of A if it is contained in convex hulls of two disjoint subsets of A. Prove that if A is in general position (no $d+1$ points affinely dependent), then its Radon point is unique. ③

10. (a) Let $X, Y \subset \mathbf{R}^2$ be finite point sets, and suppose that for every subset $S \subseteq X \cup Y$ of at most 4 points, $S \cap X$ can be separated (strictly) by a line from $S \cap Y$. Prove that X and Y are line-separable. ③
    (b) Extend (a) to sets $X, Y \subset \mathbf{R}^d$, with $|S| \leq d+2$. ⑤
    The result (b) is called Kirchberger's theorem [Kir03].

1.4 Centerpoint and Ham Sandwich

We prove an interesting result as an application of Helly's theorem.

1.4.1 Definition (Centerpoint). *Let X be an n-point set in $\mathbf{R}^d$. A point $x \in \mathbf{R}^d$ is called a* centerpoint *of X if each closed half-space containing x contains at least $\frac{n}{d+1}$ points of X.*

Let us stress that one set may generally have many centerpoints, and a centerpoint need not belong to X.

The notion of centerpoint can be viewed as a generalization of the *median* of one-dimensional data. Suppose that $x_1, \ldots, x_n \in \mathbf{R}$ are results of measurements of an unknown real parameter x. How do we estimate x from the x_i? We can use the arithmetic mean, but if one of the measurements is completely wrong (say, 100 times larger than the others), we may get quite a bad estimate. A more "robust" estimate is a *median*, i.e., a point x such that at least $\frac{n}{2}$ of the x_i lie in the interval $(-\infty, x]$ and at least $\frac{n}{2}$ of them lie in $[x, \infty)$. The centerpoint can be regarded as a generalization of the median for higher-dimensional data.

In the definition of centerpoint we could replace the fraction $\frac{1}{d+1}$ by some other parameter $\alpha \in (0, 1)$. For $\alpha > \frac{1}{d+1}$, such an "α-centerpoint" need not always exist: Take $d+1$ points in general position for X. With $\alpha = \frac{1}{d+1}$ as in the definition above, a centerpoint always exists, as we prove next.

Centerpoints are important, for example, in some algorithms of divide-and-conquer type, where they help divide the considered problem into smaller subproblems. Since no really efficient algorithms are known for finding "exact" centerpoints, the algorithms often use α-centerpoints with a suitable $\alpha < \frac{1}{d+1}$, which are easier to find.

1.4.2 Theorem (Centerpoint theorem). *Each finite point set in $\mathbf{R}^d$ has at least one centerpoint.*

Proof. First we note an *equivalent definition of a centerpoint:* x is a centerpoint of X if and only if it lies in each open half-space γ such that $|X \cap \gamma| > \frac{d}{d+1} n$.

We would like to apply Helly's theorem to conclude that all these open half-spaces intersect. But we cannot proceed directly, since we have infinitely many half-spaces and they are open and unbounded. Instead of such an open half-space γ, we thus consider the compact convex set $\operatorname{conv}(X \cap \gamma) \subset \gamma$.

$\operatorname{conv}(\gamma \cap X)$

Letting γ run through all open half-spaces γ with $|X \cap \gamma| > \frac{d}{d+1} n$, we obtain a family $\mathcal{C}$ of compact convex sets. Each of them contains more than $\frac{d}{d+1} n$ points of X, and so the intersection of any $d+1$ of them contains at least one point of X. The family $\mathcal{C}$ consists of finitely many distinct sets (since X has finitely many distinct subsets), and so $\bigcap \mathcal{C} \neq \emptyset$ by Helly's theorem. Each point in this intersection is a centerpoint. $\qquad\square$

In the definition of a centerpoint we can regard the finite set X as defining a distribution of mass in $\mathbf{R}^d$. The centerpoint theorem asserts that for some point x, any half-space containing x encloses at least $\frac{1}{d+1}$ of the total mass. It is not difficult to show that this remains valid for continuous mass distributions, or even for arbitrary Borel probability measures on $\mathbf{R}^d$ (Exercise 1).

Ham-sandwich theorem and its relatives. Here is another important result, not much related to convexity but with a flavor resembling the centerpoint theorem.

1.4.3 Theorem (Ham-sandwich theorem). *Every d finite sets in $\mathbf{R}^d$ can be simultaneously bisected by a hyperplane. A hyperplane h bisects a finite set A if each of the open half-spaces defined by h contains at most $\lfloor |A|/2 \rfloor$ points of A.*

This theorem is usually proved via continuous mass distributions using a tool from algebraic topology: the *Borsuk–Ulam theorem*. Here we omit a proof.

Note that if A_i has an odd number of points, then every h bisecting A_i passes through a point of A_i. Thus if $A_1, \ldots, A_d$ all have odd sizes and their union is in general position, then every hyperplane simultaneously bisecting them is determined by d points, one of each A_i. In particular, there are only finitely many such hyperplanes.

Again, an analogous ham-sandwich theorem holds for arbitrary d Borel probability measures in $\mathbf{R}^d$.

Center transversal theorem. There can be beautiful new things to discover even in well-studied areas of mathematics. A good example is the following recent result, which "interpolates" between the centerpoint theorem and the ham-sandwich theorem.

1.4.4 Theorem (Center transversal theorem). *Let $1 \leq k \leq d$ and let $A_1, A_2, \ldots, A_k$ be finite point sets in $\mathbf{R}^d$. Then there exists a $(k-1)$-flat f such that for every hyperplane h containing f, both the closed half-spaces defined by h contain at least $\frac{1}{d-k+2} |A_i|$ points of A_i, $i = 1, 2, \ldots, k$.*

The ham-sandwich theorem is obtained for $k = d$ and the centerpoint theorem for $k = 1$. The proof, which we again have to omit, is based on a result of algebraic topology, too, but it uses a considerably more advanced machinery than the ham-sandwich theorem. However, the weaker result with $\frac{1}{d+1}$ instead of $\frac{1}{d-k+2}$ is easy to prove; see Exercise 2.

Bibliography and remarks. The centerpoint theorem was established by Rado [Rad47]. According to Steinlein's survey [Ste85], the ham-sandwich theorem was conjectured by Steinhaus (who also invented the popular 3-dimensional interpretation, namely, that the ham, the cheese, and the bread in any ham sandwich can be simultaneously bisected by a single straight motion of the knife) and proved by Banach. The center transversal theorem was found by Doľnikov [Dol'92] and, independently, by Živaljević and Vrećica [ŽV90].

Significant effort has been devoted to efficient algorithms for finding (approximate) centerpoints and ham-sandwich cuts (i.e., hyperplanes as in the ham-sandwich theorem). In the plane, a ham-sandwich cut for two n-point sets can be computed in linear time (Lo, Matoušek, and Steiger [LMS94]). In a higher but fixed dimension, the complexity of the best exact algorithms is currently slightly better than $O(n^{d-1})$. A centerpoint in the plane, too, can be found in linear time (Jadhav and Mukhopadhyay [JM94]). Both approximate ham-sandwich cuts (in the ratio $1 : 1+\varepsilon$ for a fixed $\varepsilon > 0$) and approximate centerpoints $((\frac{1}{d+1}-\varepsilon)$-centerpoints) can be computed in time $O(n)$ for every fixed dimension d and every fixed $\varepsilon > 0$, but the constant depends exponentially on d, and the algorithms are impractical if the dimension is not quite small. A practically efficient randomized algorithm for computing approximate centerpoints in high dimensions (α-centerpoints with $\alpha \approx 1/d^2$) was given by Clarkson, Eppstein, Miller, Sturtivant, and Teng [CEM+96].

Exercises

1. (Centerpoints for general mass distributions)
   (a) Let μ be a Borel probability measure on $\mathbf{R}^d$; that is, $\mu(\mathbf{R}^d) = 1$ and each open set is measurable. Show that for each open half-space γ with $\mu(\gamma) > t$ there exists a compact set $C \subset \gamma$ with $\mu(C) > t$. [2]
   (b) Prove that each Borel probability measure in $\mathbf{R}^d$ has a centerpoint (use (a) and the infinite Helly's theorem). [2]
2. Prove that for any k finite sets $A_1, \ldots, A_k \subset \mathbf{R}^d$, where $1 \leq k \leq d$, there exists a $(k-1)$-flat such that every hyperplane containing it has at least $\frac{1}{d+1}|A_i|$ points of A_i in both of its closed half-spaces for all $i = 1, 2, \ldots, k$. [1]

2

Lattices and Minkowski's Theorem

This chapter is a quick excursion into the *geometry of numbers*, a field where number-theoretic results are proved by geometric arguments, often using properties of convex bodies in $\mathbf{R}^d$. We formulate the simple but beautiful theorem of Minkowski on the existence of a nonzero lattice point in every symmetric convex body of sufficiently large volume. We derive several consequences, concluding with a geometric proof of the famous theorem of Lagrange claiming that every natural number can be written as the sum of at most 4 squares.

2.1 Minkowski's Theorem

In this section we consider the integer lattice $\mathbf{Z}^d$, and so a lattice point is a point in $\mathbf{R}^d$ with integer coordinates. The following theorem can be used in many interesting situations to establish the existence of lattice points with certain properties.

2.1.1 Theorem (Minkowski's theorem). *Let $C \subseteq \mathbf{R}^d$ be symmetric (around the origin, i.e., $C = -C$), convex, bounded, and suppose that $\mathrm{vol}(C) > 2^d$. Then C contains at least one lattice point different from 0.*

Proof. We put $C' = \frac{1}{2}C = \{\frac{1}{2}x \colon x \in C\}$.

Claim: There exists a nonzero integer vector $v \in \mathbf{Z}^d \setminus \{0\}$ such that $C' \cap (C' + v) \neq \emptyset$; i.e., C' and a translate of C' by an integer vector intersect.

> *Proof.* By contradiction; suppose the claim is false. Let R be a large integer number. Consider the family $\mathcal{C}$ of translates of C' by the

integer vectors in the cube $[-R, R]^d$: $\mathcal{C} = \{C' + v: v \in [-R, R]^d \cap \mathbf{Z}^d\}$, as is indicated in the drawing (C is painted in gray).

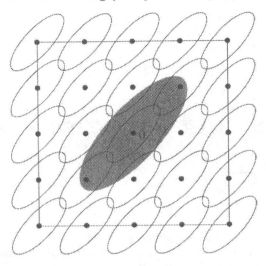

Each such translate is disjoint from C', and thus every two of these translates are disjoint as well. They are all contained in the enlarged cube $K = [-R - D, R + D]^d$, where D denotes the diameter of C'. Hence

$$\text{vol}(K) = (2R + 2D)^d \geq |\mathcal{C}|\,\text{vol}(C') = (2R + 1)^d\,\text{vol}(C'),$$

and

$$\text{vol}(C') \leq \left(1 + \frac{2D - 1}{2R + 1}\right)^d.$$

The expression on the right-hand side is arbitrarily close to 1 for sufficiently large R. On the other hand, $\text{vol}(C') = 2^{-d}\,\text{vol}(C) > 1$ is a fixed number exceeding 1 by a certain amount independent of R, a contradiction. The claim thus holds.   □

Now let us fix a $v \in \mathbf{Z}^d$ as in the claim and let us choose a point $x \in C' \cap (C' + v)$. Then we have $x - v \in C'$, and since C' is symmetric, we obtain $v - x \in C'$. Since C' is convex, the midpoint of the segment $x(v - x)$ lies in C' too, and so we have $\frac{1}{2}x + \frac{1}{2}(v - x) = \frac{1}{2}v \in C'$. This means that $v \in C$, which proves Minkowski's theorem.   □

2.1.2 Example (About a regular forest). Let K be a circle of diameter 26 (meters, say) centered at the origin. Trees of diameter 0.16 grow at each lattice point within K except for the origin, which is where you are standing. Prove that you cannot see outside this miniforest.

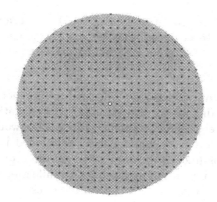

Proof. Suppose than one could see outside along some line ℓ passing through the origin. This means that the strip S of width 0.16 with ℓ as the middle line contains no lattice point in K except for the origin. In other words, the symmetric convex set $C = K \cap S$ contains no lattice points but the origin. But as is easy to calculate, $\operatorname{vol}(C) > 4$, which contradicts Minkowski's theorem.

□

2.1.3 Proposition (Approximating an irrational number by a fraction). *Let $\alpha \in (0,1)$ be a real number and N a natural number. Then there exists a pair of natural numbers m, n such that $n \leq N$ and*

$$\left| \alpha - \frac{m}{n} \right| < \frac{1}{nN}.$$

This proposition implies that there are infinitely many pairs m, n such that $|\alpha - \frac{m}{n}| < 1/n^2$ (Exercise 4). This is a basic and well-known result in elementary number theory. It can also be proved using the pigeonhole principle.

The proposition has an analogue concerning the approximation of several numbers $\alpha_1, \ldots, \alpha_k$ by fractions with a common denominator (see Exercise 5), and there a proof via Minkowski's theorem seems to be the simplest.

Proof of Proposition 2.1.3. Consider the set

$$C = \left\{ (x, y) \in \mathbf{R}^2 \colon -N - \tfrac{1}{2} \leq x \leq N + \tfrac{1}{2},\ |\alpha x - y| < \tfrac{1}{N} \right\}.$$

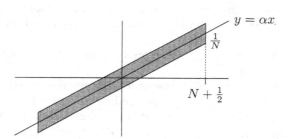

This is a symmetric convex set of area $(2N+1)\frac{2}{N} > 4$, and therefore it contains some nonzero integer lattice point (n, m). By symmetry, we may assume $n > 0$. The definition of C gives $n \le N$ and $|\alpha n - m| < \frac{1}{N}$. In other words, $|\alpha - \frac{m}{n}| < \frac{1}{nN}$. $\square$

Bibliography and remarks. The name "geometry of numbers" was coined by Minkowski, who initiated a systematic study of this field (although related ideas appeared in earlier works). He proved Theorem 2.1.1, in a more general form mentioned later on, in 1891 (see [Min96]). His first application was a theorem on simultaneously making linear forms small (Exercise 2.2.4). While geometry of numbers originated as a tool in number theory, for questions in Diophantine approximation and quadratic forms, today it also plays a significant role in several other diverse areas, such as coding theory, cryptography, the theory of uniform distribution, and numerical integration.

Theorem 2.1.1 is often called *Minkowski's first theorem*. What is, then, Minkowski's second theorem? We answer this natural question in the notes to Section 2.2, where we also review a few more of the basic results in the geometry of numbers and point to some interesting connections and directions of research.

Most of our exposition in this chapter follows a similar chapter in Pach and Agarwal [PA95]. Older books on the geometry of numbers are Cassels [Cas59] and Gruber and Lekkerkerker [GL87]. A pleasant but somewhat aged introduction is Siegel [Sie89]. The Gruber [Gru93] provides a concise recent overview.

Exercises

1. Prove: If $C \subseteq \mathbf{R}^d$ is convex, symmetric around the origin, bounded, and such that $\mathrm{vol}(C) > k2^d$, then C contains at least $2k$ lattice points. [2]

2. By the method of the proof of Minkowski's theorem, show the following result (Blichtfeld; Van der Corput): If $S \subseteq \mathbf{R}^d$ is measurable and $\mathrm{vol}(S) > k$, then there are points $s_1, s_2, \ldots, s_k \in S$ with all $s_i - s_j \in \mathbf{Z}^d$, $1 \le i, j \le k$. [3]

3. Show that the boundedness of C in Minkowski's theorem is not really necessary. [1]

4. (a) Verify the claim made after Example 2.1.3, namely, that for any irrational α there are infinitely many pairs m, n such that $|\alpha - m/n| < 1/n^2$. [1]
   (b) Prove that for $\alpha = \sqrt{2}$ there are only finitely many pairs m, n with $|\alpha - m/n| < 1/4n^2$. [2]
   (c) Show that for any algebraic irrational number α (i.e., a root of a univariate polynomial with integer coefficients) there exists a constant D such that $|\alpha - m/n| < 1/n^D$ holds for finitely many pairs (m, n) only. Conclude that, for example, the number $\sum_{i=1}^{\infty} 2^{-i^i}$ is not algebraic. [4]

5. (a) Let $\alpha_1, \alpha_2 \in (0,1)$ be real numbers. Prove that for a given $N \in \mathbf{N}$ there exist $m_1, m_2, n \in \mathbf{N}$, $n \leq N$, such that $|\alpha_i - \frac{m_i}{n}| < \frac{1}{n\sqrt{N}}$, $i = 1, 2$. [4]

(b) Formulate and prove an analogous result for the simultaneous approximation of d real numbers by rationals with a common denominator. [2] (This is a result of Dirichlet [Dir42].)

6. Let $K \subset \mathbf{R}^2$ be a compact convex set of area α and let x be a point chosen uniformly at random in $[0,1)^2$.

(a) Prove that the expected number of points of $\mathbf{Z}^2$ in the set $K + x$ equals α. [2]

(b) Show that with probability at least $1 - \alpha$, $K + x$ contains no point of $\mathbf{Z}^2$. [1]

2.2 General Lattices

Let $z_1, z_2, \ldots, z_d$ be a d-tuple of linearly independent vectors in $\mathbf{R}^d$. We define the *lattice with basis* $\{z_1, z_2, \ldots, z_d\}$ as the set of all linear combinations of the z_i with integer coefficients; that is,

$$\Lambda = \Lambda(z_1, z_2, \ldots, z_d) = \{i_1 z_1 + i_2 z_2 + \cdots + i_d z_d \colon (i_1, i_2, \ldots, i_d) \in \mathbf{Z}^d\}.$$

Let us remark that this lattice has in general many different bases. For instance, the sets $\{(0,1),(1,0)\}$ and $\{(1,0),(3,1)\}$ are both bases of the "standard" lattice $\mathbf{Z}^2$.

Let us form a $d \times d$ matrix Z with the vectors $z_1, \ldots, z_d$ as columns. We define the *determinant of the lattice* $\Lambda = \Lambda(z_1, z_2, \ldots, z_d)$ as $\det \Lambda = |\det Z|$. Geometrically, $\det \Lambda$ is the volume of the parallelepiped $\{\alpha_1 z_1 + \alpha_2 z_2 + \cdots + \alpha_d z_d \colon \alpha_1, \ldots, \alpha_d \in [0,1]\}$:

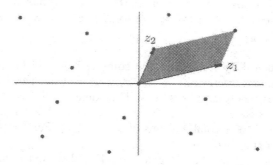

(the proof is left to Exercise 1). The number $\det \Lambda$ is indeed a property of the lattice Λ (as a point set), and it does not depend on the choice of the basis of Λ (Exercise 2). It is not difficult to show that if Z is the matrix of some basis of Λ, then the matrix of every basis of Λ has the form BU, where U is an integer matrix with determinant ± 1.

2.2.1 Theorem (Minkowski's theorem for general lattices). *Let Λ be a lattice in $\mathbf{R}^d$, and let $C \subseteq \mathbf{R}^d$ be a symmetric convex set with $\mathrm{vol}(C) > 2^d \det \Lambda$. Then C contains a point of Λ different from 0.*

Proof. Let $\{z_1, \ldots, z_d\}$ be a basis of Λ. We define a linear mapping $f\colon \mathbf{R}^d \to \mathbf{R}^d$ by $f(x_1, x_2, \ldots, x_d) = x_1 z_1 + x_2 z_2 + \cdots + x_d z_d$. Then f is a bijection and $\Lambda = f(\mathbf{Z}^d)$. For any convex set X, we have $\mathrm{vol}(f(X)) = \det(\Lambda) \mathrm{vol}(X)$. (Sketch of proof: This holds if X is a cube, and a convex set can be approximated by a disjoint union of sufficiently small cubes with arbitrary precision.) Let us put $C' = f^{-1}(C)$. This is a symmetric convex set with $\mathrm{vol}(C') = \mathrm{vol}(C)/\det \Lambda > 2^d$. Minkowski's theorem provides a nonzero vector $v \in C' \cap \mathbf{Z}^d$, and $f(v)$ is the desired point as in the theorem. $\qquad\square$

A seemingly more general definition of a lattice. What if we consider integer linear combinations of more than d vectors in $\mathbf{R}^d$? Some caution is necessary: If we take $d = 1$ and the vectors $v_1 = (1)$, $v_2 = (\sqrt{2})$, then the integer linear combinations $i_1 v_1 + i_2 v_2$ are dense in the real line (by Example 2.1.3), and such a set is not what we would like to call a lattice.

In order to exclude such pathology, we define a *discrete subgroup of* $\mathbf{R}^d$ as a set $\Lambda \subset \mathbf{R}^d$ such that whenever $x, y \in \Lambda$, then also $x - y \in \Lambda$, and such that the distance of any two distinct points of Λ is at least δ, for some fixed positive real number $\delta > 0$.

It can be shown, for instance, that if $v_1, v_2, \ldots, v_n \in \mathbf{R}^d$ are vectors with *rational* coordinates, then the set Λ of all their integer linear combinations is a discrete subgroup of $\mathbf{R}^d$ (Exercise 3). As the following theorem shows, any discrete subgroup of $\mathbf{R}^d$ whose linear span is all of $\mathbf{R}^d$ is a lattice in the sense of the definition given at the beginning of this section.

2.2.2 Theorem (Lattice basis theorem). *Let $\Lambda \subset \mathbf{R}^d$ be a discrete subgroup of $\mathbf{R}^d$ whose linear span is $\mathbf{R}^d$. Then Λ has a basis; that is, there exist d linearly independent vectors $z_1, z_2, \ldots, z_d \in \mathbf{R}^d$ such that $\Lambda = \Lambda(z_1, z_2, \ldots, z_d)$.*

Proof. We proceed by induction. For some i, $1 \leq i \leq d+1$, suppose that linearly independent vectors $z_1, z_2, \ldots, z_{i-1} \in \Lambda$ with the following property have already been constructed. If F_{i-1} denotes the $(i-1)$-dimensional subspace spanned by $z_1, \ldots, z_{i-1}$, then all points of Λ lying in F_{i-1} can be written as integer linear combinations of $z_1, \ldots, z_{i-1}$. For $i = d+1$, this gives the statement of the theorem.

So consider an $i \leq d$. Since Λ generates $\mathbf{R}^d$, there exists a vector $w \in \Lambda$ not lying in the subspace F_{i-1}. Let P be the i-dimensional parallelepiped determined by $z_1, z_2, \ldots, z_{i-1}$ and by w: $P = \{\alpha_1 z_1 + \alpha_2 z_2 + \cdots + \alpha_{i-1} z_{i-1} + \alpha_i w\colon \alpha_1, \ldots, \alpha_i \in [0,1]\}$. Among all the (finitely many) points of Λ lying in P but not in F_{i-1}, choose one nearest to F_{i-1} and call it z_i, as in the picture:

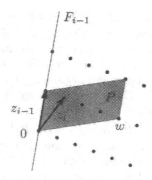

Note that if the points of $\Lambda \cap P$ are written in the form $\alpha_1 z_1 + \alpha_2 z_2 + \cdots + \alpha_{i-1} z_{i-1} + \alpha_i w$, then z_i is one with the smallest α_i. It remains to show that $z_1, z_2, \ldots, z_i$ have the required property.

So let $v \in \Lambda$ be a point lying in F_i (the linear span of $z_1, \ldots, z_i$). We can write $v = \beta_1 z_1 + \beta_2 z_2 + \cdots + \beta_i z_i$ for some real numbers $\beta_1, \ldots, \beta_i$. Let γ_j be the fractional part of β_j, $j = 1, 2, \ldots, i$; that is, $\gamma_j = \beta_j - \lfloor \beta_j \rfloor$. Put $v' = \gamma_1 z_1 + \gamma_2 z_2 + \cdots + \gamma_i z_i$. This point also lies in Λ (since v and v' differ by an integer linear combination of vectors of Λ). We have $0 \leq \gamma_j < 1$, and hence v' lies in the parallelepiped P. Therefore, we must have $\gamma_i = 0$, for otherwise, v' would be nearer to F_{i-1} than z_i. Hence $v' \in \Lambda \cap F_{i-1}$, and by the inductive hypothesis, we also get that all the other γ_j are 0. So all the β_j are in fact integer coefficients, and the inductive step is finished. $\quad\square$

Therefore, a lattice can also be defined as a full-dimensional discrete subgroup of $\mathbf{R}^d$.

Bibliography and remarks. First we mention several fundamental theorems in the "classical" geometry of numbers.

Lattice packing and the Minkowski–Hlawka theorem. For a compact $C \subset \mathbf{R}^d$, the *lattice constant* $\Delta(C)$ is defined as $\min\{\det(\Lambda): \Lambda \cap C = \{0\}\}$, where the minimum is over all lattices Λ in $\mathbf{R}^d$ (it can be shown by a suitable compactness argument, known as the *compactness theorem of Mahler*, that the minimum is attained). The ratio $\mathrm{vol}(C)/\Delta(C)$ is the smallest number $D = D(C)$ for which the Minkowski-like result holds: Whenever $\det(\Lambda) > D$, we have $C \cap \Lambda \neq \{0\}$. It is also easy to check that $2^{-d} D(C)$ equals the maximum *density of a lattice packing of C*; i.e., the fraction of $\mathbf{R}^d$ that can be filled by the set $C + \Lambda$ for some lattice Λ such that all the translates $C + v$, $v \in \Lambda$, have pairwise disjoint interiors. A basic result (obtained by an averaging argument) is the *Minkowski–Hlawka theorem*, which shows that $D \geq 1$ for all star-shaped compact sets C. If C is star-shaped and symmetric, then we have the improved lower bound (better packing) $D \geq 2\zeta(d) = 2\sum_{n=1}^{\infty} n^{-d}$. This brings us to the fascinating field of *lattice packings*, which we do not pursue in this book; a nice geometric introduction is in the first half of the book Pach and Agarwal [PA95], and an authoritative reference is Conway and Sloane [CS99]. Let us remark that the lattice constant (and hence the maximum lattice packing density) is not known in general even for Euclidean spheres, and many ingenious constructions and arguments have been developed for packing them efficiently. These problems also have close connections to *error-correcting codes*.

Successive minima and Minkowski's second theorem. Let $C \subset \mathbf{R}^d$ be a convex body containing 0 in the interior and let $\Lambda \subset \mathbf{R}^d$ be a lattice. The *ith successive minimum of C with respect to Λ*, denoted by $\lambda_i = \lambda_i(C, \Lambda)$, is the infimum of the scaling factors $\lambda > 0$ such that λC contains at least i linearly independent vectors of Λ. In particular, λ_1 is the smallest number for which $\lambda_1 C$ contains a nonzero lattice vector, and Minkowski's theorem guarantees that $\lambda_1^d \leq 2^d \det(\Lambda)/\operatorname{vol}(C)$. Minkowski's second theorem asserts $(2^d/d!) \det(\Lambda) \leq \lambda_1 \lambda_2 \cdots \lambda_d \cdot \operatorname{vol}(C) \leq 2^d \det(\Lambda)$.

The flatness theorem. If a convex body K is not required to be symmetric about 0, then it can have arbitrarily large volume without containing a lattice point. But any lattice-point free body has to be *flat*: For every dimension d there exists $c(d)$ such that any convex body $K \subseteq \mathbf{R}^d$ with $K \cap \mathbf{Z}^d = \emptyset$ has *lattice width* at most $c(d)$. The lattice width of K is defined as $\min\{\max_{x \in K} \langle x, y \rangle - \min_{x \in K} \langle x, y \rangle : y \in \mathbf{Z}^d \setminus \{0\}\}$; geometrically, we essentially count the number of hyperplanes orthogonal to y, spanned by points of $\mathbf{Z}^d$, and intersecting K. Such a result was first proved by Khintchine in 1948, and the current best bound $c(d) = O(d^{3/2})$ is due to Banaszczyk, Litvak, Pajor, and Szarek [BLPS99]; we also refer to this paper for more references.

Computing lattice points in convex bodies. Minkowski's theorem provides the existence of nonzero lattice points in certain convex bodies. Given one of these bodies, how efficiently can one actually *compute* a nonzero lattice point in it? More generally, given a convex body in $\mathbf{R}^d$, how difficult is it to decide whether it contains a lattice point, or to count all lattice points? For simplicity, we consider only the integer lattice $\mathbf{Z}^d$ here.

First, if the dimension d is considered as a constant, such problems can be solved efficiently, at least in theory. An algorithm due to Lenstra [Len83] finds in polynomial time an integer point, if one exists, in a given convex polytope in $\mathbf{R}^d$, d fixed. It is based on the flatness theorem mentioned above (the ideas are also explained in many other sources, e.g., [GLS88], [Lov86], [Sch86], [Bar97]). More recently, Barvinok [Bar93] (or see [Bar97]) provided a polynomial-time algorithm for counting the integer points in a given fixed-dimensional convex polytope. Both algorithms are nice and certainly nontrivial, and especially the latter can be recommended as a neat application of classical mathematical results in a new context.

On the other hand, if the dimension d is considered as a part of the input then (exact) calculations with lattices tend to be algorithmically difficult. Most of the difficult problems of combinatorial optimization can be formulated as instances of *integer programming*, where a given linear function should be minimized over the set of integer points in a given convex polytope. This problem is well known to be NP-hard, and so is the problem of deciding whether a given convex polytope contains an integer point (both problems are actually polynomially equivalent). For an introduction to integer programming see, e.g., Schrijver [Sch86].

Some much more special problems concerning lattices have also been shown to be algorithmically difficult. For example, finding a *shortest (nonzero) vector* in a given lattice Λ specified by a basis is NP-hard (with respect to randomized polynomial-time reductions). (In the notation introduced above, we are asking for $\lambda_1(B^d, \Lambda)$, the first successive minimum of the ball. This took quite some time to prove (Micciancio [Mic98] has obtained the strongest result to date, inapproximability up to the factor of $\sqrt{2}$, building on earlier work mainly of Ajtai), although the analogous hardness result for the shortest vector in the maximum norm (i.e., $\lambda_1([-1, 1]^d, \Lambda)$) has been known for a long time.

Basis reduction and applications. Although finding the shortest vector of a lattice Λ is algorithmically difficult, the shortest vector can be approximated in the following sense. For every $\varepsilon > 0$ there is a polynomial-time algorithm that, given a basis of a lattice Λ in $\mathbf{R}^d$, computes a nonzero vector of Λ whose length is at most $(1 + \varepsilon)^d$ times the length of the shortest vector of Λ; this was proved by Schnorr [Sch87]. The first result of this type, with a worse bound on the approximation factor, was obtained in the seminal work of Lenstra, Lenstra, and Lovász [LLL82]. The LLL algorithm, as it is called, computes not only a single short vector but a whole "short" basis of Λ.

The key notion in the algorithm is that of a *reduced basis* of Λ; intuitively, this means a basis that cannot be much improved (made significantly shorter) by a simple local transformation. There are many technically different notions of reduced bases. Some of them are classical and have been considered by mathematicians such as Gauss and Lagrange. The definition of the *Lovász-reduced basis* used in the LLL algorithm is sufficiently relaxed so that a reduced basis can be computed from any initial basis by polynomially many local improvements, and, at the same time, is strong enough to guarantee that a reduced basis is relatively short. These results are covered in many sources; the thin book by Lovász [Lov86] can still be recommended as a delightful introduction. Numerous refinements of the LLL algorithm, as well as efficient implementations, are available.

We sketch an ingenious application of the LLL algorithm for polynomial factorization (from Kannan, Lenstra, and Lovász [KLL88]; the original LLL technique is somewhat different). Assume for simplicity that we want to factor a monic polynomial $p(x) \in \mathbf{Z}[x]$ (integer coefficients, leading coefficient 1) into a product of factors irreducible over $\mathbf{Z}[x]$. By numerical methods we can compute a root α of $p(x)$ with very high precision. If we can find the minimal polynomial of α, i.e., the lowest-degree monic polynomial $q(x) \in \mathbf{Z}[x]$ with $q(\alpha) = 0$, then we are done, since $q(x)$ is irreducible and divides $p(x)$. Let us write $q(x) = x^d + a_{d-1}x^{d-1} + \cdots + a_0$. Let K be a large number and let us consider the d-dimensional lattice Λ in $\mathbf{R}^{d+1}$ with basis $(K, 1, 0, \ldots, 0)$, $(K\alpha, 0, 1, 0, \ldots, 0)$, $(K\alpha^2, 0, 0, 1, 0, \ldots, 0), \ldots, (K\alpha^d, 0, \ldots, 0, 1)$. Combining the basis vectors with the coefficients $a_0, a_1, \ldots, a_{d-1}, 1$, respectively, we obtain the vector $v_0 = (0, a_0, a_1, \ldots, a_{d-1}, 1) \in \Lambda$. It turns out that if K is sufficiently large compared to the a_i, then v_0 is the shortest nonzero vector, and moreover, every vector not much longer than v_0 is a multiple of v_0. The LLL algorithm applied to Λ thus finds v_0, and this yields $q(x)$. Of course, we do not know the degree of $q(x)$, but we can test all possible degrees one by one, and the required magnitude of K can be estimated from the coefficients of $p(x)$.

The LLL algorithm has been used for the *knapsack problem* and for the *subset sum problem*. Typically, the applications are problems where one needs to express a given number (or vector) as a linear combination of given numbers (or vectors) with small integer coefficients. For example, the subset sum problem asks, for given integers $a_1, a_2, \ldots, a_n$ and b, for a subset $I \subseteq \{1, 2, \ldots, n\}$ with $\sum_{i \in I} a_i = b$; i.e., b should be expressed as a linear combination of the a_i with $0/1$ coefficients. These and many other significant applications can be found in Grötschel, Lovász, and Schrijver [GLS88]. In *cryptography*, several cryptographic systems proposed in the literature were broken with the help of the LLL algorithm (references are listed, e.g., in [GLS88], [Dwo97]). On the other hand, lattices play a prominent role in recent constructions, mainly due to Ajtai, of new cryptographic systems. While currently the security of every known efficient cryptographic system depends on an (unproven) assumption of hardness of a certain computational problem, Ajtai's methods suffice with a considerably weaker and more plausible assumption than those required by the previous systems (see [Ajt98] or [Dwo97] for an introduction).

Exercises

1. Let $v_1, \ldots, v_d$ be linearly independent vectors in $\mathbf{R}^d$. Form a matrix A with $v_1, \ldots, v_d$ as rows. Prove that $|\det A|$ is equal to the volume of the

parallelepiped $\{\alpha_1 v_1 + \alpha_2 v_2 + \cdots + \alpha_d v_d\colon \alpha_1, \ldots, \alpha_d \in [0, 1]\}$. (You may want to start with $d = 2$.) ③

2. Prove that if $z_1, \ldots, z_d$ and $z_1', \ldots, z_d'$ are vectors in $\mathbf{R}^d$ such that $\Lambda(z_1, \ldots, z_d) = \Lambda(z_1', \ldots, z_d')$, then $|\det Z| = |\det Z'|$, where Z is the $d \times d$ matrix with the z_i as columns, and similarly for Z'. ③

3. Prove that for n rational vectors $v_1, \ldots, v_n$, the set $\Lambda = \{i_1 v_1 + i_2 v_2 + \cdots + i_n v_n \colon i_1, i_2, \ldots, i_n \in \mathbf{Z}\}$ is a discrete subgroup of $\mathbf{R}^d$. ③

4. (Minkowski's theorem on linear forms) Prove the following from Minkowski's theorem: Let $\ell_i(x) = \sum_{j=1}^d a_{ij} x_j$ be linear forms in d variables, $i = 1, 2, \ldots, d$, such that the $d \times d$ matrix $(a_{ij})_{i,j}$ has determinant 1. Let $b_1, \ldots, b_d$ be positive real numbers with $b_1 b_2 \cdots b_d = 1$. Then there exists a nonzero integer vector $z \in \mathbf{Z}^d \setminus \{0\}$ with $|\ell_i(z)| \le b_i$ for all $i = 1, 2, \ldots, d$. ③

2.3 An Application in Number Theory

We prove one nontrivial result of elementary number theory. The proof via Minkowski's theorem is one of several possible proofs. Another proof uses the pigeonhole principle in a clever way.

2.3.1 Theorem (Two-square theorem). *Each prime $p \equiv 1 \,(\mathrm{mod}\,4)$ can be written as a sum of two squares: $p = a^2 + b^2$, $a, b \in \mathbf{Z}$.*

Let $F = GF(p)$ stand for the field of residue classes modulo p, and let $F^* = F \setminus \{0\}$. An element $a \in F^*$ is called a *quadratic residue* modulo p if there exists an $x \in F^*$ with $x^2 \equiv a \,(\mathrm{mod}\,p)$. Otherwise, a is a *quadratic nonresidue*.

2.3.2 Lemma. *If p is a prime with $p \equiv 1 \,(\mathrm{mod}\,4)$ then -1 is a quadratic residue modulo p.*

Proof. The equation $i^2 = 1$ has two solutions in the field F, namely $i = 1$ and $i = -1$. Hence for any $i \ne \pm 1$ there exists exactly one $j \ne i$ with $ij = 1$ (namely, $j = i^{-1}$, the inverse element in F), and all the elements of $F^* \setminus \{-1, 1\}$ can be divided into pairs such that the product of elements in each pair is 1. Therefore, $(p-1)! = 1 \cdot 2 \cdots (p-1) \equiv -1 \,(\mathrm{mod}\,p)$.

For a contradiction, suppose that the equation $i^2 = -1$ has no solution in F. Then all the elements of F^* can be divided into pairs such that the product of the elements in each pair is -1. There are $(p-1)/2$ pairs, which is an even number. Hence $(p-1)! \equiv (-1)^{(p-1)/2} = 1$, a contradiction. $\quad\square$

Proof of Theorem 2.3.1. By the lemma, we can choose a number q such that $q^2 \equiv -1 \,(\mathrm{mod}\,p)$. Consider the lattice $\Lambda = \Lambda(z_1, z_2)$, where $z_1 = (1, q)$ and $z_2 = (0, p)$. We have $\det \Lambda = p$. We use Minkowski's theorem for general lattices (Theorem 2.2.1) for the disk $C = \{(x, y) \in \mathbf{R}^2 \colon x^2 + y^2 < 2p\}$. The area of C is $2\pi p > 4p = 4 \det \Lambda$, and so C contains a point $(a, b) \in \Lambda \setminus \{0\}$. We have $0 < a^2 + b^2 < 2p$. At the same time, $(a, b) = iz_1 + jz_2$ for some $i, j \in \mathbf{Z}$, which means that $a = i, b = iq + jp$. We calculate $a^2 + b^2 = i^2 + (iq + jp)^2 = i^2 + i^2q^2 + 2iqjp + j^2p^2 \equiv i^2(1 + q^2) \equiv 0 \pmod{p}$. Therefore $a^2 + b^2 = p$. □

Bibliography and remarks.   The fact that every prime congruent to 1 mod 4 can be written as the sum of two squares was already known to Fermat (a more rigorous proof was given by Euler). The possibility of expressing every natural number as a sum of at most 4 squares was proved by Lagrange in 1770, as a part of his work on quadratic forms. The proof indicated in Exercise 1 below is due to Davenport.

Exercises

1. (Lagrange's four-square theorem) Let p be a prime.
   (a) Show that there exist integers a, b with $a^2 + b^2 \equiv -1 \pmod{p}$. ③
   (b) Show that the set $\Lambda = \{(x, y, z, t) \in \mathbf{Z}^4 : z \equiv ax + by \pmod{p}, t \equiv bx - ay \pmod{p}\}$ is a lattice, and compute $\det(\Lambda)$. ①
   (c) Show the existence of a nonzero point of Λ in a ball of a suitable radius, and infer that p can be written as a sum of 4 squares of integers. ②
   (d) Show that any natural number can be written as a sum of 4 squares of integers. ③

3

Convex Independent Subsets

Here we consider geometric Ramsey-type results about finite point sets in the plane. Ramsey-type theorems are generally statements of the following type: Every sufficiently large structure of a given type contains a "regular" substructure of a prescribed size. In the forthcoming Erdős–Szekeres theorem (Theorem 3.1.3), the "structure of a given type" is simply a finite set of points in general position in $\mathbf{R}^2$, and the "regular substructure" is a set of points forming the vertex set of a convex polygon, as is indicated in the picture:

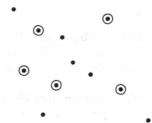

A prototype of Ramsey-type results is Ramsey's theorem itself: *For every choice of natural numbers p, r, n, there exists a natural number N such that whenever X is an N-element set and $c: \binom{X}{p} \to \{1, 2, \ldots, r\}$ is an arbitrary coloring of the system of all p-element subsets of X by r colors, then there is an n-element subset $Y \subseteq X$ such that all the p-tuples in $\binom{Y}{p}$ have the same color.* The most famous special case is with $p = r = 2$, where $\binom{X}{2}$ is interpreted as the edge set of the complete graph K_N on N vertices. Ramsey's theorem asserts that if each of the edges of K_N is colored red or blue, we can always find a complete subgraph on n vertices with all edges red or all edges blue.

Many of the geometric Ramsey-type theorems, including the Erdős–Szekeres theorem, can be derived from Ramsey's theorem. But the quantitative bound for the N in Ramsey's theorem is very large, and consequently, the size of the "regular" configurations guaranteed by proofs via Ramsey's theorem is very small. Other proofs tailored to the particular problems and using more of their geometric structure often yield much better quantitative results.

3.1 The Erdős–Szekeres Theorem

3.1.1 Definition (Convex independent set). *We say that a set $X \subseteq \mathbf{R}^d$ is* convex independent *if for every $x \in X$, we have $x \notin \mathrm{conv}(X \setminus \{x\})$.*

The phrase "in convex position" is sometimes used synonymously with "convex independent." In the plane, a finite convex independent set is the set of vertices of a convex polygon. We will discuss results concerning the occurrence of convex independent subsets in sufficiently large point sets. Here is a simple example of such a statement.

3.1.2 Proposition. *Among any 5 points in the plane in general position (no 3 collinear), we can find 4 points forming a convex independent set.*

Proof. If the convex hull has 4 or 5 vertices, we are done. Otherwise, we have a triangle with two points inside, and the two interior points together with one of the sides of the triangle define a convex quadrilateral.   □

Next, we prove a general result.

3.1.3 Theorem (Erdős–Szekeres theorem). *For every natural number k there exists a number $n(k)$ such that any $n(k)$-point set $X \subset \mathbf{R}^2$ in general position contains a k-point convex independent subset.*

First proof (using Ramsey's theorem and Proposition 3.1.2). Color a 4-tuple $T \subset X$ red if its four points are convex independent and blue otherwise. If n is sufficiently large, Ramsey's theorem provides a k-point subset $Y \subset X$ such that all 4-tuples from Y have the same color. But for $k \geq 5$ this color cannot be blue, because any 5 points determine at least one red 4-tuple. Consequently, Y is convex independent, since every 4 of its points are (Carathéodory's theorem).   □

Next, we give an inductive proof; it yields an almost tight bound for $n(k)$.

Second proof of the Erdős–Szekeres theorem. In this proof, by a set in general position we mean a set with no 3 points on a common line and no 2 points having the same x-coordinate. The latter can always be achieved by rotating the coordinate system.

Let X be a finite point set in the plane in general position. We call X a *cup* if X is convex independent and its convex hull is bounded from above by a single edge (in other words, if the points of X lie on the graph of a convex function).

Similarly, we define a *cap*, with a single edge bounding the convex hull from below.

A k-cap is a cap with k points, and similarly for an ℓ-cup.

We define $f(k, \ell)$ as the smallest number N such than any N-point set in general position contains a k-cup or an ℓ-cap. By induction on k and ℓ, we prove the following formula for $f(k, \ell)$:

$$f(k, \ell) \leq \binom{k + \ell - 4}{k - 2} + 1. \tag{3.1}$$

Theorem 3.1.3 clearly follows from this, with $n(k) \leq f(k, k)$. For $k \leq 2$ or $\ell \leq 2$ the formula holds. Thus, let $k, \ell \geq 3$, and consider a set P in general position with $N = f(k-1, \ell) + f(k, \ell-1) - 1$ points. We prove that it contains a k-cup or an ℓ-cap. This will establish the inequality $f(k, \ell) \leq f(k-1, \ell) + f(k, \ell-1) - 1$, and then (3.1) follows by induction; we leave the simple manipulation of binomial coefficients to the reader.

Suppose that there is no ℓ-cap in X. Let $E \subseteq X$ be the set of points $p \in X$ such that X contains a $(k-1)$-cup ending with p.

We have $|E| \geq N - f(k-1, \ell) + 1 = f(k, \ell-1)$, because $X \setminus E$ contains no $(k-1)$-cup and so $|X \setminus E| < f(k-1, \ell)$.

Either the set E contains a k-cup, and then we are done, or there is an $(\ell-1)$-cap. The first point p of such an $(\ell-1)$-cap is, by the definition of E, the last point of some $(k-1)$-cup in X, and in this situation, either the cup or the cap can be extended by one point:

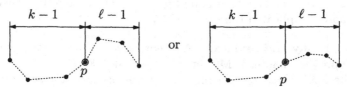

This finishes the inductive step.   □

A lower bound for sets without k-cups and ℓ-caps. Interestingly, the bound for $f(k, \ell)$ proved above is tight, not only asymptotically but exactly! This means, in particular, that there are n-point planar sets in general position where any convex independent subset has at most $O(\log n)$ points, which is somewhat surprising at first sight.

An example of a set $X_{k,\ell}$ of $\binom{k+\ell-4}{k-2}$ points in general position with no k-cup and no ℓ-cap can be constructed, again by induction on $k + \ell$. If $k \leq 2$ or $\ell \leq 2$, then $X_{k,\ell}$ can be taken as a one-point set.

Supposing both $k \geq 3$ and $\ell \geq 3$, the set $X_{k,\ell}$ is obtained from the sets $L = X_{k-1,\ell}$ and $R = X_{k,\ell-1}$ according to the following picture:

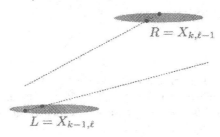

The set L is placed to the left of R in such a way that all lines determined by pairs of points in L go below R and all lines determined by pairs of points of R go above L.

Consider a cup C in the set $X_{k,\ell}$ thus constructed. If $C \cap L = \emptyset$, then $|C| \leq k-1$ by the assumption on R. If $C \cap L \neq \emptyset$, then C has at most 1 point in R, and since no cup in L has more than $k-2$ points, we get $|C| \leq k-1$ as well. The argument for caps is symmetric.

We have $|X_{k,\ell}| = |X_{k-1,\ell}| + |X_{k,\ell-1}|$, and the formula for $|X_{k,\ell}|$ follows by induction; the calculation is almost the same as in the previous proof. $\square$

Determining the exact value of $n(k)$ in the Erdős–Szekeres theorem is much more challenging. Here are the best known bounds:

$$2^{k-2} + 1 \leq n(k) \leq \binom{2k-5}{k-2} + 2.$$

The upper bound is a small improvement over the bound $f(k,k)$ derived above; see Exercise 5. The lower bound results from an inductive construction slightly more complicated than that of $X_{k,\ell}$.

Bibliography and remarks. A recent survey of the topics discussed in the present chapter is Morris and Soltan [MS00].

The Erdős–Szekeres theorem was one of the first Ramsey-type results [ES35], and Erdős and Szekeres independently rediscovered the general Ramsey's theorem at that occasion. Still another proof, also using Ramsey's theorem, was noted by Tarsi: Let the points of X be numbered $x_1, x_2, \ldots, x_n$, and color the triple $\{x_i, x_j, x_k\}$, $i < j < k$, red if we make a right turn when going from x_i to x_k via x_j, and blue if we make a left turn. It is not difficult to check that a homogeneous subset, with all triples having the same color, is in convex position.

The original upper bound of $n(k) \leq \binom{2k-4}{k-2}+1$ from [ES35] has been improved only recently and very slightly; the last improvement to the bound stated in the text above is due to Tóth[1] and Valtr [TV98].

The Erdős–Szekeres theorem was generalized to planar convex sets. The following somewhat misleading term is used: A family of pairwise disjoint convex sets is *in general position* if no set is contained in the convex hull of the union of two other sets of the family. For every k there exists n such that in any family of n pairwise disjoint convex sets in the plane in general position, there are k sets in convex position, meaning that none of them is contained in the convex hull of the union of the others. This was shown by Bisztriczky and G. Fejes Tóth [BT89] and, with a different proof and better quantitative bound, by Pach and Tóth [PT98]. The assumption of general position is necessary.

An interesting problem is the generalization of the Erdős–Szekeres theorem to $\mathbf{R}^d$, $d \geq 3$. The existence of $n_d(k)$ such that every $n_d(k)$ points in $\mathbf{R}^d$ in general position contain a k-point subset in convex position is easy to see (Exercise 4), but the order of magnitude is wide open. The current best upper bound $n_d(k) \leq \binom{2k-2d-1}{k-d}+d$ [Kár01] slightly improves the immediate bound. Füredi [unpublished] conjectured that $n_3(k) \leq e^{O(\sqrt{k})}$. If true, this would be best possible: A construction of Károlyi and Valtr [KV01] shows that for every fixed $d \geq 3$, $n_d(k) \geq e^{c_d k^{1/(d-1)}}$ with a suitable $c_d > 0$. The construction starts with a one-point set X_0, and X_{i+1} is obtained from X_i by replacing each point $x \in X_i$ by the two points $x - (\varepsilon_i^d, \varepsilon_i^{d-1}, \ldots, \varepsilon_i)$ and $x + (\varepsilon_i^d, \varepsilon_i^{d-1}, \ldots, \varepsilon_i)$, with $\varepsilon_i > 0$ sufficiently small, and then perturbing the resulting set very slightly, so that X_{i+1} is in suitable general position. We have $|X_i| = 2^i$, and the key lemma asserts that $\mathrm{mc}(X_{i+1}) \leq \mathrm{mc}(X_i) + \mathrm{mc}(\pi(X_i))$, where $\mathrm{mc}(X)$ denotes the maximum size of a convex independent subset of X and π is the projection to the hyperplane $\{x_d = 0\}$.

Another interesting generalization of the Erdős–Szekeres theorem to $\mathbf{R}^d$ is mentioned in Exercise 5.4.3.

The bounds in the Erdős–Szekeres theorem were also investigated for special point sets, namely, for the so-called *dense sets* in the plane. An n-point $X \subset \mathbf{R}^2$ is called c-dense if the ratio of the maximum and minimum distances of points in X is at most $c\sqrt{n}$. For *every* planar n-point set, this ratio is at least $c_0\sqrt{n}$ for a suitable constant $c_0 > 0$, as an easy volume argument shows, and so the dense sets are quite well spread. Improving on slightly weaker results of Alon, Katchalski, and Pulleyblank [AKP89], Valtr [Val92a] showed, by a probabilistic argument, that every c-dense n-point set in general position contains

---

[1] The reader should be warned that four mathematicians named Tóth are mentioned throughout the book. For two of them, the *surname* is actually Fejes Tóth (László and Gábor), and for the other two it is just Tóth (Géza and Csaba).

a convex independent subset of at least $c_1 n^{1/3}$ points, for some $c_1 > 0$ depending on c, and he proved that this bound is asymptotically optimal. Simplified proofs, as well as many other results on dense sets, can be found in Valtr's thesis [Val94].

Exercises

1. Find a configuration of 8 points in general position in the plane with no 5 convex independent points (thereby showing that $n(5) \geq 9$). ②
2. Prove that the set $\{(i,j);\ i = 1, 2, \ldots, m,\ j = 1, 2, \ldots, m\}$ contains no convex independent subset with no more that $Cm^{2/3}$ points (with C some constant independent of m). ④
3. Prove that for each k there exists $n(k)$ such that each $n(k)$-point set in the plane contains a k-point convex independent subset or k points lying on a common line. ③
4. Prove an Erdős–Szekeres theorem in $\mathbf{R}^d$: For every k there exists $n = n_d(k)$ such that any n points in $\mathbf{R}^d$ in general position contain a k-point convex independent subset. ②
5. (A small improvement on the upper bound on $n(k)$) Let $X \subset \mathbf{R}^d$ be a planar set in general position with $f(k, \ell)+1$ points, where f is as in the second proof of Erdős–Szekeres, and let t be the (unique) topmost point of X. Prove that X contains a *k-cup with respect to t* or an *ℓ-cap with respect to t*, where a cup with respect to t is a subset $Y \subseteq X \setminus \{t\}$ such that $Y \cup \{t\}$ is in convex position, and a cap with respect to t is a subset $Y \subseteq X \setminus \{t\}$ such that $\{x, y, z, t\}$ is not in convex position for any triple $\{x, y, z\} \subseteq Y$. Infer that $n(k) \leq f(k-1, k)+1$. ②
6. Show that the construction of $X_{k,\ell}$ described in the text can be realized on a polynomial-size grid. That is, if we let $n = |X_{k,\ell}|$, we may suppose that the coordinates of all points in $X_{k,\ell}$ are integers between 1 and n^c with a suitable constant c. (This was observed by Valtr.) ③

3.2 Horton Sets

Let X be a set in $\mathbf{R}^d$. A k-point set $Y \subseteq X$ is called a *k-hole in X* if Y is convex independent and $\operatorname{conv}(Y) \cap X = Y$. In the plane, Y determines a convex k-gon with no points of X inside. Erdős raised the question about the rather natural strengthening of the Erdős–Szekeres theorem: Is it true that for every k there exists an $n(k)$ such that any $n(k)$-point set in the plane in general position has a k-hole?

A construction due to Horton, whose streamlined version we present below, shows that this is *false* for $k \geq 7$: There are arbitrarily large sets without a 7-hole. On the other hand, a positive result holds for $k \leq 5$. For $k = 6$, the answer is not known, and this "6-hole problem" appears quite challenging.

3.2.1 Proposition (The existence of a 5-hole). *Every sufficiently large planar point set in general position contains a 5-hole.*

Proof. By the Erdős–Szekeres theorem, we may assume that there exists a 6-point convex independent subset of our set X. Consider a 6-point convex independent subset $H \subseteq X$ with the smallest possible $|X \cap \mathrm{conv}(H)|$. Let $I = \mathrm{conv}(H) \cap (X \setminus H)$ be the points inside the convex hull of H.

- If $I = \emptyset$, we have a 6-hole.
- If there is one point x in I, we consider a diagonal that partitions the hexagon into two quadrilaterals:

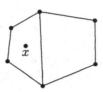

The point x lies in one of these quadrilaterals, and the vertices of the other quadrilateral together with x form a 5-hole.
- If $|I| \geq 2$, we choose an edge xy of $\mathrm{conv}(I)$. Let γ be an open half-plane bounded by the line xy and containing no points of I (it is determined uniquely unless $|I| = 2$).
If $|\gamma \cap H| \geq 3$, we get a 5-hole formed by x, y, and 3 points of $\gamma \cap H$. For $|\gamma \cap H| \leq 2$, we have one of the two cases indicated in the following picture:

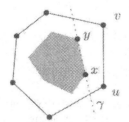 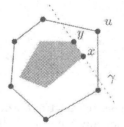

By replacing u and v by x and y in the left situation, or u by x in the right situation, we obtain a 6-point convex independent set having fewer points inside than H, which is a contradiction. $\qquad \square$

3.2.2 Theorem (Seven-hole theorem). *There exist arbitrarily large finite sets in the plane in general position without a 7-hole.*

The sets constructed in the proof have other interesting properties as well.

Definitions. Let X and Y be finite sets in the plane. We say that X is *high above* Y (and that Y is *deep below* X) if the following hold:

(i) No line determined by two points of $X \cup Y$ is vertical.
(ii) Each line determined by two points of X lies above all the points of Y.
(iii) Each line determined by two points of Y lies below all the points of X.

For a set $X = \{x_1, x_2, \ldots, x_n\}$, with no two points having equal x-coordinates and with notation chosen so that the x-coordinates of the x_i increase with i, we define the sets $X_0 = \{x_2, x_4, \ldots\}$ (consisting of the points with even indices) and $X_1 = \{x_1, x_3, \ldots\}$ (consisting of the points with odd indices).

A finite set $H \subset \mathbf{R}^2$ is a *Horton set* if $|H| \leq 1$, or the following conditions hold: $|H| > 1$, both H_0 and H_1 are Horton sets, and H_1 lies high above H_0 or H_0 lies high above H_1.

3.2.3 Lemma. *For every $n \geq 1$, an n-point Horton set exists.*

Proof. We note that one can produce a smaller Horton set from a larger one by deleting points from the right. We construct $H^{(k)}$, a Horton set of size 2^k, by induction.

We define $H^{(0)}$ as the point $(0,0)$. Suppose that we can construct a Horton set $H^{(k)}$ with 2^k points whose x-coordinates are $0, 1, \ldots, 2^k - 1$. The induction step goes as follows.

Let $A = 2H^{(k)}$ (i.e., $H^{(k)}$ expanded twice), and $B = A + (1, h_k)$, where h_k is a sufficiently large number. We set $H^{(k+1)} = A \cup B$. It is easily seen that if h_k is large enough, B lies high above A, and so $H^{(k+1)}$ is Horton as well. The set $H^{(3)}$ looks like this:

$$\cdot \quad \cdot \quad \cdot \quad \cdot \quad \cdot \quad \cdot$$

□

Closedness from above and from below. A set X in $\mathbf{R}^2$ is *r-closed from above* if for any r-cup in X there exists a point in X lying above the r-cup (i.e., above the bottom part of its convex hull).

Similarly, we define a set *r-closed from below* using r-caps.

3.2.4 Lemma. *Every Horton set is both 4-closed from above and 4-closed from below.*

Proof. We proceed by induction on the size of the Horton set. Let H be a Horton set, and assume that H_0 lies deep below H_1 (the other possible case is analogous). Let $C \subseteq H$ be a 4-cup.

If $C \subseteq H_0$ or $C \subseteq H_1$, then a point closing C from above exists by the inductive hypothesis. Thus, let $C \cap H_0 \neq \emptyset \neq C \cap H_1$.

The cup C may have at most 2 points in H_1 (the upper part): If there were 3 points, say a, b, c (in left-to-right order), then H_0 lies below the lines ab and bc, and so the remaining point of C, which was supposed to lie in H_0, cannot form a cup with $\{a, b, c\}$:

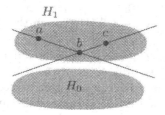

This means that C has at least 2 points, a and b, in the lower part H_0. Since the points of H_0 and H_1 alternate along the x-axis, there is a point $c \in H_1$ between a and b in the ordering by x-coordinates. This c is above the segment ab, and so it closes the cup C from above. We argue similarly for a 4-cap. $\quad\square$

3.2.5 Proposition. *No Horton set contains a 7-hole.*

Proof. (Very similar to the previous one.) For contradiction, suppose there is a 7-hole X in the considered Horton set H. If $X \subseteq H_0$ or $X \subseteq H_1$, we use induction. Otherwise, we select the part (H_0 or H_1) containing the larger portion of X; this has at least 4 points of X. If this part is, say, H_0, and it lies deep below H_1, these 4 points must form a cup in H_0, for if some 3 of them were a cap, no point of H_1 could complete them to a convex independent set. By Lemma 3.2.4, H_0 (being a Horton set) contains a point closing the 4-cup from above. Such a point must be contained in the convex hull of the 7-hole X, a contradiction. $\quad\square$

Bibliography and remarks. The existence of a 5-hole in every 10-point planar set in general position was proved by Harborth [Har79]. Horton [Hor83] constructed arbitrarily large sets without a 7-hole; we followed a presentation of his construction according to Valtr [Val92a].

The question of existence of k-holes can be generalized to point sets in $\mathbf{R}^d$. Valtr [Val92b] proved that $(2d+1)$-holes exist in all sufficiently large sets in general position in $\mathbf{R}^d$, and he constructed arbitrarily large sets without k-holes for $k \geq 2^{d-1}(P(d-1)+1)$, where $P(d-1)$ is the product of the first $d-1$ primes. We outline the construction. Let H

be a finite set in $\mathbf{R}^d$, $d \geq 2$, in general position (no $d{+}1$ on a common hyperplane and no two sharing the value of any coordinate). Let $H = \{x_1, x_2, \ldots, x_n\}$ be enumeration of H by increasing first coordinate, and let $H_{q,r} = \{x_i : i \equiv r \,(\mathrm{mod}\ q)\}$. Let $p_1 = 2, p_2 = 3, \ldots, p_{d-1}$ be the first $d{-}1$ primes, and let us write $p = p_{d-1}$ for brevity. The set H is called d-Horton if

(i) its projection on the first $d{-}1$ coordinates is a $(d{-}1)$-Horton set in $\mathbf{R}^{d-1}$ (where all sets in $\mathbf{R}^1$ are 1-Horton), and

(ii) either $|H| \leq 1$ or all the sets $H_{p,r}$ are d-Horton, $r = 0, 1, \ldots, p{-}1$, and for every subset $I \subseteq \{0, 1, \ldots, p{-}1\}$ of at least two indices, there is a partition $I = J \cup K$, $J \neq \emptyset \neq K$, such that $\bigcup_{r \in J} H_{p,r}$ lies high above $\bigcup_{r \in K} H_{p,r}$.

Here A lies high above B if every hyperplane determined by d points of A lies above B (in the direction of the dth coordinate) and vice versa. Arbitrarily large d-Horton sets can be constructed by induction: We first construct the $(d{-}1)$-dimensional projection, and then we determine the dth coordinates suitably to meet condition (ii). The nonexistence of large holes is proved using an appropriate generalization of r-closedness from above and from below.

Since large sets generally need not contain k-holes, it is natural to look for other, less special, configurations. Bialostocki, Dierker, and Voxman [BDV91] proved the existence of *k-holes modulo q*: For every q and for all $k \geq q{+}2$, each sufficiently large set X (in terms of q and k) in general position contains a k-point convex independent subset Y such that the number of points of X in the interior of $\mathrm{conv}(Y)$ is divisible by q; see Exercise 6. Károlyi, Pach, and Tóth [KPT01] obtained a similar result with the weaker condition $k \geq \frac{5}{6}q + O(1)$. They also showed that every sufficiently large *1-almost convex* set in the plane contains a k-hole, and Valtr [Val01] extended this to k-almost convex sets, where X is k-almost convex if no triangle with vertices at points of X contains more than k points of X inside.

Exercises

1. Prove that an n-point Horton set contains no convex independent subset with more than $4 \log_2 n$ points. ☐2
2. Find a configuration of 9 points in the plane in general position with no 5-hole. ☐2
3. Prove that every sufficiently large set in general position in $\mathbf{R}^3$ has a 7-hole. ☐3
4. Let H be a Horton set and let $k \geq 7$. Prove that if $Y \subseteq H$ is a k-point subset in convex position, then $|H \cap \mathrm{conv}(Y)| \geq 2^{\lfloor k/4 \rfloor}$. Thus, not only does H contain no k-holes, but each convex k-gon has even exponentially many points inside. ☐4

This result is due to Nyklová [Nyk00], who proved exact bounds for Horton sets and observed that the number of points inside each convex k-gon can be somewhat increased by replacing each point of a Horton set by a tiny copy of a small Horton set.

5. Call a set $X \subset \mathbf{R}^2$ in general position *almost convex* if no triangle with vertices at points of X contains more than 1 point of X in its interior. Let $X \subset \mathbf{R}^2$ be a finite set in general position such that no triangle with vertices at vertices of conv(X) contains more than 1 point of X. Prove that X is almost convex. ③

6. (a) Let $q \geq 2$ be an integer and let $k = mq+2$ for an integer $m \geq 1$. Prove that every sufficiently large set $X \subset \mathbf{R}^2$ in general position contains a k-point convex independent subset Y such that the number of points of X in the interior of conv(Y) is divisible by q. Use Ramsey's theorem for triples. ④

(b) Extend the result of (a) to all $k \geq q + 2$. ③

4

Incidence Problems

In this chapter we study a very natural problem of combinatorial geometry: the maximum possible number of incidences between m points and n lines in the plane. In addition to its mathematical appeal, this problem and its relatives are significant in the analysis of several basic geometric algorithms. In the proofs we encounter number-theoretic arguments, results about graph drawing, the probabilistic method, forbidden subgraphs, and line arrangements.

4.1 Formulation

Point–line incidences. Consider a set P of m points and a set L of n lines in the plane. What is the maximum possible number of their *incidences*, i.e., pairs (p, ℓ) such that $p \in P$, $\ell \in L$, and p lies on ℓ? We denote the number of incidences for specific P and L by $I(P, L)$, and we let $I(m, n)$ be the maximum of $I(P, L)$ over all choices of an m-element P and an n-element L. For example, the following picture illustrates that $I(3, 3) \geq 6$,

and it is not hard to see that actually $I(3, 3) = 6$.

A trivial upper bound is $I(m, n) \leq mn$, but it it can never be attained unless $m = 1$ or $n = 1$. In fact, if m has a similar order of magnitude as n then $I(m, n)$ is asymptotically much smaller than mn. The order of magnitude is known exactly:

4.1.1 Theorem (Szemerédi–Trotter theorem). *For all $m, n \geq 1$, we have $I(m, n) = O(m^{2/3}n^{2/3} + m + n)$, and this bound is asymptotically tight.*

We give two proofs in the sequel, one simpler and one including techniques useful in more general situations. We will mostly consider only the most interesting case $m = n$. The general case needs no new ideas but only a little more complicated calculation.

Of course, the problem of point–line incidences can be generalized in many ways. We can consider incidences between points and hyperplanes in higher dimensions, or between points in the plane and some family of curves, and so on. A particularly interesting case is that of points and unit circles, which is closely related to counting unit distances.

Unit distances and distinct distances. Let $U(n)$ denote the maximum possible number of pairs of points with unit distance in an n-point set in the plane. For $n \leq 3$ we have $U(n) = \binom{n}{2}$ (all distances can be 1), but already for $n = 4$ at most 5 of the 6 distances can be 1; i.e., $U(4) = 5$:

We are interested in the asymptotic behavior of the function $U(n)$ for $n \to \infty$.

This can also be reformulated as an incidence problem. Namely, consider an n-point set P and draw a unit circle around each point of p, thereby obtaining a set C of n unit circles. Each pair of points at unit distance contributes two point–circle incidences, and hence $U(n) \leq \frac{1}{2} I_{\text{1circ}}(n, n)$, where $I_{\text{1circ}}(m, n)$ denotes the maximum possible number of incidences between m points and n unit circles.

Unlike the case of point–line incidences, the correct order of magnitude of $U(n)$ is not known. An upper bound of $O(n^{4/3})$ can be obtained by modifying proofs of the Szemerédi–Trotter theorem. But the best known lower bound is $U(n) \geq n^{1+c_1/\log\log n}$, for some positive constant c_1; this is superlinear in n but grows more slowly than $n^{1+\varepsilon}$ for every fixed $\varepsilon > 0$.

A related quantity is the minimum possible number of distinct distances determined by n points in the plane; formally,

$$g(n) = \min_{P \subset \mathbf{R}^2 : |P| = n} |\{\text{dist}(x, y) : x, y \in P\}|.$$

Clearly, $g(n) \geq \binom{n}{2}/U(n)$, and so the bound $U(n) = O(n^{4/3})$ mentioned above gives $g(n) = \Omega(n^{2/3})$. This has been improved several times, and the current best lower bound is approximately $\Omega(n^{0.863})$. The best known upper bound is $O(n/\sqrt{\log n})$.

Arrangements of lines. We need to introduce some terminology concerning line arrangements. Consider a finite set L of lines in the plane. They divide the plane into convex subsets of various dimensions, as is indicated in the following picture with 4 lines:

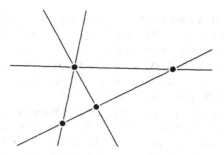

The intersections of the lines, indicated by black dots, are called the *vertices*. By removing all the vertices lying on a line $\ell \in L$, the line is split into two unbounded rays and several segments, and these parts are the *edges*. Finally, by deleting all the lines of L, the plane is divided into open convex polygons, called the *cells*. In Chapter 6 we will study arrangements of lines and hyperplanes further, but here we need only this basic terminology and (later) the simple fact that an arrangement of n lines in general position has $\binom{n}{2}$ vertices, n^2 edges, and $\binom{n}{2}+n+1$ cells. For the time being, the reader can regard this as an exercise, or wait until Chapter 6 for a proof.

Many cells in arrangements. What is the maximum total number of vertices of m distinct cells in an arrangement of n lines in the plane? Let us denote this number by $K(m,n)$. A simple construction shows that the maximum number of incidences $I(m,n)$ is asymptotically bounded from above by $K(m,n)$; more exactly, we have $I(m,n) \leq \frac{1}{2} K(m,2n)$. To see this, consider a set P of m points and a set L of n lines realizing $I(m,n)$, and replace each line $\ell \in L$ by a pair of lines ℓ', ℓ'' parallel to ℓ and lying at distance ε from ℓ:

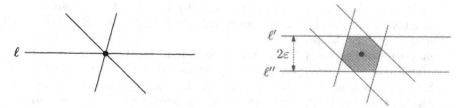

If $\varepsilon > 0$ is sufficiently small, then a point $p \in P$ incident to k lines in the original arrangement now lies in a tiny cell with $2k$ vertices in the modified arrangement.

It turns out that $K(m,n)$ has the same order of magnitude as $I(m,n)$, and the upper bound can be obtained by methods similar to those used for $I(m,n)$. In higher-dimensional problems, even determining the maximum possible complexity of a single cell can be quite challenging. For example, the maximum complexity of a single cell in an arrangement of n hyperplanes is described by the so-called upper bound theorem from the 1970s, which will be discussed in Chapter 5.

Bibliography and remarks. This chapter is partially based on a nice presentation of the discussed topics in the book by Pach and Agarwal [PA95], which we recommend as a source of additional information concerning history, bibliographic references, and various related problems. But we also include some newer results and techniques discovered since the publication of that book.

The following neat problem concerning point–line incidences was posed by Sylvester [Syl93] in 1893: Prove that it is impossible to arrange a finite number of points in the plane so that a line through every two of them passes through a third, unless they all lie on the same line. This problem remained unsolved until 1933, when it was asked again by Erdős and solved shortly afterward by Gallai. The solution shows, in particular, that it is impossible to embed the points of a finite projective plane $\mathcal{F}$ into $\mathbf{R}^2$ in such a way that points of each line of $\mathcal{F}$ lie on a straight line in $\mathbf{R}^2$. For example, the well-known drawing of the Fano plane of order 3 has to contain a curved line:

Recently Pinchasi [Pin02] proved the following conjecture of Bezdek, resembling Sylvester's problem: For every finite family of at least 5 unit circles in the plane, every two of them intersecting, there exists an intersection point common to exactly 2 of the circles.

The problems of estimating the maximum number of point–line incidences, the maximum number of unit distances, and the minimum number of distinct distances were raised by Erdős [Erd46]. For *point–line incidences*, he proved the lower bound $I(m,n) = \Omega(m^{2/3}n^{2/3} + m + n)$ (see Section 4.2) and conjectured it to be the right order of magnitude. This was first proved by Szemerédi and Trotter [ST83]. Simpler proofs were found later by Clarkson, Edelsbrunner, Guibas, Sharir, and Welzl [CEG+90], by Székely [Szé97], and by Aronov and Sharir [AS01a]; they are quite different from one another, and we discuss them all in this chapter.

Tóth [Tót01a] proved the analogy of the Szemerédi–Trotter theorem for the complex plane; he used the original Szemerédi–Trotter technique, since none of the simpler proofs seems to work there.

A beautiful application of techniques of Clarkson et al. [CEG+90] in geometric measure theory can be found in Wolff [Wol97]. This paper deals with a variation of the *Kakeya problem*: It shows that any Borel set in the plane containing a circle of every radius has Hausdorff dimension 2.

For *unit distances in the plane* Erdős [Erd46] established the lower bound $U(n) = \Omega(n^{1+c/\log\log n})$ (Section 4.2) and again conjectured it to be tight, but the best known upper bound remains $O(n^{4/3})$. This was first shown by Spencer, Szemerédi, and Trotter [SST84], and it can be re-proved by modifying each of the proofs mentioned above for point–line incidences. Further improvement of the upper bound probably needs different, more "algebraic," methods, which would use the "circularity" in a strong way, not just in the form of simple combinatorial axioms (such as that two points determine at most two unit circles).

For the analogous problem of *unit distances among n points in* $\mathbf{R}^3$, Erdős [Erd60] proved $\Omega(n^{4/3}\log\log n)$ from below and $O(n^{5/3})$ from above. The example for the lower bound is the grid $\{1, 2, \ldots, \lfloor n^{1/3}\rfloor\}^3$ appropriately scaled; the bound $\Omega(n^{4/3})$ is entirely straightforward, and the extra $\log\log n$ factor needs further number-theoretic considerations. The upper bound follows by an argument with forbidden $K_{3,3}$; similar proofs are shown in Section 4.5. The current best bound is close to $O(n^{3/2})$; more precisely, it is $n^{3/2}2^{O(\alpha^2(n))}$ [CEG+90]. Here the function $\alpha(n)$, to be defined in Section 7.2, grows extremely slowly, more slowly than $\log n$, $\log\log n$, $\log\log\log n$, etc. In *dimensions 4 and higher*, the number of unit distances can be $\Omega(n^2)$ (Exercise 2). Here even the constant at the leading term is known; see [PA95]. Among other results related to the unit-distance problems and considering point sets with various restrictions, we mention a neat construction of Erdős, Hickerson, and Pach [EHP89] showing that, for every $\alpha \in (0, 2)$, there is an n-point set on the 2-dimensional unit sphere with the distance α occurring at least $\Omega(n\log^* n)$ times (the special distance $\sqrt{2}$ can even occur $\Omega(n^{4/3})$ times), and the annoying (and still unsolved) problem of Erdős and Moser, whether the number of unit distances in an n-point planar set in convex position is always bounded by $O(n)$ (see [PA95] for partial results and references).

For *distinct distances in the plane*, the best known upper bound, due to Erdős, is $O(n/\sqrt{\log n})$. This bound is attained for the $\sqrt{n} \times \sqrt{n}$ square grid. After a series of increases of the lower bound (Moser [Mos52], Chung [Chu84], Beck [Bec83], Clarkson et al. [CEG+90], Chung, Szemerédi, and Trotter [CST92], Székely [Szé97], Solymosi and Tóth [ST01]) the current record is $\Omega(n^{4/(5-1/e)-\varepsilon})$ for every fixed $\varepsilon > 0$ (the exponent is approximately 0.863) by Tardos [Tar01], who improved a number-theoretic lemma in the Solymosi–Tóth proof. Aronov and Sharir [AS01b] obtained the lower bound of approximately $n^{0.526}$ for distinct distances in $\mathbf{R}^3$.

Another challenging quantity is the number $I_{\mathrm{circ}}(m, n)$ of *incidences of m points with n arbitrary circles* in the plane. The lower bound for point–line incidences can be converted to an example with

m points, n circles, and $\Omega(m^{2/3}n^{2/3} + m + n)$ incidences, but in the case of $I_{\text{circ}}(m, n)$, this lower bound is *not* the best possible for all m and n: Consider an example of an n-point set with $t = O(n/\sqrt{\log n})$ distinct distances and draw the t circles with these distances as radii around each point; the resulting $tn = o(n^2)$ circles have $\Omega(n^2)$ incidences with the n points. The current record in the upper bound is due to Aronov and Sharir [AS01a], and for $m = n$ it yields $I_{\text{circ}}(n, n) = O(n^{15/11+\varepsilon}) = O(n^{1.364})$. A little more about their approach is mentioned in the notes to Section 4.5, including an outline of a proof of a weaker bound $I_{\text{circ}}(n, n) = O(n^{1.4})$. Two other methods for obtaining upper bounds are indicated in Exercises 4.4.2 and 4.6.4.

More generally, one can consider $I(P, \Gamma)$, the number of incidences between an m-point $P \subset \mathbf{R}^2$ and a family Γ of n planar curves. Pach and Sharir [PS98a] proved by Székely's method that if Γ is a *family of curves with k degrees of freedom and multiplicity type s*, meaning that for any k points there are at most s curves of Γ passing through all of them and no two curves intersect in more than k points, then $|I(P, \Gamma)| = O\left(m^{k/(2k-1)}n^{1-1/(2k-1)} + m + n\right)$, with the constant of proportionality depending on k and s. Earlier [PS92], they proved the same bound with some additional technical assumptions on the family Γ by the technique of Clarkson et al. [CEG+90]. Most likely this bound is not tight for $k \geq 3$. Aronov and Sharir [AS01a] improved the bound slightly for Γ a family of graphs of univariate polynomials of degree at most k. The best known lower bound is mentioned in the notes to Section 4.2 below.

Point–plane incidences. Considering n points on a line in $\mathbf{R}^3$ and m planes containing that line, we see that the number of incidences can be mn without further assumptions on the position of the points and/or planes. Agarwal and Aronov [AA92] proved the upper bound $O(m^{3/5}n^{4/5} + m + n)$ for the number of incidences between m planes and n points in $\mathbf{R}^3$ if no 3 of the points are collinear, slightly improving on a result of Edelsbrunner, Guibas, and Sharir [EGS90]. In dimension d, the maximum number of incidences of n hyperplanes with m vertices of their arrangement is $O(m^{2/3}n^{d/3} + n^{d-1})$ [AA92], and this is tight for $m \geq n^{d-2}$ (for smaller m, the trivial $O(mn)$ bound is tight).

The complexity of *many cells in an arrangement of lines* was first studied by Canham [Can69], who proved $K(m, n) = O(m^2 + n)$, using the fact that two cells can have at most 4 lines incident to both of them (essentially a "forbidden $K_{2,5}$" argument; see Section 4.5). The tight bound $O(m^{2/3}n^{2/3} + m + n)$ was first achieved by Clarkson et al. [CEG+90]. Among results for the complexity of m cells in other types of arrangements we mention the bound $O(m^{2/3}n^{2/3} + n\alpha(n) + n\log m)$ for segments by Aronov, Edelsbrunner, Guibas, and Sharir [AEGS92], $O(m^{2/3}n^{2/3}\alpha(n)^{1/3} + n)$ for unit circles [CEG$^+$90] (improved to $O(m^{2/3}n^{2/3}) + n$) by Agarwal, Aronov, and Sharir [AAS01]), $O(m^{3/5}n^{4/5}2^{0.4\alpha(n)} + n)$ for arbitrary circles [CEG$^+$90] (also improved in [AAS01]; see the notes to Section 4.5), $O(m^{2/3}n + n^2)$ for planes in $\mathbf{R}^3$ by Agarwal and Aronov [AA92] (which is tight), and $O(m^{1/2}n^{d/2}(\log n)^{(\lfloor d/2\rfloor - 1)/2})$ for hyperplanes in $\mathbf{R}^d$ by Aronov, Matoušek, and Sharir [AMS94]. If one counts only facets of m cells in an arrangement of n hyperplanes in $\mathbf{R}^d$, then the tight bound is $O(m^{2/3}n^{d/3} + n^{d-1})$ [AA92]. A few more references on this topic can be found in Agarwal and Sharir [AS00a].

The number of similar copies of a configuration. The problem of unit distances can be rephrased as follows. Let K denote a set consisting of two points in the plane with unit distance. What is the maximum number of congruent copies of K that can occur in an n-point set in the plane? This reformulation opens the way to various interesting generalizations, where one can vary K, or one can consider homothetic or similar copies of K, and so on. Elekes's survey [Ele01] nicely describes these problems, their relation to the incidence bounds, and other connections. Here we sketch some of the main developments.

Beautiful results were obtained by Laczkovich and Ruzsa [LR97], who investigated the maximum number of *similar* copies of a given finite configuration K that can be contained in an n-point set in the plane. Earlier, Elekes and Erdős [EE94] proved that this number is $\Omega(n^{2-(\log n)^{-c}})$ for all K, where $c > 0$ depends on K, and it is $\Omega(n^2)$ whenever all the coordinates of the points in K are algebraic numbers. Building on these results, Laczkovich and Ruzsa proved that the maximum number of similar copies of K is $\Omega(n^2)$ if and only if the *cross-ratio* of every 4 points of K is algebraic, where the cross-ratio of points $a, b, c, d \in \mathbf{R}^2$ equals $\frac{c-a}{c-b} \cdot \frac{d-b}{d-a}$, with a, b, c, d interpreted as complex numbers in this formula.

Their proof makes use of very nice results from the additive theory of numbers, most notably a theorem of Freiman [Fre73] (also see Ruzsa [Ruz94]): If A is a set of n integers such that $|A + A| \leq cn$, where $A + A = \{a + b : a, b \in A\}$ and $c > 0$ is a constant, then A is contained in a d-dimensional generalized arithmetic progression of size at most Cn, with C and d depending on c only. Here a d-dimensional *generalized arithmetic progression* is a set of integers of the form $\{z_0 + i_1 q_1 + i_2 q_2 + \cdots + i_d q_d : i_1 = 0, 1, \ldots, n_1, i_2 = 0, 1, \ldots, n_2, \ldots, i_d = 0, 1, \ldots, n_d\}$ for some integers z_0 and $q_1, q_2, \ldots, q_d$. It is easy to see that $|A + A| \leq C_d|A|$ for every d-dimensional generalized arithmetic progression, and Freiman's theorem is a sort of converse statement: If $|A + A| = O(|A|)$, then A is not too far from a generalized arithmetic progression. (Freiman's theorem has also been used for incidence-related problems by Erdős, Füredi, Pach, and Ruzsa [EFPR93], and

Gowers's paper [Gow98] is an impressive application of results of this type in combinatorial number theory.)

Polynomials attaining $O(n)$ values on Cartesian products. Interesting results related to those of Freiman, as well as to incidence problems, were obtained in a series of papers by Elekes and his coworkers (they are described in the already mentioned survey [Ele01]). Perhaps even more significant than the particular results is the direction of research opened by them, combining algebraic and combinatorial tools. Let us begin with a conjecture of Purdy proved by Elekes and Rónyai [ER00] as a consequence of their theorems. Let P be a set of n distinct points lying on a line $u \subset \mathbf{R}^2$, let Q be a set of n distinct points lying on a line $v \subset \mathbf{R}^2$, and let $\mathrm{Dist}(P,Q) = \{\|p - q\| : p \in P, q \in Q\}$. If, for example, u and v are parallel and if both P and Q are placed with equal spacing along their lines, then $|\mathrm{Dist}(P,Q)| \le 2n$. Another such case is $P = \{(\sqrt{i}, 0) : i = 1, 2, \ldots, n\}$ and $Q = \{(0, \sqrt{j}) : j = 1, 2, \ldots, n\}$: This time u and v are perpendicular, and again $|\mathrm{Dist}(P,Q)| \le 2n$. According to Purdy's conjecture, these are the only possible positions of u and v if the number of distances is linear: For every $C > 0$ there is an n_0 such that if $n \ge n_0$ and $|\mathrm{Dist}(P,Q)| \le Cn$, then u and v are parallel or perpendicular.

If we parameterize the line u by a real parameter x, and v by y, and denote the cosine of the angle of u and v by λ, then Purdy's conjecture can be reformulated in algebraic terms as follows: Whenever $X, Y \subset \mathbf{R}$ are n-point sets such that the polynomial $F(x,y) = x^2 + y^2 + 2\lambda xy$ attains at most Cn distinct values on $X \times Y$, i.e., $|\{F(x,y) : x \in X, y \in Y\}| \le Cn$, then necessarily $\lambda = 0$ or $\lambda = \pm 1$, provided that $n \ge n_0(C)$.

Elekes and Rónyai [ER00] characterized all bivariate polynomials $F(x,y)$ that attain only $O(n)$ values on Cartesian products $X \times Y$. For every C, d there exists an n_0 such that if $F(x,y)$ is a bivariate polynomial of degree at most d and $X, Y \subset \mathbf{R}$ are n-point sets, $n \ge n_0$, such that $F(x,y)$ attains at most Cn distinct values on $X \times Y$, then $F(x,y)$ has one of the two special forms $f(g(x) + h(y))$ or $f(g(x)h(y))$, where f, g, h are univariate polynomials. In fact, we need not consider the whole $X \times Y$; it suffices to assume that F attains at most Cn values on an arbitrary subset of δn^2 pairs from $X \times Y$ (with n_0 depending on δ, too). A similar result holds for a bivariate *rational function* $F(x,y)$, with one more special form to consider, namely, $F(x,y) = f((g(x) + h(y))/(1 - g(x)h(y)))$.

We indicate a proof only for the special case of the polynomial $F(x,y) = x^2 + y^2 + 2\lambda xy$ from Purdy's conjecture (following Elekes [Ele99]); the basic idea of the general case is similar, but several more tools are needed, especially from elementary algebraic geometry. So let $Z = F(X,Y)$ be the set of values attained by F on $X \times Y$. For each $y_i \in Y$, put $f_i(x) = F(x, y_i)$, and define the family $\Gamma = \{\gamma_{ij} : i, j =$

$1, 2, \ldots, n$, $i \neq j\}$ of planar curves by $\gamma_{ij} = \{(f_i(t), f_j(t)): t \in \mathbf{R}\}$ (this is the key trick). Each γ_{ij} contains at least $\frac{n}{2}$ points of $Z \times Z$, since among the n points $(f_i(x_k), f_j(x_k))$, $x_k \in X$, no 3 can coincide, because the f_i are quadratic polynomials. Moreover, a straightforward (although lengthy) calculation using resultants verifies that for $\lambda \notin \{0, \pm 1\}$, at most 8 distinct curves γ_{ij} can pass through any two given distinct points $a, b \in \mathbf{R}^2$. Consequently, Γ contains at least $\frac{1}{9}n^2$ distinct curves. Using the bound of Pach and Sharir [PS92], [PS98a] on the number of incidences between points and algebraic curves mentioned above, with $Z \times Z$ as the points and the at least $\frac{1}{9}n^2$ distinct curves of Γ as the curves, we obtain that $|Z| = \Omega(n^{5/4})$. So there is even a significant gap: Either $\lambda \in \{0, \pm 1\}$, and then $F(X, Y)$ can have only $2n$ distinct elements for suitable X, Y, or $\lambda \notin \{0, \pm 1\}$ and then $|F(X, Y)| = \Omega(n^{5/4})$ for all X, Y.

Perhaps this latter bound can be improved to $\Omega(n^{2-\varepsilon})$ for every $\varepsilon > 0$ (so there would be an almost-dichotomy: either the number of values of F can be linear, or it has to be always near-quadratic). On the other hand, it is known that the polynomial $x^2 + y^2 + xy$ attains only $O(n^2/\sqrt{\log n})$ distinct values for x, y ranging over $\{1, 2, \ldots, n\}$, and so the bound need not always be linear or quadratic. It seems likely that in the general case of the Elekes–Rónyai theorem the number of values attained by F should be near-quadratic unless F is one of the special forms.

Further generalizations of the Elekes–Rónyai theorem were obtained by Elekes and Szabó; see [Ele01].

Exercises

1. Let $I_{1\text{circ}}(m, n)$ be the maximum number of incidences of m points with n unit circles and let $U(n)$ be the maximum number of unit distances for an n-point set.
   (a) Prove that $I_{1\text{circ}}(2n, 2n) = O(I_{1\text{circ}}(n, n))$. ☐1
   (b) We have seen that $U(n) \leq \frac{1}{2}I_{1\text{circ}}(n, n)$. Prove that $I_{1\text{circ}}(n, n) = O(U(n))$. ☐2
2. Show that an n-point set in $\mathbf{R}^4$ may determine $\Omega(n^2)$ unit distances. ☐4
3. Prove that if $X \subset \mathbf{R}^d$ is a set where every two points have distance 1, then $|X| \leq d+1$. ☐3
4. What can be said about the maximum possible number of incidences of n lines in $\mathbf{R}^3$ with m points? ☐2
5. Use the Szemerédi–Trotter theorem to show that n points in the plane determine at most
   (a) $O(n^{7/3})$ triangles of unit area, ☐3
   (b) $O(n^{7/3})$ triangles with a given fixed angle α. ☐2

The result in (a) was first proved by Erdős and Purdy [EP71]. As for (b), Pach and Sharir [PS92] proved the better bound $O(n^2 \log n)$; also see [PA95].

6. (a) Using the Szemerédi–Trotter theorem, show that the maximum possible number of distinct lines such that each of them contains at least k points of a given m-point set P in the plane is $O(m^2/k^3 + m/k)$. [2]
    (b) Prove that such lines have at most $O(m^2/k^2 + m)$ incidences with P. [3]

7. (Many points on a line or many lines)
    (a) Let P be an m-point set in the plane and let $k \le \sqrt{m}$ be an integer parameter. Prove (using Exercise 6, say) that at most $O(m^2/k)$ pairs of points of P lie on lines containing at least k and at most $\sqrt{m}$ points of P. [3]
    (b) Similarly, for $K \ge \sqrt{m}$, the number of pairs lying on lines with at least $\sqrt{m}$ and at most K points is $O(Km)$. [3]
    (c) Prove the following theorem of Beck [Bec83]: There is a constant $c > 0$ such that for any n-point $P \subseteq \mathbf{R}^2$, at least cn^2 distinct lines are determined by P or there is a line containing at least cn points of P. [2]
    (d) Derive that there exists a constant $c > 0$ such that for every n-point set P in the plane that does not lie on a single line there exists a point $p \in P$ lying on at least cn distinct lines determined by points of P. [1]
    Part (d) is a weak form of the *Dirac–Motzkin conjecture*; the full conjecture, still unsolved, is the same assertion with $c = \frac{1}{2}$.

8. (Many distinct radii)
    (a) Assume that $I_{\mathrm{circ}}(m, n) = O(m^\alpha n^\beta + m + n)$ for some constants $\alpha < 1$ and $\beta < 1$, where $I_{\mathrm{circ}}(m, n)$ is the maximum number of incidences of m points with n circles in the plane. In analogy with to Exercise 7, derive that there is a constant $c > 0$ such that for any n-point set $P \subset \mathbf{R}^2$, there are at least cn^3 distinct circles containing at least 3 points of P each or there is a circle or line containing at least cn points of P. [3]
    (b) Using (a), prove the following result of Elekes (an answer to a question of Balog): For any n-point set $P \subset \mathbf{R}^2$ not lying on a common circle or line, the circles determined by P (i.e., those containing 3 or more points of P) have $\Omega(n)$ distinct radii. [4]
    (c) Find an example of an n-point set with only $O(n)$ distinct radii. [3]

9. (Sums and products cannot both be few) Let $A \subset \mathbf{R}$ be a set of n distinct real numbers and let $S = A + A = \{a + b: a, b \in A\}$ and $P = A \cdot A = \{ab: a, b \in A\}$.
    (a) Check that each of the n^2 lines $\{(x, y) \in \mathbf{R}^2: y = a(x - b)\}$, $a, b \in A$, contains at least n distinct points of $S \times P$. [1]
    (b) Conclude using Exercise 6 that $|S \times P| = \Omega(n^{5/2})$, and consequently, $\max(|S|, |P|) = \Omega(n^{5/4})$; i.e., the set of sums and the set of products can never both have almost linear size. [2] (This is a theorem of Elekes [Ele97] improving previous results on a problem raised by Erdős and Szemerédi.)

10. (a) Find n-point sets in the plane that contain $\Omega(n^2)$ similar copies of the vertex set of an equilateral triangle. ☐1

(b) Verify that the following set P_m has $n = O(m^4)$ points and contains $\Omega(n^2)$ similar copies of the vertex set of a regular pentagon: Identify $\mathbf{R}^2$ with the complex plane $\mathbf{C}$, let $\omega = e^{2\pi i/5}$ denote a primitive 5th root of unity, and put

$$P_m = \{i_0 + i_1\omega + i_2\omega^2 + i_3\omega^3 \colon i_0, i_1, i_2, i_3 \in \mathbf{Z}, \ |i_j| \leq m\}.$$

☐4

The example in (b) is from Elekes and Erdős [EE94], and the set P_∞ is called a *pentagonal pseudolattice*. The following picture shows P_2:

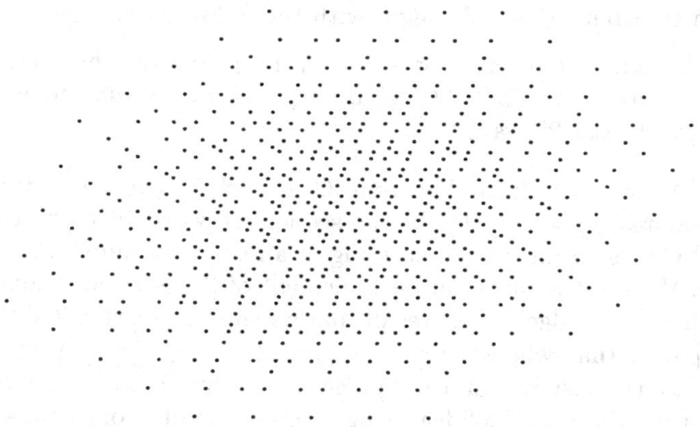

4.2 Lower Bounds: Incidences and Unit Distances

4.2.1 Proposition (Many point–line incidences). *We have* $I(n,n) = \Omega(n^{4/3})$, *and so the upper bound for the maximum number of incidences of n points and n lines in the plane in the Szemerédi–Trotter theorem is asymptotically optimal.*

It is not easy to come up with good constructions "by hand." Small cases do not seem to be helpful for discovering a general pattern. Surprisingly, an asymptotically optimal construction is quite simple. The appropriate lower bound for $I(m,n)$ with $n \neq m$ is obtained similarly (Exercise 1).

Proof. For simplicity, we suppose that $n = 4k^3$ for a natural number k. For the point set P, we choose the $k \times 4k^2$ grid; i.e., we set $P = \{(i,j) \colon i = 0, 1, 2, \ldots, k-1, \ j = 0, 1, \ldots, 4k^2-1\}$. The set L consists of all the lines with equations $y = ax + b$, where $a = 0, 1, \ldots, 2k-1$ and $b = 0, 1, \ldots, 2k^2-1$. These are n lines, as it should be. For $x \in [0, k)$, we have $ax + b < ak + b <$

$2k^2 + 2k^2 = 4k^2$. Therefore, for each $i = 0, 1, \ldots, k-1$, each line of L contains a point of P with the x-coordinate equal to i, and so $I(P, L) \geq k \cdot |L| = \frac{1}{4} n^{4/3}$.

             □

Next, we consider unit distances, where the construction is equally simple but the analysis uses considerable number-theoretic tools.

4.2.2 Theorem (Many unit distances). *For all $n \geq 2$, there exist configurations of n points in the plane determining at least $n^{1+c_1/\log\log n}$ unit distances, with a positive constant c_1.*

A configuration with the asymptotically largest known number of unit distances is a $\sqrt{n} \times \sqrt{n}$ regular grid with a suitably chosen step. Here unit distances are related to the number of possible representations of an integer as a sum of two squares. We begin with the following claim:

4.2.3 Lemma. *Let $p_1 < p_2 < \cdots < p_r$ be primes of the form $4k+1$, and put $M = p_1 p_2 \cdots p_r$. Then M can be expressed as a sum of two squares of integers in at least 2^r ways.*

Proof. As we know from Theorem 2.3.1, each p_j can be written as a sum of two squares: $p_j = a_j^2 + b_j^2$. In the sequel, we work with the ring $\mathbf{Z}[\mathrm{i}]$, the so-called Gaussian integers, consisting of all complex numbers $u + \mathrm{i}v$, where $u, v \in \mathbf{Z}$. We use the fact that each element of $\mathbf{Z}[\mathrm{i}]$ can be uniquely factored into primes. From algebra, we recall that a *prime* in the ring $\mathbf{Z}[\mathrm{i}]$ is an element $\gamma \in \mathbf{Z}[\mathrm{i}]$ such that whenever $\gamma = \gamma_1 \gamma_2$ with $\gamma_1, \gamma_2 \in \mathbf{Z}[\mathrm{i}]$, then $|\gamma_1| = 1$ or $|\gamma_2| = 1$. Both existence and uniqueness of prime factorization follows from the fact that $\mathbf{Z}[\mathrm{i}]$ is a Euclidean ring (see an introductory course on algebra for an explanation of these notions).

Let us put $\alpha_j = a_j + \mathrm{i} b_j$, and let $\bar{\alpha}_j = a_j - \mathrm{i} b_j$ be the complex conjugate of α_j. We have $\alpha_j \bar{\alpha}_j = (a_j + \mathrm{i} b_j)(a_j - \mathrm{i} b_j) = a_j^2 + b_j^2 = p_j$. Let us choose an arbitrary subset $J \subseteq I = \{1, 2, \ldots, r\}$ and define $A_J + \mathrm{i} B_J = \left(\prod_{j \in J} \alpha_j \right) \left(\prod_{j \in I \setminus J} \bar{\alpha}_j \right)$. Then $A_J - \mathrm{i} B_J = \left(\prod_{j \in J} \bar{\alpha}_j \right) \left(\prod_{j \in I \setminus J} \alpha_j \right)$, and hence $M = (A_J + \mathrm{i} B_J)(A_J - \mathrm{i} B_J) = A_J^2 + B_J^2$. This gives one expression of the number M as a sum of two squares. It remains to prove that for two sets $J \neq J'$, $A_J + \mathrm{i} B_J \neq A_{J'} + \mathrm{i} B_{J'}$. To this end, it suffices to show that all the α_j and $\bar{\alpha}_j$ are primes in $\mathbf{Z}[\mathrm{i}]$. Then the numbers $A_J + \mathrm{i} B_J$ and $A_{J'} + \mathrm{i} B_{J'}$ are distinct, since they have distinct prime factorizations. (No α_j or $\bar{\alpha}_j$ can be obtained from another one by multiplying it by a unit of the ring $\mathbf{Z}[\mathrm{i}]$: The units are only the elements $1, -1, \mathrm{i},$ and $-\mathrm{i}$.)

So suppose that $\alpha_j = \gamma_1 \gamma_2$, $\gamma_1, \gamma_2 \in \mathbf{Z}[\mathrm{i}]$. We have $p_j = \alpha_j \bar{\alpha}_j = \gamma_1 \gamma_2 \bar{\gamma}_1 \bar{\gamma}_2 = |\gamma_1|^2 |\gamma_2|^2$. Now, $|\gamma_1|^2$ and $|\gamma_2|^2$ are both integers, and since p_j is a prime, we get that $|\gamma_1| = 1$ or $|\gamma_2| = 1$.

             □

Next, we need to know that the primes of the form $4k+1$ are sufficiently dense. First we recall the well-known *prime number theorem*: If $\pi(n)$ denotes the number of primes not exceeding n, then

$$\pi(n) = (1 + o(1))\frac{n}{\ln n} \quad \text{as } n \to \infty.$$

Proofs of this fact are quite complicated; on the other hand, it is not so hard to prove weaker bounds $cn/\log n < \pi(n) < Cn/\log n$ for suitable positive constants c, C.

We consider primes in the arithmetic progression $1, 5, 9, \ldots, 4k+1, \ldots$. A famous theorem of Dirichlet asserts that every arithmetic progression contains infinitely many primes unless this is impossible for a trivial reason, namely, unless all the terms have a nontrivial common divisor. The following theorem is still stronger:

4.2.4 Theorem. *Let d and a be relatively prime natural numbers, and let $\pi_{d,a}(n)$ be the number of primes of the form $a + kd$ ($k = 0, 1, 2, \ldots$) not exceeding n. We have*

$$\pi_{d,a}(n) = (1 + o(1))\frac{1}{\varphi(d)} \cdot \frac{n}{\ln n},$$

where φ denotes the Euler function: $\varphi(d)$ is the number of integers between 1 and d that are relatively prime to d.

For every $d \geq 2$, there are $\varphi(d)$ residue classes modulo d that can possibly contain primes. The theorem shows that the primes are quite uniformly distributed among these residue classes.

The proof of the theorem is not simple, and we omit it, but it is very nice, and we can only recommend to the reader to look it up in a textbook on number theory.

Proof of the lower bound for unit distances (Theorem 4.2.2). Let us suppose that n is a square. For the set P we choose the points of the $\sqrt{n} \times \sqrt{n}$ grid with step $1/\sqrt{M}$, where M is the product of the first $r-1$ primes of the form $4k+1$, and r is chosen as the largest number such that $M \leq \frac{n}{4}$.

It is easy to see that each point of the grid participates in at least as many unit distances as there are representations of M as a sum of two squares of nonnegative integers. Since one representation by a sum of two squares of nonnegative integers corresponds to at most 4 representations by a sum of two squares of arbitrary integers (the signs can be chosen in 4 ways), we have at least $2^{r-1}/16$ unit distances by Lemma 4.2.3.

By the choice of r, we have $4p_1 p_2 \cdots p_{r-1} \leq n < 4p_1 p_2 \cdots p_r$, and hence $2^r \leq n$ and $p_r > (\frac{n}{4})^{1/r}$. Further, we obtain, by Theorem 4.2.4, $r = \pi_{4,1}(p_r) \geq (\frac{1}{2} - o(1))p_r/\log p_r \geq \sqrt{p_r} \geq n^{1/3r}$ for sufficiently large n, and thus $r^{3r} \geq n$. Taking logarithms, we have $3r \log r \geq \log n$, and hence $r \geq \log n/(3 \log r) \geq \log n/(3 \log \log n)$. The number of unit distances is at least $n\, 2^{r-4} \geq n^{1+c_1/\log \log n}$, as Theorem 4.2.2 claims. Let us remark that for sufficiently large n the constant c_1 can be made as close to 1 as desired. $\square$

Bibliography and remarks. Proposition 4.2.1 is due to Erdős [Erd46]. His example is outlined in Exercise 2 (also see [PA95]); the

analysis requires a bit of number theory. The simpler example in the
text is from Elekes [Ele01]. Its extension provides the best known
lower bound for the number of incidences between m points and $n \geq$
$m^{(k-1)/2}$ curves with k degrees of freedom: For a parameter $t \leq m^{1/k}$,
let $P = \{(i,j): 0 \leq i < t, 0 \leq j < \frac{n}{t}\}$, and let Γ consist of the
graphs of the polynomials $\sum_{\ell=0}^{k-1} a_\ell x^\ell$ with $a_\ell = 0, 1, \ldots, \lfloor \frac{n}{kt^{\ell+1}} \rfloor$, $\ell =$
$0, 1, \ldots, k-1$.

Theorem 4.2.2 is due to Erdős [Erd46], and the proof uses ingredients well known in number theory. The prime number theorem (and
also Theorem 4.2.4) was proved in 1896, by de la Valée Poussin and
independently by Hadamard (see Narkiewicz [Nar00]).

Exercises

1. By extending the example in the text, prove that for all m, n with $n^2 \leq m$
   and $m^2 \leq n$, we have $I(m,n) = \Omega(n^{2/3}m^{2/3})$. ③
2. (Another example for incidences) Suppose that $n = 4t^6$ for an integer
   $t \geq 1$ and let $P = \{(i,j): 0 \leq i,j < \sqrt{n}\}$. Let $S = \{(a,b), a,b =$
   $1, 2, \ldots, t, \gcd(a,b) = 1\}$, where $\gcd(a,b)$ denotes the greatest common
   divisor of a and b. For each point $p \in P$, consider the lines passing
   through p with slope a/b, for all pairs $(a,b) \in S$. Let L be the union of
   all the lines thus obtained for all points $p \in P$.
   (a) Check that $|L| \leq n$. ②
   (b) Prove that $|S| \geq ct^2$ for a suitable positive constant $c > 0$, and infer
   that $I(P,L) = \Omega(nt^2) = \Omega(n^{4/3})$. ④

4.3 Point–Line Incidences via Crossing Numbers

Here we present a very simple proof of the Szemerédi–Trotter theorem based
on a result concerning graph drawing. We need the notion of the crossing
number of a graph G; this is the minimum possible number of edge crossings
in a drawing of G. To make this rigorous, let us first recall a formal definition
of a drawing.

An *arc* is the image of a continuous injective map $[0,1] \to \mathbf{R}^2$. A *drawing*
of a graph G is a mapping that assigns to each vertex of G a point in the plane
(distinct vertices being assigned distinct points) and to each edge of G an
arc connecting the corresponding two (images of) vertices and not incident
to any other vertex. We do not insist that the drawing be planar, so the
arcs are allowed to cross. A *crossing* is a point common to at least two arcs
but distinct from all vertices. In this section we will actually deal only with
drawings where each edge is represented by a straight segment.

Let G be a graph (or multigraph). The *crossing number* of a drawing of
G in the plane is the number of crossings in the considered drawing, where a
crossing incident to $k \geq 2$ edges is counted $\binom{k}{2}$ times. So a drawing is planar

if and only if its crossing number is 0. The *crossing number* of the graph G is the smallest possible crossing number of a drawing of G; we denote it by $\mathrm{cr}(G)$. For example, $\mathrm{cr}(K_5) = 1$:

As is well known, for $n > 2$, a planar graph with n vertices has at most $3n-6$ edges. This can be rephrased as follows: If the number of edges is at least $3n-5$ then $\mathrm{cr}(G) > 0$. The following theorem can be viewed as a generalization of this fact.

4.3.1 Theorem (Crossing number theorem). *Let $G = (V, E)$ be a simple graph (no multiple edges). Then*

$$\mathrm{cr}(G) \geq \frac{1}{64} \cdot \frac{|E|^3}{|V|^2} - |V|$$

(the constant $\frac{1}{64}$ can be improved by a more careful calculation).

The lower bound in this theorem is asymptotically tight; i.e., there exist graphs with n vertices, m edges, and crossing number $O(m^3/n^2)$; see Exercise 1. The assumption that the graph is simple cannot be omitted.

For a proof of this theorem, we need a simple lemma:

4.3.2 Lemma. *The crossing number of any simple graph $G = (V, E)$ is at least $|E| - 3|V|$.*

Proof. If $|E| > 3|V|$ and some drawing of the graph had fewer than $|E|-3|V|$ crossings, then we could delete one edge from each crossing and obtain a planar graph with more than $3|V|$ edges. □

Proof of Theorem 4.3.1. Consider some drawing of a graph $G = (V, E)$ with n vertices, m edges, and crossing number x. We may assume $m \geq 4n$, for otherwise, the claimed bound is negative. Let $p \in (0, 1)$ be a parameter; later on we set it to a suitable value. We choose a random subset $V' \subseteq V$ by including each vertex $v \in V$ into V' independently with probability p. Let G' be the subgraph of G induced by the subset V'. Put $n' = |V'|$, $m' = |E(G')|$, and let x' be the crossing number of the graph G' in the drawing "inherited" from the considered drawing of G. The expectation of n' is $\mathbf{E}[n'] = np$. The probability that a given edge appears in $E(G')$ is p^2, and hence $\mathbf{E}[m'] = mp^2$, and similarly we get $\mathbf{E}[x'] = xp^4$. At the same time, by Lemma 4.3.2 we always have $x' \geq m' - 3n'$, and so this relation holds for the expectations as well: $\mathbf{E}[x'] \geq \mathbf{E}[m'] - 3\mathbf{E}[n']$. So we have $xp^4 \geq mp^2 - 3np$. Setting $p = \frac{4n}{m}$ (which is at most 1, since we assume $m \geq 4n$), we calculate that

$$x \geq \frac{1}{64} \frac{m^3}{n^2}.$$

The crossing number theorem is proved.   □

Proof of the Szemerédi–Trotter theorem (Theorem 4.1.1). We consider a set P of m points and a set L of n lines in the plane realizing the maximum number of incidences $I(m,n)$. We define a certain topological graph $G = (V, E)$, that is, a graph together with its drawing in the plane. Each point $p \in P$ becomes a vertex of G, and two points $p, q \in P$ are connected by an edge if they lie on a common line $\ell \in L$ next to one another. So we have a drawing of G where the edges are straight segments. This is illustrated below, with G drawn thick:

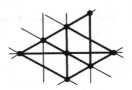

If a line $\ell \in L$ contains $k \geq 1$ points of P, then it contributes $k-1$ edges to P, and hence $I(m,n) = |E| + n$. Since the edges are parts of the n lines, at most $\binom{n}{2}$ pairs may cross: $\mathrm{cr}(G) \leq \binom{n}{2}$. On the other hand, from the crossing number theorem we get $\mathrm{cr}(G) \geq \frac{1}{64} \cdot |E|^3/m^2 - m$. So $\frac{1}{64} \cdot |E|^3/m^2 - m \leq \mathrm{cr}(G) \leq \binom{n}{2}$, and a calculation gives $|E| = O(n^{2/3}m^{2/3} + m)$. This proves the Szemerédi–Trotter theorem.   □

The best known upper bound on the number of unit distances, $U(n) = O(n^{4/3})$, can be proved along similar lines; see Exercise 2.

Bibliography and remarks. The presented proof of the Szemerédi–Trotter theorem is due to Székely [Szé97].

The crossing number theorem was proved by Ajtai, Chvátal, Newborn, and Szemerédi [ACNS82] and independently by Leighton [Lei84]. This result belongs to the theory of *geometric graphs*, which studies the properties of graphs drawn in the plane (most often with edges drawn as straight segments). A nice introduction to this area is given in Pach and Agarwal [PA95], and a newer survey is Pach [Pac99]. In the rest of this section we mention mainly some of the more recent results.

Pach and Tóth [PT97] improved the constant $\frac{1}{64}$ in Theorem 4.3.1 to approximately 0.0296, which is already within a factor of 2.01 of the best known upper bound (obtained by connecting all pairs of points of distance at most d in a regular $\sqrt{n} \times \sqrt{n}$ grid, for a suitable d). The improvement is achieved by establishing a better version of Lemma 4.3.2, namely, $\mathrm{cr}(G) \geq 5|E| - 25|V|$ for $|E| > 7|V| - 14$.

Pach, Spencer, and Tóth [PST00] proved that for graphs with certain forbidden subgraphs, the bound can be improved substantially: For example, if G has n vertices, m edges, and contains no cycle of length 4, then $\mathrm{cr}(G) = \Omega(m^4/n^3)$ for $m \geq 400n$, which is asymptotically tight. Generally, let $\mathcal{G}$ be a class of graphs that is monotone (closed under adding edges) and such that any n-vertex graph in $\mathcal{G}$ has at most $O(n^{1+\alpha})$ edges, for some $\alpha \in (0,1)$. Then $\mathrm{cr}(G) \geq cm^{2+1/\alpha}/n^{1+1/\alpha}$ for any $G \in \mathcal{G}$ with n vertices and $m \geq Cn\log^2 n$ edges, with suitable constants $C, c > 0$ depending on $\mathcal{G}$. The proof applies a generally useful lower bound on the crossing number, which we outline next. Let $\mathrm{bw}(G)$ denote the *bisection width* of G, i.e., the minimum number of edges connecting V_1 and V_2, over all partitions (V_1, V_2) of $V(G)$ with $|V_1|, |V_2| \geq \frac{1}{3}|V(G)|$. Leighton [Lei83] proved that $\mathrm{cr}(G) = \Omega(\mathrm{bw}(G)^2) - |V(G)|$ for any graph G of maximum degree bounded by a constant. Pach, Shahrokhi, and Szegedy [PSS96], and independently Sýkora and Vrťo [SV94], extended this to graphs with arbitrary degrees:

$$\mathrm{cr}(G) = \Omega\big(\mathrm{bw}(G)^2\big) - \frac{1}{16}\sum_{v \in V(G)} \deg_G(v)^2, \qquad (4.1)$$

where $\deg_G(v)$ is the degree of v in G. The proof uses the following version, due to Gazit and Miller [GM90], of the well-known Lipton–Tarjan separator theorem for planar graphs: For any planar graph H and any nonnegative weight function $w\colon V(H) \to [0, \frac{2}{3}]$ with $\sum_{v \in V(H)} w(v) = 1$, one can delete at most $1.58\sqrt{\sum_{v \in V(H)} \deg_H(v)^2}$ edges in such a way that the total weight of vertices in each component of the resulting graph is at most $\frac{2}{3}$. To deduce (4.1), consider a drawing of G with the minimum number of crossings, replace each crossing by a vertex of degree 4, assign weight 0 to these vertices and weight $\frac{1}{|V(G)|}$ to the original vertices, and apply the separator theorem (see, e.g., [PA95] for a more detailed account). Djidjev and Vrťo [DV02] have recently strengthened (4.1), replacing $\mathrm{bw}(G)$ by the *cutwidth* of G. To define the cutwidth, we consider an injective mapping $f\colon V(G) \to \mathbf{R}$. Each edge corresponds to a closed interval, and we find the maximum number of these intervals with a common interior point. The cutwidth is the minimum of this quantity over all f.

To derive the result of Pach et al. [PST00] on the crossing number of graphs with forbidden subgraphs mentioned above from (4.1), we consider a graph $G \in \mathcal{G}$ with n vertices and m edges. If $\mathrm{cr}(G)$ is small, then the bisection width is small, so G can be cut into two parts of almost equal size by removing not too many edges. For each of these parts, we bisect again, and so on, until parts of some suitable size s (depending on n and m) are reached. By the assumption on $\mathcal{G}$, each of the resulting parts has $O(s^{1+\alpha})$ edges, and so there are $O(ns^\alpha)$ edges within the parts. This number of edges plus the number of edges deleted in the bisections add up to m, and this provides an inequality relating $\mathrm{cr}(G)$ to n and m; see [PST00] for the calculations.

The notion of crossing number is a subtle one. Actually, one can give several natural definitions; a study of various notions and of their relations was made by Pach and Tóth [PT00]. Besides counting the crossings, as we did in the definition of $\mathrm{cr}(G)$, one can count the number of (unordered) pairs of edges that cross; the resulting notion is called the *pairwise crossing number* in [PT00], and we denote it by pair-$\mathrm{cr}(G)$. We always have pair-$\mathrm{cr}(G) \leq \mathrm{cr}(G)$, but since two edges (arcs) are allowed to cross several times, it is not clear whether pair-$\mathrm{cr}(G) = \mathrm{cr}(G)$ for all graphs G, and currently this seems to be a challenging open problem (see Exercise 4 for a typical false attempt at a proof). A simple argument shows that $\mathrm{cr}(G) \leq 2\,\mathrm{pair\text{-}cr}(G)^2$ (Exercise 4(c)). A stronger claim, proved in [PT00], is $\mathrm{cr}(G) \leq 2\,\mathrm{odd\text{-}cr}(G)^2$, where odd-$\mathrm{cr}(G)$ is the *odd-crossing number* of G, counting the number of pairs of edges that cross an odd number of times. An inspiration for their proof is a theorem of Hanani and Tutte claiming that a graph G is planar if and only if odd-$\mathrm{cr}(G) = 0$. In a drawing of G, call an edge e *even* if there is no edge crossed by e an odd number of times. Pach and Tóth show, by a somewhat complicated proof, that if we consider a drawing of G and let E_0 be the set of the even edges, then there is another drawing of G in which the edges of E_0 are involved in no crossings at all. The inequality $\mathrm{cr}(G) \leq 2\,\mathrm{odd\text{-}cr}(G)^2$ then follows by an argument similar to that in Exercise 4(c).

Finally, let us remark that if we consider rectilinear drawings (where each edge is drawn as a straight segment), then the resulting *rectilinear crossing number* can be much larger than any of the crossing numbers considered above: Graphs are known with $\mathrm{cr}(G) = 4$ and arbitrarily large rectilinear crossing numbers (Bienstock and Dean [BD93]).

Exercises

1. Show that for any n and m, $5n \leq m \leq \binom{n}{2}$, there exist graphs with n vertices, m edges, and crossing number $O(m^3/n^2)$. ②

2. In a manner similar to the above proof for point–line incidences, prove the bound $I_{1\mathrm{circ}}(n,n) = O(n^{4/3})$, where $I_{1\mathrm{circ}}(m,n)$ denotes the maximum possible number of incidences between m points and n unit circles in the plane (be careful in handling possible multiple edges in the considered topological graph!). ③

3. Let $K(n,m)$ denote the maximum total number of edges of m distinct cells in an arrangement of n lines in the plane. Prove $K(n,m) = O(n^{2/3}m^{2/3} + n + m)$ using the method of the present section (it may be

convenient to classify edges into top and bottom ones and bound each type separately). [3]
4. (a) Prove that in a drawing of G with the smallest possible number of crossings, no two edges cross more than once. [2]
(b) Explain why the result in (a) does not imply that pair-cr$(G) =$ cr(G) (where pair-cr(G) is the minimum number of pairs of crossing edges in a drawing of G). [2]
(c) Prove that if G is a graph with pair-cr$(G) = k$, then cr$(G) \leq \binom{2k}{2}$. [4]

4.4 Distinct Distances via Crossing Numbers

Here we use the methods from the preceding sections to establish a lower bound on the number of distinct distances determined by an n-point set in the plane. We do not go for the best known bound, whose proof is too complicated for our purposes, but in the notes below we indicate how the improvement is achieved.

4.4.1 Proposition (Distinct distances in R²). *The minimum number $g(n)$ of distinct distances determined by an n-point set in the plane satisfies $g(n) = \Omega(n^{4/5})$.*

Proof. Fix an n-point set P, and let t be the number of distinct distances determined by P. This means that for each point $p \in P$, all the other points are contained in t circles centered at p (the radii correspond to the t distances appearing in P).

These tn circles obtained for all the n points of P have $n(n-1)$ incidences with the points of P. The first idea is to bound this number of incidences from above in terms of n and t, in a way similar to the proof of the Szemerédi–Trotter theorem in the preceding section, which yields a lower bound for t.

First we delete all circles with at most 2 points on them (the innermost circle and the second outermost circle in the above picture). We have destroyed at most $2nt$ incidences, and so still almost n^2 incidences remain (we may assume that t is much smaller than n, for otherwise, there is nothing to prove). Now we define a graph G: The vertices are the points of P and the edges are the arcs of the circles between the points. This graph has n vertices, almost n^2 edges, and there are at most t^2n^2 crossings because every two circles intersect in at most 2 points.

Now *if* we could apply the crossing number theorem to this graph, we would get that with n vertices and n^2 edges there must be at least $\Omega(n^6/n^2) = \Omega(n^4)$ crossings, and so $t = \Omega(n)$ would follow. This, of course, is too good to be true, and indeed we cannot use the crossing number theorem directly because our graph may have multiple edges: Two points can be connected by several arcs.

A multigraph can have arbitrarily many edges even if it is planar. But if we have a bound on the maximum edge multiplicity, we can still infer a lower bound on the crossing number:

4.4.2 Lemma. *Let $G = (V, E)$ be a multigraph with maximum edge multiplicity k. Then*

$$\mathrm{cr}(G) = \Omega\left(\frac{|E|^3}{k|V|^2}\right) - O(k^2|V|).$$

We defer the proof to the end of this section.

In the graph G defined above, it appears that the maximum edge multiplicity can be as high as t. If we used Lemma 4.4.2 with $k = t$ in the manner indicated above, we would get only the estimate $t = \Omega(n^{2/3})$.

The next idea is to deal with the edges of very high multiplicity separately. Namely, we observe that if a pair $\{u, v\}$ of points is connected by k arcs, then the centers of these arcs lie on the symmetry axis ℓ_{uv} of the segment uv:

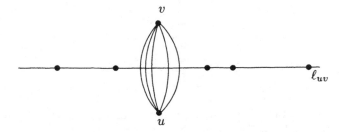

So the line ℓ_{uv} has at least k incidences with the points of P. But the Szemerédi–Trotter theorem tells us that there cannot be too many distinct lines, each incident to many points of P. Let us make this precise.

By a consequence of the Szemerédi–Trotter theorem stated in Exercise 4.1.6(b), lines containing at least k points of P each have altogether no more than $O(n^2/k^2 + n)$ incidences with P.

Let M be the set of pairs $\{u, v\}$ of vertices of G connected by at least k edges in G, and let E be the set of edges (arcs) connecting these pairs. Each edge in E connecting the pair $\{u, v\}$ contributes one incidence of the bisecting line ℓ_{uv} with a point $p \in P$. On the other hand, one incidence of such p with some ℓ_{uv} can correspond to at most $2t$ edges of E, because at most t circles are centered at p, and so ℓ_{uv} intersects at most $2t$ arcs with center p. So we have $|E| = O(tn^2/k^2 + tn)$.

Let us set k as large as possible but so that $|E| \leq \frac{1}{2}n^2$, i.e., $k = C\sqrt{t}$ for a sufficiently large constant C. If we delete all edges of E, the remaining graph still has $\Omega(n^2)$ edges, but the maximum multiplicity is now below k. We can finally apply Lemma 4.4.2: With n vertices, $\Omega(n^2)$ edges, and edge multiplicity at most $k = O(\sqrt{t})$, we have at least $\Omega(n^4/\sqrt{t})$ crossings. This number must be below $t^2 n^2$, which yields $t = \Omega(n^{4/5})$ as claimed. □

Proof of Lemma 4.4.2. Consider a fixed drawing of G. We choose a subgraph G' of G by the following random experiment. In the first stage, we consider each edge of G independently, and we delete it with probability $1 - \frac{1}{k}$. In the second stage, we delete all the remaining multiple edges, and this gives G', which has n vertices, m' edges, and x' crossing pairs of edges. Consider the probability p_e that a fixed edge $e \in E$ remains in G'. Clearly, $p_e \leq \frac{1}{k}$. On the other hand, if e was one of $k' \leq k$ edges connecting the same pair of vertices, then the probability that e survives the first stage while all the other edges connecting its two vertices are deleted is

$$\frac{1}{k}\left(1 - \frac{1}{k}\right)^{k'-1} \geq \frac{1}{3k}$$

(since $(1-1/k)^{k-1} \geq \frac{1}{3}$). We get $\mathbf{E}[m'] \geq |E|/3k$ and $\mathbf{E}[x'] \leq x/k^2$. Applying the crossing number theorem for the graph G' and taking expectations, we have

$$\mathbf{E}[x'] \geq \frac{1}{64} \cdot \frac{\mathbf{E}[m'^3]}{n^2} - n.$$

By convexity (Jensen's inequality), we have $\mathbf{E}[m'^3] \geq (\mathbf{E}[m'])^3 = \Omega(|E|^3/k^3)$. Plugging this plus the bound $\mathbf{E}[x'] \leq x/k^2$ into the above formula, we get

$$\frac{x}{k^2} = \Omega\left(\frac{|E|^3}{k^3 n^2}\right) - O(n),$$

and the lemma follows. □

Bibliography and remarks. The proof presented above is, with minor modifications, that of Székely [Szé97]. The bound has subsequently been improved by Solymosi and Tóth [ST01] to $\Omega(n^{6/7})$ and then by Tardos [Tar01] to (approximately) $\Omega(n^{0.863})$.

The weakest point of the proof shown above seems to be the lower bound on the number of incidences between the points of P and the "rich" bisectors ℓ_{uv} ($\{u, v\}$ being the pairs connected by k or more edges). We counted as if each such incidence could be responsible for as many as t edges. While this does not look geometrically very plausible, it seems hard to exclude such a possibility directly. Instead, Solymosi and Tóth prove a better lower bound for the number of incidences of P with the rich bisectors differently; they show that if there are many edges with multiplicity at least k, then each of $\Omega(n)$ suitable points is incident to many (namely $\Omega(n/t^{3/2})$ in their proof) rich bisectors. We outline this argument.

We need to modify the definition of the graph G. The new definition uses an auxiliary parameter r (a constant, with $r = 3$ in the original Solymosi–Tóth proof). First, we note that by the theorem of Beck mentioned in Exercise 4.1.7, there is a subset $P' \subseteq P$ of $\Omega(n)$ points such that each $p \in P'$ sees the other points of P in $\Omega(n)$ distinct directions. For each $p \in P'$, we draw the t circles around p. If several points of P are visible from p in the same direction, we temporarily delete all of them but one. Then, on each circle, we group the remaining points into groups by r consecutive points, and on each circle we delete the at most $r-1$ leftover points fitting in no such group. This still leaves $\Omega(n)$ r-point groups on the circles centered at p.

Next, we consider one such r-point group and all the $\binom{r}{2}$ bisecting lines of its points. If at least one of these bisectors, call it ℓ_{uv}, contains fewer than k points of P (k being a suitable threshold), then we add the arc connecting u and v as an edge of G:

If this bisector has at most k points of P, then the arc $\{u, v\}$ is added to G.

(This is not quite in agreement with our definition of a graph drawing, since the arc may pass though other vertices of G, but it is easy to check that if we permit arcs through vertices and modify the definition of the crossing number appropriately, Lemma 4.4.2 remains valid.) The groups where every bisector contains at least k points of P (call them *rich* groups) do not contribute any edges of G.

Setting $k = \alpha n^2/t^2$ for a small constant α, we argue by Lemma 4.4.2 that G has at most βn^2 edges for a small $\beta = \beta(\alpha) > 0$. It follows that most of the r-point groups must be rich, and so there is a subset $P'' \subseteq P'$ of $\Omega(n)$ points, each of them possessing $\Omega(n)$ rich groups on its circles. It remains to prove that each point $p \in P''$ is incident to many rich bisectors. We divide the plane around p into angular sectors such that each sector contains about $3rt$ points (of the $\Omega(n)$ points in the rich groups belonging to p). Each sector contains at least t complete rich groups (since there are t circles, and so the sector's boundaries cut through at most $2t$ groups), and we claim that it has to contain many rich bisectors. This leads to the following number-theoretic problem: we have tr distinct real numbers (corresponding to the angles of the points in the sector as seen from p), arranged into

t groups by r numbers, and we form all the $\binom{r}{2}$ arithmetic averages of the pairs in each group (corresponding to the rich bisectors of the group). This yields $t\binom{r}{2}$ real numbers, and we want to know how many of them must be distinct.

It is not hard to see that for $r = 3$, there must be at least $\Omega(t^{1/3})$ distinct numbers, because the three averages $(a+b)/2$, $(a+c)/2$, and $(b+c)/2$ determine the numbers a, b, c uniquely. It follows, still for $r = 3$, that each of the $\Omega(\frac{n}{t})$ sectors has $\Omega(t^{1/3})$ distinct bisectors, and so each point in P'' has $\Omega(n/t^{2/3})$ incidences with the rich lines. Applying Szemerédi–Trotter now yields the Solymosi–Tóth bound of $t = \Omega(n^{6/7})$ distinct distances.

Tardos [Tar01] considered the number-theoretic problem above for larger r, and he proved, by a complicated argument, that for r large but fixed, the number of distinct pairwise averages is $\Omega(t^{1/e+\varepsilon})$, with $\varepsilon \to 0$ as $r \to \infty$. Plugging this into the proof leads to the current best bound mentioned above. An example by Ruzsa shows that the number of distinct pairwise averages can be $O(\sqrt{t})$ for any fixed r, and it follows that the Solymosi–Tóth method as is cannot provide a bound better than $\Omega(n^{8/9})$. But surely one can look forward to the further continuation of the adventure of distinct distances.

Exercises

1. Let $I_{\mathrm{circ}}(m, n)$ be the maximum number of incidences between m points and n arbitrary circles in the plane. Fill in the details of the following approach to bounding $I_{\mathrm{circ}}(n, n)$. Let K be a set of n circles, C the set of their centers, and P a set of n points.

   (a) First, assume that the centers of the circles are mutually distinct, i.e., $|C| = |K|$. Proceed as in the proof in the text: Remove circles with at most 2 incidences, and let the others define a drawing of a multigraph G with vertex set P and arcs of the circles as edges. Handle the edges with multiplicity k or larger via Szemerédi–Trotter, using the incidences of the bisectors with the set C, and those with multiplicity $< k$ by Lemma 4.4.2. Balance k suitably. What bound is obtained for the total number of incidences? ③

   (b) Extend the argument to handle concentric circles too. ③

2. This exercise provides another bound for $I_{\mathrm{circ}}(n, n)$, the maximum possible number of incidences between n arbitrary circles and n points in the plane. Let K be the set of circles and P the set of points. Let P_i be the points with at least $d_i = 2^i$ and fewer than 2^{i+1} incidences; we will argue for each P_i separately.

   Define the multigraph G on P_i as usual, with arcs of circles of K connecting neighboring points of P_i (the circles with at most 2 incidences with P_i are deleted). Let E be the set of edges of G. For a point $u \in P_i$, let $N(u)$ be the set of its neighboring points, and for a $v \in N(u)$, let

$\mu(u,v)$ be the number of edges connecting u and v. For an edge e, define its *partner edge* as the edge following after e clockwise around its circle.

(a) Show that for each $u \in P_i$, $|\{v \in N(u): \mu(u,v) \geq 4\sqrt{d_i}\}| < \sqrt{d_i}/2$. ①

(b) Let $E_h \subseteq E$ be the edges of multiplicity at least $4\sqrt{d_i}$. Argue that for at least $\frac{1}{4}$ of the edges in E_h, their partner edges do not belong to E_h, and hence $|E \setminus E_h| = \Omega(|E|)$. ③

(c) Delete the edges of E_h from the graph, and apply Lemma 4.4.2 to bound $|E \setminus E_h|$. What overall bound does all this give for $I_{\text{circ}}(n,n)$? ②

A similar proof appears in Pach and Sharir [PS98a] (for the more general case of curves mentioned in the notes to Section 4.1).

4.5 Point–Line Incidences via Cuttings

Here we explain another proof of the upper bound $I(n,n) = O(n^{4/3})$ for point–line incidences. The technique is quite different. It leads to an efficient algorithm and seems more generally applicable than the one with the crossing number theorem.

4.5.1 Lemma (A worse but useful bound).

$$I(m,n) = O(n\sqrt{m} + m), \qquad (4.2)$$
$$I(m,n) = O(m\sqrt{n} + n). \qquad (4.3)$$

Proof. There are at most $\binom{n}{2}$ crossing pairs of lines in total. On the other hand, a point $p_i \in P$ with d_i incidences "consumes" $\binom{d_i}{2}$ crossing pairs (their intersections all lie at p_i). Therefore, $\sum_{i=1}^{m} \binom{d_i}{2} \leq \binom{n}{2}$.

We want to bound $\sum_{i=1}^{m} d_i$ from above. Since points with no incidences can be deleted from P in advance, we may assume $d_i \geq 1$ for all i, and then we have $\binom{d_i}{2} \geq (d_i-1)^2/2$. By the Cauchy–Schwarz inequality,

$$\sum_{i=1}^{m}(d_i-1) \leq \sqrt{m \sum_{i=1}^{m}(d_i-1)^2} \leq \sqrt{2\binom{n}{2}m},$$

and hence $\sum d_i = O(n\sqrt{m} + m)$.

The other inequality in the lemma can be proved similarly by looking at pairs of points on each line. Alternatively, the equality $I(n,m) = I(m,n)$ for all m,n follows using the geometric duality introduced in Section 5.1. $\square$

Forbidden subgraph arguments. For integers $r,s \geq 1$, let $K_{r,s}$ denote the complete bipartite graph on $r+s$ vertices; the picture shows $K_{3,4}$:

The above proof can be expressed using graphs with forbidden $K_{2,2}$ as a subgraph and thus put into the context of extremal graph theory.

A typical question in extremal graph theory is the maximum possible number of edges of a (simple) graph on n vertices that does not contain a given forbidden subgraph, such as $K_{2,2}$. Here the subgraph is understood in a noninduced sense: For example, the complete graph K_4 does contain $K_{2,2}$ as a subgraph. More generally, one can forbid all subgraphs from a finite or infinite family $\mathcal{F}$ of graphs, or consider "containment" relations other than being a subgraph, such as "being a minor."

If the forbidden subgraph H is not bipartite, then, for example, the complete bipartite graph $K_{n,n}$ has $2n$ vertices, n^2 edges, and no subgraph isomorphic to H. This shows that forbidding a nonbipartite H does not reduce the maximum number of edges too significantly, and the order of magnitude remains quadratic.

On the other hand, forbidding $K_{r,s}$ with some fixed r and s decreases the exponent of n, and forbidden bipartite subgraphs are the key to many estimates in incidence problems and elsewhere.

4.5.2 Theorem (Kővári–Sós–Turán theorem). *Let $r \le s$ be fixed natural numbers. Then any graph on n vertices containing no $K_{r,s}$ as a subgraph has at most $O(n^{2-1/r})$ edges.*

If G is a bipartite graph with color classes of sizes m and n containing no subgraph $K_{r,s}$ with the r vertices in the class of size m and the s vertices in the class of size n, then

$$|E(G)| = O\left(\min(mn^{1-1/r} + n, m^{1-1/s}n + m)\right).$$

(In both parts, the constant of proportionality depends on r and s.)

Note that in the second part of the theorem, the situation is not symmetric: By forbidding the "reverse" placement of $K_{r,s}$, we get a different bound in general.

The upper bound in the theorem is suspected to be tight, but a matching lower bound is known only for some special values of r and s, in particular for $r \le 3$ (and all $s \ge r$).

To see the relevance of forbidden $K_{2,2}$ to the point–line incidences, we consider a set P of points and a set L of lines and we define a bipartite graph with vertex set $P \cup L$ and with edges corresponding to incidences. An edge $\{p, \ell\}$ means that the point p lies on the line ℓ. So the number of incidences equals the number of edges. Since two points determine a line, this graph contains no $K_{2,2}$ as a subgraph: Its presence would mean that two distinct lines both contain the same two distinct points. The Kővári–Sós–Turán theorem thus immediately implies Lemma 4.5.1, and the above proof of this lemma is the usual proof of that theorem, for the special case $r = s = 2$, rephrased in terms of points and lines.

As was noted above, for arbitrary bipartite graphs with forbidden $K_{2,2}$, not necessarily being incidence graphs of points and lines in the plane, the bound in the Kővári–Sós–Turán theorem cannot be improved. So, in order to do better for point–line incidences, one has to use some more geometry than just the excluded $K_{2,2}$. In fact, this was one of the motivations of the problem of point–line incidences: In a finite projective plane of order q, we have $n = q^2+q+1$ points, n lines, and $(q+1)n \approx n^{3/2}$ incidences, and so the Szemerédi–Trotter theorem strongly distinguishes the Euclidean plane from finite projective planes in a combinatorial sense.

Proof of the Szemerédi–Trotter theorem (Theorem 4.1.1) for $m = n$. The bound from Lemma 4.5.1 is weaker than the tight Szemerédi–Trotter bound, but it is tight if $n^2 \leq m$ or $m^2 \leq n$. The idea of the present proof is to convert the "balanced" case (n points and n lines) into a collection of "unbalanced" subproblems, for which Lemma 4.5.1 is optimal. We apply the following important result:

4.5.3 Lemma (Cutting lemma). *Let L be a set of n lines in the plane, and let r be a parameter, $1 < r < n$. Then the plane can be subdivided into t generalized triangles (this means intersections of three half-planes) $\Delta_1, \Delta_2, \ldots, \Delta_t$ in such a way that the interior of each Δ_i is intersected by at most $\frac{n}{r}$ lines of L, and we have $t \leq Cr^2$ for a certain constant C independent of n and r.*

Such a collection $\Delta_1, \ldots, \Delta_t$ may look like this, for example:

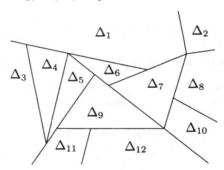

The lines of L are not shown.

In order to express ourselves more economically, we introduce the following terminology. A *cutting* is a subdivision of the plane into finitely many generalized triangles. (We sometimes omit the adjective "generalized" in the sequel.) A given cutting is a $\frac{1}{r}$-*cutting for* a set L of n lines if the interior of each triangle of the cutting is intersected by at most $\frac{n}{r}$ lines of L.

Proofs of the cutting lemma will be discussed later, and now we continue the proof of the Szemerédi–Trotter theorem.

Let P be the considered n-point set, L the set of n lines, and $I(P, L)$ the number of their incidences. We fix a "magic" value $r = n^{1/3}$, and we divide the plane into $t = O(r^2) = O(n^{2/3})$ generalized triangles $\Delta_1, \ldots, \Delta_t$ so that the interior of each Δ_i is intersected by at most $n/r = n^{2/3}$ lines of L, according to the cutting lemma .

Let P_i denote the points of P lying inside Δ_i or on its boundary but not at the vertices of Δ_i, and let L_i be the set of lines of L intersecting the interior of Δ_i. The pairs (L_i, P_i) define the desired "unbalanced" subproblems. We have $|L_i| \le n^{2/3}$, and while the size of the P_i may vary, the average $|P_i|$ is $\frac{n}{t} \approx n^{1/3}$, which is about the square root of the size of L_i.

We have to be a little careful, since not all incidences of L and P are necessarily included among the incidences of some L_i and P_i. One exceptional case is a point $p \in P$ not appearing in any of the P_i.

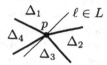

Such a point has to be the vertex of some Δ_i, and so there are no more than $3t$ such exceptional points. These points have at most $I(n, 3t)$ incidences with the lines of L. Another exceptional case is a line of L containing a side of Δ_i but not intersecting its interior and therefore not included in L_i, although it may be incident with some points on the boundary of Δ_i.

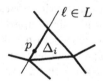

There are at most $3t$ such exceptional lines, and they have at most $I(3t, n)$ incidences with the points of P. So we have

$$I(L, P) \le I(n, 3t) + I(3t, n) + \sum_{i=1}^{t} I(L_i, P_i).$$

By Lemma 4.5.1, $I(n, 3t)$ and $I(3t, n)$ are both bounded by $O(t\sqrt{n} + n) = O(n^{7/6}) \ll n^{4/3}$, and it remains to estimate the main term. We have $|L_i| \le n^{2/3}$ and $\sum_{i=1}^{t} |P_i| \le 2n$, since each point of P goes into at most two P_i. Using the bound (4.2) for each $I(L_i, P_i)$ we obtain

$$\sum_{i=1}^{t} I(L_i, P_i) \le \sum_{i=1}^{t} I(n^{2/3}, |P_i|) = \sum_{i=1}^{t} O(|P_i| n^{1/3} + n^{2/3})$$

$$= O(n^{1/3}) \left(\sum_{i=1}^{t} |P_i| \right) + O(t n^{2/3}) = O(n^{4/3}).$$

This finally shows that $I(n,n) = O(n^{4/3})$.   □

Bibliography and remarks. The bound in Lemma 4.5.1 using excluded $K_{2,2}$ is due to Erdős [Erd46].

Determining the maximum possible number of edges in a $K_{r,s}$-free bipartite graph with given sizes of the two color classes is known as the *Zarankiewicz problem*. The general upper bound given in the text was shown by Kővári, Sós, and Turán [KST54]. For a long time, matching lower bounds (constructions) were known only for $r \leq 3$ and all $s \geq r$ (in these cases, even the constant in the leading term is known exactly; see Füredi [Für96] for some of these results and references). In particular, $K_{2,2}$-free graphs on n vertices with $\Omega(n^{3/2})$ edges are provided by incidence graphs of finite projective planes, and $K_{3,3}$-free graphs on n vertices with $\Omega(n^{5/3})$ edges were obtained by Brown [Bro66]. His construction is the "distance-k graph" in the 3-dimensional affine space over finite fields of order $q \equiv -1 \mod 4$, for a suitable $k = k(q)$. Recently, Kollár, Rónyai, and Szabó [KRS96] constructed asymptotically optimal $K_{r,s}$-free graphs for s very large compared to r, namely $s \geq r!+1$, using results of algebraic geometry. This was slightly improved by Alon, Rónyai, and Szabó [ARS99] to $s \geq (r-1)!+1$. They also obtained an alternative to Brown's construction of $K_{3,3}$-free graphs with a better constant, and asymptotically tight lower bounds for some "asymmetric" instances of the Zarankiewicz problem, where one wants a $K_{r,s}$-free bipartite graph with color classes of sizes n and m (with the "orientation" of the forbidden $K_{r,s}$ fixed).

The approach to incidence problems using cuttings first appeared in a seminal paper of Clarkson, Edelsbrunner, Guibas, Sharir, and Welzl [CEG+90], based on probabilistic methods developed in computational geometry ([Cla87], [HW87], and [CS89] are among the most influential papers in this development). Clarkson et al. did not use cuttings in our sense but certain "cuttings on the average." Namely, if n_i is the number of lines intersecting the interior of Δ_i, then their cuttings have $t = O(r^2)$ triangles and satisfy $\sum_{i=1}^{t} n_i^c \leq C(c) \cdot r^2 (\frac{n}{r})^c$, where $c \geq 1$ is an integer constant, which can be selected as needed for each particular application, and $C(c)$ is a constant depending on c. This means that the cth degree average of the n_i is, up to a constant, the same as if all the n_i were $O(\frac{n}{r})$. Technically, these "cuttings on the average" can replace the optimal $\frac{1}{r}$-cuttings in most applications. Clarkson et al. [CEG+90] proved numerous results on various incidence problems and many-cells problems by this method; see the notes to Section 4.1.

The cutting lemma was first proved by Chazelle and Friedman [CF90] and, independently, by Matoušek [Mat90a]. The former proof yields an optimal cutting lemma in every fixed dimension and will be discussed in Section 6.5, while the latter proof applies only to planar cutting and is presented in Section 4.7. A third, substantially different, proof was discovered by Chazelle [Cha93a].

Yet another proof of the Szemerédi–Trotter theorem was recently found by Aronov and Sharir (it is a simplification of the techniques in [AS01a]). It is based on the case $d = 2$ of the following partition theorem of Matoušek [Mat92]: *For every n-point set $X \subset \mathbf{R}^d$, d fixed, and every r, $1 \le r \le n$, there exists a partition $X = X_1 \cup X_2 \cup \cdots \cup X_t$, $t = O(r)$, such that $\frac{n}{r} \le |X_i| \le \frac{2n}{r}$ for all i and no hyperplane h crosses more than $O(r^{1-1/d})$ of the sets X_i*. Here h crossing X_i means that X_i is not completely contained in one of the open half-spaces defined by h or in h itself.[1] This result is proved using the d-dimensional cutting lemma (see Section 4.6). The bound $O(r^{1-1/d})$ is asymptotically the best possible in general.

To use this result for bounding $I(L, P)$, where L is a set of n lines and P a set of n points in the plane, we let $X = \mathcal{D}_0(L)$ be the set of points dual to the lines of L (see Section 5.1). We apply the partition theorem to X with $r = n^{2/3}$ and dualize back, which yields a partition $L = L_1 \cup L_2 \cup \cdots \cup L_t$, $t = O(r)$, with $|L_i| \approx \frac{n}{r} = n^{1/3}$. The crossing condition implies that no point p is incident to lines from more than $O(\sqrt{r})$ of the L_i, not counting the pathological L_i where p is common to all the lines of L_i.

We consider the incidences of a point $p \in P$ with the lines of L_i. The i where p lies on at most one line of L_i contribute at most $O(\sqrt{r})$ incidences, which gives a total of $O(n\sqrt{r}) = O(n^{4/3})$ for all $p \in P$. On the other hand, if p lies on at least two lines of L_i then it is a vertex of the arrangement of L_i. As is easy to show, the number of incidences of k lines with the vertices of their arrangement is $O(k^2)$ (Exercise 6.1.6), and so the total contribution from these cases is $O(\sum |L_i|^2) = O(n^2/r) = O(n^{4/3})$. This proves the balanced case of Szemerédi–Trotter, and the unbalanced case works in the same way with an appropriate choice of r. Unlike the previous proofs, this one does not directly apply with pseudolines instead of lines.

Improved point–circle incidences. A similar method also proves that $I_{\mathrm{circ}}(n, n) = O(n^{1.4})$ (see Exercise 4.4.2 for another proof). Circles are dualized to points and points to surfaces of cones in $\mathbf{R}^3$, and the appropriate partition theorem holds as well, with no surface of a cone crossing more than $O(r^{2/3})$ of the subsets X_i.

Aronov and Sharir [AS01a] improved the bound to $I_{\mathrm{circ}}(m, n) = O(m^{2/3}n^{2/3} + m)$ for large m, namely $m \ge n^{(5-3\varepsilon)/(4-9\varepsilon)}$, and to $I_{\mathrm{circ}}(m, n) = O(m^{(6+3\varepsilon)/11}n^{(9-\varepsilon)/11} + n)$ for the smaller m (here, as usual, $\varepsilon > 0$ can be chosen arbitrarily small, influencing the constants

---

[1] A slightly stronger result is proved in [Mat92]: For every X_i we can choose a relatively open simplex $\sigma_i \supseteq X_i$, and no h crosses more than $O(r^{1-1/d})$ of the σ_i.

of proportionality). Agarwal et al. [AAS01] obtained almost the same bounds for the maximum complexity of m cells in an arrangement of n circles.

A key ingredient in the Aronov–Sharir proof are results on the following question of independent interest. Given a family of n curves in the plane, into how many pieces ("pseudosegments") must we cut them, in the worst case, so that no two pieces intersect more than once? This problem, first studied by Tamaki and Tokuyama [TT98], will be briefly discussed in the notes to Section 11.1. For the curves being circles, Aronov and Sharir [AS01a] obtained the estimate $O(n^{3/2+\varepsilon})$, improving on several previous results.

To bound the number $I(P,C)$ of incidences of an m-point set P and a set C of n circles, we delete the circles containing at most 2 points, we cut the circles into $O(n^{3/2+\varepsilon})$ pieces as above, and we define a graph with vertex set P and with edges being the circle arcs that connect consecutive points along the pieces. The number of edges is at least $I(P,C) - O(n^{3/2+\varepsilon})$. The crossing number theorem applies (since the graph is simple) and yields $I(P,C) = O(m^{2/3}n^{2/3} + n^{3/2+\varepsilon})$, which is tight for m about $n^{5/4}$ and larger. For smaller m, Aronov and Sharir use the method with partition in the dual space outlined above to divide the original problem into smaller subproblems, and for these they use the bound just mentioned.

Exercises

1. Let $I_{1\mathrm{circ}}(m,n)$ be the maximum number of incidences between m points and n unit circles in the plane. Prove that $I_{1\mathrm{circ}}(m,n) = O(m\sqrt{n}+n)$ by the method of Lemma 4.5.1. ②
2. Let $I_{\mathrm{circ}}(m,n)$ be the maximum possible number of incidences between m points and n arbitrary circles in the plane. Prove that $I_{\mathrm{circ}}(m,n) = O(n\sqrt{m}+n)$ and $I_{\mathrm{circ}}(m,n) = O(mn^{2/3}+n)$. ②

4.6 A Weaker Cutting Lemma

Here we prove a version of the cutting lemma (Lemma 4.5.3) with a slightly worse bound on the number of the Δ_i. The proof uses the probabilistic method and the argument is very simple and general. We will improve on it later and obtain tight bounds in a more general setting in Section 6.5. In Section 4.7 below we give another, self-contained, elementary geometric proof of the planar cutting lemma .

Here we are going to prove that every set of n lines admits a $\frac{1}{r}$-cutting consisting of $O(r^2 \log^2 n)$ triangles. But first let us see why at least $\Omega(r^2)$ triangles are necessary.

A lower bound. Consider n lines in general position. Their arrangement has, as we know, $\binom{n}{2}+n+1 \geq n^2/2$ cells. On the other hand, considering a triangle Δ_i whose interior is intersected by $k \leq \frac{n}{r}$ lines ($k \geq 1$), we see that Δ_i is divided into at most $\binom{k}{2}+k+1 \leq 2k^2$ cells. Since each cell of the arrangement has to show up in the interior of at least one triangle Δ_i, the number of triangles is at least $n^2/4k^2 = \Omega(r^2)$. Hence the cutting lemma is asymptotically optimal for $r \to \infty$.

Proof of a weaker version of the cutting lemma (Lemma 4.5.3). We select a random sample $S \subseteq L$ of the given lines. We make s independent random draws, drawing a random line from L each time. These are draws with replacement: One line can be selected several times, and so S may have fewer than s lines.

Consider the arrangement of S. Partition the cells that are not (generalized) triangles by adding some suitable diagonals, as illustrated below:

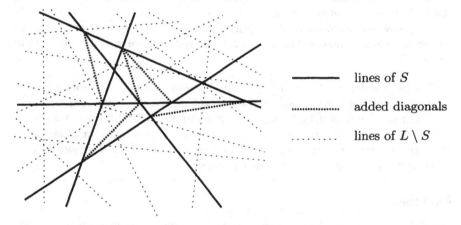

lines of S

added diagonals

lines of $L \setminus S$

This creates (generalized) triangles $\Delta_1, \Delta_2, \ldots, \Delta_t$ with $t = O(s^2)$ (since we have a drawing of a planar graph with $\binom{s}{2}+1$ vertices; also see Exercise 2).

4.6.1 Lemma. *For $s = 6r \ln n$, the following holds with a positive probability: The Δ_i form a $\frac{1}{r}$-cutting for L; that is, the interior of no Δ_i is intersected by more than $\frac{n}{r}$ lines of L.*

This implies the promised weaker version of the cutting lemma: Since the probability of the sample S being good is positive, there exists at least one good S that yields the desired collection of triangles.

Proof of Lemma 4.6.1. Let us say that a triangle T is *dangerous* if its interior is intersected by at least $k = \frac{n}{r}$ lines of L. We fix some arbitrary dangerous triangle T. What is the probability that no line of the sample S intersects the interior of T? We select a random line s times. The probability that we never hit one of the k lines intersecting the interior of T is at most

$(1 - k/n)^s$. Using the well-known inequality $1+x \leq e^x$, we can bound this probability by $e^{-ks/n} = e^{-6 \ln n} = n^{-6}$.

Call a triangle T *interesting* (for L) if it can appear in a triangulation for some sample $S \subseteq L$. Any interesting triangle has vertices at some three vertices of the arrangement of L, and hence there are fewer than n^6 interesting triangles.[2] Therefore, with a positive probability, a random sample S intersects the interiors of all the dangerous interesting triangles simultaneously. In particular, none of the triangles Δ_i appearing in the triangulation of such a sample S can be dangerous. This proves Lemma 4.6.1.   □

More sophisticated probabilistic reasoning shows that it is sufficient to choose $s = \text{const} \cdot r \log r$ in Lemma 4.6.1, instead of $\text{const} \cdot r \log n$, and still, with a positive probability no interesting dangerous triangle is missed by S (see Section 6.5 and also Exercise 10.3.4). This improvement is important for r small, say constant: It shows that the number of triangles in a $\frac{1}{r}$-cutting can be bounded independent of n.

To prove the asymptotically tight bound $O(r^2)$ by a random sampling argument seems considerably more complicated and we will discuss this in Section 6.5.

Bibliography and remarks. The ideas in the above proof of the weaker cutting lemma can be traced back at least to early papers of Clarkson (such as [Cla87]) on random sampling in computational geometry. The presented proof was constructed ex post facto for didactic purposes; the cutting lemma was first proved, as far as I know, in a stronger form (with $\log r$ instead of $\log n$).

Exercises

1. Calculate the exact expected size of S, a sample drawn from n elements by s independent random draws with replacements. ③
2. Calculate the number of (generalized) triangles arising by triangulating an arrangement of n lines in the plane in general position. (First, specify how exactly the unbounded cells are triangulated.) ②
3. (A cutting lemma for circles) Consider a set K of n circles in the plane. Select a sample $S \subseteq K$ by s independent random draws with replacement. Consider the arrangement of S, and construct its *vertical decomposition*; that is, from each vertex extend vertical segments upwards and downwards until they hit a circle of S (or all the way to infinity). Similarly extend vertical segments from the leftmost and rightmost points of each circle.

---

[2] The unbounded triangles have only 1 or 2 vertices, but they are completely determined by their two unbounded rays, and so their number is at most n^2.

(a) Show that this partitions the plane into $O(s^2)$ "circular trapezoids" (shapes bounded by at most two vertical segments and at most two circular arcs). [2]

(b) Show that for $s = Cr \ln n$ with a sufficiently large constant C, there is a positive probability that the sample S intersects all the dangerous interesting circular trapezoids, where "dangerous" and "interesting" are defined analogously to the definitions in the proof of the weaker version of the cutting lemma . [3]

4. Using Exercises 3 and 4.5.1, show that the number of unit distances determined by n points in the plane is $O(n^{4/3} \log^{2/3} n)$. [3]

5. Using Exercises 3 and 4.5.2, show that $I_{\mathrm{circ}}(n, n) = O(n^{1.4} \log^c n)$ (for some constant c), where $I_{\mathrm{circ}}(m, n)$ is the maximum possible number of incidences between m points and n arbitrary circles in the plane. [3]

4.7 The Cutting Lemma: A Tight Bound

Here we prove the cutting lemma in full strength. The proof is simple and elementary, but it does not seem to generalize to higher-dimensional situations.

For simplicity, we suppose that the given set L of n lines is in general position. (If not, perturb it slightly to get general position, construct the $\frac{1}{r}$-cutting, and perturb back; this gives a $\frac{1}{r}$-cutting for the original collection of lines; we omit the details.) First we need some definitions and observations concerning levels.

Levels and their simplifications. Let L be a fixed finite set of lines in the plane; we assume that no line of L is vertical. The *level* of a point $x \in \mathbf{R}^2$ is defined as the number of lines of L lying strictly below x.

We note that the level of all points of an (open) cell of the arrangement of L is the same, and similarly for a (relatively open) edge. On the other hand, the level of an edge can differ from the levels of its endpoints, for example.

We define the *level k* of the arrangement of L, where $0 \le k < n$, as the set E_k of all edges of the arrangement of L having level exactly k. These edges plus their endpoints form an x-monotone polygonal line, where x-monotone means that each vertical line intersects it at exactly one point.

It is easy to see that the level k makes a turn at each endpoint of its edges; it looks somewhat like this:

The level k is drawn thick, while the thin segments are pieces of lines of L and they do not belong to the level k.

Let $e_0, e_1, \ldots, e_t$ be the edges of E_k numbered from left to right; e_0 and e_t are the unbounded rays. Let us fix a point p_i in the interior of e_i. For an integer parameter $q \geq 2$, we define the *q-simplification of the level k* as the monotone polygonal line containing the left part of e_0 up to the point p_0, the segments $p_0 p_q$, $p_q p_{2q}, \ldots, p_{\lfloor (t-1)/q \rfloor q} p_t$, and the part of e_t to the right of p_t. Thus, the q-simplification has at most $\frac{t}{q} + 2$ edges. Here is an illustration for $t = 9$, $q = 4$:

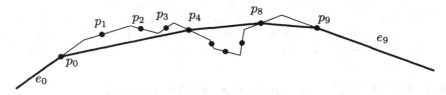

(We could have defined the q-simplification by connecting every qth vertex of the level, but the present way makes some future considerations a little simpler.)

4.7.1 Lemma.

(i) *The portion Π of the level k (considered as a polygonal line) between the points p_j and p_{j+q} is intersected by at most $q+1$ lines of L.*

(ii) *The segment $p_j p_{j+q}$ is intersected by at most $q+1$ lines of L.*

(iii) *The q-simplification of the level k is contained in the strip between the levels $k - \lceil q/2 \rceil$ and $k + \lceil q/2 \rceil$.*

Proof. Part (i) is obvious: Each line of L intersecting Π contains one of the edges $e_j, e_{j+1}, \ldots, e_{j+q}$. As for (ii), Π is connected, and hence all lines intersecting its convex hull must intersect Π itself as well. The segment $p_j p_{j+q}$ is contained in conv(Π).

Concerning (iii), imagine walking along some segment $p_j p_{j+q}$ of the q-simplification. We start at an endpoint, which has level k. Our current level may change only as we cross lines of L. Moreover, having traversed the whole segment we must be back to level k. Thus, to get from level k to $k + i$ and back to k we need to cross at least $2i$ lines on the way. From this and (ii), $2i \leq q+1$, and hence $i \leq \lfloor (q+1)/2 \rfloor = \lceil q/2 \rceil$.   □

Proof of the cutting lemma for lines in general position. Let r be the given parameter. If $r = \Omega(n)$, then it suffices to produce a 0-cutting of size $O(n^2)$ by simply triangulating the arrangement of L. Hence we may assume that r is much smaller than n.

Set $q = \lceil n/10r \rceil$. Divide the levels $E_0, E_1, \ldots, E_{n-1}$ into q groups: The ith group contains all E_j with j congruent to i modulo q ($i = 0, 1, \ldots, q-1$). Since the total number of edges in the arrangement is n^2, there is an i such that the ith group contains at most n^2/q edges. We fix one such i; from now on, we consider only the levels $i, q+i, 2q+i, \ldots$, and we construct the desired $\frac{1}{r}$-cutting from them.

Let P_j be the q-simplification of the level $jq+i$. If E_{jq+i} has m_j edges, then P_j has at most $m_j/q + 3$ edges, and the total number of edges of the P_j, $j = 0, 1, \ldots, \lfloor(n-1)/q\rfloor$, can be estimated by $n^2/q^2 + 3(n/q+1) = O(n^2/q^2)$. We note that the polygonal chains P_j never intersect properly: If they did, a vertex of some P_j, which has level $qj + i$, would be above P_{j+1}, and this is ruled out by Lemma 4.7.1(iii).

We form the vertical decomposition for the P_j; that is, we extend vertical segments from each vertex of P_j upwards and downwards until they hit P_{j-1} and P_{j+1}:

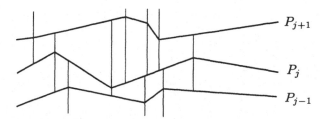

This subdivides the plane into $O(n^2/q^2) = O(r^2)$ trapezoids.

We claim that each such trapezoid is intersected by at most $\frac{n}{r}$ lines of L. We look at a trapezoid in the strip between P_j and P_{j+1}. By Lemma 4.7.1(iii), it lies between the levels $qj+i - \lceil q/2 \rceil$ and $q(j+1)+i+\lceil q/2 \rceil$, and therefore, each of its vertical sides is intersected by no more than $3q$ lines. The bottom side is a part of an edge of P_j, and consequently, it is intersected by no more than $q+1$ lines; similarly for the top side. Hence the number of lines intersecting the considered trapezoid is certainly at most $10q \leq \frac{n}{r}$. (A more careful analysis shows that one trapezoid is in fact intersected by at most $2q + O(1)$ lines; see Exercise 1.)

Finally, a $\frac{1}{r}$-cutting can be obtained by subdividing each trapezoid into two triangles by a diagonal. But let us remark that for applications of $\frac{1}{r}$-cuttings, trapezoids are usually as good as triangles.   □

Bibliography and remarks. The basic ideas of the presented proof are from [Mat90a], and the presentation basically follows [Mat98]. The latter paper provides some estimates for the number of triangles or trapezoids in a $\frac{1}{r}$-cutting, as $r \to \infty$: For example, at least $2.54(1 - o(1))r^2$ trapezoids are sometimes necessary, and $8(1+o(1))r^2$ trapezoids always suffice. The notion of levels and their simplifications, as well as Lemma 4.7.1, are due to Edelsbrunner and Welzl [EW86].

Exercises

1. (a) Verify that each trapezoid arising in the described construction is intersected by at most $2.5q + O(1)$ lines. Setting q appropriately, show that the plane can subdivided into $12.5r^2 + O(r)$ trapezoids, each intersected by at most $\frac{n}{r}$ lines, assuming $1 \ll r \ll n$. ②
(b) Improve the bounds from (a) to $2q + O(1)$ and $8r^2 + O(r)$, respectively. ④

5

Convex Polytopes

Convex polytopes are convex hulls of finite point sets in $\mathbf{R}^d$. They constitute the most important class of convex sets with an enormous number of applications and connections.

Three-dimensional convex polytopes, especially the regular ones, have been fascinating people since the antiquity. Their investigation was one of the main sources of the theory of planar graphs, and thanks to this well-developed theory they are quite well understood. But convex polytopes in dimension 4 and higher are considerably more challenging, and a surprisingly deep theory, mainly of algebraic nature, was developed in attempts to understand their structure.

A strong motivation for the study of convex polytopes comes from practically significant areas such as combinatorial optimization, linear programming, and computational geometry. Let us look at a simple example illustrating how polytopes can be associated with combinatorial objects. The 3-dimensional polytope in the picture

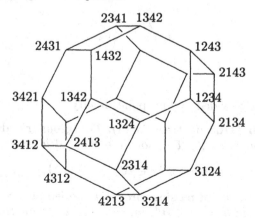

is called the *permutahedron*. Although it is 3-dimensional, it is most naturally defined as a subset of $\mathbf{R}^4$, namely, the convex hull of the 24 vectors obtained by permuting the coordinates of the vector $(1, 2, 3, 4)$ in all possible ways. In the picture, the (visible) vertices are labeled by the corresponding permutations. Similarly, the d-dimensional permutahedron is the convex hull of the $(d+1)!$ vectors in $\mathbf{R}^{d+1}$ arising by permuting the coordinates of $(1, 2, \ldots, d+1)$. One can observe that the edges of the polytope connect exactly pairs of permutations differing by a transposition of two adjacent numbers, and a closer examination reveals other connections between the structure of the permutahedron and properties of permutations.

There are many other, more sophisticated, examples of convex polytopes assigned to combinatorial and geometric objects such as graphs, partially ordered sets, classes of metric spaces, or triangulations of a given point set. In many cases, such convex polytopes are a key tool for proving hard theorems about the original objects or for obtaining efficient algorithms. Two impressive examples are discussed in Chapter 12, and several others are scattered in other chapters.

The present chapter should convey some initial working knowledge of convex polytopes for a nonpolytopist. It is just a small sample of an extensive theory. A more comprehensive modern introduction is the book by Ziegler [Zie94].

5.1 Geometric Duality

First we discuss geometric duality, a simple technical tool indispensable in the study of convex polytopes and handy in many other situations. We begin with a simple motivating question.

How can we visualize the set of all lines intersecting a convex pentagon as in the picture?

A suitable way is provided by line–point duality.

5.1.1 Definition (Duality transform). *The (geometric) duality transform is a mapping denoted by $\mathcal{D}_0$. To a point $a \in \mathbf{R}^d \setminus \{0\}$ it assigns the hyperplane*

$$\mathcal{D}_0(a) = \{x \in \mathbf{R}^d : \langle a, x \rangle = 1\},$$

and to a hyperplane h not passing through the origin, which can be uniquely written in the form $h = \{x \in \mathbf{R}^d : \langle a, x \rangle = 1\}$, it assigns the point $\mathcal{D}_0(h) = a \in \mathbf{R}^d \setminus \{0\}$.

Here is the geometric meaning of the duality transform. If a is a point at distance δ from 0, then $\mathcal{D}_0(a)$ is the hyperplane perpendicular to the line $0a$ and intersecting that line at distance $\frac{1}{\delta}$ from 0, in the direction from 0 towards a.

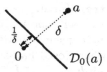

A nice interpretation of duality is obtained by working in $\mathbf{R}^{d+1}$ and identifying the "primal" $\mathbf{R}^d$ with the hyperplane $\pi = \{x \in \mathbf{R}^{d+1}: x_{d+1} = 1\}$ and the "dual" $\mathbf{R}^d$ with the hyperplane $\rho = \{x \in \mathbf{R}^{d+1}: x_{d+1} = -1\}$. The hyperplane dual to a point $a \in \pi$ is produced as follows: We construct the hyperplane in $\mathbf{R}^{d+1}$ perpendicular to $0a$ and containing 0, and we intersect it with ρ. Here is an illustration for $d = 2$:

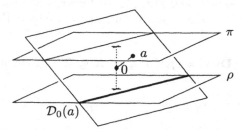

In this way, the duality $\mathcal{D}_0$ can be naturally extended to k-flats in $\mathbf{R}^d$, whose duals are $(d-k-1)$-flats. Namely, given a k-flat $f \subset \pi$, we consider the $(k+1)$-flat F through 0 and f, we construct the orthogonal complement of F, and we intersect it with ρ, obtaining $\mathcal{D}_0(f)$.

Let us consider the pentagon drawn above and place it so that the origin lies in the interior. Let $v_i = \mathcal{D}_0(\ell_i)$, where ℓ_i is the line containing the side $a_i a_{i+1}$. Then the points dual to the lines intersecting the pentagon $a_1 a_2 \ldots a_5$ fill exactly the exterior of the convex pentagon $v_1 v_2 \ldots v_5$:

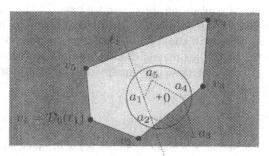

This follows easily from the properties of duality listed below (of course, there is nothing special about a pentagon here). Thus, the considered set of lines can be nicely described in the dual plane. A similar passage from lines to points or back is useful in many geometric or computational problems.

Properties of the duality transform. Let p be a point of $\mathbf{R}^d$ distinct from the origin and let h be a hyperplane in $\mathbf{R}^d$ not containing the origin. Let h^- stand for the closed half-space bounded by h and containing the origin, while h^+ denotes the other closed half-space bounded by h. That is, if $h = \{x \in \mathbf{R}^d \colon \langle a, x \rangle = 1\}$, then $h^- = \{x \in \mathbf{R}^d \colon \langle a, x \rangle \leq 1\}$.

5.1.2 Lemma (Duality preserves incidences).

(i) $p \in h$ *if and only if* $\mathcal{D}_0(h) \in \mathcal{D}_0(p)$.
(ii) $p \in h^-$ *if and only if* $\mathcal{D}_0(h) \in \mathcal{D}_0(p)^-$.

Proof. (i) Let $h = \{x \in \mathbf{R}^d \colon \langle a, x \rangle = 1\}$. Then $p \in h$ means $\langle a, p \rangle = 1$. Now, $\mathcal{D}_0(h)$ is the point a, and $\mathcal{D}_0(p)$ is the hyperplane $\{y \in \mathbf{R}^d \colon \langle y, p \rangle = 1\}$, and hence $\mathcal{D}_0(h) = a \in \mathcal{D}_0(p)$ also means just $\langle a, p \rangle = 1$. Part (ii) is proved similarly. □

5.1.3 Definition (Dual set). *For a set* $X \subseteq \mathbf{R}^d$, *we define the set dual to* X, *denoted by* X^*, *as follows:*

$$X^* = \left\{ y \in \mathbf{R}^d \colon \langle x, y \rangle \leq 1 \text{ for all } x \in X \right\}.$$

Another common name used for the duality is *polarity*; the dual set would then be called the *polar set*. Sometimes it is denoted by X°.

Geometrically, X^* is the intersection of all half-spaces of the form $\mathcal{D}_0(x)^-$ with $x \in X$. Or in other words, X^* consists of the origin plus all points y such that $X \subseteq \mathcal{D}_0(y)^-$. For example, if X is the pentagon $a_1 a_2 \ldots a_5$ drawn above, then X^* is the pentagon $v_1 v_2 \ldots v_5$.

For any set X, the set X^* is obviously closed and convex and contains the origin. Using the separation theorem (Theorem 1.2.4), it is easily shown that for any set $X \subseteq \mathbf{R}^d$, the set $(X^*)^*$ is the closure $\mathrm{conv}(X \cup \{0\})$. In particular, for a closed convex set containing the origin we have $(X^*)^* = X$ (Exercise 3).

For a hyperplane h, the dual set h^* is different from the point $\mathcal{D}_0(h)$.[1]

For readers familiar with the duality of planar graphs, let us remark that it is closely related to the geometric duality applied to convex polytopes in $\mathbf{R}^3$. For example, the next drawing illustrates a planar graph and its dual graph (dashed):

---

[1] In the literature, however, the "star" notation is sometimes also used for the dual point or hyperplane, so for a point p, the hyperplane $\mathcal{D}_0(p)$ would be denoted by p^*, and similarly, h^* may stand for $\mathcal{D}_0(h)$.

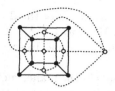

Later we will see that these are graphs of the 3-dimensional cube and of the regular octahedron, which are polytopes dual to each other in the sense defined above. A similar relation holds for all 3-dimensional polytopes and their graphs.

Other variants of duality. The duality transform $\mathcal{D}_0$ defined above is just one of a class of geometric transforms with similar properties. For some purposes, other such transforms (dualities) are more convenient. A particularly important duality, denoted by $\mathcal{D}$, corresponds to moving the origin to the "minus infinity" of the x_d-axis (the x_d-axis is considered vertical). A formal definition is as follows.

5.1.4 Definition (Another duality). *A nonvertical hyperplane h can be uniquely written in the form $h = \{x \in \mathbf{R}^d \colon x_d = a_1 x_1 + \cdots + a_{d-1} x_{d-1} - a_d\}$. We set $\mathcal{D}(h) = (a_1, \ldots, a_{d-1}, a_d)$. Conversely, the point $a = (a_1, \ldots, a_{d-1}, a_d)$ maps back to h.*

The property (i) of Lemma 5.1.2 holds for this $\mathcal{D}$, and an analogue of (ii) is:

(ii′) A point p lies above a hyperplane h if and only if the point $\mathcal{D}(h)$ lies above the hyperplane $\mathcal{D}(p)$.

Exercises

1. Let $C = \{x \in \mathbf{R}^d \colon |x_1| + \cdots + |x_d| \leq 1\}$. Show that C^* is the d-dimensional cube $\{x \in \mathbf{R}^d \colon \max |x_i| \leq 1\}$. Picture both bodies for $d = 3$. ②
2. Prove the assertion made in the text about the lines intersecting a convex pentagon. ②
3. Show that for any $X \subseteq \mathbf{R}^d$, $(X^*)^*$ equals the closure of $\mathrm{conv}(X \cup \{0\})$, where X^* stands for the dual set to X. ③
4. Let $C \subseteq \mathbf{R}^d$ be a convex set. Prove that C^* is bounded if and only if 0 lies in the interior of C. ②
5. Show that $C = C^*$ if and only if C is the unit ball centered at the origin. ②
6. (a) Let $C = \mathrm{conv}(X) \subseteq \mathbf{R}^d$. Prove that $C^* = \bigcap_{x \in X} \mathcal{D}_0(x)^-$. ②
   (b) Show that if $C = \bigcap_{h \in H} h^-$, where H is a collection of hyperplanes not passing through 0, and if C is bounded, then $C^* = \mathrm{conv}\{\mathcal{D}_0(h) \colon h \in H\}$. ②
   (c) What is the right analogue of (b) if C is unbounded? ②
7. What is the dual set h^* for a hyperplane h, and what about h^{**}? ③

8. Verify the geometric interpretation of the duality $\mathcal{D}_0$ outlined in the text (using the embeddings of $\mathbf{R}^d$ into $\mathbf{R}^{d+1}$). ②

9. (a) Let s be a segment in the plane. Describe the set of all points dual to lines intersecting s. ①

   (b) Consider $n \geq 3$ segments in the plane, such that none of them contains 0 but they all lie on lines passing through 0. Show that if every 3 among such segments can be intersected by a single line, then all the segments can be simultaneously intersected by a line. ③

   (c) Show that the assumption in (b) that the extensions of the segments pass through 0 is essential: For each $n \geq 1$, construct $n+1$ pairwise disjoint segments in the plane that cannot be simultaneously intersected by a line but every n of them can (such an example was first found by Hadwiger and Debrunner). ④

5.2 *H*-Polytopes and *V*-Polytopes

A convex polytope in the plane is a convex polygon. Famous examples of convex polytopes in $\mathbf{R}^3$ are the Platonic solids: regular tetrahedron, cube, regular octahedron, regular dodecahedron, and regular icosahedron. A convex polytope in $\mathbf{R}^3$ is a convex set bounded by finitely many convex polygons. Such a set can be regarded as a convex hull of a finite point set, or as an intersection of finitely many half-spaces. We thus define two types of convex polytopes, based on these two views.

5.2.1 Definition (*H*-polytope and *V*-polytope). *An H-polyhedron is an intersection of finitely many closed half-spaces in some $\mathbf{R}^d$. An H-polytope is a bounded H-polyhedron.*

   A V-polytope is the convex hull of a finite point set in $\mathbf{R}^d$.

A basic theorem about convex polytopes claims that from the mathematical point of view, *H*-polytopes and *V*-polytopes are equivalent.

5.2.2 Theorem. *Each V-polytope is an H-polytope. Each H-polytope is a V-polytope.*

This is one of the theorems that may look "obvious" and whose proof needs no particularly clever idea but does require some work. In the present case, we do not intend to avoid it. Actually, we have quite a neat proof in store, but we postpone it to the end of this section.

Although *H*-polytopes and *V*-polytopes are mathematically equivalent, there is an enormous difference between them from the computational point of view. That is, it matters a lot whether a convex polytope is given to us as a convex hull of a finite set or as an intersection of half-spaces. For example, given a set of n points specifying a *V*-polytope, how do we find its representation as an *H*-polytope? It is not hard to come up with some algorithm, but the problem is to find an efficient algorithm that would allow one to handle large real-world problems. This algorithmic question is not yet satisfactorily solved. Moreover, in some cases the number of required half-spaces may be astronomically large compared to the number n of points, as we will see later in this chapter.

As another illustration of the computational difference between V-polytopes and H-polytopes, we consider the maximization of a given linear function over a given polytope. For V-polytopes it is a trivial problem, since it suffices to substitute all points of V into the given linear function and select the maximum of the resulting values. But maximizing a linear function over the intersection of a collection of half-spaces is the basic problem of linear programming, and it is certainly nontrivial.

Terminology. The usual terminology does not distinguish V-polytopes and H-polytopes. A *convex polytope* means a point set in $\mathbf{R}^d$ that is a V-polytope (and thus also an H-polytope). An arbitrary, possibly unbounded, H-polyhedron is called a *convex polyhedron*. All polytopes and polyhedra considered in this chapter are convex, and so the adjective "convex" is often omitted.

The *dimension* of a convex polyhedron is the dimension of its affine hull. It is the smallest dimension of a Euclidean space containing a congruent copy of P.

Basic examples. One of the easiest classes of polytopes is that of *cubes*. The d-dimensional cube as a point set is the Cartesian product $[-1, 1]^d$.

$$d = 1 \qquad d = 2 \qquad d = 3$$

As a V-polytope, the d-dimensional cube is the convex hull of the set $\{-1, 1\}^d$ (2^d points), and as an H-polytope, it can be described by the inequalities $-1 \le x_i \le 1$, $i = 1, 2, \ldots, d$, i.e., by $2d$ half-spaces. We note that it is also the unit ball of the maximum norm $\|x\|_\infty = \max_i |x_i|$.

Another important example is the class of *crosspolytopes* (or generalized octahedra). The d-dimensional crosspolytope is the convex hull of the "coordinate cross," i.e., $\mathrm{conv}\{e_1, -e_1, e_2, -e_2, \ldots, e_d, -e_d\}$, where $e_1, \ldots, e_d$ are the vectors of the standard orthonormal basis.

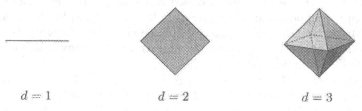

$$d = 1 \qquad d = 2 \qquad d = 3$$

It is also the unit ball of the ℓ_1-norm $\|x\|_1 = \sum_{i=1}^d |x_i|$. As an H-polytope, it can be expressed by the 2^d half-spaces of the form $\langle \sigma, \leq \rangle 1$, where σ runs through all vectors in $\{-1, 1\}^d$.

The polytopes with the smallest possible number of vertices (for a given dimension) are called simplices.

5.2.3 Definition (Simplex). *A simplex is the convex hull of an affinely independent point set in some* $\mathbf{R}^d$.

A d-dimensional simplex in $\mathbf{R}^d$ can also be represented as an intersection of $d+1$ half-spaces, as is not difficult to check.

A *regular* d-dimensional simplex is the convex hull of $d+1$ points with all pairs of points having equal distances.

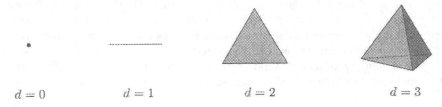

$d = 0$ $d = 1$ $d = 2$ $d = 3$

Unlike cubes and crosspolytopes, d-dimensional regular simplices do not have a very nice coordinate representation in $\mathbf{R}^d$. The simplest and most useful representation lives one dimension higher: The convex hull of the $d+1$ vectors $e_1, \ldots, e_{d+1}$ of the standard orthonormal basis in $\mathbf{R}^{d+1}$ is a d-dimensional regular simplex with side length $\sqrt{2}$.

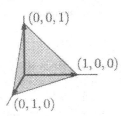

$(0,0,1)$

$(1,0,0)$

$(0,1,0)$

Proof of Theorem 5.2.2 (equivalence of H-polytopes and V-polytopes). We first show that any H-polytope is also a V-polytope. We proceed by induction on d. The case $d = 1$ being trivial, we suppose that $d \geq 2$.

So let Γ be a finite collection of closed half-spaces in $\mathbf{R}^d$ such that $P = \bigcap \Gamma$ is nonempty and bounded. For each $\gamma \in \Gamma$, let $F_\gamma = P \cap \partial\gamma$ be the intersection of P with the bounding hyperplane of γ. Each nonempty F_γ is an H-polytope of dimension at most $d-1$ (correct?), and so it is the convex hull of a finite set $V_\gamma \subset F_\gamma$ by the inductive hypothesis.

We claim that $P = \mathrm{conv}(V)$, where $V = \bigcup_{\gamma \in \Gamma} V_\gamma$. Let $x \in P$ and let ℓ be a line passing through x. The intersection $\ell \cap P$ is a segment; let y and z be its endpoints. There are $\alpha, \beta \in \Gamma$ such that $y \in F_\alpha$ and $z \in F_\beta$ (if y were not on the boundary of any $\gamma \in \Gamma$, we could continue along ℓ a little further within P).

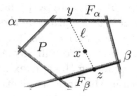

We have $y \in \operatorname{conv}(V_\alpha)$ and $z \in \operatorname{conv}(V_\beta)$, and thus $x \in \operatorname{conv}(V_\alpha \cup V_\beta) \subseteq \operatorname{conv}(V)$.

We have proved that any H-polytope is a V-polytope, and it remains to show that a V-polytope can be expressed as the intersection of finitely many half-spaces. This follows easily by duality (and implicitly uses the separation theorem).

Let $P = \operatorname{conv}(V)$ with V finite, and assume that 0 is an interior point of P. By Exercise 5.1.6(a), the dual body P^* equals $\bigcap_{v \in V} \mathcal{D}_0(v)^-$, and by Exercise 5.1.4 it is bounded. By what we have already proved, P^* is a V-polytope, and by Exercise 5.1.6(a) again, $P = (P^*)^*$ is the intersection of finitely many half-spaces.   □

Bibliography and remarks. The theory of convex polytopes is a well-developed area covered in numerous books and surveys, such as the already recommended recent monograph [Zie94] (with addenda and updates on the web page of its author), the very influential book by Grünbaum [Grü67], the chapters on polytopes in the handbooks of discrete and computational geometry [GO97], of convex geometry [GW93], and of combinatorics [GGL95], or the books McMullen and Shephard [MS71] and Brønsted [Brø83], concentrating on questions about the numbers of faces. Recent progress in combinatorial and computational polytope theory is reflected in the collection [KZ00]. For analyzing examples, one should be aware of (free) software systems for manipulating convex polytopes, such as **polymake** by Gawrilow and Joswig [GJ00].

Interesting discoveries about 3-dimensional convex polytopes were already made in ancient Greece. The treatise by Schläfli [Sch01] written in 1850–52 is usually mentioned as the beginning of modern theory, and several books were published around the turn of the century. We refer to Grünbaum [Grü67], Schrijver [Sch86], and to the other sources mentioned above for historical accounts.

The permutahedron mentioned in the introduction to this chapter was considered by Schoute [Sch11], and it arises by at least two other quite different and natural constructions (see [Zie94]).

There are several ways of proving the equivalence of H-polytopes and V-polytopes. Ours is inspired by a proof by Edmonds, as presented in Fukuda's lecture notes (ETH Zürich). A classical algorithmic proof is provided by the *Fourier–Motzkin elimination procedure*, which proceeds by projections on coordinate hyperplanes; see [Zie94] for a detailed exposition. The *double-description method* is a similar algorithm formulated in the dual setting, and it is still one of the most efficient known computational methods. We will say a little more about the algorithmic problem of expressing the convex hull of a finite set as the intersection of half-spaces in the notes to Section 5.5.

One may ask, What is a "vertex description" of an *unbounded H-polyhedron*? Of course, it is not the convex hull of a finite set, but it can be expressed as the Minkowski sum $P + C$, where P is a V-polytope and C is a convex cone described as the convex hull of finitely many rays emanating from 0.

Exercises

1. Verify that a d-dimensional simplex in $\mathbf{R}^d$ can be expressed as the intersection of $d+1$ half-spaces. $\boxed{2}$
2. (a) Show that every convex polytope in $\mathbf{R}^d$ is an orthogonal projection of a simplex of a sufficiently large dimension onto the space $\mathbf{R}^d$ (which we consider embedded as a d-flat in some $\mathbf{R}^n$). $\boxed{3}$
   (b) Prove that every convex polytope P symmetric about 0 (i.e., with $P = -P$) is the affine image of a crosspolytope of a sufficiently large dimension. $\boxed{3}$

5.3 Faces of a Convex Polytope

The surface of the 3-dimensional cube consists of 8 "corner" points called vertices, 12 edges, and 6 squares called *facets*. According to the perhaps more usual terminology in 3-dimensional geometry, the facets would be called faces. But in the theory of convex polytopes, the word face has a slightly different meaning, defined below. For the cube, not only the squares but also the vertices and the edges are all called *faces* of the cube.

5.3.1 Definition (Face). *A* face *of a convex polytope P is defined as*

- *either P itself, or*
- *a subset of P of the form $P \cap h$, where h is a hyperplane such that P is fully contained in one of the closed half-spaces determined by h.*

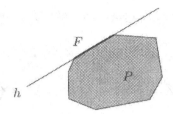

We observe that each face of P is a convex polytope. This is because P is the intersection of finitely many half-spaces and h is the intersection of two half-spaces, so the face is an H-polyhedron, and moreover, it is bounded.

If P is a polytope of dimension d, then its faces have dimensions -1, 0, $1, \ldots, d$, where -1 is, by definition, the dimension of the empty set. A face of dimension j is also called a j-face.

Names of faces. The 0-faces are called *vertices*, the 1-faces are called *edges*, and the $(d-1)$-faces of a d-dimensional polytope are called *facets*. The $(d-2)$-faces of a d-dimensional polytope are *ridges*; in the familiar 3-dimensional situation, edges = ridges. For example, the 3-dimensional cube has 28 faces in total: the empty face, 8 vertices, 12 edges, 6 facets, and the whole cube.

The following proposition shows that each V-polytope is the convex hull of its vertices, and that the faces can be described combinatorially: They are the convex hulls of certain subsets of vertices. This includes some intuitive facts such as that each edge connects two vertices.

A helpful notion is that of an *extremal point* of a set: For a set $X \subseteq \mathbf{R}^d$, a point $x \in X$ is extremal if $x \notin \operatorname{conv}(X \setminus \{x\})$.

5.3.2 Proposition. *Let $P \subset \mathbf{R}^d$ be a (bounded) convex polytope.*

(i) *("Vertices are extremal") The extremal points of P are exactly its vertices, and P is the convex hull of its vertices.*

(ii) *("Face of a face is a face") Let F be a face of P. The vertices of F are exactly those vertices of P that lie in F. More generally, the faces of F are exactly those faces of P that are contained in F.*

The proof is not essential for our further considerations, and it is given at the end of this section (but Exercise 9 below illustrates that things are not quite as simple as it might perhaps seem). The proposition has an appropriate analogue for polyhedra, but in order to avoid technicalities, we treat the bounded case only.

Graphs of polytopes. Each 1-dimensional face, or edge, of a convex polytope has exactly two vertices. We can thus define the *graph* $G(P)$ of a polytope P in the natural way: The vertices of the polytope are vertices of the graph, and two vertices are connected by an edge in the graph if they are vertices of the same edge of P. (The terms "vertices" and "edges" for graphs actually come from the corresponding notions for 3-dimensional convex polytopes.)

Here is an example of a 3-dimensional polytope, the regular octahedron, with its graph:

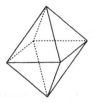 

For polytopes in $\mathbf{R}^3$, the graph is always planar: Project the polytope from its interior point onto a circumscribed sphere, and then make a "cartographic map" of this sphere, say by stereographic projection. Moreover, it can be shown that the graph is vertex 3-connected. (A graph G is called *vertex k-connected* if $|V(G)| \geq k+1$ and deleting any at most $k-1$ vertices leaves G connected.) Nicely enough, these properties characterize graphs of convex 3-polytopes:

5.3.3 Theorem (Steinitz theorem). *A finite graph is isomorphic to the graph of a 3-dimensional convex polytope if and only if it is planar and vertex 3-connected.*

We omit a proof of the considerably harder "if" part (exhibiting a polytope for every vertex 3-connected planar graph); all known proofs are quite complicated.

Graphs of higher-dimensional polytopes probably have no nice description comparable to the 3-dimensional case, and it is likely that the problem of deciding whether a given graph is isomorphic to a graph of a 4-dimensional convex polytope is NP-hard. It is known that the graph of every d-dimensional polytope is vertex d-connected (*Balinski's theorem*), but this is only a necessary condition.

Examples. A d-dimensional simplex has been defined as the convex hull of a $(d+1)$-point affinely independent set V. It is easy to see that each subset of V determines a face of the simplex. Thus, there are $\binom{d+1}{k+1}$ faces of dimension k, $k = -1, 0, \ldots, d$, and 2^{d+1} faces in total.

The d-dimensional crosspolytope has $V = \{e_1, -e_1, \ldots, e_d, -e_d\}$ as the vertex set. A proper subset $F \subset V$ determines a face if and only if there is no i such that both $e_i \in F$ and $-e_i \in F$ (Exercise 2). It follows that there are $3^d + 1$ faces, including the empty one and the whole crosspolytope.

The nonempty faces of the d-dimensional cube $[-1, 1]^d$ correspond to vectors $v \in \{-1, 1, 0\}^d$. The face corresponding to such v has the vertex set $\{u \in \{-1, 1\}^d \colon u_i = v_i \text{ for all } i \text{ with } v_i \neq 0\}$. Geometrically, the vector v is the center of gravity of its face.

The face lattice. Let $\mathcal{F}(P)$ be the set of all faces of a (bounded) convex polytope P (including the empty face $\emptyset$ of dimension -1). We consider the partial ordering of $\mathcal{F}(P)$ by inclusion.

5.3.4 Definition (Combinatorial equivalence). *Two convex polytopes P and Q are called* combinatorially equivalent *if $\mathcal{F}(P)$ and $\mathcal{F}(Q)$ are isomorphic as partially ordered sets.*

We are going to state some properties of the partially ordered set $\mathcal{F}(P)$ without proofs. These are not difficult and can be found in [Zie94].

It turns out that $\mathcal{F}(P)$ is a lattice (a partially ordered set satisfying additional axioms). We recall that this means the following two conditions:

- *Meets condition:* For any two faces $F, G \in \mathcal{F}(P)$, there exists a face $M \in \mathcal{F}(P)$, called the *meet* of F and G, that is contained in both F and G and contains all other faces contained in both F and G.
- *Joins condition:* For any two faces $F, G \in \mathcal{F}(P)$, there exists a face $J \in \mathcal{F}(P)$, called the *join* of F and G, that contains both F and G and is contained in all other faces containing both F and G.

The meet of two faces is their geometric intersection $F \cap G$.

For verifying the joins and meets conditions, it may be helpful to know that for a finite partially ordered set possessing the minimum element and the maximum element, the meets condition is equivalent to the joins condition, and so it is enough to check only one of the conditions.

Here is the face lattice of a 3-dimensional pyramid:

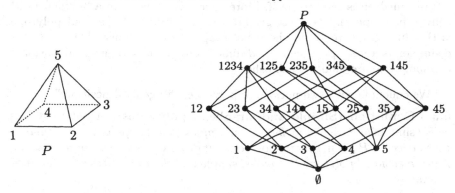

The vertices are numbered 1–5, and the faces are labeled by the vertex sets.

The face lattice is *graded*, meaning that every maximal chain has the same length (the rank of a face F is $\dim(F)+1$). Quite obviously, it is *atomic*: Every face is the join of its vertices. A little less obviously, it is *coatomic*; that is, every face is the meet (intersection) of the facets containing it. An important consequence is that combinatorial type of a polytope is determined by the vertex–facet incidences. More precisely, if we know the dimension and all subsets of vertices that are vertex sets of facets (but without knowing the coordinates of the vertices, of course), we can uniquely reconstruct the whole face lattice in a simple and purely combinatorial way.

Face lattices of convex polytopes have several other nice properties, but no full algebraic characterization is known, and the problem of deciding whether a given lattice is a face lattice is algorithmically difficult (even for 4-dimensional polytopes).

The face lattice can be a suitable representation of a convex polytope in a computer. Each j-face is connected by pointers to its $(j-1)$-faces and to the $(j+1)$-faces containing it. On the other hand, it is a somewhat redundant representation: Recall that the vertex–facet incidences already contain the full information, and for some applications, even less data may be sufficient, say the graph of the polytope.

The dual polytope. Let P be a convex polytope containing the origin in its interior. Then the dual set P^* is also a polytope; we have verified this in the proof of Theorem 5.2.2.

5.3.5 Proposition. *For each $j = -1, 0, \ldots, d$, the j-faces of P are in a bijective correspondence with the $(d-j-1)$-faces of P^*. This correspondence also reverses inclusion; in particular, the face lattice of P^* arises by turning the face lattice of P upside down.*

Again we refer to the reader's diligence or to [Zie94] for a proof. Let us examine a few examples instead.

Among the five regular Platonic solids, the cube and the octahedron are dual to each other, the dodecahedron and the icosahedron are also dual, and the tetrahedron is dual to itself. More generally, if we have a 3-dimensional convex polytope and G is its graph, then the graph of the dual polytope is the dual graph to G, in the usual graph-theoretic sense. The dual of a d-simplex is a d-simplex, and the d-dimensional cube and the d-dimensional crosspolytope are dual to each other.

We conclude with two notions of polytopes "in general position."

5.3.6 Definition (Simple and simplicial polytopes). *A polytope P is called* simplicial *if each of its facets is a simplex (this happens, in particular, if the vertices of P are in general position, but general position is not necessary). A d-dimensional polytope P is called* simple *if each of its vertices is contained in exactly d facets.*

The faces of a simplex are again simplices, and so each proper face of a simplicial polytope is a simplex. Among the five Platonic solids, the tetrahedron, the octahedron, and the icosahedron are simplicial; and the tetrahedron, the cube, and the dodecahedron are simple. Crosspolytopes are simplicial, and cubes are simple. An example of a polytope that is neither simplicial nor simple is the 4-sided pyramid used in the illustration of the face lattice.

The dual of a simple polytope is simplicial, and vice versa. For a simple d-dimensional polytope, a small neighborhood of each vertex looks combinatorially like a neighborhood of a vertex of the d-dimensional cube. Thus, for each vertex v of a d-dimensional simple polytope, there are d edges emanating from v, and each k-tuple of these edges uniquely determines one k-face incident to v. Consequently, v belongs to $\binom{d}{k}$ k-faces, $k = 0, 1, \ldots, d$.

Proof of Proposition 5.3.2. In (i) ("vertices are extremal"), we assume that P is the convex hull of a finite point set. Among all such sets, we fix one that is inclusion-minimal and call it V_0. Let V_v be the vertex set of P, and let V_e be the set of all extremal points of P. We prove that $V_0 = V_v = V_e$, which gives (i). We have $V_e \subseteq V_0$ by the definition of an extremal point.

Next, we show that $V_v \subseteq V_e$. If $v \in V_v$ is a vertex of P, then there is a hyperplane h with $P \cap h = \{v\}$, and all of $P \setminus \{v\}$ lies in one of the open half-spaces defined by h. Hence $P \setminus \{v\}$ is convex, which means that v is an extremal point of P, and so $V_v \subseteq V_e$.

Finally we verify $V_0 \subseteq V_v$. Let $v \in V_0$; by the inclusion-minimality of V_0, we get that $v \notin C = \mathrm{conv}(V_0 \setminus \{v\})$. Since C and $\{v\}$ are disjoint compact convex sets, they can be strictly separated by a hyperplane h. Let h_v be the hyperplane parallel to h and containing v; this h_v has all points of $V_0 \setminus \{v\}$ on one side.

We want to show that $P \cap h_v = \{v\}$ (then v is a vertex of P, and we are done). The set $P \setminus h_v = \mathrm{conv}(V_0) \setminus h_v$, being the intersection of a convex set with an open half-space, is convex. Any segment vx, where $x \in P \setminus h_v$, shares only the point v with the hyperplane h_v, and so $(P \setminus h_v) \cup \{v\}$ is convex as well. Since this set contains V_0 and is convex, it contains $P = \mathrm{conv}(V_0)$, and so $P \cap h_v = \{v\}$ indeed.

As for (ii) ("face of a face is a face"), it is clear that a face G of P contained in F is a face of F too (use the same witnessing hyperplane). For the reverse direction, we begin with the case of vertices. By a consideration similar to that at the end of the proof of (i), we see that $F = \mathrm{conv}(V) \cap h = \mathrm{conv}(V \cap h)$. Hence all the extremal points of F, which by (i) are exactly the vertices of F, are in V.

Finally, let F be a face of P defined by a hyperplane h, and let $G \subset F$ be a face of F defined by a hyperplane g within h; that is, g is a $(d-2)$-dimensional affine subspace of h with $G = g \cap F$ and with all of F on one side. Let γ be the closed half-space bounded by h with $P \subset \gamma$. We start rotating the boundary h of γ around g in the direction such that the rotated half-space γ' still contains F.

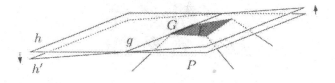

If we rotate by a sufficiently small amount, then all the vertices of P not lying in F are still in the interior of γ'. At the same time, the interior of γ' contains all the vertices of F not lying in G, while all the vertices of G remain on the boundary h' of γ'. So h' defines a face of P (since all of P is on one side), and this face has the same vertex set as G, and so it equals G by the first part of (ii) proved above. $\square$

Bibliography and remarks. Most of the material in this section is quite old, and we restrict ourselves to a few comments and remarks on recent developments.

Graphs of polytopes. The Steinitz theorem was published in [Ste22]. A proof (of the harder implication) can be found in [Zie94]. In this type of proof, one starts with the planar graph K_4, which is obviously realizable as a graph of a 3-dimensional polytope, and creates the given 3-connected planar graph by a sequence of suitable elementary operations, the so-called ΔY transformations, which are shown to preserve the realizability. Another type of proof first finds a suitable straight edge planar drawing of the given graph G and then shows that the vertices of such a drawing can be lifted to $\mathbf{R}^3$ to form the appropriate polytope. The drawings needed here are "rubber band" drawings: Pin down the vertices of an outer face and think of the edges as rubber bands of various strengths, which left alone would contract to points. Then the equilibrium position, where the forces at every inner vertex add up to 0, specifies the drawing (see, e.g., Richter-Gebert [RG97] for a presentation). These ideas go back to Maxwell; the result about the equilibrium position specifying straight edge drawing for every 3-connected planar graph was proved by Tutte [Tut60]. Very interesting related results about graphs with higher connectivity are due to Linial, Lovász, and Wigderson [LW88]. Another way of obtaining suitable drawings is via *Koebe's representation theorem* (see, e.g., [PA95] for an exposition): Every planar graph G can be represented by touching circles; that is, every vertex $v \in V(G)$ can be assigned a circular disk in the plane in such a way that the disks have pairwise disjoint interiors and two of them touch if and only if their two vertices are connected by an edge.

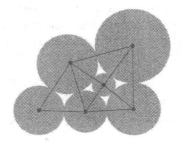

On the other hand, Koebe's theorem follows easily from a stronger version of the Steinitz theorem due to Andreev: Every 3-connected planar graph has a *cage representation*, i.e., as the graph of a 3-dimensional convex polytope P whose edges are all tangent to the unit sphere (each vertex of P can see a cap of the unit sphere, and a suitable stereographic projection of these caps yields the disks as in Koebe's theorem). These beautiful results, as well as several others along these lines, would certainly deserve to be included in a book like this, but here they are not for space and time reasons.

A result of Blind and Mani-Levitska, with a beautiful simple new proof by Kalai [Kal88], shows that a simple polytope is determined by its dimension and its graph; that is, if two d-dimensional simple polytopes P and Q have isomorphic graphs, then they are combinatorially equivalent.

One of the most challenging problems about graphs of convex polytopes is the *Hirsch conjecture*. In its basic form, it states that the graph of any d-dimensional polytope with n facets has diameter at most $n-d$; i.e., every two vertices can be connected by a path of at most $n-d$ edges. This conjecture is implied by its special case with $n = 2d$, the so-called *d-step conjecture*. There are several variants of the Hirsch conjecture. Some of them are known to be false, such as the Hirsch conjecture for d-dimensional polyhedra with n-facets; their graph can have diameter at least $n-d+\lfloor d/5 \rfloor$. But even here the conjecture fails just by a little, while the crucial and wide open question is whether the diameter of the graph can be bounded by a fixed polynomial in d and n.

The Hirsch conjecture is motivated by linear programming (and it was published in Dantzig's book [Dan63]), since the running time of all variants of the simplex algorithm is bounded from below by the number of edges that must be traversed in order to get from the starting vertex of the polyhedron of admissible solutions to the optimum vertex.

The best upper bound is due to Kalai. He published several papers on this subject, successively improving and simplifying his arguments, and this sequence is concluded with [Kal92]. He proves the following: *Let P be a convex polyhedron in $\mathbf{R}^d$ with n facets. Assume that no edge of P is horizontal and that P has a (unique) topmost vertex w. Then from every vertex v of P there is a path to w consisting of at most $f(d,n) \leq 2n\binom{d+\lfloor \log_2 n \rfloor -1}{d-1} \leq 2n^{\log_2 d+1}$ edges and going upward all the time.* The proof is quite short and uses only very simple properties of polytopes (also see [Zie94] or [Kal97]).

Kalai [Kal92] also discovered a randomized variant of the simplex algorithm for linear programming for which the expected number of pivot steps, for every linear program with n constraints in $\mathbf{R}^d$, is bounded by a subexponential function of n and d, namely by $n^{O(\sqrt{d})}$. All the previous worst-case bounds were exponential. Interestingly, essentially the same algorithm (in a dual setting) was found by Sharir and Welzl and a little later analyzed in [MSW96], independent of Kalai's work and at almost the same time, but coming from a quite different direction. The Sharir–Welzl algorithm is formulated in an abstract framework, and it can be used for many other optimization problems besides linear programming.

Realizations of polytopes. By a *realization* of a d-dimensional polytope P we mean any polytope $Q \subseteq \mathbf{R}^d$ that is combinatorially equivalent to P. The proof of Steinitz's theorem shows that every 3-dimensional polytope has a realization whose vertices have integer coordinates. For 3-polytopes with n vertices, Richter-Gebert [RG97] proved that the vertex coordinates can be chosen as positive integers no larger than 2^{18n^2}, and if the polytope has at least one triangular facet, the upper bound becomes 43^n (a previous, slightly worse, estimate was given by Onn and Sturmfels [OS94]). No nontrivial lower bounds seem to be known. Let us remark that for straight edge drawings of planar graphs, the vertices of every n-vertex graph can be placed on a grid with side $O(n)$. This was first proved by de Fraysseix, Pach, and Pollack [dFPP90] with the $(2n-4) \times (n-2)$ grid, and re-proved by Schnyder [Sch90] by a different method, with the $(n-1) \times (n-1)$ grid; see also Kant [Kan96] for more recent results in this direction.

For higher-dimensional polytopes, the situation is strikingly different. Although all simple polytopes and all simplicial polytopes can be realized with integer vertex coordinates, there are 4-dimensional polytopes for which every realization requires irrational coordinates (we will see an 8-dimensional example in Section 5.6). There are also 4-dimensional n-vertex polytopes for which every realization with integer coordinates uses doubly exponential coordinates, of order $2^{2^{\Omega(n)}}$. There are numerous other results indicating that the polytopes of dimension 4 and higher are complicated. For example, the problem of deciding whether a given finite lattice is isomorphic to the face lattice of a 4-dimensional polytope is algorithmically difficult; it is polynomially equivalent to the problem of deciding whether a system of polynomial inequalities with integer coefficients in n variables has a solution. This latter problem is known to be NP-hard, but most likely it is even harder; the best known algorithm needs exponential time and polynomial space. An overview of such results, and references to previous work on which they are built, can be found in Richter-Gebert [RG99], and detailed proofs in [RG97]. Section 6.2 contains a few more remarks on realizability (see, in particular, Exercise 6.2.3).

Exercises

1. Verify that if $V \subset \mathbf{R}^d$ is affinely independent, then each subset $F \subseteq V$ determines a face of the simplex $\mathrm{conv}(V)$. ②

2. Verify the description of the faces of the cube and of the crosspolytope given in the text. ③

3. Consider the $(n-1)$-dimensional permutahedron as defined in the introduction to this chapter.
   (a) Verify that it really has $n!$ vertices corresponding to the permutations of $\{1, 2, \ldots, n\}$. ②
   (b) Describe all faces of the permutahedron combinatorially (what sets of permutations are vertex sets of faces?). ③
   (c) Determine the dimensions of the faces found in (b). In particular, show that the facets correspond to ordered partitions (A, B) of $\{1, 2, \ldots, n\}$, $A, B \neq \emptyset$, and count them. ③

4. Let $P \subset \mathbf{R}^4 = \mathrm{conv}\{\pm e_i \pm e_j \colon i, j = 1, 2, 3, 4, \, i \neq j\}$, where $e_1, \ldots, e_4$ is the standard basis (this P is called the 24-*cell*). Describe the face lattice of P and prove that P is combinatorially equivalent to P^* (in fact, P can be obtained from P^* by an isometry and scaling). ③

5. Using Proposition 5.3.2, prove the following:
   (a) If F is a face of a convex polytope P, then F is the intersection of P with the affine hull of F. ②
   (b) If F and G are faces of a convex polytope P, then $F \cap G$ is a face, too. ①

6. Let P be a convex polytope in $\mathbf{R}^3$ containing the origin as an interior point, and let F be a j-face of P, $j = 0, 1, 2$.
   (a) Give a precise definition of the face F' of the dual polytope P^* corresponding to F (i.e., describe F' as a subset of $\mathbf{R}^3$). ②
   (b) Verify that F' is indeed a face of P^*. ②

7. Let $V \subset \mathbf{R}^d$ be the vertex set of a convex polytope and let $U \subset V$. Prove that U is the vertex set of a face of $\mathrm{conv}(V)$ if and only if the affine hull of U is disjoint from $\mathrm{conv}(V \setminus U)$. ③

8. Prove that the graph of any 3-dimensional convex polytope is 3-connected; i.e., removing any 2 vertices leaves the graph connected. ⑤

9. Let C be a convex set. Call a point $x \in C$ *exposed* if there is a hyperplane h with $C \cap h = \{x\}$ and all the rest of C on one side. For convex polytopes, exposed points are exactly the vertices, and we have shown that any extremal point is also exposed. Find an example of a compact convex set $C \subset \mathbf{R}^2$ with an extremal point that is not exposed. ③

10. (On extremal points) For a set $X \subseteq \mathbf{R}^d$, let $\mathrm{ex}(X) = \{x \in X \colon x \notin \mathrm{conv}(X \setminus \{x\})\}$ denote the set of extremal points of X.
    (a) Find a convex set $C \subseteq \mathbf{R}^d$ with $C \neq \mathrm{conv}(\mathrm{ex}(C))$. ①
    (b) Find a compact convex $C \subseteq \mathbf{R}^3$ for which $\mathrm{ex}(C)$ is not closed. ④

(c) By modifying the proof of Theorem 5.2.2, prove that $C = \mathrm{conv}(\mathrm{ex}(C))$ for every compact convex $C \subset \mathbf{R}^d$ (this is a finite-dimensional version of the well known *Krein–Milman theorem*). [4]

5.4 Many Faces: The Cyclic Polytopes

A convex polytope P can be given to us by the list of vertices. How difficult is it to recover the full face lattice, or, more modestly, a representation of P as an intersection of half-spaces? The first question to ask is how large the face lattice or the collection of half-spaces can be, compared to the number of vertices. That is, what is the maximum total number of faces, or the maximum number of facets, of a convex polytope in $\mathbf{R}^d$ with n vertices? The dual question is, of course, the maximum number of faces or vertices of a bounded intersection of n half-spaces in $\mathbf{R}^d$.

Let $f_j = f_j(P)$ denote the number of j-faces of a polytope P. The vector $(f_0, f_1, \ldots, f_d)$ is called the *f-vector* of P. We thus assume $f_0 = n$ and we are interested in estimating the maximum value of f_{d-1} and of $\sum_{k=0}^{d} f_k$.

In dimensions 2 and 3, the situation is simple and favorable. For $d = 2$, our polytope is a convex polygon with n vertices and n edges, and so $f_0 = f_1 = n$, $f_2 = 1$. The f-vector is even determined uniquely.

A 3-dimensional polytope can be regarded as a drawing of a planar graph, in our case with n vertices. By well-known results for planar graphs, we have $f_1 \leq 3n-6$ and $f_2 \leq 2n-4$. Equalities hold if and only if the polytope is simplicial (all facets are triangles).

In both cases the total number of faces is linear in n. But as the dimension grows, polytopes become much more complicated. First of all, even the total number of faces of the most innocent convex polytope, the d-dimensional simplex, is *exponential* in d. But here we consider d fixed and relatively small, and we investigate the dependence on the number of vertices n.

Still, as we will see, for every $n \geq 5$ there is a 4-dimensional convex polytope with n vertices and with every two vertices connected by an edge, i.e., with $\binom{n}{2}$ edges! This looks counterintuitive, but our intuition is based on the 3-dimensional case. In any fixed dimension d, the number of facets can be of order $n^{\lfloor d/2 \rfloor}$, which is rather disappointing for someone wishing to handle convex polytopes efficiently. On the other hand, complete desperation is perhaps not appropriate: Certainly not all polytopes exhibit this very bad behavior. For example, it is known that if we choose n points uniformly at random in the unit ball B^d, then the expected number of faces of their convex hull is only $o(n)$, for every fixed d.

It turns out that the number of faces for a given dimension and number of vertices is the largest possible for so-called *cyclic polytopes*, to be introduced next. First we define a very useful curve in $\mathbf{R}^d$.

5.4.1 Definition (Moment curve). *The curve* $\gamma = \{(t, t^2, \ldots, t^d) \colon t \in \mathbf{R}\}$ *in* $\mathbf{R}^d$ *is called the* moment curve.

5.4.2 Lemma. *Any hyperplane* h *intersects the moment curve* γ *in at most* d *points. If there are* d *intersections, then* h *cannot be tangent to* γ, *and thus at each intersection,* γ *passes from one side of* h *to the other.*

Proof. A hyperplane h can be expressed by the equation $\langle a, x \rangle = b$, or in coordinates $a_1 x_1 + a_2 x_2 + \cdots + a_d x_d = b$. A point of γ has the form $(t, t^2, \ldots, t^d)$, and if it lies in h, we obtain $a_1 t + a_2 t^2 + \cdots + a_d t^d - b = 0$. This means that t is a root of a nonzero polynomial $p_h(t)$ of degree at most d, and hence the number of intersections of h with γ is at most d. If there are d distinct roots, then they must be all simple. At a simple root, the polynomial $p_h(t)$ changes sign, and this means that the curve γ passes from one side of h to the other. $\square$

As a corollary, we see that every d points of the moment curve are affinely independent, for otherwise, we could pass a hyperplane through them plus one more point of γ. So the moment curve readily supplies *explicit* examples of point sets in general position.

5.4.3 Definition (Cyclic polytope). *The convex hull of finitely many points on the moment curve is called a* cyclic polytope.

How many facets does a cyclic polytope have? Each facet is determined by a d-tuple of vertices, and distinct d-tuples determine distinct facets. Here is a criterion telling us exactly which d-tuples determine facets.

5.4.4 Proposition (Gale's evenness criterion). *Let* V *be the vertex set of a cyclic polytope* P *considered with the linear ordering* $\leq$ *along the moment curve (larger vertices have larger values of the parameter* t*). Let* $F = \{v_1, v_2, \ldots, v_d\} \subseteq V$ *be a* d-tuple of vertices of P, where $v_1 < v_2 < \cdots < v_d$. *Then* F *determines a facet of* P *if and only if for any two vertices* $u, v \in V \setminus F$, *the number of vertices* $v_i \in F$ *with* $u < v_i < v$ *is even.*

Proof. Let h_F be the hyperplane affinely spanned by F. Then F determines a facet if and only if all the points of $V \setminus F$ lie on the same side of h_F.

Since the moment curve γ intersects h_F in exactly d points, namely at the points of F, it is partitioned into $d+1$ pieces, say $\gamma_0, \ldots, \gamma_d$, each lying completely in one of the half-spaces, as is indicated in the drawing:

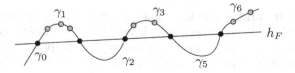

Hence, if the vertices of $V \setminus F$ are all contained in the odd-numbered pieces $\gamma_1, \gamma_3, \ldots$, as in the picture, or if they are all contained in the even-numbered pieces $\gamma_0, \gamma_2, \ldots$, then F determines a facet. This condition is equivalent to Gale's criterion. $\square$

Now we can count the facets.

5.4.5 Theorem. *The number of facets of a d-dimensional cyclic polytope with n vertices $(n \geq d+1)$ is*

$$\binom{n - \lfloor d/2 \rfloor}{\lfloor d/2 \rfloor} + \binom{n - \lfloor d/2 \rfloor - 1}{\lfloor d/2 \rfloor - 1} \text{ for } d \text{ even, and}$$

$$2\binom{n - \lfloor d/2 \rfloor - 1}{\lfloor d/2 \rfloor} \text{ for } d \text{ odd.}$$

For fixed d, this has the order of magnitude $n^{\lfloor d/2 \rfloor}$.

Proof. The number of facets equals the number of ways of placing d black circles and $n - d$ white circles in a row in such a way that we have an even number of black circles between each two white circles.

Let us say that an arrangement of black and white circles is *paired* if any contiguous segment of black circles has an even length (the arrangements permitted by Gale's criterion need not be paired because of the initial and final segments). The number of paired arrangements of $2k$ black circles and $n - 2k$ white circles is $\binom{n-k}{k}$, since by deleting every second black circle we get a one-to-one correspondence with selections of the positions of k black circles among $n - k$ possible positions.

Let us return to the original problem, and first consider an *odd* $d = 2k+1$. In a valid arrangement of circles, we must have an odd number of consecutive black circles at the beginning or at the end (but not both). In the former case, we delete the initial black circle, and we get a paired arrangement of $2k$ black and $n-1-2k$ white circles. In the latter case, we similarly delete the black circle at the end and again get a paired arrangement as in the first case. This establishes the formula in the theorem for odd d.

For *even* $d = 2k$, the number of initial consecutive black circles is either odd or even. In the even case, we have a paired arrangement, which contributes $\binom{n-k}{k}$ possibilities. In the odd case, we also have an odd number of consecutive black circles at the end, and so by deleting the first and last black circles we obtain a paired arrangement of $2(k-1)$ black circles and $n-2k$ white circles. This contributes $\binom{n-k-2}{k-1}$ possibilities. $\square$

> **Bibliography and remarks.** The convex hull of the moment curve
> was studied by by Carathéodory [Car07]. In the 1950s, Gale con-
> structed neighborly polytopes by induction. Cyclic polytopes and the
> evenness criterion appear in Gale [Gal63]. The moment curve is an
> important object in many other branches besides the theory of convex

polytopes. For example, in elementary algebraic topology it is used for proving that every (at most countable) d-dimensional simplicial complex has a geometric realization in $\mathbf{R}^{2d+1}$.

Convex hulls of random sets. Bárány [Bár89] proved that if n points are chosen uniformly and independently at random from a fixed d-dimensional convex polytope K (for example, the unit cube), then the number of k-dimensional faces of their convex hull has the order $(\log n)^{d-1}$ for every fixed d and k, $0 \leq k \leq d-1$ (the constant of proportionality depending on d, k, and K). If K is a smooth convex body (such as the unit ball), then the order of magnitude is $n^{(d-1)/(d+1)}$, again with d, k, and K fixed. For more references and wider context see, e.g., Weil and Wieacker [WW93].

Exercises

1. (a) Show that if V is a finite subset of the moment curve, then all the points of V are extreme in $\text{conv}(V)$; that is, they are vertices of the corresponding cyclic polytope. ☐2

   (b) Show that any two cyclic polytopes in $\mathbf{R}^d$ with n vertices are combinatorially the same: They have isomorphic face lattices. Thus, we can speak of *the* cyclic polytope. ☐3

2. (Another curve like γ) Let $\beta \subset \mathbf{R}^d$ be the curve $\{(\frac{1}{t+1}, \frac{1}{t+2}, \ldots, \frac{1}{t+d}) : t \in \mathbf{R}, t > 0\}$. Show that any hyperplane intersects β in at most d points (and if there are d intersections, then there is no tangency), and conclude that any n distinct points on β form the vertex set of a polytope combinatorially isomorphic to the cyclic polytope. ☐4 (Let us remark that many other curves have these properties as well; the moment curve is just the most convenient example.)

3. (Universality of the cyclic polytope)

   (a) Let $x_1, \ldots, x_n$ be points in $\mathbf{R}^d$. Let y_i denote the vector arising by appending 1 as the $(d+1)$st component of x_i. Show that if the determinants of all matrices with columns $y_{i_1}, \ldots, y_{i_{d+1}}$, for all choices of indices $i_1 < i_2 < \cdots < i_{d+1}$, have the same nonzero sign, then $x_1, \ldots, x_n$ form the vertex set of a convex polytope combinatorially equivalent to the n-vertex cyclic polytope in $\mathbf{R}^d$. ☐4

   (b) Show that for any integers n and d there exists N such that among any N points in $\mathbf{R}^d$ in general position, one can choose n points forming the vertex set of a convex polytope combinatorially equivalent to the n-vertex cyclic polytope. ☐3 (This can be seen as a d-dimensional generalization of the Erdős–Szekeres theorem.)

4. Prove that if n is sufficiently large in terms of d, then for every set of n points in $\mathbf{R}^d$ in general position, one can choose $d+1$ simplices of dimension d with vertices at some of these points such that any hyperplane avoids at least one of these simplices. Use Exercise 3. ☐2

This exercise is a special case of a problem raised by Lovász, and it was communicated to me by Bárány. A detailed solution can be found in [BVS+99].

5. Show that for cyclic polytopes in dimensions 4 and higher, every pair of vertices is connected by an edge. For dimension 4 and two arbitrary vertices, write out explicitly the equation of a hyperplane intersecting the cyclic polytope exactly in this edge. ③

6. Determine the f-vector of a cyclic polytope with n vertices in dimensions 4, 5, and 6. ③

5.5 The Upper Bound Theorem

The upper bound theorem, one of the earlier major achievements of the theory of convex polytopes, claims that the cyclic polytope has the largest possible number of faces.

5.5.1 Theorem (Upper bound theorem). *Among all d-dimensional convex polytopes with n vertices, the cyclic polytope maximizes the number of faces of each dimension.*

In this section we prove only an approximate result, which gives the correct order of magnitude for the maximum number of facets.

5.5.2 Proposition (Asymptotic upper bound theorem). *A d-dimensional convex polytope with n vertices has at most $2\binom{n}{\lfloor d/2 \rfloor}$ facets and no more than $2^{d+1}\binom{n}{\lfloor d/2 \rfloor}$ faces in total. For d fixed, both quantities thus have the order of magnitude $n^{\lfloor d/2 \rfloor}$.*

First we establish this proposition for simplicial polytopes, in the following form.

5.5.3 Proposition. *Let P be a d-dimensional simplicial polytope. Then*

(a) $f_0(P) + f_1(P) + \cdots + f_d(P) \leq 2^d f_{d-1}(P)$, and
(b) $f_{d-1}(P) \leq 2f_{\lfloor d/2 \rfloor - 1}(P)$.

This implies Proposition 5.5.2 for simplicial polytopes, since the number of $(\lfloor d/2 \rfloor - 1)$-faces is certainly no bigger than $\binom{n}{\lfloor d/2 \rfloor}$, the number of all $\lfloor d/2 \rfloor$-tuples of vertices.

Proof of Proposition 5.5.3. We pass to the dual polytope P^*, which is simple. Now we need to prove $\sum_{k=0}^d f_k(P^*) \leq 2^d f_0(P^*)$ and $f_0(P^*) \leq 2f_{\lceil d/2 \rceil}(P^*)$.

Each face of P^* has at least one vertex, and every vertex of a simple d-polytope is incident to 2^d faces, which gives the first inequality.

We now bound the number of vertices in terms of the number of $\lceil d/2 \rceil$-faces. This is the heart of the proof, and it shows where the mysterious exponent $\lfloor d/2 \rfloor$ comes from.

Let us rotate the polytope P^* so that no two vertices share the x_d-coordinate (i.e., no two vertices have the same vertical level).

Consider a vertex v with the d edges emanating from it. By the pigeonhole principle, there are at least $\lceil d/2 \rceil$ edges directed downwards or at least $\lceil d/2 \rceil$ edges directed upwards. In the former case, every $\lceil d/2 \rceil$-tuple of edges going up determines a $\lceil d/2 \rceil$-face for which v is the lowest vertex. In the latter case, every $\lceil d/2 \rceil$-tuple of edges going down determines a $\lceil d/2 \rceil$-face for which v is the highest vertex. Here is an illustration, unfortunately for the not too interesting 3-dimensional case, showing a situation with 2 edges going up and the corresponding 2-dimensional face having v as the lowest vertex:

We have exhibited at least one $\lceil d/2 \rceil$-face for which v is the lowest vertex or the highest vertex. Since the lowest vertex and the highest vertex are unique for each face, the number of vertices is no more than twice the number of $\lceil d/2 \rceil$-faces. $\square$

Warning. For simple polytopes, the total combinatorial complexity is proportional to the number of vertices, and for simplicial polytopes it is proportional to the number of facets (considering the dimension fixed, that is). For polytopes that are neither simple nor simplicial, the number of faces of intermediate dimensions can have larger order of magnitude than both the number of facets and the number of vertices; see Exercise 1.

Nonsimplicial polytopes. To prove the asymptotic upper bound theorem, it remains to deal with nonsimplicial polytopes. This is done by a perturbation argument, similar to numerous other results where general position is convenient for the proof but where we want to show that the result holds in degenerate cases as well. In most instances in this book, the details of perturbation arguments are omitted, but here we make an exception, since the proof seems somewhat nontrivial.

5.5.4 Lemma. *For any d-dimensional convex polytope P there exists a d-dimensional simplicial polytope Q with $f_0(P) = f_0(Q)$ and $f_k(Q) \geq f_k(P)$ for all $k = 1, 2, \ldots, d$.*

Proof. The basic idea is very simple: Move (perturb) every vertex of P by a very small amount, in such a way that the vertices are in general position, and show that each k-face of P gives rise to at least one k-face of the perturbed polytope. There are several ways of doing this proof.

We process the vertices one by one. Let V be the vertex set of P and let $v \in V$. The operation of ε-*pushing* v is as follows: We choose a point v' lying in the interior of P, at distance at most ε from v, and on no hyperplane determined by the points of V, and we set $V' = (V \setminus \{v\}) \cup \{v'\}$. If we successively ε_v-push each vertex v of the polytope, the resulting vertex set is in general position and we have a simple polytope.

It remains to show that for any polytope P with vertex set V and any $v \in V$, there is an $\varepsilon > 0$ such that ε-pushing v does not decrease the number of faces.

Let $U \subset V$ be the vertex set of a k-face of P, $0 \leq k \leq d-1$, and let V' arise from V by ε-pushing v. If $v \notin U$, then no doubt, U determines a face of conv(V'), and so we assume that $v \in U$. First suppose that v lies in the affine hull of $U \setminus \{v\}$; we claim that then $U \setminus \{v\}$ determines a k-face of conv(V'). This follows easily from the criterion in Exercise 5.3.7: A subset $U \subset V$ is the vertex set of a face of conv(V) if and only if the affine hull of U is disjoint from conv($V \setminus U$). We leave a detailed argument to the reader (one must use the fact that v is pushed inside).

If v lies outside of the affine hull of $U \setminus \{v\}$, then we want to show that $U' = (U \setminus \{v\}) \cup \{v'\}$ determines a k-face of conv(V'). The affine hull of U is disjoint from the compact set conv($V \setminus U$). If we move v continuously by a sufficiently small amount, the affine hull of U moves continuously, and so there is an $\varepsilon > 0$ such that if we move v within ε from its original position, the considered affine hull and conv($V \setminus U$) remain disjoint. $\square$

The h-vector and such. Here we introduce some notions extremely useful for deeper study of the f-vectors of convex polytopes. In particular, they are crucial in proofs of the (exact) upper bound theorem.

Let us go back to the setting of the proof of Proposition 5.5.3. There we considered a simple polytope that used to be called P^* but now, for simplicity, let us call it P. It is positioned in $\mathbf{R}^d$ in such a way that no edge is horizontal, and so for each vertex v, there are some i_v edges going upwards and $d - i_v$ edges going downwards.

The central definition is this: The h-*vector* of P is $(h_0, h_1, \ldots, h_d)$, where h_i is the number of vertices v with exactly i edges going upwards. So, for example, we have $h_0 = h_d = 1$.

Next, we relate the h-vector to the f-vector. Each vertex v is the lowest vertex for exactly $\binom{i_v}{k}$ faces of dimension k, and each k-face has exactly one lowest vertex, and so

$$f_k = \sum_{i=0}^{d} \binom{i}{k} h_i \qquad (5.1)$$

(for $i < k$ we have $\binom{i}{k} = 0$). So the h-vector determines the f-vector. Less obviously, the h-vector can be uniquely reconstructed from the f-vector! A quick way of seeing this is via generating functions: If $f(x)$ is the polynomial $\sum_{k=0}^{d} f_k x^k$ and $h(x) = \sum_{i=0}^{d} h_i x^i$, then (5.1) translates to $f(x) = h(x+1)$, and therefore $h(x) = f(x-1)$. Explicitly, we have

$$h_i = \sum_{k=0}^{d} (-1)^{i-k} \binom{k}{i} f_k. \tag{5.2}$$

We have defined the h-vector using one particular choice of the vertical direction, but now we know that it is determined by the f-vector and thus independent of the chosen direction. By turning P upside down, we see that

$$h_i = h_{d-i} \quad \text{for all } i = 0, 1, \ldots, d.$$

These equalities are known as the *Dehn–Sommerville relations*. They include the usual Euler formula $f_0 + f_2 = f_1 + 2$ for 3-dimensional polytopes.

Let us stress once again that all we have said about h-vectors concerns only *simple polytopes*. For a *simplicial polytope* P, the h-vector can now be defined as the h-vector of the dual simple polytope P^*. Explicitly,

$$h_j = \sum_{k=0}^{j} (-1)^{j-k} \binom{d-k}{d-j} f_{k-1}.$$

The upper bound theorem has the following neat reformulation in terms of h-vectors: For any d-dimensional simplicial polytope with $f_0 = n$ vertices, we have

$$h_i \leq \binom{n-d+i-1}{i}, \quad i = 0, 1, \ldots, \lfloor d/2 \rfloor. \tag{5.3}$$

Proving the upper bound theorem is not one of our main topics, but an outline of a proof *can* be found in this book. It starts in the next section and finishes in Exercise 11.3.6, and it is not among the most direct possible proofs. Deriving the upper bound theorem from (5.3) is a pure and direct calculation, verifying that the h-vector of the cyclic polytope satisfies (5.3) with equality. We omit this part.

Bibliography and remarks. The upper bound theorem was conjectured by Motzkin in 1957 and proved by McMullen [McM70]. Many partial results have been obtained in the meantime. Perhaps most notably, Klee [Kle64] found a simple proof for polytopes with not too few vertices (at least about d^2 vertices in dimension d). That proof applies to simplicial complexes much more general than the boundary complexes of simplicial polytopes: It works for Eulerian pseudomanifolds and, in particular, for all *simplicial spheres*, i.e., simplicial complexes homeomorphic to S^{d-1}. Presentations of McMullen's proof and Klee's proof can be found in Ziegler's book [Zie94]. A nice variation was described by Alon and Kalai [AK85]. Another proof, based on linear programming duality and results on hyperplane arrangements, was given by Clarkson [Cla93]. An elegant presentation of similar ideas,

using the Gale transform discussed below in Section 5.6, can be found
in Welzl [Wel01] and in Exercises 11.3.5 and 11.3.6. Our exposition of
the asymptotic upper bound theorem is based on Seidel [Sei95].

The ordering of the vertices of a simple polytope P by their height
in the definition of the h-vector corresponds to a linear ordering of the
facets of P^*. This ordering of the facets is a *shelling*. Shelling, even
in the strictly peaceful mathematical sense, is quite important, also
beyond the realm of convex polytopes. Let $\mathcal{K}$ be a finite cell complex
whose cells are convex polytopes (such as the boundary complex of a
convex polytope), and suppose that all maximal cells have the same
dimension k. Such $\mathcal{K}$ is called *shellable* if $k = 0$ or $k \geq 1$ and $\mathcal{K}$ has
a *shelling*. A shelling of $\mathcal{K}$ is an enumeration $F_1, F_2, \ldots, F_n$ of the
facets (maximum-dimension cells) of $\mathcal{K}$ such that (i) the boundary
complex of F_1 is shellable, and (ii) for every $i > 1$, there is a shelling
of the complex $F_i \cap \bigcup_{j=1}^{i-1} F_j$ that can be extended to a shelling of the
boundary complex of F_i. The boundary complex of a convex polytope
is homeomorphic to a sphere, and a shelling builds the sphere in such
a way that each new cell is glued by contractible part of its boundary
to the previously built part, except for the last cell, which closes the
remaining hole.

McMullen's proof of the upper bound theorem does not generalize
to simplicial spheres (i.e., finite simplicial complexes homeomorphic
to spheres), for example because they need not be shellable, counterintuitive as this may look. The upper bound theorem for them was
proved by Stanley [Sta75] using much heavier algebraic and algebraic-
topological tools.

An interesting extension of the upper bound theorem was found
by Kalai [Kal91]. Let P be a simplicial d-dimensional polytope. All
proper faces of P are simplices, and so the boundary is a simplicial
complex. Let K be any subcomplex of the boundary (a subset of the
proper faces of P such that if $F \in K$, then all faces of F also lie in
K). The *strong upper bound theorem*, as Kalai's result is called, asserts
that if K has at least as many $(d-1)$-faces as the d-dimensional cyclic
polytope on n vertices, then K has at least as many k-faces as that
cyclic polytope, for all $k = 0, 1, \ldots, d-1$. (Note that we do not assume
that P has n vertices!) The proof uses methods developed for the
proof of the g-theorem mentioned below as well as Kalai's technique
of *algebraic shifting*.

Another major achievement concerning the f-vectors of polytopes
is the so-called *g-theorem*. The inventive name *g-vector* of a d-dimen-
sional simple polytope refers to the vector $(g_0, g_1, \ldots, g_{\lfloor d/2 \rfloor})$, where
$g_0 = h_0$ and $g_i = h_i - h_{i-1}$, $i = 1, 2, \ldots, \lfloor d/2 \rfloor$. The g-theorem characterizes all possible integer vectors that can appear as the g-vector
of a d-dimensional simple (or simplicial) polytope. Since the g-vector

uniquely determines the f-vector, we have a complete characterization of f-vectors of simple polytopes. In particular, the g-theorem guarantees that all the components of the g-vector are always non-negative (this fact is known as the *generalized lower bound theorem*), and therefore the h-vector is unimodal: We have $h_0 \leq h_1 \leq \cdots \leq h_{\lfloor d/2 \rfloor} = h_{\lceil d/2 \rceil} \geq \cdots \geq h_d$. (On the other hand, the f-vector of a simple polytope need *not* be unimodal; more exactly, it is unimodal in dimensions up to 19, and there are 20-dimensional nonunimodal examples.) We again refer to [Zie94] for a full statement of the g-theorem. The proof has two independent parts; one of them, due to Billera and Lee [BL81], constructs suitable polytopes, and the other part, first proved by Stanley [Sta80], shows certain inequalities for all simple polytopes. For studying the most elementary proof of the second part currently available, one can start with McMullen [McM96] and continue with [McM93].

For nonsimple (and nonsimplicial) polytopes, a characterization of possible f-vectors remains elusive. It seems, anyway, that the *flag vector* might be a more appropriate parameter for nonsimple polytopes. The flag vector counts, for every $k = 1, 2, \ldots, d$ and for every $i_1 < i_2 < \cdots < i_k$, the number of chains $F_1 \subset F_2 \subset \cdots \subset F_k$, where $F_1, \ldots, F_k$ are faces with $\dim(F_j) = i_j$ (such a chain is called a *flag*).

No analogue of the upper bound theorem is known for centrally symmetric polytopes. A few results concerning their face counts, obtained by methods quite different from the ones for arbitrary polytopes, will be mentioned in Section 14.5.

The proof of Lemma 5.5.4 by pushing vertices inside is similar to an argument in Klee [Kle64], but he proves more and presents the proof in more detail.

Convex hull computation. What does it mean to compute the convex hull of a given n-point set $V \subset \mathbf{R}^d$? One possible answer, briefly touched upon in the notes to Section 5.2, is to express $\operatorname{conv}(V)$ as the intersection of half-spaces and to compute the vertex sets of all facets. (As we know, the face lattice can be reconstructed from this information purely combinatorially; see Kaibel and Pfetsch [KP01] for an efficient algorithm.) Of course, for some applications it may be sufficient to know much less about the convex hull, say only the graph of the polytope or only the list of its vertices, but here we will discuss only algorithms for computing all the vertex–facet incidences or the whole face lattice. For a more detailed overview of convex hull algorithms see, e.g., Seidel [Sei97].

For the dimension d considered fixed, there is a quite simple and practical randomized algorithm. It computes the convex hull of n points in $\mathbf{R}^d$ in expected time $O(n^{\lfloor d/2 \rfloor} + n \log n)$ (Seidel [Sei91], simplifying Clarkson and Shor [CS89]), and also a very complicated but deterministic algorithm with the same asymptotic running time (Chazelle [Cha93b]; somewhat simplified in Brönnimann, Chazelle, and Matoušek [BCM99]). This is worst-case optimal, since an n-vertex polytope may have about $n^{\lfloor d/2 \rfloor}$ facets. There are also output-sensitive algorithms, whose running time depends on the total number f of faces of the resulting polytope. Recent results in this direction, including an algorithm that computes the convex hull of n points in general position in $\mathbf{R}^d$ (d fixed) in time $O(n \log f + (nf)^{1-1/(\lfloor d/2 \rfloor + 1)}(\log n)^{c(d)})$, can be found in Chan [Cha00b].

Still, none of the known algorithms is theoretically fully satisfactory, and practical computation of convex hulls even in moderate dimensions, say 10 or 20, can be quite challenging. Some of the algorithms are too complicated and with too large constants hidden in the asymptotic notation to be of practical value. Algorithms requiring general position of the points are problematic for highly degenerate point configurations (which appear in many applications), since small perturbations used to achieve general position often increase the number of faces tremendously. Some of the randomized algorithms compute intermediate polytopes that can have many more faces than the final result. Often we are interested just in the vertex–facet incidences, but many algorithms compute all faces, whose number can be much larger, or even a triangulation of every face, which may again increase the complexity. Such problems of existing algorithms are discussed in Avis, Bremner, and Seidel [ABS97].

For actual computations, simple and theoretically suboptimal algorithms are often preferable. One of them is the double-description method mentioned earlier, and another algorithm that seems to behave well in many difficult instances is the *reverse search* of Avis and Fukuda [AF92]. It enumerates the vertices of the intersection of a given set H of half-spaces one by one, using quite small storage. Conceptually, one thinks of optimizing a generic linear function over $\bigcap H$ by a simplex algorithm with Bland's rule. This defines a spanning tree in the graph of the polytope, and this tree is searched depth-first starting from the optimum vertex, essentially by running the simplex algorithm "backwards." The main problem of this algorithm is with degenerate vertices of high degree, which may correspond to an enormous number of bases in the simplex algorithm.

Also, it sometimes helps if one knows some special properties of the convex hull in a particular problem, say many symmetries. For example, very extensive computations of convex hulls were performed by Deza, Fukuda, Pasechnik, and Sato [DFPS00], who studied the *metric polytope*. Before we define this interesting polytope, let us first introduce the *metric cone* M_n. This is a set in $\mathbf{R}^{\binom{n}{2}}$ representing all metrics on $\{1, 2, \ldots, n\}$, where the coordinate $x_{\{i,j\}}$ specifies the distance of

i to j, $1 \leq i < j \leq n$. So M_n is defined by the triangle inequalities $x_{\{i,j\}} + x_{\{j,k\}} \leq x_{\{i,k\}}$, where i, j, k are three distinct indices. The metric polytope m_n is the subset of M_n defined by the additional inequalities saying that the perimeter of each triangle is at most 2, namely $x_{\{i,j\}} + x_{\{j,k\}} + x_{\{i,k\}} \leq 2$. Deza et al. were able to enumerate all the approximately $1.5 \cdot 10^9$ vertices of the 28-dimensional polytope m_8; this may give some idea of the extent of these computational problems. Without using many symmetries of m_n, a polytope of this size would currently be out of reach. Such computations might provide insight into various conjectures concerning the metric polytope, which are important for combinatorial optimization problems (see, e.g., Deza and Laurent [DL97] for background).

Exercises

1. (a) Let P be a k-dimensional convex polytope in $\mathbf{R}^k$, and Q an ℓ-dimensional convex polytope in $\mathbf{R}^\ell$. Show that the Cartesian product $P \times Q \subset \mathbf{R}^{k+\ell}$ is a convex polytope of dimension $k + \ell$. ②

   (b) If F is an i-face of P, and G is a j-face of Q, $i, j \geq 0$, then $F \times G$ is an $(i+j)$-face of $P \times Q$. Moreover, this yields all the nonempty faces of $P \times Q$. ③

   (c) Using the product of suitable polytopes, find an example of a "fat-lattice" polytope, i.e., a polytope for which the total number of faces has a larger order of magnitude than the number of vertices plus the number of facets together (the dimension should be a constant). ③

   (d) Show that the following yields a 5-dimensional fat-lattice polytope: The convex hull of two regular n-gons whose affine hulls are skew 2-flats in $\mathbf{R}^5$. ③

   For recent results on fat-lattice polytopes see Eppstein, Kuperberg, and Ziegler [EKZ01].

5.6 The Gale Transform

On a very general level, the Gale transform resembles the duality transform defined in Section 5.1. Both convert a (finite) geometric configuration into another geometric configuration, and they may help uncover some properties of the original configuration by making them more apparent, or easier to visualize, in the new configuration. The Gale transform is more complicated to explain and probably more difficult to get used to, but it seems worth the effort. It was invented for studying high-dimensional convex polytopes, and recently it has been used for solving problems about point configurations by relating them to advanced theorems on convex polytopes. It is also closely related to the duality of linear programming (see Section 10.1), but we will not elaborate on this connection here.

The Gale transform assigns to a sequence $\boldsymbol{a} = (a_1, a_2, \ldots, a_n)$ of $n \geq d+1$ points in $\mathbf{R}^d$ another sequence $\bar{\boldsymbol{g}} = (\bar{g}_1, \bar{g}_2, \ldots, \bar{g}_n)$ of n points. The points $\bar{g}_1, \bar{g}_2, \ldots, \bar{g}_n$ live in a different dimension, namely in $\mathbf{R}^{n-d-1}$. For example, n points in the plane are transformed to n points in $\mathbf{R}^{n-3}$ and vice versa. In the literature one finds many results about k-dimensional polytopes with $k+3$ or $k+4$ vertices; this is because their vertex sets have a low-dimensional Gale transform.

Let us stress that the Gale transform operates on *sequences*, not individual points: We cannot say what $\bar{g}_1$ is without knowing all of $a_1, a_2, \ldots, a_n$. We also require that the affine hull of the a_i be the whole $\mathbf{R}^d$; otherwise, the Gale transform is not defined. (On the other hand, we do not need any sort of general position, and some of the a_i may even coincide.)

The reader might wonder why the points of the Gale transform are written with bars. This is to indicate that they should be interpreted as vectors in a vector space, rather than as points in an affine space. As we will see, "affine" properties of the sequence $\boldsymbol{a}$, such as affine dependencies, correspond to "linear" properties of the Gale transform, such as linear dependencies.

In order to obtain the Gale transform of $\boldsymbol{a}$, we first convert the a_i into $(d+1)$-dimensional vectors: $\bar{a}_i \in \mathbf{R}^{d+1}$ is obtained from a_i by appending a $(d+1)$st coordinate equal to 1. This is the embedding $\mathbf{R}^d \to \mathbf{R}^{d+1}$ often used for relating affine notions in $\mathbf{R}^d$ to linear notions in $\mathbf{R}^{d+1}$; see Section 1.1.

Let A be the $d \times n$ matrix with $\bar{a}_i$ as the ith column. Since we assume that there are $d+1$ affinely independent points in $\boldsymbol{a}$, the matrix A has rank $d+1$, and so the vector space V generated by the *rows* of A is a $(d+1)$-dimensional subspace of $\mathbf{R}^n$. We let $V^\perp$ be the orthogonal complement of V in $\mathbf{R}^n$; that is, $V^\perp = \{w \in \mathbf{R}^n : \langle v, w \rangle = 0 \text{ for all } v \in V\}$. We have $\dim(V^\perp) = n-d-1$. Let us choose some basis $(b_1, b_2, \ldots, b_{n-d-1})$ of $V^\perp$, and let B be the $(n-d-1) \times n$ matrix with b_j as the jth row. Finally, we let $\bar{g}_i \in \mathbf{R}^{n-d-1}$ be the ith column of B. The sequence $\bar{\boldsymbol{g}} = (\bar{g}_1, \bar{g}_2, \ldots, \bar{g}_n)$ is the *Gale transform* of $\boldsymbol{a}$. Here is a pictorial summary:

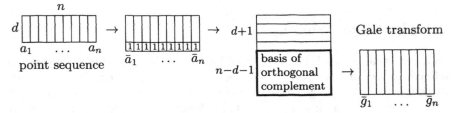

5.6.1 Observation.

(i) *(The Gale transform is determined up to linear isomorphism) In the construction of $\bar{\boldsymbol{g}}$, we can choose an arbitrary basis of $V^\perp$. Choosing a different basis corresponds to multiplying the matrix B from the left by a nonsingular $(n-d-1) \times (n-d-1)$ matrix T (Exercise 1), and this means transforming $(\bar{g}_1, \ldots, \bar{g}_n)$ by a linear isomorphism of $\mathbf{R}^{n-d-1}$.*

(ii) *A sequence $\bar{g}$ in $\mathbf{R}^{n-d-1}$ is the Gale transform of some a if and only if it spans $\mathbf{R}^{n-d-1}$ and has 0 as the center of gravity: $\sum_{i=1}^{n} \bar{g}_i = 0$.*

(iii) *Let us consider a sequence $\bar{g}$ in $\mathbf{R}^{n-d-1}$ satisfying the condition in (ii). If we interpret it as a point sequence (breaking the convention that the result of the Gale transform should be thought of as a sequence of vectors), apply the Gale transform to it, again consider the result as a point sequence, and apply the Gale transform the second time, we recover the original $\bar{g}$, up to linear isomorphism (Exercise 5).*

Two ways of probing a configuration. We would like to set up a dictionary for translating between geometric properties of a sequence a and those of its Gale transform. First we discuss how some familiar geometric properties of a configuration of points or vectors are reflected in the values of affine or linear functions on the configuration, and how they manifest themselves in affine or linear dependencies. For a sequence $\bar{a} = (\bar{a}_1, \ldots, \bar{a}_n)$ of vectors in $\mathbf{R}^{d+1}$, we define two vector subspaces of $\mathbf{R}^n$:

$\mathsf{LinVal}(\bar{a}) = \{(f(\bar{a}_1), f(\bar{a}_2), \ldots, f(\bar{a}_n)): f\colon \mathbf{R}^{d+1} \to \mathbf{R} \text{ is a linear function}\}$,
$\mathsf{LinDep}(\bar{a}) = \{\alpha \in \mathbf{R}^n: \alpha_1 \bar{a}_1 + \alpha_2 \bar{a}_2 + \cdots + \alpha_n \bar{a}_n = 0\}$.

For a point sequence $a = (a_1, \ldots, a_n)$, we then let $\mathsf{AffVal}(a) = \mathsf{LinVal}(\bar{a})$ and $\mathsf{AffDep}(a) = \mathsf{LinDep}(\bar{a})$, where $\bar{a}$ is obtained from a as above, by appending 1's. Another description is

$\mathsf{AffVal}(a) = \{(f(a_1), f(a_2), \ldots, f(a_n)): f\colon \mathbf{R}^d \to \mathbf{R} \text{ is an affine function}\}$,
$\mathsf{AffDep}(a) = \{\alpha \in \mathbf{R}^n: \alpha_1 a_1 + \cdots + \alpha_n a_n = 0, \alpha_1 + \cdots + \alpha_n = 0\}$.

The knowledge of $\mathsf{LinVal}(\bar{a})$ tells us a lot about $\bar{a}$, and we only have to learn to decode the information. As usual, we assume that $\bar{a}$ linearly spans all of $\mathbf{R}^{d+1}$.

Each nonzero linear function $f\colon \mathbf{R}^{d+1} \to \mathbf{R}$ determines the linear hyperplane $h_f = \{x \in \mathbf{R}^{d+1}: f(x) = 0\}$ (by a *linear hyperplane* we mean a hyperplane passing through 0). This h_f is oriented (one of its half-spaces is positive and the other negative), and the sign of $f(\bar{a}_i)$ determines whether $\bar{a}_i$ lies on h_f, on its positive side, or on its negative side.

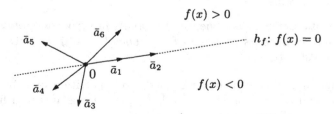

We begin our decoding of the properties of $\bar{a}$ with the property "spanning a linear hyperplane." That is, we choose our favorite index set $I \subseteq$

$\{1, 2, \ldots, n\}$, and we ask whether the points of the subsequence $\bar{a}_I = (\bar{a}_i \colon i \in I)$ span a linear hyperplane. First, we observe that they lie in a common linear hyperplane if and only if there is a nonzero $\varphi \in \mathsf{LinVal}(\bar{a})$ such that $\varphi_i = 0$ for all $i \in I$. It could still happen that all of $\bar{a}_I$ lies in a lower-dimensional linear subspace. Using the assumption that $\bar{a}$ spans $\mathbf{R}^{d+1}$, it is not difficult to see that $\bar{a}_I$ spans a linear hyperplane if and only if all $\varphi \in \mathsf{LinVal}(\bar{a})$ that vanish on $\bar{a}_I$ have identical zero sets; that is, the set $\{i \colon \varphi_i = 0\}$ is the same for all such φ. If we know that $\bar{a}_I$ spans a linear hyperplane, we can also see how the other vectors in $\bar{a}$ are distributed with respect to this linear hyperplane.

Analogously, knowing $\mathsf{AffVal}(a)$, we can determine which subsequences of a span (affine) hyperplanes and how the other points are partitioned by these hyperplanes. For example, we can tell whether there are some $d{+}1$ points on a common hyperplane, and so we know whether a is in general position. As a more complicated example, let $P = \mathrm{conv}(a)$. We can read off from $\mathsf{AffVal}(a)$ which of the a_i are the vertices of P, and also the whole face lattice of P (Exercise 6).

Similar information can be inferred from $\mathsf{AffDep}(a)$ (exactly the same information, in fact, since $\mathsf{AffDep}(a) = \mathsf{AffVal}(a)^\perp$; see Exercise 7). For an $\alpha \in \mathsf{AffDep}(a)$ let $I_+(\alpha) = \{i \in \{1, 2, \ldots, n\} \colon \alpha_i > 0\}$ and $I_-(\alpha) = \{i \in \{1, 2, \ldots, n\} \colon \alpha_i < 0\}$. As we learned in the proof of Radon's lemma (Lemma 1.3.1), $I_+ = I_+(\alpha)$ and $I_- = I_-(\alpha)$ correspond to Radon partitions of a. Namely, $\sum_{i \in I_+} \alpha_i a_i = \sum_{i \in I_-} (-\alpha_i) a_i$, and dividing by $\sum_{i \in I_+} \alpha_i = \sum_{i \in I_-} (-\alpha_i)$, we have convex combinations on both sides, and so $\mathrm{conv}(a_{I_+}) \cap \mathrm{conv}(a_{I_-}) \neq \emptyset$. Conversely, if I_1 and I_2 are disjoint index sets with $\mathrm{conv}(a_{I_1}) \cap \mathrm{conv}(a_{I_2}) \neq \emptyset$, then there is a nonzero $\alpha \in \mathsf{AffDep}(a)$ with $I_+(\alpha) \subseteq I_1$ and $I_-(\alpha) \subseteq I_2$. For example, a_i is a vertex of $\mathrm{conv}(a)$ if and only if there is no $\alpha \in \mathsf{AffDep}(a)$ with $I_+(\alpha) = \{i\}$.

For a sequence $\bar{a}$ of vectors, linear dependencies correspond to expressing 0 as a convex combination. Namely, for disjoint index sets I_1 and I_2, we have $0 \in \mathrm{conv}(\{\bar{a}_i \colon i \in I_1\} \cup \{-\bar{a}_i \colon i \in I_2\})$ if and only if there is a nonzero $\alpha \in \mathsf{LinDep}(\bar{a})$ with $I_+(\alpha) \subseteq I_1$ and $I_-(\alpha) \subseteq I_2$.

Together with these geometric interpretations of $\mathsf{LinVal}(\bar{a})$, $\mathsf{AffVal}(a)$, $\mathsf{LinDep}(\bar{a})$, and $\mathsf{AffDep}(a)$, the following lemma (whose proof is left to Exercise 8) allows us to translate properties of point configurations to those of their Gale transforms.

5.6.2 Lemma. *Let a be a sequence of n points in $\mathbf{R}^d$ whose points affinely span $\mathbf{R}^d$, and let $\bar{g}$ be its Gale transform. Then $\mathsf{LinVal}(\bar{g}) = \mathsf{AffDep}(a)$ and $\mathsf{LinDep}(\bar{g}) = \mathsf{AffVal}(a)$.* $\qquad\square$

So a Radon partition of a corresponds to a partition of $\bar{g}$ by a linear hyperplane, and a partition of a by a hyperplane translates to a linear dependence (i.e., a "linear Radon partition") of $\bar{g}$.

Let us list several interesting connections, again leaving the simple but instructive proofs to the reader.

5.6.3 Corollary (Dictionary of the Gale transform).

(i) (*Lying in a common hyperplane*) *For every* $(d+1)$-*point index set* $I \subseteq \{1, 2, \ldots, n\}$, *the points* a_i *with* $i \in I$ *lie in a common hyperplane if and only if all the vectors* $\bar{g}_j$ *with* $j \notin I$ *lie in a common linear hyperplane.*

(ii) (*General position*) *In particular, the points of* $\boldsymbol{a}$ *are in general position (no* $d+1$ *on a common hyperplane) if and only if every* $n-d-1$ *vectors among* $\bar{g}_1, \ldots, \bar{g}_n$ *span* $\mathbf{R}^{n-d-1}$ *(which is a natural condition of general position for vectors).*

(iii) (*Faces of the convex hull*) *The points* a_i *with* $i \in I$ *are contained in a common facet of* $P = \mathrm{conv}(\boldsymbol{a})$ *if and only if* $0 \in \mathrm{conv}\{\bar{g}_j \colon j \notin I\}$. *In particular, if* P *is a simplicial polytope, then its* k-*faces exactly correspond to complements of the* $(n-k-1)$-*element subsets of* $\bar{g}$ *containing* 0 *in the convex hull.*

(iv) (*Convex independence*) *The* a_i *form a convex independent set if and only if there is no oriented linear hyperplane with exactly one of the* $\bar{g}_j$ *on the positive side.*

Here is, finally, a picture of a 3-dimensional convex polytope with 6 vertices and the (planar) Gale transform of its vertex set:

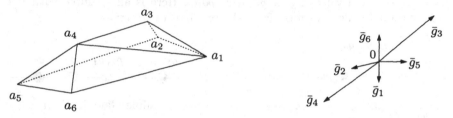

For example, the facet $a_1 a_2 a_5 a_6$ is reflected by the complementary pair $\bar{g}_3, \bar{g}_4$ of parallel oppositely oriented vectors, and so on.

Signs suffice. As was noted above, in order to find out whether some a_i is a vertex of $\mathrm{conv}(\boldsymbol{a})$, we ask whether there is an $\alpha \in \mathsf{AffDep}(\boldsymbol{a})$ with $I_+(\alpha) = \{i\}$. Only the signs of the vectors in $\mathsf{AffDep}(\boldsymbol{a})$ are important here, and this is the case with all the combinatorial-geometric information about point sequences or vector sequences in Corollary 5.6.3. For such purposes, the knowledge of $\mathrm{sgn}(\mathsf{AffDep}(\boldsymbol{a})) = \{(\mathrm{sgn}(\alpha_1), \ldots, \mathrm{sgn}(\alpha_n)) \colon \alpha \in \mathsf{AffDep}(\boldsymbol{a})\}$ is as good as the knowledge of $\mathsf{AffDep}(\boldsymbol{a})$.

We can thus declare two sequences $\boldsymbol{a}$ and $\boldsymbol{b}$ combinatorially isomorphic if $\mathrm{sgn}(\mathsf{AffDep}(\boldsymbol{a})) = \mathrm{sgn}(\mathsf{AffDep}(\boldsymbol{b}))$ and $\mathrm{sgn}(\mathsf{AffVal}(\boldsymbol{a})) = \mathrm{sgn}(\mathsf{AffVal}(\boldsymbol{b}))$.[2] We will hear a little more about this notion of combinatorial isomorphism in Section 9.3 when we discuss order types, and also in the notes to Section 6.2 in connection with oriented matroids.

---

[2] It is nontrivial but true that either of these equalities implies the other one.

Here we need only one very special case: If $\bar{g} = (\bar{g}_1, \ldots, \bar{g}_n)$ is a sequence of vectors, $t_1, \ldots, t_n > 0$ are positive real numbers, and $\bar{g}' = (t_1\bar{g}_1, \ldots, t_n\bar{g}_n)$, then clearly,

$$\mathsf{sgn}(\mathsf{LinVal}(\bar{g})) = \mathsf{sgn}(\mathsf{LinVal}(\bar{g}')) \quad \text{and} \quad \mathsf{sgn}(\mathsf{LinDep}(\bar{g})) = \mathsf{sgn}(\mathsf{LinDep}(\bar{g}')),$$

and so $\bar{g}$ and $\bar{g}'$ are combinatorially isomorphic vector configurations.

Affine Gale diagrams. We have seen a certain asymmetry of the Gale transform: While the sequence $\boldsymbol{a}$ is interpreted affinely, as a point sequence, its Gale transform needs to be interpreted linearly, as a sequence of vectors (with 0 playing a special role). Could one reduce the dimension of $\bar{g}$ by 1 and pass to an "affine version" of the Gale transform? This is indeed possible, but one has to distinguish "positive" and "negative" points in the affine version.

Let $\bar{g}$ be the Gale transform of some $\boldsymbol{a}$, $\bar{g}_1, \ldots, \bar{g}_n \in \mathbf{R}^{n-d-1}$. Let us assume for simplicity that all the $\bar{g}_i$ are nonzero. We choose a hyperplane h not parallel to any of the $\bar{g}_i$ and not passing through 0, and we project the $\bar{g}_i$ centrally from 0 into h, obtaining points $g_1, \ldots, g_n \in h \cong \mathbf{R}^{n-d-2}$. If g_i lies on the same side of 0 as $\bar{g}_i$, i.e., if $g_i = t_i\bar{g}_i$ with $t_i > 0$, we set $\sigma_i = +1$, and call g_i a *positive point*. For g_i lying on the other side of 0 than $\bar{g}_i$ we let $\sigma_i = -1$, and we call g_i a *negative point*. Here is an example with the 2-dimensional Gale transform from the previous drawing:

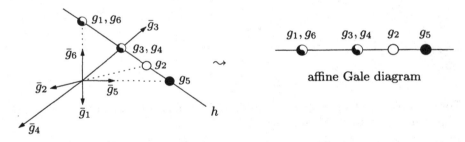

The positive g_i are marked by full circles, the negative ones by empty circles, and we have borrowed the (incomplete) yin–yang symbol for marking the positions shared by one positive and one negative point. This sequence $\boldsymbol{g}$ of positive and negative points in $\mathbf{R}^{n-d-2}$, or more formally the pair $(\boldsymbol{g}, \sigma)$, is called an *affine Gale diagram* of $\boldsymbol{a}$. It conveys the same combinatorial information as $\bar{g}$, although we cannot reconstruct $\boldsymbol{a}$ from it up to linear isomorphism, as was the case with $\bar{g}$. (For this reason, we speak of Gale *diagram* rather than Gale *transform*.) One has to get used to interpreting the positive and negative points properly. If we put

$$\mathsf{AffVal}(\boldsymbol{g}, \sigma) = \{(\sigma_1 f(g_1), \ldots, \sigma_n f(g_n)) : f : \mathbf{R}^{n-d-2} \to \mathbf{R} \text{ affine}\},$$

$$\mathsf{AffDep}(\boldsymbol{g}, \sigma) = \Big\{\alpha \in \mathbf{R}^n : \textstyle\sum_{i=1}^{n} \sigma_i \alpha_i g_i = 0, \ \sum_{i=1}^{n} \sigma_i \alpha_i = 0\Big\},$$

then, as is easily checked,

$$\mathrm{sgn}(\mathsf{AffDep}(g,\sigma)) = \mathrm{sgn}(\mathsf{LinDep}(\bar{g})) \quad\text{and}\quad \mathrm{sgn}(\mathsf{AffVal}(g,\sigma)) = \mathrm{sgn}(\mathsf{LinVal}(\bar{g})).$$

Here is a reinterpretation of Corollary 5.6.3 in terms of the affine Gale diagram.

5.6.4 Proposition (Dictionary of affine Gale diagrams). *Let a be a sequence of n points in $\mathbf{R}^d$, let $\bar{g}$ be the Gale transform of a, and assume that all the $\bar{g}_i$ are nonzero. Let (g,σ) be an affine Gale diagram of a in $\mathbf{R}^{n-d-2}$.*

(i) *A subsequence a_I lies in a common facet of $\mathrm{conv}(a)$ if and only if $\mathrm{conv}(\{g_j\colon j \notin I, \sigma_j = 1\}) \cap \mathrm{conv}(\{g_j\colon j \notin I, \sigma_j = -1\}) \neq \emptyset$.*

(ii) *The points of a are in convex position if and only if for every oriented hyperplane in $\mathbf{R}^{n-d-2}$, the number of positive points of g on its positive side plus the number of negative points of g on its negative side is at least 2.* □

So far we have assumed that $\bar{g}_i \neq 0$ for all i. This need not hold in general, and points $\bar{g}_i = 0$ need a special treatment in the affine Gale diagram: They are called the *special points*, and for a full specification of the affine Gale diagram, we draw the positive and negative points and give the number of special points. It is easy to find out how the presence of special points influences the conditions in the previous proposition.

A nonrational polytope. Configurations of $k+4$ points in $\mathbf{R}^k$ have planar affine Gale diagrams. This leads to many interesting constructions of k-dimensional convex polytopes with $k+4$ vertices. Here we give just one example: an 8-dimensional polytope with 12 vertices that cannot be realized with rational coordinates; that is, no polytope with isomorphic face lattice has all vertex coordinates rational. First one has to become convinced that if 9 distinct points are placed in $\mathbf{R}^2$ so that they are not all collinear and there are collinear triples and 4-tuples as is marked by segments in the left drawing below,

 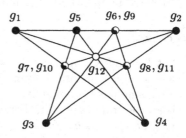

then not all coordinates of the points can be rational. We omit the proof, which has little to do with the Gale transform or convex polytopes.

Next, we declare some points negative, some positive, and some both positive and negative, as in the right drawing, obtaining 12 points. These points have a chance of being an affine Gale diagram of the vertex set of an 8-dimensional convex polytope, since condition (ii) in Proposition 5.6.4

is satisfied. How do we construct such a polytope? For $g_i = (x_i, y_i)$, we put $\bar{g}_i = (t_i x_i, t_i y_i, t_i) \in \mathbf{R}^3$, choosing $t_i > 0$ for positive g_i and $t_i < 0$ for negative t_i, in such a way that $\sum_{i=1}^{12} \bar{g}_i = 0$. Then the Gale transform of $\bar{g}$ is the vertex set of the desired convex polytope P (see Observation 5.6.1(ii) and (iii)).

Let P' be some convex polytope with an isomorphic face lattice and let (g', σ') be an affine Gale diagram of its vertex set a'. We have, for example, $g'_7 = g'_{10}$ because $\{a'_i: i \neq 7, 10\}$ form a facet of P', and similarly for the other point coincidences. The triple g'_1, g'_{12}, g'_8 (where g'_8 is positive) is collinear, because $\{a'_i: i \neq 1, 8, 12\}$ is a facet. In this way, we see that the point coincidences and collinearities are preserved, and so no affine Gale diagram of P' can have all coordinates rational. At the same time, by checking the definition, we see that a point sequence with rational coordinates has at least one affine Gale diagram with rational coordinates. Thus, P cannot be realized with rational coordinates.

Bibliography and remarks. Gale diagrams and the Gale transform emerged from the work of Gale [Gal56] and were further developed by Perles, as is documented in [Grü67] (also see, e.g., [MS71]). Our exposition essentially follows Ziegler's book [Zie94] (his treatment is combined with an introduction to oriented matroids). We aim at concreteness, and so, for example, the Gale transform is defined using the orthogonal complement, although it might be mathematically more elegant to work with the annihilator in the dual space $(\mathbf{R}^n)^*$, and so on. The construction of an irrational 8-polytope is due to Perles.

In Section 11.3 (Exercise 6) we mention an interpretation of the h-vector of a simplicial convex polytope via the Gale transform. Using this correspondence, Wagner and Welzl [WW01] found an interesting continuous analogue of the upper bound theorem, which speaks about probability distributions in $\mathbf{R}^d$. For other recent applications of a similar correspondence see the notes to Section 11.3.

Exercises

1. Let B be a $k \times n$ matrix of rank $k \leq n$. Check that for any $k \times n$ matrix B' whose rows generate the same vector space as the rows of B, there exists a nonsingular $k \times k$ matrix T with $B' = TB$. Infer that if $\bar{g} = (\bar{g}_1, \ldots, \bar{g}_n)$ is a Gale transform of a, then any other Gale transform of a has the form $(T\bar{g}_1, T\bar{g}_2, \ldots, T\bar{g}_n)$ for a nonsingular square matrix T. ③

2. Let a be a sequence of $d+1$ affinely independent points in $\mathbf{R}^d$. What is the Gale transform of a, and what are $\mathsf{AffVal}(a)$ and $\mathsf{AffDep}(a)$? ①

3. Let $\bar{g}$ be a Gale transform of the vertex set of a convex polytope $P \subset \mathbf{R}^d$, and let $\bar{h}$ be obtained from $\bar{g}$ by appending the zero vector. Check that $\bar{h}$ is again a Gale transform of a convex independent set. What is the relation of this set to P? ③

4. Using affine Gale diagrams, count the number of classes of combinatorial equivalence of d-dimensional convex polytopes with $d+2$ vertices. How many of them are simple, and how many simplicial? [3]

5. Verify the characterization in Observation 5.6.1(ii) of sequences $\bar{g}$ in $\mathbf{R}^{n-d-1}$ that are Gale transforms of some a, and check that if the Gale transform is applied twice to such $\bar{g}$, we obtain $\bar{g}$ up to linear isomorphism. [3]

6. Let $a = (a_1, \ldots, a_n)$ be a point sequence in $\mathbf{R}^d$ whose affine hull is all of $\mathbf{R}^d$, and let $P = \operatorname{conv}\{a_1, \ldots, a_n\}$.
Given $\mathsf{AffVal}(a)$, explain how we can determine which of the a_i are the vertices of P and how we reconstruct the face lattice of P. [2]

7. Let $\bar{a}$ be a sequence of n vectors in $\mathbf{R}^{d+1}$ that spans $\mathbf{R}^{d+1}$.
   (a) Find $\dim \mathsf{LinVal}(\bar{a})$ and $\dim \mathsf{LinDep}(\bar{a})$. [2]
   (b) Check that $\mathsf{LinVal}(\bar{a})$ is the orthogonal complement of $\mathsf{LinDep}(\bar{a})$. [2]

8. Prove Lemma 5.6.2. [3]

9. Verify Corollary 5.6.3. [3]

5.7 Voronoi Diagrams

Consider a finite set $P \subset \mathbf{R}^d$. For each point $p \in P$, we define a region $reg(p)$, which is the "sphere of influence" of the point p: It consists of the points $x \in \mathbf{R}^d$ for which p is the closest point among the points of P. Formally,

$$reg(p) = \{x \in \mathbf{R}^d\colon \operatorname{dist}(x,p) \le \operatorname{dist}(x,q) \text{ for all } q \in P\},$$

where $\operatorname{dist}(x,y)$ denotes the Euclidean distance of the points x and y. The *Voronoi diagram* of P is the set of all regions $reg(p)$ for $p \in P$. (More precisely, it is the cell complex induced by these regions; that is, every intersection of a subset of the regions is a face of the Voronoi diagram.) Here an example of the Voronoi diagram of a point set in the plane:

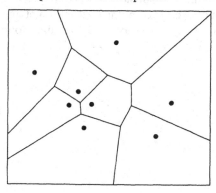

(Of course, the Voronoi diagram is clipped by a rectangle so that it fits into a finite page.) The points of P are traditionally called the *sites* in the context of Voronoi diagrams.

5.7.1 Observation. *Each region reg(p) is a convex polyhedron with at most* $|P|-1$ *facets.*

Indeed,

$$reg(p) = \bigcap_{q \in P \setminus \{p\}} \{x \colon \mathrm{dist}(x,p) \le \mathrm{dist}(x,q)\}$$

is an intersection of $|P| - 1$ half-spaces. □

For $d = 2$, a Voronoi diagram of n points is a subdivision of the plane into n convex polygons (some of them are unbounded). It can be regarded as a drawing of a planar graph (with one vertex at the infinity, say), and hence it has a linear combinatorial complexity: n regions, $O(n)$ vertices, and $O(n)$ edges.

In the literature the Voronoi diagram also appears under various other names, such as the *Dirichlet tessellation*.

Examples of applications. Voronoi diagrams have been reinvented and used in various branches of science. Sometimes the connections are surprising. For instance, in archaeology, Voronoi diagrams help study cultural influences. Here we mention a few applications, mostly algorithmic.

- ("Post office problem" or nearest neighbor searching) Given a point set P in the plane, we want to construct a data structure that finds the point of P nearest to a given query point x as quickly as possible. This problem arises directly in some practical situations or, more significantly, as a subroutine in more complicated problems. The query can be answered by determining the region of the Voronoi diagram of P containing x. For this problem (point location in a subdivision of the plane), efficient data structures are known; see, e.g., the book [dBvKOS97] or other introductory texts on computational geometry.
- (Robot motion planning) Consider a disk-shaped robot in the plane. It should pass among a set P of point obstacles, getting from a given start position to a given target position and touching none of the obstacles.

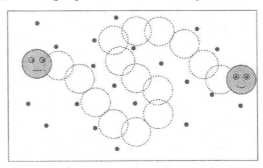

If such a passage is possible at all, the robot can always walk along the edges of the Voronoi diagram of P, except for the initial and final segments of the tour. This allows one to reduce the robot motion problem to a graph search problem: We define a subgraph of the Voronoi diagram consisting of the edges that are passable for the robot.

- (A nice triangulation: the Delaunay triangulation) Let $P \subset \mathbf{R}^2$ be a finite point set. In many applications one needs to construct a triangulation of P (that is, to subdivide conv(P) into triangles with vertices at the points of P) in such a way that the triangles are not too skinny. Of course, for some sets, some skinny triangles are necessary, but we want to avoid them as much as possible. One particular triangulation that is usually very good, and provably optimal with respect to several natural criteria, is obtained as the dual graph to the Voronoi diagram of P. Two points of P are connected by an edge if and only if their Voronoi regions share an edge.

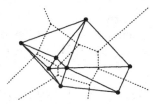

If no 4 points of P lie on a common circle then this indeed defines a triangulation, called the *Delaunay triangulation*[3] of P; see Exercise 5. The definition extends to points sets in $\mathbf{R}^d$ in a straightforward manner.

- (Interpolation) Suppose that $f: \mathbf{R}^2 \to \mathbf{R}$ is some smooth function whose values are known to us only at the points of a finite set $P \subset \mathbf{R}^2$. We would like to interpolate f over the whole polygon conv(P). Of course, we cannot really tell what f looks like outside P, but still we want a reasonable interpolation rule that provides a nice smooth function with the given values at P. Multidimensional interpolation is an extensive semiempirical discipline, which we do not seriously consider here; we explain only one elegant method based on Voronoi diagrams. To compute the interpolated value at a point $x \in$ conv(P), we construct the Voronoi diagram of P, and we overlay it with the Voronoi diagram of $P \cup \{x\}$.

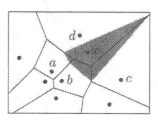

---

[3] Being a transcription from Russian, the spelling of Delaunay's name varies in the literature. For example, in crystallography literature he is usually spelled "Delone."

The region of the new point x cuts off portions of the regions of some of the old points. Let w_p be the area of the part of $reg(p)$ in the Voronoi diagram of P that belongs to $reg(x)$ after inserting x. The interpolated value $f(x)$ is

$$f(x) = \sum_{p \in P} \frac{w_p}{\sum_{q \in P} w_q} f(p).$$

An analogous method can be used in higher dimensions, too.

Relation of Voronoi diagrams to convex polyhedra. We now show that Voronoi diagrams in $\mathbf{R}^d$ correspond to certain convex polyhedra in $\mathbf{R}^{d+1}$.

First we define the *unit paraboloid* in $\mathbf{R}^{d+1}$:

$$U = \{x \in \mathbf{R}^{d+1}: x_{d+1} = x_1^2 + x_2^2 + \cdots + x_d^2\}.$$

For $d = 1$, U is a parabola in the plane.

In the sequel, let us imagine the space $\mathbf{R}^d$ as the hyperplane $x_{d+1} = 0$ in $\mathbf{R}^{d+1}$. For a point $p = (p_1, \ldots, p_d) \in \mathbf{R}^d$, let $\mathsf{e}(p)$ denote the hyperplane in $\mathbf{R}^{d+1}$ with equation

$$x_{d+1} = 2p_1 x_1 + 2p_2 x_2 + \cdots + 2p_d x_d - p_1^2 - p_2^2 - \cdots - p_d^2.$$

Geometrically, $\mathsf{e}(p)$ is the hyperplane tangent to the paraboloid U at the point $u(p) = (p_1, p_2, \ldots, p_d, p_1^2 + \cdots + p_d^2)$ lying vertically above p. It is perhaps easier to remember this geometric definition of $\mathsf{e}(p)$ and derive its equation by differentiation when needed. On the other hand, in the forthcoming proof we start out from the equation of $\mathsf{e}(p)$, and as a by-product, we will see that $\mathsf{e}(p)$ is the tangent to U at $u(p)$ as claimed.

5.7.2 Proposition. *Let $p, x \in \mathbf{R}^d$ be points and let $u(x)$ be the point of U vertically above x. Then $u(x)$ lies above the hyperplane $\mathsf{e}(p)$ or on it, and the vertical distance of $u(x)$ to $\mathsf{e}(p)$ is δ^2, where $\delta = \mathrm{dist}(x, p)$.*

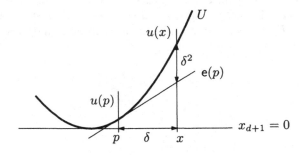

Proof. We just substitute into the equations of U and of $\mathsf{e}(p)$. The x_{d+1}-coordinate of $u(x)$ is $x_1^2 + \cdots + x_d^2$, while the x_{d+1}-coordinate of the point of $e(p)$ above x is $2p_1x_1 + \cdots + 2p_dx_d - p_1^2 - \cdots - p_d^2$. The difference is $(x_1 - p_1)^2 + \cdots + (x_d - p_d)^2 = \delta^2$. $\qquad\square$

Let $\mathcal{E}(p)$ denote the half-space lying above the hyperplane $e(p)$. Consider an n-point set $P \subset \mathbf{R}^d$. By Proposition 5.7.2, $x \in reg(p)$ holds if and only if $e(p)$ is vertically closest to U at x among all $e(q)$, $q \in P$. Here is what we have derived:

5.7.3 Corollary. *The Voronoi diagram of P is the vertical projection of the facets of the polyhedron $\bigcap_{p \in P} \mathcal{E}(p)$ onto the hyperplane $x_{d+1} = 0$.* $\qquad\square$

Here is an illustration for a planar Voronoi diagram:

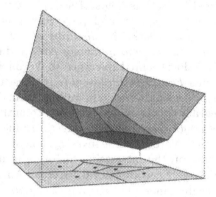

5.7.4 Corollary. *The maximum total number of faces of all regions of the Voronoi diagram of an n-point set in $\mathbf{R}^d$ is $O(n^{\lceil d/2 \rceil})$.*

Proof. We know that the combinatorial complexity of the Voronoi diagram equals the combinatorial complexity of an H-polyhedron with at most n facets in $\mathbf{R}^{d+1}$. By intersecting this H-polyhedron with a large simplex we can obtain a bounded polytope with at most $n+d+2$ facets, and we have not decreased the number of faces compared to the original H-polyhedron. Then the dual version of the asymptotic upper bound theorem (Theorem 5.5.2) implies that the total number of faces is $O(n^{\lceil d/2 \rceil})$, since $\lfloor (d+1)/2 \rfloor = \lceil d/2 \rceil$. $\qquad\square$

The convex polyhedra in $\mathbf{R}^{d+1}$ obtained from Voronoi diagrams in $\mathbf{R}^d$ by the above construction are rather special, and so a lower bound for the combinatorial complexity of convex polytopes cannot be automatically transferred to Voronoi diagrams. But it turns out that the number of vertices of a Voronoi diagram on n points in $\mathbf{R}^d$ can really be of order $n^{\lceil d/2 \rceil}$ (Exercise 2).

Let us remark that the trick used for transforming Voronoi diagrams to convex polyhedra is an example of a more general technique, called *linearization* or *Veronese mapping*, which will be discussed a little more in

Section 10.3. This method sometimes allows us to convert a problem about algebraic curves or surfaces of bounded degree to a problem about k-flats in a suitable higher-dimensional space.

The farthest-point Voronoi diagram. The projection of the H-poly-hedron $\bigcap_{p \in P} \mathcal{E}(p)^{\mathrm{op}}$, where γ^{op} denotes the half-space opposite to γ, forms the *farthest-neighbor Voronoi diagram*, in which each point $p \in P$ is assigned the regions of points for which it is the farthest point. It can be shown that all nonempty regions of this diagram are unbounded and they correspond precisely to the points appearing on the surface of $\mathrm{conv}(P)$.

Bibliography and remarks. The concept of Voronoi diagrams independently emerged in various fields of science, for example as the *medial axis transform* in biology and physiology, the *Wigner–Seitz zones* in chemistry and physics, the *domains of action* in crystallography, and the *Thiessen polygons* in meteorology and geography. Apparently, the earliest documented reference to Voronoi diagrams is a picture in the famous *Principia Philosopiae* by Descartes from 1644 (that picture actually seems to show a power diagram, a generalization of the Voronoi diagram to sites with different strengths of influence). Mathematically, Voronoi diagrams were first introduced by Dirichlet [Dir50] and by Voronoi [Vor08] for the investigation of quadratic forms. For more information on the interesting history and a surprising variety of applications we refer to several surveys: Aurenhammer and Klein [AK00], Aurenhammer [Aur91], and the book Okabe, Boots, and Sugihara [OBS92]. Every computational geometry textbook also has at least a chapter devoted to Voronoi diagrams, and most papers on this subject appear in computational geometry.

The *Delaunay triangulation* (or, more correctly, the Delaunay tessellation, since it need not be a triangulation in general) was first considered by Voronoi as the dual to the Voronoi diagram, and later by Delaunay [Del34] with the definition given in Exercise 5(b) below. The Delaunay triangulation of a planar point set P optimizes several quality measures among all triangulations of P: It maximizes the minimum angle occurring in any triangle, minimizes the maximum circumradius of the triangles, maximizes the sum of inradii, and so on (see [AK00] for references). Such optimality properties can usually be proved by *local flipping*. We consider an arbitrary triangulation $\mathcal{T}$ of a given finite $P \subset \mathbf{R}^2$ (say with no 4 cocircular points). If there is a 4-point $Q \subseteq P$ such that $\mathrm{conv}(Q)$ is a quadrilateral triangulated by two triangles of $\mathcal{T}$ but in such a way that these two triangles are not the Delaunay triangulation of Q, then the diagonal of Q can be flipped:

not locally
Delaunay $\rightsquigarrow$

locally
Delaunay

It can be shown that every sequence of such local flips is finite and finishes with the Delaunay triangulation of P (Exercise 7). This procedure has an analogue in higher dimensions, where it gives a simple and practically successful algorithm for computing Delaunay triangulations (and Voronoi diagrams); see, e.g., Edelsbrunner and Shah [ES96].

Generalizations of Voronoi diagrams. The example in the text with robot motion planning, as well as other applications, motivates various notions of generalized Voronoi diagrams. First, instead of the Euclidean distance, one can take various other distance functions, say the ℓ_p-metrics. Second, instead of the spheres of influence of points, we can consider the spheres of influence of other sites, such as disjoint polygons (this is what we get if we have a circular robot moving amidst polygonal obstacles). We do not attempt to survey the numerous results concerning such generalizations, again referring to [AK00]. Results on the combinatorial complexity of Voronoi diagrams under non-Euclidean metrics and/or for nonpoint sites will be mentioned in the notes to Section 7.7.

In another, very general, approach to Voronoi diagrams, one takes the Voronoi diagram induced by two objects as a primitive notion. So for every two objects we are given a partition of space into two regions separated by a *bisector*, and Voronoi diagrams for more than two objects are built using the 2-partitions for all pairs. If one postulates a few geometric properties of the bisectors, one gets a reasonable theory of Voronoi diagrams (the so-called *abstract Voronoi diagrams*), including efficient algorithms. So, for example, we do not even need a notion of distance at this level of generality. Abstract Voronoi diagrams (in the plane) were suggested by Klein [Kle89].

A geometrically significant generalization of the Euclidean Voronoi diagram is the *power diagram*: Each point $p \in P$ is assigned a real weight $w(p)$, and $reg(P) = \{x \in \mathbf{R}^d : \|x - p\|^2 - w(p) \le \|x - q\|^2 - w(q)$ for all $q \in P\}$. While Voronoi diagrams in $\mathbf{R}^d$ are projections of *certain* convex polyhedra in $\mathbf{R}^{d+1}$, the projection into $\mathbf{R}^d$ of every intersection of finitely many nonvertical upper half-spaces in $\mathbf{R}^{d+1}$ is a power diagram. Moreover, a hyperplane section of a power diagram is again a power diagram. Several other generalized Voronoi diagrams in $\mathbf{R}^d$ (for example, with multiplicative weights of the sites) can be obtained by intersecting a suitable power diagram in $\mathbf{R}^{d+1}$ with a simple surface and projecting into $\mathbf{R}^d$, which yields fast algorithms; see Aurenhammer and Imai [AI88].

Another generalization are *higher-order Voronoi diagrams*. The kth-order Voronoi diagram of a finite point set P assigns to each k-point $T \subseteq P$ the region $reg(T)$ consisting of all $x \in \mathbf{R}^d$ for which the points of T are the k nearest neighbors of x in P. The usual Voronoi diagram arises for $k = 1$, and the farthest-point Voronoi diagram for $k = |P| - 1$. The kth-order Voronoi diagram of $P \subset \mathbf{R}^d$ is the projection of the kth level facets in the arrangement of the hyperplanes $e(p)$, $p \in P$ (see Chapter 6 for these notions). Lee [Lee82] proved that the kth-order Voronoi diagram of n points in the plane has combinatorial complexity $O(k(n-k))$; this is better than the maximum possible complexity of level k in an arrangement of n arbitrary planes in $\mathbf{R}^3$.

Applications of Voronoi diagrams are too numerous to be listed here, and we add only a few remarks to those already mentioned in the text. Using point location in Voronoi diagrams as in the post office problem, several basic computational problems in the plane can be solved efficiently, such as finding the *closest pair* in a point set or the *largest disk* contained in a given polygon and not containing any of the given points.

Besides providing good triangulations, the Delaunay triangulation contains several other interesting graphs as subgraphs, such as a minimum spanning tree of a given point set (Exercise 6). In the plane, this leads to an $O(n \log n)$ algorithm for the minimum spanning tree. In $\mathbf{R}^3$, subcomplexes of the Delaunay triangulation, the so-called α-complexes, have been successfully used in molecular modeling (see, e.g., Edelsbrunner [Ede98]); they allow one to quickly answer questions such as, "how many tunnels and voids are there in the given molecule?"

Robot motion planning using Voronoi diagrams (or, more generally, the *retraction approach*, where the whole free space for the robot is replaced by some suitable low-dimensional skeleton) was first considered by Ó'Dúnlaig and Yap [ÓY85]. Algorithmic motion planning is an extensive discipline with innumerable variants of the problem. For a brief introduction from the computational-geometric point of view see, e.g., [dBvKOS97]; among several monographs we mention Laumond and Overmars [LO96] and Latombe [Lat91].

The spatial interpolation of functions using Voronoi diagrams was considered by Sibson [Sib81].

Exercises

1. Prove that the region $reg(p)$ of a point p in the Voronoi diagram of a finite point set $P \subset \mathbf{R}^d$ is unbounded if and only if p lies on the surface of $conv(P)$. ②

2. (a) Show that the Voronoi diagram of the $2n$-point set $\{(\frac{i}{n}, 0, 0): i = 1, 2, \ldots, n\} \cup \{(0, 1, \frac{j}{n}): j = 1, 2, \ldots, n\}$ in $\mathbf{R}^3$ has $\Omega(n^2)$ vertices. ③

   (b) Let $d = 2k+1$ be odd, let $e_1, \ldots, e_d$ be vectors of the standard orthonormal basis in $\mathbf{R}^d$, and let e_0 stand for the zero vector. For $i = 0, 1, \ldots, k$ and $j = 1, 2, \ldots, n$, let $p_{i,j} = e_{2i} + \frac{i}{n} e_{2i+1}$. Prove that for every choice of $j_0, j_1, \ldots, j_k \in \{1, 2, \ldots, n\}$, there is a point in $\mathbf{R}^d$ for which the nearest points among the $p_{i,j}$ are exactly $p_{0,j_0}, p_{1,j_1}, \ldots, p_{k,j_k}$. Conclude that the Voronoi diagram of the $p_{i,j}$ has combinatorial complexity $\Omega(n^k) = \Omega(n^{\lceil d/2 \rceil})$. ③

3. (Voronoi diagram of flats) Let $\varepsilon_1, \ldots, \varepsilon_{d-1}$ be small distinct positive numbers and for $i = 1, 2, \ldots, d-1$ and $j = 1, 2, \ldots, n$, let $F_{i,j}$ be the $(d-2)$-flat $\{x \in \mathbf{R}^d: x_i = j, x_d = \varepsilon_i\}$. For every choice of $j_1, j_2, \ldots, j_{d-1} \in \{1, 2, \ldots, n\}$, find a point in $\mathbf{R}^d$ for which the nearest sites (under the Euclidean distance) among the $F_{i,j}$ are exactly $F_{1,j_1}, F_{2,j_2}, \ldots, F_{d-1,j_{d-1}}$. Conclude that the Voronoi diagram of the $F_{i,j}$ has combinatorial complexity $\Omega(n^{d-1})$. ③

   This example is from Aronov [Aro00].

4. For a finite point set in the plane, define the farthest-point Voronoi diagram as indicated in the text, verify the claimed correspondence with a convex polyhedron in $\mathbf{R}^3$, and prove that all nonempty regions are unbounded. ③

5. (Delaunay triangulation) Let P be a finite point set in the plane with no 3 points collinear and no 4 points cocircular.

   (a) Prove that the dual graph of the Voronoi diagram of P, where two points $p, q \in P$ are connected by a straight edge if and only if the boundaries of $reg(p)$ and $reg(q)$ share a segment, is a plane graph where the outer face is the complement of $conv(P)$ and every inner face is a triangle. ③

   (b) Define a graph on P as follows: Two points p and q are connected by an edge if and only if there exists a circular disk with both p and q on the boundary and with no point of P in its interior. Prove that this graph is the same as in (a), and so we have an alternative definition of the Delaunay triangulation. ③

6. (Delaunay triangulation and minimum spanning tree) Let $P \subset \mathbf{R}^2$ be a finite point set with no 3 points collinear and no 4 cocircular. Let T be a spanning tree of minimum total edge length in the complete graph with the vertex set P, where the length of an edge is just its Euclidean length. Prove that all edges of T are also edges of the Delaunay triangulation of P. ③

7. (Delaunay triangulation by local flipping) Let $P \subset \mathbf{R}^2$ be an n-point set with no 3 points collinear and no 4 cocircular. Let $\mathcal{T}$ be an arbitrary triangulation of $conv(P)$. Suppose that triangulations $\mathcal{T}_1, \mathcal{T}_2, \ldots$ are obtained from $\mathcal{T}$ by successive local flips as described in the notes above (in each step, we select a convex quadrilateral in the current triangulation

partitioned into two triangles in a way that is not the Delaunay triangulation of the four vertices and we flip the diagonal of the quadrilateral).
(a) Prove that the sequence of triangulations is always finite (and give as good an estimate for its maximum length as you can). ③
(b) Show that if no local flipping is possible, then the current triangulation is the Delaunay triangulation of P. ④

8. Consider a finite set of disjoint segments in the plane. What types of curves may bound the regions in their Voronoi diagram? The region of a given segment is the set of points for which this segment is a closest one. ②

9. Let A and B be two finite point sets in the plane. Choose $a_0 \in A$ arbitrarily. Having defined $a_0, \ldots, a_i$ and $b_1, \ldots, b_{i-1}$, define b_{i+1} as a point of $B \setminus \{b_1, \ldots, b_i\}$ nearest to a_i, and a_{i+1} as a point of $A \setminus \{a_0, \ldots, a_i\}$ nearest to b_{i+1}. Continue until one of the sets becomes empty. Prove that at least one of the pairs (a_i, b_{i+1}), (b_{i+1}, a_{i+1}), $i = 0, 1, 2, \ldots$, realizes the shortest distance between a point of A and a point of B. (This was used by Eppstein [Epp95] in some dynamical geometric algorithms.) ③

10. (a) Let C be any circle in the plane $x_3 = 0$ (in $\mathbf{R}^3$). Show that there exists a half-space h such that C is the vertical projection of the set $h \cap U$ onto $x_3 = 0$, where $U = \{x \in \mathbf{R}^3 : x_3 = x_1^2 + x_2^2\}$ is the unit paraboloid. ①
(b) Consider n arbitrary circular disks $K_1, \ldots, K_n$ in the plane. Show that there exist only $O(n)$ intersections of their boundaries that lie inside no other K_i (this means that the boundary of the union of the K_i consists of $O(n)$ circular arcs). ③

11. Define a "spherical polytope" as an intersection of n balls in $\mathbf{R}^3$ (such an object has facets, edges, and vertices similar to an ordinary convex polytope).
(a) Show that any such spherical polytope in $\mathbf{R}^3$ has $O(n^2)$ faces. You may assume that the spheres are in general position. ④
(b) Find an example of an intersection of n balls having quadratically many vertices. ③
(c) Show that the intersection of n *unit* balls has $O(n)$ complexity only. ④

6

Number of Faces in Arrangements

Arrangements of lines in the plane and their higher-dimensional generalization, arrangements of hyperplanes in $\mathbf{R}^d$, are a basic geometric structure whose significance is comparable to that of convex polytopes. In fact, arrangements and convex polytopes are quite closely related: A cell in a hyperplane arrangement is a convex polyhedron, and conversely, each hyperplane arrangement in $\mathbf{R}^d$ corresponds canonically to a convex polytope in $\mathbf{R}^{d+1}$ of a special type, the so-called zonotope. But as is often the case with different representations of the same mathematical structure, convex polytopes and arrangements of hyperplanes emphasize different aspects of the structure and lead to different questions.

Whenever we have a problem involving a finite point set in $\mathbf{R}^d$ and partitions of the set by hyperplanes, we can use geometric duality, and we obtain a problem concerning a hyperplane arrangement. Arrangements appear in many other contexts as well; for example, some models of molecules give rise to arrangements of spheres in $\mathbf{R}^3$, and automatic planning of the motion of a robot among obstacles involves, implicitly or explicitly, arrangements of surfaces in higher-dimensional spaces.

Arrangements of hyperplanes have been investigated for a long time from various points of view. In several classical areas of mathematics one is mainly interested in topological and algebraic properties of the whole arrangement. Hyperplane arrangements are related to such marvelous objects as Lie algebras, root systems, and Coxeter groups. In the theory of *oriented matroids* one studies the systems of sign vectors associated to hyperplane arrangements in an abstract axiomatic setting.

We are going to concentrate on estimating the combinatorial complexity (number of faces) in arrangements and neglect all the other directions.

General probabilistic techniques for bounding the complexity of geometric configurations constitute the second main theme of this chapter. These methods have been successful in attacking many more problems than can even be mentioned in this book. We begin with a simple but powerful sampling argument in Section 6.3 (somewhat resembling the proof of the crossing number theorem), add more tricks in Section 6.4, and finish with quite a sophisticated method, demonstrated on a construction of optimal $\frac{1}{r}$-cuttings, in Section 6.5.

6.1 Arrangements of Hyperplanes

We recall from Section 4.1 that for a finite set H of lines in the plane, the arrangement of H is a partition of the plane into relatively open convex subsets, the *faces* of the arrangement. In this particular case, the faces are the vertices (0-faces), the edges (1-faces), and the cells (2-faces).[1]

An arrangement of a finite set H of hyperplanes in $\mathbf{R}^d$ is again a partition of $\mathbf{R}^d$ into relatively open convex faces. Their dimensions are 0 through d. As in the plane, the 0-faces are called vertices, the 1-faces edges, and the d-faces *cells*. Sometimes the $(d-1)$-faces are referred to as *facets*.

The cells are the connected components of $\mathbf{R}^d \setminus \bigcup H$. To obtain the facets, we consider the $(d-1)$-dimensional arrangements induced in the hyperplanes of H by their intersections with the other hyperplanes. That is, for each $h \in H$ we take the connected components of $h \setminus \bigcup_{h' \in H: h' \neq h} h'$. To obtain k-faces, we consider every possible k-flat L defined as the intersection of some $d-k$ hyperplanes of H. The k-faces of the arrangement lying within L are the connected components of $L \setminus \bigcup(H \setminus H_L)$, where $H_L = \{h \in H: L \subseteq h\}$.

Remark on sign vectors. A face of the arrangement of H can be described by its *sign vector*. First we need to fix the *orientation* of each hyperplane $h \in H$. Each $h \in H$ partitions $\mathbf{R}^d$ into three regions: h itself and the two open half-spaces determined by it. We choose one of these open half-spaces as positive and denote it by $h^\oplus$, and we let the other one be negative, denoted by $h^\ominus$.

Let F be a face of the arrangement of H. We define the *sign vector* of F (with respect to the chosen orientations of the hyperplanes) as $\sigma(F) = (\sigma_h: h \in H)$, where

$$\sigma_h = \begin{cases} +1 & \text{if } F \subseteq h^\oplus, \\ 0 & \text{if } F \subseteq h, \\ -1 & \text{if } F \subseteq h^\ominus. \end{cases}$$

The sign vector determines the face F, since we have $F = \bigcap_{h \in H} h^{\sigma_h}$, where $h^0 = h$, $h^1 = h^\oplus$, and $h^{-1} = h^\ominus$. The following drawing shows the sign

---

[1] This terminology is not unified in the literature. What we call faces are sometimes referred to as cells (0-cells, 1-cells, and 2-cells).

vectors of the marked faces in a line arrangement. Only the signs are shown, and the positive half-planes lie above their lines.

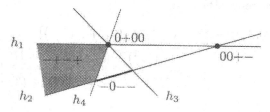

Of course, not all possible sign vectors correspond to nonempty faces. For n lines, there are 3^n sign vectors but only $O(n^2)$ faces, as we will derive below.

Counting the cells in a hyperplane arrangement. We want to count the maximum number of faces in an arrangement of n hyperplanes in $\mathbf{R}^d$. As we will see, this is much simpler than the similar task for convex polytopes!

If a set H of hyperplanes is in general position, which means that the intersection of every k hyperplanes is $(d-k)$-dimensional, $k = 2, 3, \ldots, d+1$, the arrangement of H is called *simple*. For $|H| \geq d+1$ it suffices to require that every d hyperplanes intersect at a single point and no $d+1$ have a common point.

Every d-tuple of hyperplanes in a simple arrangement determines exactly one vertex, and so a simple arrangement of n hyperplanes has exactly $\binom{n}{d}$ vertices. We now calculate the number of cells; it turns out that the order of magnitude is also n^d for d fixed.

6.1.1 Proposition. *The number of cells (d-faces) in a simple arrangement of n hyperplanes in $\mathbf{R}^d$ equals*

$$\Phi_d(n) = \binom{n}{0} + \binom{n}{1} + \cdots + \binom{n}{d}. \tag{6.1}$$

First proof. We proceed by induction on the dimension d and the number of hyperplanes n. For $d = 1$ we have a line and n points in it. These divide the line into $n+1$ one-dimensional pieces, and formula (6.1) holds. (The formula is also correct for $n = 0$ and all $d \geq 1$, since the whole space, with no hyperplanes, is a single cell.)

Now suppose that we are in dimension d, we have $n-1$ hyperplanes, and we insert another one. Since we assume general position, the $n-1$ previous hyperplanes divide the newly inserted hyperplane h into $\Phi_{d-1}(n-1)$ cells by the inductive hypothesis. Each such $(d-1)$-dimensional cell within h partitions one d-dimensional cell into exactly two new cells. The total increase in the number of cells caused by inserting h is thus $\Phi_{d-1}(n-1)$, and so

$$\Phi_d(n) = \Phi_d(n-1) + \Phi_{d-1}(n-1).$$

Together with the initial conditions (for $d = 1$ and for $n = 0$), this recurrence determines all values of Φ, and so it remains to check that formula (6.1) satisfies the recurrence. We have

$$\begin{aligned}
\Phi_d(n-1) + \Phi_{d-1}(n-1) &= \binom{n-1}{0} + \left[\binom{n-1}{1} + \binom{n-1}{0}\right] \\
&\quad + \left[\binom{n-1}{2} + \binom{n-1}{1}\right] + \cdots + \left[\binom{n-1}{d} + \binom{n-1}{d-1}\right] \\
&= \binom{n-1}{0} + \binom{n}{1} + \binom{n}{2} + \cdots + \binom{n}{d} = \Phi_d(n).
\end{aligned}$$

$\square$

Second proof. This proof looks simpler, but a complete rigorous presentation is perhaps somewhat more demanding.

We proceed by induction on d, the case $d = 0$ being trivial. Let H be a set of n hyperplanes in $\mathbf{R}^d$ in general position; in particular, we assume that no hyperplane of H is horizontal and no two vertices of the arrangement have the same vertical level (x_d-coordinate).

Let g be an auxiliary horizontal hyperplane lying below all the vertices. A cell of the arrangement of H either is bounded from below, and in this case it has a unique lowest vertex, or is not bounded from below, and then it intersects g. The number of cells of the former type is the same as the number of vertices, which is $\binom{n}{d}$. The cells of the latter type correspond to the cells in the $(d-1)$-dimensional arrangement induced within g by the hyperplanes of H, and their number is thus $\Phi_{d-1}(n)$. $\square$

What is the number of faces of the intermediate dimensions $1, 2, \ldots, d-1$ in a simple arrangement of n hyperplanes? This is not difficult to calculate using Proposition 6.1.1 (Exercise 1); the main conclusion is that the total number of faces is $O(n^d)$ for a fixed d.

What about nonsimple arrangements? It turns out that a simple arrangement of n hyperplanes maximizes the number of faces of each dimension among arrangements of n hyperplanes. This can be verified by a perturbation argument, which is considerably simpler than the one for convex polytopes (Lemma 5.5.4), and which we omit.

Bibliography and remarks. The paper of Steiner [Ste26] from 1826 gives formulas for the number of faces in arrangements of lines, circles, planes, and spheres. Of course, his results have been extended in many ways since then (see, e.g., Zaslavsky [Zas75]). An early monograph on arrangements is Grünbaum [Grü72].

The questions considered in the subsequent sections, such as the combinatorial complexity of certain parts of arrangements, have been studied mainly in the last twenty years or so. A recent survey discussing a large part of the material of this chapter and providing many more facts and references is Agarwal and Sharir [AS00a].

The algebraic and topological investigation of hyperplane arrangements (both in real and complex spaces) is reflected in the book Orlik and Terao [OT91]. Let us remark that in these areas, one usually considers *central arrangements* of hyperplanes, where all the hyperplanes pass through the origin (and so they are linear subspaces of the underlying vector space). If such a central arrangement in $\mathbf{R}^d$ is intersected with a generic hyperplane not passing through the origin, one obtains a $(d-1)$-dimensional "affine" arrangement such as those considered by us. The correspondence is bijective, and so these two views of arrangements are not very different, but for many results, the formulation with central arrangements is more elegant.

The correspondence of arrangements to zonotopes is thoroughly explained in Ziegler [Zie94].

Exercises

1. (a) Count the number of faces of dimensions 1 and 2 for a simple arrangement of n planes in $\mathbf{R}^3$. ②
    (b) Express the number of k-faces in a simple arrangement of n hyperplanes in $\mathbf{R}^d$. ②
2. Prove that the number of *unbounded* cells in an arrangement of n hyperplanes in $\mathbf{R}^d$ is $O(n^{d-1})$ (for a fixed d). ②
3. (a) Check that an arrangement of d or fewer hyperplanes in $\mathbf{R}^d$ has no bounded cell. ②
    (b) Prove that an arrangement of $d+1$ hyperplanes in general position in $\mathbf{R}^d$ has exactly one bounded cell. ③
4. How many d-dimensional cells are there in the arrangement of the $\binom{d}{2}$ hyperplanes in $\mathbf{R}^d$ with equations $\{x_i = x_j\}$, where $1 \le i < j \le d$? ③
5. How many d-dimensional cells are there in the arrangement of the hyperplanes in $\mathbf{R}^d$ with the equations $\{x_i - x_j = 0\}$, $\{x_i - x_j = 1\}$, and $\{x_i - x_j = -1\}$, where $1 \le i < j \le d$? ⑤
6. (Flags in arrangements)
    (a) Let H be a set of n lines in the plane, and let V be the set of vertices of their arrangement. Prove that the number of pairs (v, h) with $v \in V$, $h \in H$, and $v \in h$, i.e., the number of incidences $I(V, H)$, is bounded by $O(n^2)$. (Note that this is trivially true for simple arrangements.) ②
    (b) Prove that the maximum number of d-tuples $(F_0, F_1, \ldots, F_d)$ in an arrangement of n hyperplanes in $\mathbf{R}^d$, where F_i is an i-dimensional face and F_{i-1} is contained in the closure of F_i, is $O(n^d)$ (d fixed). Such d-tuples are sometimes called *flags* of the arrangement. ③
7. Let $P = \{p_1, \ldots, p_n\}$ be a point set in the plane. Let us say that points x, y have the *same view* of P if the points of P are visible in the same cyclic order from them. If rotating light rays emanate from x and from y, the points of P are lit in the same order by these rays. We assume that

neither x nor y is in P and that neither of them can see two points of P in occlusion.

(a) Show that the maximum possible number of points with mutually distinct views of P is $O(n^4)$. ☐2

(b) Show that the bound in (a) cannot be improved in general. ☐4

6.2 Arrangements of Other Geometric Objects

Arrangements can be defined not only for hyperplanes but also for other geometric objects. For example, what is the arrangement of a finite set H of segments in the plane? As in the case of lines, it is a decomposition of the plane into faces of dimension $0, 1, 2$: the vertices, the edges, and the cells, respectively. The vertices are the intersections of the segments, the edges are the portions of the segments after removing the vertices, and the cells (2-faces) are the connected components of $\mathbf{R}^2 \setminus \bigcup H$. (Note that the endpoints of the segments are *not* included among the vertices.) While the cells of line arrangements are convex polygons, those in arrangements of segments can be complicated regions, even with holes:

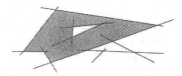

It is almost obvious that the total number of faces of the arrangement of n segments is at most $O(n^2)$. What is the maximum number of edges on the boundary of a single cell in such an arrangements? This seemingly innocuous question is surprisingly difficult, and most of Chapter 7 revolves around it.

Let us now present the definition of the arrangement for arbitrary sets $A_1, A_2, \ldots, A_n \subseteq \mathbf{R}^d$. The arrangement is a subdivision of space into connected pieces again called the *faces*. Each face is an inclusion-maximal connected set that "crosses no boundary." More precisely, first we define an equivalence relation $\approx$ on $\mathbf{R}^d$: We put $x \approx y$ whenever x and y lie in the same subcollection of the A_i, that is, whenever $\{i : x \in A_i\} = \{i : y \in A_i\}$. So for each $I \subseteq \{1, 2, \ldots, n\}$, we have one possible equivalence class, namely $\{x \in \mathbf{R}^d : x \in A_i \Leftrightarrow i \in I\}$ (this is like a field in the Venn diagram of the A_i). But in typical geometric situations, most of the classes are empty. The faces of the arrangement of the A_i are the connected components of the equivalence classes. The reader is invited to check that for both hyperplane arrangements and arrangements of segments this definition coincides with the earlier ones.

Arrangements of algebraic surfaces.   Quite often one needs to consider arrangements of the zero sets of polynomials. Let $p_1(x_1, x_2, \ldots, x_d), \ldots,$ $p_n(x_1, x_2, \ldots, x_d)$ be polynomials with real coefficients in d variables, and let $Z_i = \{x \in \mathbf{R}^d : p_i(x) = 0\}$ be the zero set of p_i. Let D denote the maximum of the degrees of the p_i; when speaking of the arrangement of $Z_1, \ldots, Z_n$, one usually assumes that D is bounded by some (small) constant. Without a bound on D, even a single Z_i can have arbitrarily many connected components.

In many cases, the Z_i are algebraic surfaces, such as ellipsoids, paraboloids, etc., but since we are in the real domain, sometimes they need not look like surfaces at all. For example, the zero set of the polynomial $p(x_1, x_2) = x_1^2 + x_2^2$ consists of the single point $(0, 0)$. Although it is sometimes convenient to think of the Z_i as surfaces, the results stated below apply to zero sets of arbitrary polynomials of bounded degree.

It is known that if both d and D are considered as constants, the maximum number of faces in the arrangement of $Z_1, Z_2, \ldots, Z_n$ as above is at most $O(n^d)$. This is one of the most useful results about arrangements, with many surprising applications (a few are outlined below and in the exercises). In the literature one often finds a (formally weaker) version dealing with *sign patterns* of the polynomials p_i. A vector $\sigma \in \{-1, 0, +1\}^n$ is called a sign pattern of $p_1, p_2, \ldots, p_n$ if there exists an $x \in \mathbf{R}^d$ such that the sign of $p_i(x)$ is σ_i, for all $i = 1, 2, \ldots, n$. Trivially, the number of sign patterns for any n polynomials is at most 3^n. For $d = 1$, it is easy to see that the actual number of sign patterns is much smaller, namely at most $2nD + 1$ (Exercise 1). It is not so easy to prove, but still true, that there are at most $C(d, D) \cdot n^d$ sign patterns in dimension d. This result is generally called the *Milnor–Thom theorem* (and it was apparently first proved by Oleinik and Petrovskiĭ, which fits the usual pattern in the history of mathematics). Here is a more precise (and more recent) version of this result, where the dependence on D and d is specified quite precisely.

6.2.1 Theorem (Number of sign patterns). *Let $p_1, p_2, \ldots, p_n$ be d-variate real polynomials of degree at most D. The number of faces in the arrangement of their zero sets $Z_1, Z_2, \ldots, Z_n \subseteq \mathbf{R}^d$, and consequently the number of sign patterns of $p_1, \ldots, p_n$ as well is at most $2(2D)^d \sum_{i=0}^d 2^i \binom{4n+1}{i}$. For $n \geq d \geq 2$, this expression is bounded by*

$$\left(\frac{50Dn}{d} \right)^d .$$

Proofs of these results are not included here because they would require at least one more chapter. They belong to the field of *real algebraic geometry*. The classical, deep, and extremely extensive field of *algebraic geometry* mostly studies algebraic varieties over algebraically closed fields, such as the complex numbers (and the questions of combinatorial complexity in our sense are not among its main interests). Real algebraic geometry investigates algebraic varieties and related concepts over the real numbers or other real-closed fields; the presence of ordering and the missing roots of polynomials makes its flavor distinctly different.

Arrangements of pseudolines. An arrangement of pseudolines is a natural generalization of an arrangement of lines. Lines are replaced by curves, but we insist that these curves behave, in a suitable sense, like lines: For example, no two of them intersect more than once. This kind of generalization is quite different from, say, arrangements of planar algebraic curves, and so it perhaps does not quite belong to the present section. But besides mentioning pseudoline arrangements as a useful and interesting concept, we also need them for a (typical) example of application of Theorem 6.2.1, and so we kill two birds with one stone by discussing them here.

An *(affine) arrangement of pseudolines* can be defined as the arrangement of a finite collection of curves in the plane that satisfy the following conditions:

(i) Each curve is x-monotone and unbounded in both directions; in other words, it intersects each vertical line in exactly one point.

(ii) Every two of the curves intersect in exactly one point and they cross at the intersection. (We do not permit "parallel" pseudolines, for they would complicate the definition unnecessarily.)[2]

The curves are called *pseudolines*, but while "being a line" is an absolute notion, "being a pseudoline" makes sense only with respect to a given collection of curves.

Here is an example of a (simple) arrangement of 5 pseudolines:

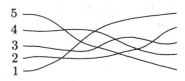

Much of what we have proved for arrangements of lines is true for arrangements of pseudolines as well. This holds for the maximum number of vertices, edges, and cells, but also for more sophisticated results like the Szemerédi–Trotter theorem on the maximum number of incidences of m points and n lines; these results have proofs that do not use any properties of straight lines not shared by pseudolines.

One might be tempted to say that pseudolines are curves that behave topologically like lines, but as we will see below, in at least one sense this is

---

[2] This "affine" definition is a little artificial, and we use it only because we do not want to assume the reader's familiarity with the topology of the projective plane. In the literature one usually considers arrangements of pseudolines in the projective plane, where the definition is very natural: Each pseudoline is a closed curve whose removal does not disconnect the projective plane, and every two pseudolines intersect exactly once (which already implies that they cross at the intersection point). Moreover, one often adds the condition that the curves do not form a single *pencil*; i.e., not all of them have a common point, since otherwise, one would have to exclude the case of a pencil in the formulation of many theorems. But here we are not going to study pseudoline arrangements in any depth.

profoundly wrong. The correct statement is that every two of them behave topologically like two lines, but arrangements of pseudolines are more general than arrangements of lines.

We should first point out that there is no problem with the "local" structure of the pseudolines, since each pseudoline arrangement can be redrawn equivalently (in a sense defined precisely below) by polygonal lines, as a *wiring diagram*:

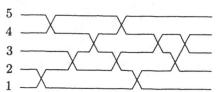

The difference between pseudoline arrangements and line arrangements is of a more global nature.

The arrangement of 5 pseudolines drawn above can be realized by straight lines:

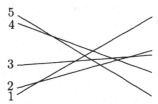

What is the meaning of "realization by straight lines"? To this end, we need a suitable notion of equivalence of two arrangements of pseudolines. There are several technically different possibilities; we again use an "affine" notion, one that is very simple to state but not the most common. Let H be a collection of n pseudolines. We number the pseudolines $1, 2, \ldots, n$ in the order in which they appear on the left of the arrangement, say from the bottom to the top. For each i, we write down the numbers of the other pseudolines in the order they are encountered along the pseudoline i from left to right. For a simple arrangement we obtain a permutation π_i of $\{1, 2, \ldots, n\} \setminus \{i\}$ for each i. For the arrangement in the pictures, we have $\pi_1 = (2, 3, 5, 4)$, $\pi_2 = (1, 5, 4, 3)$, $\pi_3 = (1, 5, 4, 2)$, $\pi_4 = (5, 1, 3, 2)$, and $\pi_5 = (4, 1, 3, 2)$. For a nonsimple arrangement, some of the π_i are linear *quasiorderings*, meaning that several consecutive numbers can be chunked together. We call two arrangements *affinely isomorphic* if they yield the same $\pi_1, \ldots, \pi_n$, i.e., if each pseudoline meets the others in the same (quasi)order as the corresponding pseudoline in the other arrangement. Two affinely isomorphic pseudoline arrangements can be converted one to another by a suitable homeomorphism of the plane.[3]

---

[3] The more usual notion of *isomorphism* of pseudoline arrangements is defined for arrangements in the projective plane. The arrangement of H is isomorphic to the

An arrangement of pseudolines is *stretchable* if it is affinely isomorphic to an arrangement of straight lines.[4] It turns out that all arrangements of 8 or fewer pseudolines are stretchable, but there exists a nonstretchable arrangement of 9 pseudolines:

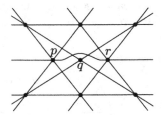

The proof of nonstretchability is based on the *Pappus theorem* in projective geometry, which states that if 8 straight lines intersect as in the drawing, then the points p, q, and r are collinear. By modifying this arrangement suitably, one can obtain a simple nonstretchable arrangement of 9 pseudolines as well.

Next, we show that most of the simple pseudoline arrangements are nonstretchable. The following construction shows that the number of isomorphism classes of simple arrangements of n pseudolines is at least $2^{\Omega(n^2)}$:

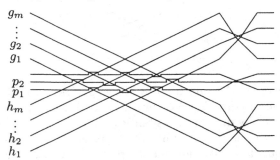

We have $m \approx \frac{n}{3}$, and the lines $h_1, \ldots, h_m$ and $g_1, \ldots, g_m$ form a regular grid. Each of the about $\frac{n}{3}$ pseudolines p_i in the middle passes near $\Omega(n)$ vertices of

arrangement of H' if there exists a homeomorphism φ of the projective plane onto itself such that each pseudoline $h \in H$ is mapped to a pseudoline $\varphi(h) \in H'$. For affinely isomorphic arrangements in the affine plane, the corresponding arrangements in the projective plane are isomorphic, but the isomorphism in the projective plane also allows for mirror reflection and for "relocating the infinity." Combinatorially, the isomorphism in the projective plane can be described using the (quasi)orderings $\pi_1, \ldots, \pi_n$ as well. Here the π_i have to agree only up to a possible reversal and cyclic shift for each i, and also the numbering of the pseudolines by $1, 2, \ldots, n$ is not canonical.

We also remark that two arrangements of *lines* are isomorphic if and only if the dual point configurations have the same order type, up to a mirror reflection of the whole configuration (order types are discussed in Section 9.3).

[4] For isomorphism in the projective plane, one gets an equivalent notion of stretchability.

this grid, and for each such vertex it has a choice of going below it or above. This gives $2^{\Omega(n^2)}$ possibilities in total.

Now we use Theorem 6.2.1 to estimate the number of nonisomorphic simple arrangements of n straight lines. Let the lines be $\ell_1, \ldots, \ell_n$, where ℓ_i has the equation $y = a_i x + b_i$ and $a_1 > a_2 > \cdots > a_n$. The x-coordinate of the intersection $\ell_i \cap \ell_j$ is $\frac{b_i - b_j}{a_j - a_i}$. To determine the ordering π_i of the intersections along ℓ_i, it suffices to know the ordering of the x-coordinates of these intersections, and this can be inferred from the signs of the polynomials $p_{ijk}(a_i, b_i, a_j, b_j, a_k, b_k) = (b_i - b_j)(a_k - a_i) - (b_i - b_k)(a_j - a_i)$. So the number of nonisomorphic arrangements of n lines is no larger than the number of possible sign patterns of the $O(n^3)$ polynomials p_{ijk} in the $2n$ variables $a_1, b_1, \ldots, a_n, b_n$, and Theorem 6.2.1 yields the upper bound of $2^{O(n \log n)}$. For large n, this is a negligible fraction of the total number of simple pseudoline arrangements. (Similar considerations apply to nonsimple arrangements as well.)

The problem of deciding the stretchability of a given pseudoline arrangement has been shown to be algorithmically difficult (at least NP-hard). One can easily encounter this problem when thinking about line arrangements and drawing pictures: What we draw by hand are really pseudolines, not lines, and even with the help of a ruler it may be almost impossible to decide experimentally whether a given arrangement can really be drawn with straight lines. But there are computational methods that can decide stretchability in reasonable time at least for moderate numbers of lines.

Bibliography and remarks. A comprehensive account of real algebraic geometry is Bochnak, Coste, and Roy [BCR98]. Among the many available introductions to the "classical" algebraic geometry we mention the lively book Cox, Little, and O'Shea [CLO92].

The original bounds on the number of sign patterns, less precise than Theorem 6.2.1 but still implying the $O(n^d)$ bound for fixed d, were given independently by Oleinik and Petrovskiĭ [OP49], Milnor [Mil64], and Thom [Tho65]. Warren [War68] proved that the number of d-dimensional cells in the arrangement as in Theorem 6.2.1, and consequently the number of sign patterns consisting of ± 1's only, is at most $2(2D)^d \sum_{i=0}^{d} 2^i \binom{n}{i}$. The extension to faces of all dimensions, and to sign patterns including 0's, was obtained by Pollack and Roy [PR93].

Sometimes we have polynomials in many variables, but we are interested only in sign patterns attained at points that satisfy some additional algebraic conditions. Such a situation is covered by a result of Basu, Pollack, and Roy [BPR96]: The number of sign patterns attained by n polynomials of degree at most D on a k-dimensional algebraic variety $V \subseteq \mathbf{R}^d$, where V can be defined by polynomials of degree at most D, is at most $\binom{n}{k} O(D)^d$.

While bounding the number of sign patterns of multivariate polynomials appears complicated, there is a beautiful short proof of an almost tight bound on the number of *zero patterns*, due to Rónyai, Babai, and Ganapathy [RBG01], which we now sketch (in the simplest form, giving a slightly suboptimal result). A vector $\zeta \in \{0,1\}^n$ is a zero pattern of d-variate polynomials $p_1, \ldots, p_n$ with coefficients in a field F if there exists an $x = x(\zeta) \in F^d$ with $p_i(x) = 0$ exactly for the i with $\zeta_i = 0$. We show that if all the p_i have degree at most D, then the number of zero patterns cannot exceed $\binom{Dn+d}{d}$. For each zero pattern ζ, let q_ζ be the polynomial $\prod_{i:\, \zeta_i \neq 0} p_i$. We have $\deg q_\zeta \leq Dn$. Let us consider the q_ζ as elements of the vector space L of all d-variate polynomials over F of degree at most Dn. Using the basis of L consisting of all monomials of degree at most Dn, we obtain $\dim L \leq \binom{Dn+d}{d}$. It remains to verify that the q_ζ are linearly independent (assuming that no p_i is identically 0). Suppose that $\sum_\zeta \alpha_\zeta q_\zeta = 0$ with $\alpha_\zeta \in F$ not all 0. Choose a zero pattern ξ with $\alpha_\xi \neq 0$ and with the largest possible number of 0's, and substitute $x(\xi)$ into $\sum_\zeta \alpha_\zeta q_\zeta$. This yields $\alpha_\xi = 0$, a contradiction.

Pseudoline arrangements. The founding paper is Levi [Lev26], where, among others, the nonstretchable arrangement of 9 lines drawn above was presented. A concise survey was written by Goodman [Goo97].

Pseudoline arrangements, besides being very natural, have also turned out to be a fruitful generalization of line arrangements. Some problems concerning line arrangements or point configurations were first solved only in the more general setting of pseudoline arrangements, and certain algorithms for line arrangements, the so-called *topological sweep* methods, use an auxiliary pseudoline to speed up the computation; see [Goo97].

Infinite families of pseudolines have been considered as well, and even *topological planes*, which are analogues of the projective plane but made of pseudolines. It is known that every finite configuration of pseudolines can be extended to a topological plane, and there are uncountably many distinct topological planes; see Goodman, Pollack, Wenger, and Zamfirescu [GPWZ94].

Oriented matroids. The possibility of representing each pseudoline arrangement by a wiring diagram makes it clear that a pseudoline arrangement can also be considered as a purely combinatorial object. The appropriate combinatorial counterpart of a pseudoline arrangement is called an oriented matroid of rank 3. More generally, similar to arrangements of pseudolines, one can define arrangements of *pseudohyperplanes* in $\mathbf{R}^d$, and these are combinatorially captured by oriented matroids of rank $d+1$. Here the rank is one higher than the space dimension, because an oriented matroid of rank d is usually viewed as a combinatorial abstraction of a *central* arrangement of hyperplanes in $\mathbf{R}^d$ (with all hyperplanes passing through 0).

There are several different but equivalent definitions of an oriented matroid. We present a definition in the so-called *covector form*. An *oriented matroid* is a set $\mathcal{V}^* \subseteq \{-1, 0, 1\}^n$ that is symmetric ($v \in \mathcal{V}^*$ implies $-v \in \mathcal{V}^*$), contains the zero vector, and satisfies the following two more complicated conditions:

- (Closed under composition) If $u, v \in \mathcal{V}^*$, then $u \circ v \in \mathcal{V}^*$, where $(u \circ v)_i = u_i$ if $u_i \neq 0$ and $(u \circ v)_i = v_i$ if $u_i = 0$.

- (Admits elimination) If $u, v \in \mathcal{V}^*$ and $j \in S(u, v) = \{i: u_i = -v_i \neq 0\}$, then there exists $w \in \mathcal{V}^*$ such that $w_j = 0$ and $w_i = (u \circ v)_i$ for all $i \notin S(u, v)$.

The *rank* of an oriented matroid $\mathcal{V}^*$ is the largest r such that there is an increasing chain $v_1 \prec v_2 \prec \cdots \prec v_r$, $v_i \in \mathcal{V}^*$, where $u \preceq v$ means $u_i \preceq v_i$ for all i and where $0 \prec 1$ and $0 \prec -1$. At first sight, all this may look quite mysterious, but it becomes much clearer if one thinks of a basic example, where $\mathcal{V}^*$ is the set of sign vectors of all faces of a central arrangement of hyperplanes in $\mathbf{R}^d$.

It turns out that every oriented matroid of rank 3 corresponds to an arrangement of pseudolines. More generally, *Lawrence's representation theorem* asserts that every oriented matroid of rank d comes from some central arrangement of pseudohyperplanes in $\mathbf{R}^d$, and so the purely combinatorial notion of oriented matroid corresponds, essentially uniquely, to the topological notion of a (central) arrangement of pseudohyperplanes.[5]

Oriented matroids are also naturally obtained from configurations of points or vectors. In the notation of Section 5.6 (Gale transform), if $\bar{a}$ is a sequence of n vectors in $\mathbf{R}^r$, then both the sets $\mathrm{sgn}(\mathsf{LinVal}(\bar{a}))$ and $\mathrm{sgn}(\mathsf{LinDep}(\bar{a}))$ are oriented matroids in the sense of the above definition. The first one has rank r, and the second, rank $n-r$.

We are not going to say much more about oriented matroids, referring to Ziegler [Zie94] for a quick introduction and to Björner, Las Vergnas, Sturmfels, White, and Ziegler [BVS+99] for a comprehensive account.

Stretchability. The following results illustrate the surprising difficulty of the stretchability problem for pseudoline arrangements. They are analogous to the statements about realizability of 4-dimensional convex polytopes mentioned in Section 5.3, and they were actually found much earlier.

---

[5] The correspondence need not really be one-to-one. For example, the oriented matroids of two projectively isomorphic pseudoline arrangements agree only up to reorientation.

Certain (simple) stretchable arrangements of n pseudolines require coefficients with $2^{\Omega(n)}$ digits in the equations of the lines, in every straight-line realization (Goodman, Pollack, and Sturmfels [GPS90]). Deciding the stretchability of a given pseudoline arrangement is NP-hard (Shor [Sho91] has a relatively simple proof), and in fact, it is polynomially equivalent to the problem of solvability of a system of polynomial inequalities with integer coefficients. This follows from results of Mnëv, published in Russian in 1985 (proofs were only sketched; see [Mne89] for an English version). This work went unnoticed in the West for some time, and so some of the results were rediscovered by other authors.

Although detailed proofs of such theorems are technically demanding, the principle is rather simple. Given two real numbers, suitably represented by geometric quantities, one can produce their sum and their product by classical geometric constructions by ruler. (Since ruler constructions are invariant under projective transformations, the numbers are represented as cross-ratios.) By composing such constructions, one can express the solvability of $p(x_1, \ldots, x_n) = 0$, for a given n-variate polynomial p with integer coefficients, by the stretchability of a suitable arrangement in the projective plane. Dealing with inequalities and passing to simple arrangements is somewhat more complicated, but the idea is similar.

Practical algorithms for deciding stretchability have been studied extensively by Bokowski and Sturmfels [BS89] and by Richter-Gebert (see, e.g., [RG99]).

Mnëv [Mne89] was mainly interested in the *realization spaces* of arrangements. Let H be a fixed stretchable arrangement. Each straight-line arrangement H' affinely isomorphic to H can be represented by a point in $\mathbf{R}^{2n}$, with the $2n$ coordinates specifying the coefficients in the equations of the lines of H'. Considering all possible H' for a given H, we obtain a subset of $\mathbf{R}^{2n}$. For some time it was conjectured that this set, the realization space of H, has to be path-connected, which would mean that one straight-line realization could be converted to any other by a continuous motion while retaining the affine isomorphism type.[6] Not only is this false, but the realization space can have arbitrarily many components. In a suitable sense, it can even have arbitrary topological type. Whenever $A \subseteq \mathbf{R}^n$ is a set definable by a formula involving finitely many polynomial inequalities with integer coefficients, Boolean connectives, and quantifiers, there is a line arrangement whose realization space S is homotopy equivalent to A (Mnëv's main result actually talks about the stronger notion of *sta-*

---

[6] In fact, these questions have been studied mainly for the isomorphism of arrangements in the projective plane. There one has to be a little careful, since a mirror reflection can easily make the realization space disconnected, and so the mirror reflection (or the whole action of the general linear group) is factored out first.

ble equivalence of S and A; see, e.g., [Goo97] or [BVS$^+$99]). Similar
theorems were proved by Richter-Gebert for the realization spaces of
4-dimensional polytopes [RG99], [RG97].

These results for arrangements and polytopes can be regarded as
instances of a vague but probably quite general principle: *"Almost
none of the combinatorially imaginable geometric configurations are
geometrically realizable, and it is difficult to decide which ones are."*
Of course, there are exceptions, such as the graphs of 3-dimensional
convex polytopes.

Encoding pseudoline arrangements. The lower bound $2^{\Omega(n^2)}$ for the
number of isomorphism classes of pseudoline arrangements is asymptotically tight. Felsner [Fel97] found a nice encoding of such an arrangement by an $n \times n$ matrix of 0's and 1's, from which the isomorphism
type can be reconstructed: The entry (i, j) of the matrix is 1 iff the jth
leftmost crossing along the pseudoline number i is with a pseudoline
whose number k is larger than i.

Exercises

1. Let $p_1(x), \ldots, p_n(x)$ be univariate real polynomials of degree at most D.
   Check that the number of sign patterns of the p_i is at most $2nD+1$. [2]
2. (Intersection graphs) Let S be a set of n line segments in the plane. The
   intersection graph of S is the graph on n vertices, which correspond to
   the segments of S, with two vertices connected by an edge if and only if
   the corresponding two segments intersect.
   (a) Prove that the graph obtained from K_5 by subdividing each edge
   exactly once is not the intersection graph of segments in the plane (and
   not even the intersection graph of any arcwise connected sets in the
   plane). [4]
   (b) Use Theorem 6.2.1 to prove that *most* graphs are not intersection
   graphs of segments: While the total number of graphs on n given vertices
   is $2^{\binom{n}{2}} = 2^{n^2/2+O(n)}$, only $2^{O(n \log n)}$ of them are intersection graphs of
   segments (be careful about collinear segments!). [3]
   (c) Show that the number of (isomorphism classes of) intersection graphs
   of planar arcwise connected sets, and even of planar convex sets, on n
   vertices *cannot* be bounded by $2^{O(n \log n)}$. (The right order of magnitude
   does not seem to be known for either of these classes of intersection
   graphs.) [4]
3. (Number of combinatorially distinct simplicial convex polytopes) Use
   Theorem 6.2.1 to prove that for every dimension $d \geq 3$ there exists $C_d > 0$
   such that the number of combinatorial types of *simplicial* polytopes in
   $\mathbf{R}^d$ with n vertices is at most $2^{C_d n \log n}$. (The combinatorial equivalence
   means isomorphic face lattices; see Definition 5.3.4.) [4]

Such a result was proved by Alon [Alo86b] and by Goodman and Pollack [GP86].

4. (Sign patterns of matrices and rank) Let A be a real $n \times n$ matrix. The *sign matrix* $\sigma(A)$ is the $n \times n$ matrix with entries in $\{-1, 0, +1\}$ given by the signs of the corresponding entries in A.

(a) Check that A has rank at most q if and only if there exist $n \times q$ matrices U and V with $A = UV^T$. [3]

(b) Estimate the number of distinct sign matrices of rank q using Theorem 6.2.1, and conclude that there exists an $n \times n$ matrix S containing only entries $+1$ and -1 such that any real matrix A with $\sigma(A) = S$ has rank at least cn, with a suitable constant $c > 0$. [3]

The result in (b) is from Alon, Frankl, and Rödl [AFR85] (for another application see [Mat96b]).

5. (Extendible pseudosegments) A *family of pseudosegments* is a finite collection $S = \{s_1, s_2, \ldots, s_n\}$ of curves in the plane such that each s_i is x-monotone and its vertical projection on the x-axis is a closed interval, every two curves in the family intersect at most once, and whenever they intersect they cross (tangential contacts are not allowed). Such an S is called *extendible* if there is a family $L = \{\ell_1, \ldots, \ell_n\}$ of pseudolines such that $s_i \subseteq \ell_i$, $i = 1, 2, \ldots, n$.

(a) Find an example of a nonextendible family of 3 pseudosegments. [1]

(b) Define an oriented graph G with vertex set S and with an edge from s_i to s_j if $s_i \cap s_j \neq \emptyset$ and s_i is below s_j on the left of their intersection. Check that if S is extendible, then G is acyclic. [1]

(c) Prove that, conversely, if G is acyclic, then S is extendible. Extend the pseudosegments one by one, maintaining the acyclicity of G. [3]

(d) Let I_i be the projection of s_i on the x-axis. Show that if for every $i < j$, $I_i \cap I_j = \emptyset$ or $I_i \subseteq I_j$ or $I_j \subseteq I_i$, then G is acyclic, and hence S is extendible. [2]

(e) Given a family of closed intervals $I_1, \ldots, I_n \subseteq \mathbf{R}$, show that each interval in the family can be partitioned into at most $O(\log n)$ subintervals in such a way that the resulting family of subintervals has the property as in (d). This implies that an arbitrary family of n pseudosegments can be cut into a family of $O(n \log n)$ extendible pseudosegments. [3]

These notions and results are from Chan [Cha00a].

6.3 Number of Vertices of Level at Most k

In this section and the next one we investigate the maximum number of faces in certain naturally defined portions of hyperplane arrangements. We consider only simple arrangements, and we omit the (usually routine) perturbation arguments showing that simple arrangements maximize the investigated quantity.

Let H be a finite set of hyperplanes in $\mathbf{R}^d$, and assume that none of them is vertical, i.e., parallel to the x_d-axis. The *level* of a point $x \in \mathbf{R}^d$ is the number of hyperplanes of H lying strictly below x (the hyperplanes passing through x, if any, are not counted). This extends the definition for lines from Section 4.7.

We are interested in the maximum possible number of vertices of level at most k in a simple arrangement of n hyperplanes. The following drawing shows the region of all points of level at most 2 in an arrangement of lines; we want to count the vertices lying in the region or on its boundary.

The vertices of level 0 are the vertices of the cell lying below all the hyperplanes, and since this cell is the intersection of at most n half-spaces, it has at most $O(n^{\lfloor d/2 \rfloor})$ vertices, by the asymptotic upper bound theorem (Theorem 5.5.2). From this result we derive a bound on the maximum number of vertices of level at most k. The elegant probabilistic technique used in the proof is generally applicable and probably more important than the particular result itself.

6.3.1 Theorem (Clarkson's theorem on levels). *The total number of vertices of level at most k in an arrangement of n hyperplanes in $\mathbf{R}^d$ is at most*

$$O(n^{\lfloor d/2 \rfloor}(k+1)^{\lceil d/2 \rceil}),$$

with the constant of proportionality depending on d.

We are going to prove the theorem for *simple* arrangements only. The general case can be derived from the result for simple arrangements by a standard perturbation argument. But let us stress that the simplicity of the arrangement is essential for the forthcoming proof.

For all k ($0 \leq k \leq n - d$), the bound is tight in the worst case. To see this for $k \geq 1$, consider a set of $\frac{n}{k}$ hyperplanes such that the lower unbounded cell in their arrangement is a convex polyhedron with $\Omega((\frac{n}{k})^{\lfloor d/2 \rfloor})$ vertices, and replace each of the hyperplanes by k very close parallel hyperplanes. Then each vertex of level 0 in the original arrangement gives rise to $\Omega(k^d)$ vertices of level at most k in the new arrangement.

A much more challenging problem is to estimate the maximum possible number of vertices of level *exactly* k. This will be discussed in Chapter 11.

One of the main motivations that led to Clarkson's theorem on levels was an algorithmic problem. Given an n-point set $P \subset \mathbf{R}^d$, we want to construct a data structure for fast answering of queries of the following type: For a query point $x \in \mathbf{R}^d$ and an integer t, report the t points of P that lie nearest to x.

Clarkson's theorem on levels is needed for bounding the maximum amount of memory used by a certain efficient algorithm. The connection is not entirely simple. It uses the lifting transform described in Section 5.7, relating the algorithmic problem in $\mathbf{R}^d$ to the complexity of levels in $\mathbf{R}^{d+1}$, and we do not discuss it here.

Proof of Theorem 6.3.1 for $d = 2$. First we demonstrate this special case, for which the calculations are somewhat simpler.

Let H be a set of n lines in general position in the plane. Let p denote a certain suitable number in the interval $(0, 1)$ whose value will be determined at the end of the proof. Let us imagine the following random experiment. We choose a subset $R \subseteq H$ at random, by including each line $h \in H$ into R with probability p, the choices being independent for distinct lines h.

Let us consider the arrangement of R, temporarily discarding all the other lines, and let $f(R)$ denote the number of vertices of level 0 in the arrangement of R. Since R is random, f is a random variable. We estimate the expectation of f, denoted by $\mathbf{E}[f]$, in two ways.

First, we have $f(R) \leq |R|$ for any specific set R, and hence $\mathbf{E}[f] \leq \mathbf{E}[|R|] = pn$.

Now we estimate $\mathbf{E}[f]$ differently: We bound it from below using the number of vertices of the arrangement of H of level at most k. For each vertex v of the arrangement of H, we define an event A_v meaning "v becomes one of the vertices of level 0 in the arrangement of R." That is, A_v occurs if v contributes 1 to the value of f. The event A_v occurs if and only if the following two conditions are satisfied:

- Both lines determining the vertex v lie in R.
- None of the lines of H lying below v falls into R.

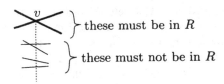

We deduce that $\mathrm{Prob}[A_v] = p^2(1 - p)^{\ell(v)}$, where $\ell(v)$ denotes the level of the vertex v.

Let V be the set of all vertices of the arrangement of H, and let $V_{\leq k} \subseteq V$ be the set of vertices of level at most k, whose cardinality we want to estimate. We have

$$\mathbf{E}[f] = \sum_{v \in V} \mathrm{Prob}[A_v] \geq \sum_{v \in V_{\leq k}} \mathrm{Prob}[A_v]$$

$$= \sum_{v \in V_{\leq k}} p^2(1-p)^{\ell(v)} \geq \sum_{v \in V_{\leq k}} p^2(1-p)^k = |V_{\leq k}| \cdot p^2(1-p)^k.$$

Altogether we have derived $np \geq \mathbf{E}[f] \geq |V_{\leq k}| \cdot p^2(1-p)^k$, and so

$$|V_{\leq k}| \leq \frac{n}{p(1-p)^k}.$$

Let us now choose the number p so as to minimize the right-hand side. A convenient value is $p = \frac{1}{k+1}$; it does not yield the exact minimum, but it comes close. We have $\left(1 - \frac{1}{k+1}\right)^k \geq e^{-1} > \frac{1}{3}$ for all $k \geq 1$. This leads to $|V_{\leq k}| \leq 3(k+1)n$. $\square$

Proof for an arbitrary dimension. The idea of the proof is the same as above. As for the technical realization, there are at least two possible routes. The first is to retain the same probability distribution for selecting the sample R (picking each hyperplane of the given set H independently with probability p); in this case, most of the proof remains as before, but we need a lemma showing that $\mathbf{E}\big[|R|^{\lfloor d/2 \rfloor}\big] = O((pn)^{\lfloor d/2 \rfloor})$. This is not difficult to prove, either from a Chernoff-type inequality or by elementary calculations (see Exercises 6.5.2 and 6.5.3).

The second possibility, which we use here, is to change the probability distribution. Namely, we define an integer parameter r and choose a random r-element subset $R \subseteq H$, with all the $\binom{n}{r}$ subsets being equally probable.

With this new way of choosing R, we proceed as in the proof for $d = 2$. We define $f(R)$ as the number of vertices of level 0 in the arrangement of R and estimate $\mathbf{E}[f]$ in two ways. On the one hand, we have $f(R) = O(r^{\lfloor d/2 \rfloor})$ for all R, and so

$$\mathbf{E}[f] = O(r^{\lfloor d/2 \rfloor}).$$

The notation V for the set of all vertices of the arrangement of H, $V_{\leq k}$ for the vertices of level at most k, and A_v for the event "v is a vertex of level 0 in the arrangement of R," is as in the previous proof. The conditions for A_v are

- All the d hyperplanes defining the vertex v fall into R.
- None of the hyperplanes of H lying below v fall into R.

So if $\ell = \ell(v)$ is the level of v, then

$$\text{Prob}[A_v] = \frac{\binom{n-d-\ell}{r-d}}{\binom{n}{r}}.$$

For brevity, we denote this quantity by $P(\ell)$. We note that it is a *decreasing* function of ℓ. Therefore,

$$\mathbf{E}[f] = \sum_{v \in V} \text{Prob}[A_v] \geq |V_{\leq k}| \cdot P(k).$$

Combining with $\mathbf{E}[f] = O(r^{\lfloor d/2 \rfloor})$ derived earlier, we obtain

$$|V_{\leq k}| \leq \frac{O(r^{\lfloor d/2 \rfloor})}{P(k)}. \tag{6.2}$$

An appropriate value for the parameter r is $r = \lfloor \frac{n}{k+1} \rfloor$. (This is not surprising, since in the previous proof, the size of R was concentrated around $pn = \frac{n}{k+1}$.) Then we have the following estimate:

6.3.2 Lemma. *Suppose that $1 \leq k \leq \frac{n}{2d} - 1$, which implies $2d \leq r \leq \frac{n}{2}$. Then*

$$P(k) \geq c_d(k+1)^{-d}$$

for a suitable $c_d > 0$ depending only on d.

We postpone the proof of the lemma a little and finish the proof of Theorem 6.3.1. We want to substitute the bound from the lemma into (6.2). In order to meet the assumptions of the lemma, we must restrict the range of k somewhat. But if, say, $k \geq \frac{n}{2d}$, then the bound claimed by the theorem is of order n^d and thus trivial, and for $k = 0$ we already know that the theorem holds. So we may assume $1 \leq k \leq \frac{n}{2d} - 1$, and we have

$$|V_{\leq k}| \leq O(r^{\lfloor d/2 \rfloor})(k + 1)^{-d} = O\left(n^{\lfloor d/2 \rfloor}(k + 1)^{\lceil d/2 \rceil}\right).$$

This establishes the theorem.   □

Proof of Lemma 6.3.2.

$$\begin{aligned}
P(k) &= \frac{\binom{n-d-k}{r-d}}{\binom{n}{r}} \\
&= \frac{(n-d-k)(n-d-k-1)\cdots(n-k-r+1)}{n(n-1)\cdots(n-r+1)} \cdot r(r-1)\cdots(r-d+1) \\
&= \frac{r(r-1)\cdots(r-d+1)}{n(n-1)\cdots(n-d+1)} \cdot \frac{n-d-k}{n-d} \cdot \frac{n-d-k-1}{n-d-1} \cdots \frac{n-k-r+1}{n-r+1} \\
&\geq \left(\frac{r}{2n}\right)^d \left(1 - \frac{k}{n-d}\right)\left(1 - \frac{k}{n-d-1}\right) \cdots \left(1 - \frac{k}{n-r+1}\right) \\
&\geq \left(\frac{r}{2n}\right)^d \left(1 - \frac{k}{n-r+1}\right)^r.
\end{aligned}$$

Now, $\frac{r}{n} \geq (\frac{n}{k+1} - 1)/n \geq \frac{1}{2(k+1)}$ (since $k < \frac{n}{2}$, say) and $1 - \frac{k}{n-r+1} \geq 1 - \frac{2k}{n}$ (a somewhat finer calculation actually gives $1 - \frac{k+1}{n}$ here). Since $k \leq \frac{n}{4}$, we can use the inequality $1-x \geq e^{-2x}$ valid for $x \in [0, \frac{1}{2}]$, and we arrive at

$$P(k) \geq (\tfrac{r}{dn})^d e^{-4kr/n} \geq c_d(k + 1)^{-d}.$$

Lemma 6.3.2 is proved.   □

Levels in arrangements. Besides vertices, we can consider all faces of level at most k, where the level of a face is the (common) level of all of its points. Using Theorem 6.3.1, it is not hard to prove that the number of all faces of level at most k in an arrangement of n hyperplanes is $O(n^{\lfloor d/2 \rfloor}(k+1)^{\lceil d/2 \rceil})$.

In the literature one often speaks about the *level* k in an arrangement of hyperplanes, meaning the boundary of the region of all points of level at most k. This is a polyhedral surface and each vertical line intersects it in exactly one point. It is a subcomplex of the arrangement; note that it may also contain faces of level different from k. In Section 4.7 we considered such levels in arrangements of lines.

Bibliography and remarks. Clarkson's theorem on levels was first proved in Clarkson [Cla88a] (see Clarkson and Shor [CS89] for the journal version). The elegant proof technique has many other applications, and we will meet it several more times, combined with additional tricks into sophisticated arguments. The theorem can be formulated in an abstract framework outlined in the notes to Section 6.5. New variations on the basic method were noted by Sharir [Sha01] (see Exercises 4 and 5).

In the planar case, the $O(nk)$ bound on the complexity of levels 0 through k was known before Clarkson's paper, apparently first proved by Goodman and Pollack [GP84]. Alon and Győri [AG86] determined the exact constant of proportionality (which Clarkson's proof in the present form cannot provide). Welzl [Wel01] proved an exact upper bound in $\mathbf{R}^3$; see the notes to Section 11.3 for a little more about his method. Several other related references can be found, e.g., in Agarwal and Sharir [AS00a].

Exercises

1. Show that for n hyperplanes in $\mathbf{R}^d$ in general position, the total number of vertices of levels $k, k+1, \ldots, n-d$ is at most $O(n^{\lfloor d/2 \rfloor}(n-k)^{\lceil d/2 \rceil})$. ☑

2. (a) Consider n lines in the plane in general position (their arrangement is simple). Call a vertex v of their arrangement an *extreme* if one of its defining lines has a positive slope and the other one has a negative slope. Prove that there are at most $O((k+1)^2)$ extremes of level at most k. Imitate the proof of Clarkson's theorem on levels. ☑

   (b) Show that the bound in (a) cannot be improved in general. ☑

3. Let $K_1, \ldots, K_n$ be circular disks in the plane. Show that the number of intersections of their boundary circles that are contained in at most k disks is bounded by $O(nk)$. Use the result of Exercise 5.7.10 and assume general position if convenient. ☒

4. Let L be a set of n nonvertical lines in the plane in general position.
   (a) Let W be an arbitrary subset of vertices of the arrangement of L, and let X_W be the number of pairs (v, ℓ), where $v \in W$, $\ell \in L$, and

ℓ goes (strictly) below v. For every real number $p \in (0,1)$, prove that $X_W \geq p^{-1}|W| - p^{-2}n$. ③

(b) Let W be a set of vertices in the arrangement of L such that no line of L lies strictly below more than k vertices of W, where $k \geq 1$. Use (a) to prove $|W| = O(n\sqrt{k})$. ②

(c) Check that the bound in (b) is tight for all $k \leq \frac{n}{2}$. ②

This exercise and the next one are from Sharir [Sha01].

5. Let P be an n-point set in the plane in general position (no 4 points on a common circle). Let C be a set of circles such that each circle in C passes through 3 points of P and contains no more than k points of P in its interior. Prove that $|C| \leq O(nk^{2/3})$, by an approach analogous to that of Exercise 4. ③

6.4 The Zone Theorem

Let H be a set of n hyperplanes in $\mathbf{R}^d$, and let g be a hyperplane that may or may not lie in H. The *zone* of g is the set of the faces of the arrangement of H that can see g. Here we imagine that the hyperplanes of H are opaque, and so we say that a face F can see the hyperplane g if there are points $x \in F$ and $y \in g$ such that the open segment xy is not intersected by any hyperplane of H (the face F is considered relatively open). Let us note that it does not matter which point $x \in F$ we choose: Either all of them can see g or none can. The picture shows the zone in a line arrangement:

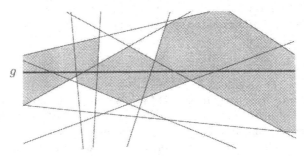

The following result bounds the maximum complexity of the zone. In the proof we will meet another interesting random sampling technique.

6.4.1 Theorem (Zone theorem). *The number of faces in the zone of any hyperplane in an arrangement of n hyperplanes in $\mathbf{R}^d$ is $O(n^{d-1})$, with the constant of proportionality depending on d.*

We prove the result only for simple arrangements; the general case follows, as usual, by a perturbation argument. Let us also assume that $g \notin H$ and that $H \cup \{g\}$ is in general position.

It is clear that the zone has $O(n^{d-1})$ *cells*, because each $(d-1)$-dimensional cell of the $(d-1)$-dimensional arrangement within g is intersects only one d-dimensional cell of the zone. On the other hand, this information is not sufficient to conclude that the total number of vertices of these cells is $O(n^{d-1})$: For example, as we know from Chapter 4, n arbitrarily chosen cells in an arrangement of n lines in the plane can together have as many as $\Omega(n^{4/3})$ vertices.

Proof. We proceed by induction on the dimension d. The base case is $d = 2$; it requires a separate treatment and does not follow from the trivial case $d = 1$ by the inductive argument shown below.

The case $d = 2$. (For another proof see Exercise 7.1.5.) Let H be a set of n lines in the plane in general position. We consider the zone of a line g. Since a convex polygon has the same number of vertices and edges, it suffices to bound the total number of 1-faces (edges) visible from the line g.

Imagine g drawn horizontally. We count the number of visible edges lying above g. Among those, at most n intersect the line g, since each line of H gives rise to at most one such edge. The others are disjoint from g.

Consider an edge uv disjoint from g and visible from a point of g. Let $h \in H$ be the line containing uv, and let a be the intersection of h with g:

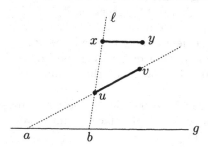

Let the notation be chosen in such a way that u is closer to a than v, and let $\ell \in H$ be the second line (besides h) defining the vertex u. Let b denote the intersection $\ell \cap g$. Let us call the edge uv a *right edge of the line ℓ* if the point b lies to the right of a, and a *left edge of the line ℓ* if b lies to the left of a.

We show that for each line ℓ there exists at most one right edge. If it were not the case, there would exist two edges, uv and xy, where u lies lower than x, which would both be right edges of ℓ, as in the above drawing. The edge xy should see some point of the line g, but the part of g lying to the right of a is obscured by the line h, and the part left of a is obscured by the line ℓ. This contradiction shows that the total number of right edges is at most n.

Symmetrically, we see that the number of left edges in the zone is at most n. The same bounds are obtained for edges of the zone lying below g. Altogether we have at most $O(n)$ edges in the zone, and the 2-dimensional case of the zone theorem is proved.

The case $d > 2$. Here we make the inductive step from $d-1$ to d. We assume that the total number of faces of a zone in $\mathbf{R}^{d-1}$ is $O(n^{d-2})$, and we want to bound the total number of zone faces in $\mathbf{R}^d$.

The first idea is to proceed by induction on n, bounding the maximum possible number of new faces created by adding a new hyperplane to $n-1$ given ones. However, it is easy to find examples showing that the number of faces can increase roughly by n^{d-1}, and so this straightforward approach fails.

In the actual proof, we use a clever averaging argument. First, we demonstrate the method for the slightly simpler case of counting only the facets (i.e., $(d-1)$-faces) of the zone.

Let $f(n)$ denote the maximum possible number of $(d-1)$-faces in the zone in an arrangement of n hyperplanes in $\mathbf{R}^d$ (the dimension d is not shown in the notation in order to keep it simple). Let H be an arrangement and g a base hyperplane such that $f(n)$ is attained for them.

We consider the following random experiment. Color a randomly chosen hyperplane $h \in H$ red and the other hyperplanes of H blue. We investigate the expected number of *blue* facets of the zone, where a facet is blue if it lies in a blue hyperplane.

On the one hand, any facet has probability $\frac{n-1}{n}$ of becoming blue, and hence the expected number of blue facets is $\frac{n-1}{n}f(n)$.

We bound the expected number of blue facets in a different way. First, we consider the arrangement of blue hyperplanes only; it has at most $f(n-1)$ blue facets in the zone by the inductive hypothesis. Next, we add the red hyperplane, and we look by how much the number of blue facets in the zone can increase.

A new blue facet can arise by adding the red hyperplane only if the red hyperplane slices some existing blue facet F into two parts F_1 and F_2, as is indicated in the picture:

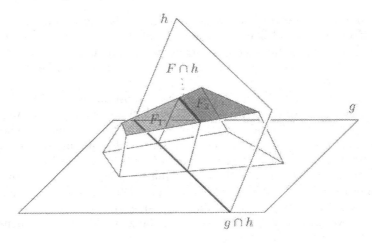

This increases the number of blue facets in the zone only if both F_1 and F_2 are visible from g. In such a case we look at the situation within the hyperplane h; we claim that $F \cap h$ is visible from $g \cap h$.

Let C be a cell of the zone in the arrangement of the blue hyperplanes having F on the boundary. We want to exhibit a segment connecting $F \cap h$ to $g \cap h$ within C. If $x_1 \in F_1$ sees a point $y_1 \in g$ and $x_2 \in F_2$ sees $y_2 \in g$, then the whole interior of the tetrahedron $x_1 x_2 y_1 y_2$ is contained in C. The intersection of this tetrahedron with the hyperplane h contains a segment witnessing the visibility of $g \cap h$ from $F \cap h$.

If we intersect all the blue hyperplanes and the hyperplane g with the red hyperplane h, we get a $(d-1)$-dimensional arrangement, in which $F \cap h$ is a facet in the zone of the $(d-2)$-dimensional hyperplane $g \cap h$. By the inductive hypothesis, this zone has $O(n^{d-2})$ facets. Hence, adding h increases the number of blue facets of the zone by $O(n^{d-2})$, and so the total number of blue facets after h has been added is never more than $f(n-1) + O(n^{d-2})$.

We have derived the following inequality:

$$\frac{n-1}{n} f(n) \le f(n-1) + O(n^{d-2}).$$

It implies $f(n) = O(n^{d-1})$, as we will demonstrate later for a slightly more general recurrence.

The previous considerations can be generalized for $(d-k)$-faces, where $1 \le k \le d-2$. Let $f_j(n)$ denote the maximum possible number of j-faces in the zone for n hyperplanes in dimension d. Let H be a collection of n hyperplanes where $f_{d-k}(n)$ is attained.

As before, we color one randomly chosen hyperplane $h \in H$ red and the others blue. A $(d-k)$-face is blue if its relative interior is disjoint from the red hyperplane. Then the probability of a fixed $(d-k)$-face being blue is $\frac{n-k}{n}$, and the expected number of blue $(d-k)$-faces in the zone is at most $\frac{n-k}{n} f_{d-k}(n)$.

On the other hand, we find that by adding the red hyperplane, the number of blue $(d-k)$-faces can increase by at most $O(n^{d-2})$, by the inductive hypothesis and by an argument similar to the case of facets. This yields the recurrence

$$\frac{n-k}{n} f_{d-k}(n) \le f_{d-k}(n-1) + O(n^{d-2}).$$

We use the substitution $\varphi(n) = \frac{f_{d-k}(n)}{n(n-1)\cdots(n-k+1)}$, which transforms our recurrence to $\varphi(n) \le \varphi(n-1) + O(n^{d-k-2})$. We assume $k < d-1$ (so the considered faces must not be edges or vertices). Then the last recurrence yields $\varphi(n) = O(n^{d-k-1})$, and hence $f_{d-k}(n) = O(n^{d-1})$.

For the case $k = d-1$ (edges), we would get only the bound $f_1(n) = O(n^{d-1} \log n)$ by this method. So the number of edges and vertices must be bounded by a separate argument, and we also have to argue separately for the planar case.

We are going to show that the number of vertices of the zone is at most proportional to the number of the 2-faces of the zone. Every vertex is contained in some 3-face of the zone. Within each such 3-face, the number of vertices is at most 3 times the number of 2-faces, because the 3-face is a 3-dimensional convex polyhedron. Since our arrangement is simple, each 2-face is contained in a bounded number of 3-faces. It follows that the total number of vertices is at most proportional to $f_2(n) = O(n^{d-1})$. The analogous bound for edges follows immediately from the bound for vertices.   □

Zones in other arrangements. The maximum complexity of a zone can be investigated for objects other than hyperplanes. We can consider two classes Z and $\mathcal{A}$ of geometric objects in $\mathbf{R}^d$ and ask for the maximum complexity of the zone of a $\zeta \in Z$ in the arrangement of n objects $a_1, a_2, \ldots, a_n \in \mathcal{A}$. This leads to a wide variety of problems. For some of them, interesting results have been obtained by extending the technique shown above.

Most notably, if ζ is a k-flat in $\mathbf{R}^d$, $0 \leq k \leq d$, or more generally, a *k-dimensional algebraic variety* in $\mathbf{R}^d$ of degree bounded by a constant, then the zone of ζ in an arrangement of n *hyperplanes* has complexity at most

$$O\left(n^{\lfloor (d+k)/2 \rfloor} (\log n)^\beta\right),$$

where $\beta = 1$ for $d + k$ odd and $\beta = 0$ for $d + k$ even. (The logarithmic factor seems likely to be superfluous in this bound; perhaps a more sophisticated proof could eliminate it.) With ζ being a k-flat, this result can be viewed as an interpolation between the asymptotic upper bound theorem and the zone theorem: For $k = 0$, with ζ being a single point, we consider the complexity of a single cell, while for $k = d-1$, we have the zone of a hyperplane. The key ideas of the proof are outlined in the notes below; for a full proof we refer to the literature.

A simple trick relates the zone problem to another question, the maximum complexity of a single cell in an arrangement. For example, what is the complexity of the zone of a segment ζ in an arrangement of n *line segments*? On the one hand, ζ can be chosen as a single point, and so the maximum zone complexity is at least the maximum possible complexity of a cell in an arrangement of n segments. On the other hand, the complexity of the zone of ζ is no more than the maximum cell complexity in an arrangement of $2n$ segments, since we can split each segment by making a tiny hole near the intersection with ζ:

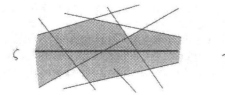
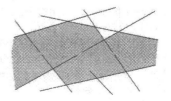

A similar reduction works for the zone of a triangle in an arrangement of triangles in $\mathbf{R}^3$ and in many other cases. Results presented in Section 7.6 will show that under quite general assumptions, the zone complexity in dimension d is no more than $O(n^{d-1+\varepsilon})$, for an arbitrarily small (but fixed) $\varepsilon > 0$.

Bibliography and remarks. The two-dimensional zone theorem was established by Chazelle, Guibas, and Lee [CGL85], with the proof shown above, and independently by Edelsbrunner, O'Rourke, and Seidel [EOS86] by a different method. The first correct proof of the general d-dimensional case, essentially the one presented here, is due to Edelsbrunner, Seidel, and Sharir [ESS93]. The main ingredients of the technique were previously developed by Sharir and his coauthors in several papers.

Bern, Eppstein, Plassman, and Yao [BEPY91] determined the best constant in the planar zone theorem: The zone of a line in an arrangement of n lines has at most $5.5n$ edges. They also showed that the zone of a convex k-gon has complexity $O(n + k^2)$.

The extension of the zone theorem to the zone of a k-dimensional algebraic variety in a hyperplane arrangement, as mentioned in the text, was proved by Aronov, Pellegrini, and Sharir [APS93]. They also obtained the same bound with ζ being the relative boundary of a $(k+1)$-dimensional convex set in $\mathbf{R}^d$.

The problem with the zone of a curved surface that did not exist for the zone of a hyperplane is that a face F of the zone of ζ can be split by a newly inserted hyperplane h into two subfaces F_1 and F_2, both of them lying in the zone, without $h \cap F$ being in the zone of $\zeta \cap h$, as is illustrated below:

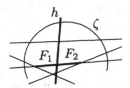

It turns out that each face F split by h in this way is adjacent to a facet in h that can be seen from ζ from both sides; such a facet is called a *popular facet* of the zone. In order to set up a suitable recurrence for the number of faces in the zone, one needs to bound the total complexity of all popular facets. This is again done by a technique similar to the proof of the zone theorem in the text. The concept of popular facet needs to be generalized to a *popular j-face*, which is a j-dimensional face F that can be seen from ζ in all the 2^{d-j} "sectors" determined by the $d - j$ hyperplanes defining F. The key observation is that if a blue popular j-face is split into two new popular j-faces by the new red hyperplane, then this can be charged to a popular $(j-1)$-face within h, as the following picture illustrates for $j = 1$:

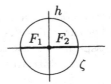

This is used to set up recurrences for the numbers of popular j-faces.

Exercises

1. (Sum of squares of cell complexities)
   (a) Let $\mathcal{C}$ be the set of all cells of an arrangement of a set H of n hyperplanes in $\mathbf{R}^d$. For $d = 2, 3$, prove that $\sum_{C \in \mathcal{C}} f_0(C)^2 = O(n^d)$, where $f_0(C)$ is the number of vertices of the cell C. [3]
   (b) Use the technique explained in this section to prove $\sum_{C \in \mathcal{C}} f_0(C)^2 = O(n^d (\log n)^{\lfloor d/2 \rfloor - 1})$ for every fixed $d \geq 3$ (or a similar bound with a larger constant in the exponent of $\log n$ if it helps). [5]
   The result in (b) is from Aronov, Matoušek, and Sharir [AMS94].
2. Define the $(\leq k)$-zone of a hyperplane g in an arrangement of hyperplanes as the collection of all faces for which some point x of their relative interior can be connected to some point $y \in g$ so that the interior of the segment xy intersects at most k hyperplanes.
   (a) By the technique of Section 6.3 (Clarkson's theorem on levels), show that the number of vertices of the $(\leq k)$-zone is $O(n^{d-1}k)$. [3]
   (b) Show that the bound in (a) cannot be improved in general. [2]
3. In this exercise we aim at bounding $K(n, n)$, the maximum total number of edges of n distinct cells in an arrangement of n lines in the plane, using the cutting lemma as in Section 4.5 (this proof is due to Clarkson, Edelsbrunner, Guibas, Sharir, and Welzl [CEG+90]). Let L be a set of n lines in general position.
   (a) Prove the bound $K(n, m) = O(n\sqrt{m} + m)$. [3]
   (b) Prove $K(n, n) = O(n^{4/3})$ using the cutting lemma. [4]
4. Consider a set H of n planes in $\mathbf{R}^3$ in general position and a sphere S (the surface of a ball).
   (a) Show that S intersects at most $O(n^2)$ cells of the arrangement of H. [2]
   (b) Using (a) and Exercise 1, prove that the zone of S in the arrangement of H has at most $O(n^{5/2})$ vertices. [4] (This is just an upper bound; the correct order of magnitude is about n^2.)

6.5 The Cutting Lemma Revisited

Here we present the most advanced version of the random sampling technique. It combines the approach to the weak cutting lemma (Section 4.6) with ingredients from the proof of Clarkson's theorem on levels and additional ideas.

We are going to re-prove the cutting lemma 4.5.3: *For every set H of n lines in the plane and every $r > 1$ there exists a $\frac{1}{r}$-cutting for H of size $O(r^2)$,* i.e., a subdivision of the plane into $O(r^2)$ generalized triangles $\Delta_1, \ldots, \Delta_t$ such that the interior of each Δ_i is intersected by at most $\frac{n}{r}$ lines of H. The proof uses random sampling, and unlike the elementary proof in Section 4.7, it can be generalized to higher dimensions without much trouble. We first give a complete proof for the planar case and then we discuss the generalizations.

Throughout this section we assume that H is in general position. A perturbation argument mentioned in Section 4.7 can be used to derive the cutting lemma for an arbitrary H.

The first idea is as in the proof of a weaker cutting lemma by random sampling in Section 4.6: We pick a random sample S of a suitable size and triangulate its arrangement.

The subsequent calculations become simpler and more elegant if we choose S by independent Bernoulli trials. That is, instead of picking s random lines with repetitions as in Section 4.6, we fix a probability $p = \frac{s}{n}$ and we include each line $h \in H$ into S with probability p, the decisions being mutually independent (this is as in the proof of the planar case of Clarkson's theorem on levels). These two ways of random sampling (by s random draws with repetitions and by independent trials with success probability $\frac{s}{n}$) can usually be thought of as nearly the same; although the actual calculations differ significantly, their results tend to be similar.

Sampling and triangulation alone do not work. Considerations similar to those in Section 4.6 show that with probability close to 1, none of the triangles in the triangulation for the random sample S as above is intersected by more than $C\frac{n}{s}\log n$ lines of H, for a suitable constant C. Later we will see that a similar statement is true with $C\frac{n}{s}\log s$ instead of $C\frac{n}{s}\log n$. But it is *not* generally true with $C\frac{n}{s}$, for any C independent of s and n. So the most direct road to an optimal $\frac{1}{r}$-cutting, namely choosing $const \cdot r$ random lines and triangulating their arrangement, is impassable.

To see this, consider a 1-dimensional situation, where $H = \{h_1, \ldots, h_n\}$ is a set of n points in $\mathbf{R}$ (or if you prefer, look at the part of a 2-dimensional arrangement along one of the lines). For simplicity, let us set $s = \frac{n}{2}$; then $p = \frac{1}{2}$, and we can imagine that we toss a fair coin n times and we include h_i into S if the ith toss is heads. The picture illustrates the result of 30 tosses, with black dots indicating heads:

We are interested in the length of the longest consecutive run of tails (empty circles). For k is fixed, it is very likely that k consecutive tails show up in a sequence of n tosses for n sufficiently large. Indeed, if we divide the tosses into blocks of length k (suppose for simplicity that n is divisible by k),

OOOOOOOOOOOOOOOOOOOOOOOOOOOOOOOOOO
|____|____|____|____|____|____|____|____|____|

then in each block, we have probability 2^{-k} of receiving all tails. The blocks are mutually independent, and so the probability of not obtaining all tails in any of the $\frac{n}{k}$ blocks is $(1 - 2^{-k})^{n/k}$. For k fixed and $n \to \infty$ this goes to 0, and a more careful calculation shows that for $k = \lfloor \frac{1}{2} \log_2 n \rfloor$ we have exponentially small probability of not receiving any block of k consecutive tails (Exercise 1). So a sequence of n tosses is very likely to contain about $\log n$ consecutive tails. (Sequences produced by humans that are intended to look random usually do not have this property; they tend to be "too uniform.") Similarly, for a smaller s, if we make a circle black with probability $\frac{s}{n}$, then the longest run typically has about $\frac{n}{s} \log s$ consecutive empty circles.

Of course, in the one-dimensional situation one can define much more uniform samples, say by making every $\frac{n}{s}$th circle black. But it is not clear how one could produce such "more uniform" samples for lines in the plane or for hyperplanes in $\mathbf{R}^d$.

The strategy: a two-level decomposition. Instead of trying to select better samples we construct a $\frac{1}{r}$-cutting for H in two stages. First we take a sample S with probability $p = \frac{r}{n}$ and triangulate the arrangement, obtaining a collection $\mathcal{T}$ of triangles. (The expected number of triangles is $O(r^2)$, as we will verify later.) Typically, $\mathcal{T}$ is not yet a $\frac{1}{r}$-cutting. Let $I(\Delta)$ denote the set of lines of H intersecting the interior of a triangle $\Delta \in \mathcal{T}$ and let $n_\Delta = |I(\Delta)|$. We define the *excess* of a triangle $\Delta \in \mathcal{T}$ as $t_\Delta = n_\Delta \cdot \frac{r}{n}$.

If $t_\Delta \leq 1$, then $n_\Delta \leq \frac{n}{r}$ and Δ is a good citizen: It can be included into the final $\frac{1}{r}$-cutting as is. On the other hand, if $t_\Delta > 1$, then Δ needs further treatment: We subdivide it into a collection of finer triangles such that each of them is intersected by at most $\frac{n}{r}$ lines of H. We do it in a seemingly naive way: We consider the whole arrangement of $I(\Delta)$, temporarily ignoring Δ, and we construct a $\frac{1}{t_\Delta}$-cutting for it. Then we intersect the triangles of this $\frac{1}{t_\Delta}$-cutting with Δ, which can produce triangles but also quadrilaterals, pentagons, and hexagons. Each of these convex polygons is further subdivided into triangles, as is illustrated below:

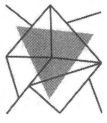
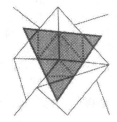

 $I(\Delta)$ a $\frac{1}{t_\Delta}$-cutting   restrict to Δ and triangulate

Note that each triangle in the $\frac{1}{t_\Delta}$-cutting is intersected by at most $\frac{n_\Delta}{t_\Delta} = \frac{n}{r}$ lines of $I(\Delta)$. Therefore, the triangles obtained within Δ are valid triangles of a $\frac{1}{r}$-cutting for H. The final $\frac{1}{r}$-cutting for H is constructed by subdividing each $\Delta \in \mathcal{T}$ with excess greater than 1 in the indicated manner and taking all the resulting triangles together.

How do we make the required $\frac{1}{t_\Delta}$-cuttings for the $I(\Delta)$? We do not yet have any suitable way of doing this unless we use the cutting lemma itself, which we do not want, of course. Fortunately, as a by-product of the subsequent considerations, we obtain a method for directly constructing slightly suboptimal cuttings:

6.5.1 Lemma (A suboptimal cutting lemma). *For every finite collection of lines and any $u > 1$, there exists a $\frac{1}{u}$-cutting consisting of at most $K(u\log(u+1))^2$ triangles, where K is a suitable constant.*

If we employ this lemma for producing the $\frac{1}{t_\Delta}$-cuttings, we can estimate the number of triangles in the resulting $\frac{1}{r}$-cutting in terms of the excesses of the triangles in $\mathcal{T}$: The total number of triangles is bounded by

$$\sum_{\Delta \in \mathcal{T}} \max\left\{1, 4K(t_\Delta \log(t_\Delta + 1))^2\right\}. \tag{6.3}$$

The key insight for the proof of the cutting lemma is that although we typically do have triangles $\Delta \in \mathcal{T}$ with excess as large as about $\log r$, they are very few. More precisely, we show that under suitable assumptions, the expected number of triangles in $\mathcal{T}$ with excess t or larger decreases exponentially as a function of t. This will take care of both estimating (6.3) by $O(r^2)$ and establishing Lemma 6.5.1.

Good and bad triangulations. Our collection $\mathcal{T}$ of triangles is obtained by triangulating the cells in the arrangement of the random sample S. Now is the time to specify how exactly the cells are triangulated, since not every triangulation works. To see this, consider a set H of n lines, each of them touching the unit circle, and let S be a random sample, again for simplicity with probability $p = \frac{1}{2}$. We have learned that such a sample is very likely to leave a gap of about $\log n$ unselected lines (as we go along the unit circle). If we maliciously triangulate the central cell in the arrangement of S by the diagonals from the vertex near such a large gap,

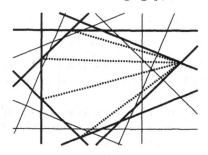

all these about $\frac{n}{2}$ triangles have excess about $\log n$; this is way too large for our purposes.

The triangulation thus cannot be quite arbitrary. For the subsequent proof, it has to satisfy simple axioms. In the planar case, it is actually technically easier not to triangulate but to construct the *vertical decomposition* of the arrangement of S. We erect vertical segments upwards and downwards from each vertex in the arrangement of S and extend them until they meet another line (or all the way to infinity):

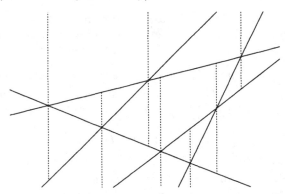

So far we have been speaking of triangles, and now we have trapezoids, but the difference is immaterial, since we can always split each trapezoid into two triangles if we wish. Let $\mathcal{T}(S)$ denote the set of (generalized) trapezoids in the vertical decomposition of S. As before, $I(\Delta)$ is the set of lines of H intersecting the interior of a trapezoid Δ.

6.5.2 Proposition (Trapezoids with large excess are rare). *Let H be a fixed set of n lines in general position, let $p = \frac{r}{n}$, where $1 \leq r \leq \frac{n}{2}$, let S be a random sample drawn from H by independent Bernoulli trials with success probability p, and let $t \geq 0$ be a real parameter. Let $\mathcal{T}(S)_{\geq t}$ denote the set of trapezoids in $\Delta \in \mathcal{T}(S)$ with excess at least t, i.e., with $|I(\Delta)| \geq t\frac{n}{r}$. Then the expected number of trapezoids in $\mathcal{T}(S)_{\geq t}$ is bounded as follows:*

$$\mathbf{E}[|\mathcal{T}(S)_{\geq t}|] \leq C \cdot 2^{-t} r^2$$

for a suitable absolute constant C.

First let us see how this result can be applied.

Proof of the suboptimal cutting lemma 6.5.1. To obtain a $\frac{1}{u}$-cutting for H, we set $r = Au \log(u+1)$ for a sufficiently large constant A and choose a sample S as in Proposition 6.5.2.

By that proposition with $t = 0$, we have $\mathbf{E}[|\mathcal{T}(S)|] \leq Br^2$ for a suitable constant B. By the same proposition with $t = A\log(u+1)$, we have $\mathbf{E}[|\mathcal{T}(S)_{\geq t}|] \leq \frac{1}{3}$ if A is sufficiently large. By linearity of expectation, we obtain

$$\mathbf{E}\left[\frac{1}{3Br^2}\,|\mathcal{T}(S)| + |\mathcal{T}(S)_{\geq t}|\right] \leq \frac{2}{3}.$$

So there exists a sample S with both $|\mathcal{T}(S)| \leq 2Br^2$ and $|\mathcal{T}(S)_{\geq t}| = 0$. This means that we have a $\frac{1}{u}$-cutting into $O(r^2) = O((u\log(u+1))^2)$ trapezoids. $\qquad\square$

For an alternative proof of Lemma 6.5.1 see Exercise 10.3.4.

Proof of the cutting lemma (Lemma 4.5.3). Most of the proof has already been described. To produce a $\frac{1}{r}$-cutting, we pick a random sample S with probability $p = \frac{r}{n}$, we let $\mathcal{T} = \mathcal{T}(S)$ be its vertical decomposition, and we refine each trapezoid $\Delta \in \mathcal{T}$ with excess $t_\Delta > 1$ using an auxiliary $\frac{1}{t_\Delta}$-cutting. The size of the resulting $\frac{1}{r}$-cutting is bounded by (6.3). So it suffices to estimate the expected value of that expression using Proposition 6.5.2:

$$\mathbf{E}\left[\sum_{\Delta \in \mathcal{T}(S)} \max\left\{1, 4K(t_\Delta \log(t_\Delta+1))^2\right\}\right]$$

$$\leq \mathbf{E}\left[\sum_{\Delta \in \mathcal{T}(S)} \max\left\{1, 4Kt_\Delta^4\right\}\right] \qquad\qquad (\text{as } \log(t_\Delta+1) \leq t_\Delta)$$

$$\leq \mathbf{E}\left[|\mathcal{T}(S)| + \sum_{i=0}^{\infty} \sum_{\substack{\Delta \in \mathcal{T}(S) \\ 2^i \leq t_\Delta < 2^{i+1}}} 4Kt_\Delta^4\right]$$

$$\leq \mathbf{E}[|\mathcal{T}(S)|] + \sum_{i=0}^{\infty} \mathbf{E}\left[|\mathcal{T}(S)|_{\geq 2^i}\right] \cdot O(2^{4(i+1)})$$

$$\leq O(r^2) + \sum_{i=0}^{\infty} C \cdot 2^{-2^i} r^2 \cdot O(2^{4(i+1)})$$

$$= O(r^2).$$

The cutting lemma is proved. $\qquad\qquad\qquad\qquad\qquad\qquad\qquad\qquad\square$

Note that it was not important that the suboptimal cutting lemma is near-optimal: *Any bound subexponential in u for the size of a $\frac{1}{u}$-cutting would do.* In particular, for any fixed $c \geq 1$, the expected cth-degree average of the excess is only a constant.

For the proof of Proposition 6.5.2, we need several definitions and some simple properties of the vertical decomposition. Let H be a fixed set of lines in general position, and let $Reg = \bigcup_{S \subseteq H} \mathcal{T}(S)$ be the set of all trapezoids that can ever appear in the vertical decomposition for some $S \subseteq H$ (including $S = \emptyset$). For a trapezoid $\Delta \in Reg$, let $D(\Delta)$ be the set of the lines of H incident to at least one vertex of Δ. By the general-position assumption, we have $|D(\Delta)| \leq 4$ for all Δ. The various possible cases, up to symmetry, are drawn below; the picture shows the lines of $D(\Delta)$ with Δ marked in gray:

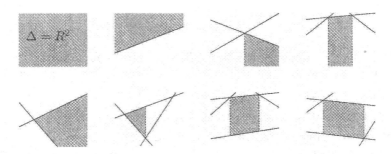

The set $D(\Delta)$ is called the *defining set* of Δ. Note that the same defining set can belong to several trapezoids.

Now we list the properties required for the proof; some of them are obvious or have already been noted.

(C0) We have $|D(\Delta)| \leq 4$ for all $\Delta \in Reg$. Moreover, any set $S_0 \subseteq H$ is the defining set for at most a constant number of $\Delta \in Reg$ (certainly no more than the maximum of $|\mathcal{T}(S_0)|$ for $|S_0| \leq 4$).

(C1) For any $\Delta \in \mathcal{T}(S)$, we have $D(\Delta) \subseteq S$ (the defining set must be present) and $S \cap I(\Delta) = \emptyset$ (no intersecting line may be present).

(C2) For any $\Delta \in Reg$ and any $S \subseteq H$ such that $D(\Delta) \subseteq S$ and $I(\Delta) \cap S = \emptyset$, we have $\Delta \in \mathcal{T}(S)$.

(C3) For every $S \subseteq H$, we have $|\mathcal{T}(S)| = O(|S|^2 + 1)$. To see this, think of adding the vertical segments to the arrangement of S one by one. Each of them splits an existing region in two.

The most interesting condition is (C2), which says that the vertical decomposition is defined "locally." It implies, in particular, that Δ is one of the trapezoids in the vertical decomposition of its defining set. More generally, it says that $\Delta \in Reg$ is present in $\mathcal{T}(S)$ whenever it is not excluded for simple local reasons (which can be checked by looking only at Δ). Checking (C2) in our situation is easy, and we leave it to the reader. Also note that it is (C2) that is generally violated for the mischievous triangulation considered earlier.

Proof of Proposition 6.5.2. First we prove that if $S \subseteq H$ is a random sample drawn with probability $p = \frac{r}{n}$, $0 < r < n$, then

$$\mathbf{E}[|\mathcal{T}(S)|] = O(r^2 + 1). \qquad (6.4)$$

This takes care of the case $t \leq 1$ in the proposition. By (C3), we have $|\mathcal{T}(S)| = O(|S|^2 + 1)$ for every fixed S, and so it suffices to show that $\mathbf{E}[|S|^2] = O(r^2 + 1)$. Now, $|S|$ is the sum of independent random variables, each of them attaining value 1 with probability p and value 0 with probability $1 - p$, and it is easy to check that $\mathbf{E}[|S|^2] \leq r^2 + r$ (Exercise 2(a)).

Next, we assume $t \geq 1$. Let $S \subseteq H$ be a random sample drawn with probability p. We observe that the conditions (C1) and (C2) allow us to express the probability $p(\Delta)$ that a certain trapezoid $\Delta \in Reg$ appears in the vertical decomposition $\mathcal{T}(S)$: Since Δ appears if and only if all lines of $D(\Delta)$ are selected into S and none of $I(\Delta)$ is selected, we have

$$p(\Delta) = p^{|D(\Delta)|}(1 - p)^{|I(\Delta)|}.$$

(An analogous formula appeared in the proof of the planar Clarkson's theorem on levels, and one can say that the technique of that proof is developed one step further in the present proof.) If we write $Reg_{\geq t} = \{\Delta \in Reg: |I(\Delta)| \geq t\frac{n}{r}\}$ for the set of all potential trapezoids with excess at least t, the expected number of trapezoids in $\mathcal{T}(S)_{\geq t}$ can be written as

$$\mathbf{E}[|\mathcal{T}(S)_{\geq t}|] = \sum_{\Delta \in Reg_{\geq t}} p(\Delta). \tag{6.5}$$

It seems difficult to estimate this sum directly; the trick is to compare it with a similar sum obtained for the expected number of trapezoids for another sample.

We define another probability $\tilde{p} = \frac{p}{t}$, and we let $\tilde{S}$ be a sample drawn from H by Bernoulli trials with success probability $\tilde{p}$. On the one hand, we have $\mathbf{E}\left[|\mathcal{T}(\tilde{S})|\right] = O(r^2/t^2 + 1)$ by (6.4). On the other hand, setting $\tilde{p}(\Delta) = \tilde{p}^{|D(\Delta)|}(1 - \tilde{p})^{|I(\Delta)|}$ we have, in analogy to (6.5),

$$\mathbf{E}\left[|\mathcal{T}(\tilde{S})|\right] = \sum_{\Delta \in Reg} \tilde{p}(\Delta) \geq \sum_{\Delta \in Reg_{\geq t}} \tilde{p}(\Delta)$$

$$= \sum_{\Delta \in Reg_{\geq t}} p(\Delta) \cdot \frac{\tilde{p}(\Delta)}{p(\Delta)} \geq \mathbf{E}[|\mathcal{T}(S)_{\geq t}|] \cdot R, \tag{6.6}$$

where

$$R = \min\left\{\frac{\tilde{p}(\Delta)}{p(\Delta)}: \Delta \in Reg_{\geq t}\right\}.$$

Now R can be bounded from below. For every $\Delta \in Reg_{\geq t}$, we have $|I(\Delta)| \geq t\frac{r}{n}$ and $|D(\Delta)| \leq 4$, and so

$$\frac{\tilde{p}(\Delta)}{p(\Delta)} = \left(\frac{\tilde{p}}{p}\right)^{|D(\Delta)|}\left(\frac{1 - \tilde{p}}{1 - p}\right)^{|I(\Delta)|} \geq t^{-4}\left(\frac{1 - \tilde{p}}{1 - p}\right)^{tn/r}.$$

We use $1 - p \leq e^{-p}$ (this holds for all real p) and $1 - \tilde{p} \geq e^{-2\tilde{p}}$ (this is true for all $\tilde{p} \in [0, \frac{1}{2}]$, and we have $\tilde{p} \leq p \leq \frac{1}{2}$). Therefore $R \geq t^{-4}e^{t-2}$. Substituting into (6.6), we finally derive

$$\mathbf{E}[|\mathcal{T}(S)_{\geq t}|] \leq \frac{1}{R} \cdot \mathbf{E}\left[|\mathcal{T}(\tilde{S})|\right] \leq t^4 e^{-(t-2)} \cdot O\left(\frac{r^2}{t^2} + 1\right) \leq C \cdot 2^{-t}r^2$$

for a sufficiently large constant C (the proposition assumes $r \geq 1$). Proposition 6.5.2 is proved. $\qquad \square$

The following can be proved by the same technique:

6.5.3 Theorem (Cutting lemma for arbitrary dimension). *Let $d \geq 1$ be a fixed integer, let H be a set of n hyperplanes in $\mathbf{R}^d$, and let r be a parameter, $1 < r \leq n$. Then there exists a $\frac{1}{r}$-cutting for H of size $O(r^d)$; that is, a subdivision of $\mathbf{R}^d$ into $O(r^d)$ generalized simplices such that the interior of each simplex is intersected by at most $\frac{n}{r}$ hyperplanes of H.*

The only new part of the proof is the construction of a suitable triangulation scheme that plays the role of $\mathcal{T}(S)$. A vertical decomposition does not work. More precisely, it is not known whether the vertical decomposition of an arrangement of n hyperplanes in $\mathbf{R}^d$ always has at most $O(n^d)$ cells (prisms); this would be needed as the analogue of condition (C3). Instead one can use the *bottom-vertex triangulation*, which we define next.

First we specify the bottom-vertex triangulation of a k-dimensional convex polytope $P \subset \mathbf{R}^d$, $1 \leq k \leq d$, by induction on k. For $k = 1$, P is a line segment, and the triangulation consists of P itself. For $k > 1$, we let v be the vertex of P with the smallest last coordinate (the "bottom vertex"); ties can be broken by lexicographic ordering of the coordinate vectors. We triangulate all proper faces of P inductively, and we add the simplices obtained by erecting the cone with apex v over all simplices in the triangulations of the faces not containing v.

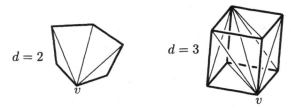

It is not difficult to check that this yields a triangulation of P (even a simplicial complex, although this is not needed in the present proof), and that if P is a simple polytope, then the total number of simplices in this triangulation is at most proportional to the number of vertices of P (with the constant of proportionality depending on d); see Exercise 4.

All the bounded cells of the arrangement of S are triangulated in this way. Some care is needed for the unbounded cells, and several ways are available. One of the simplest is to intersect the arrangement with a sufficiently large box containing all the vertices and construct the $\frac{1}{r}$-cutting only inside that box. This is sufficient for most applications of $\frac{1}{r}$-cuttings. Alternatively (and almost equivalently), we can consider the whole arrangement in the projective d-space instead of $\mathbf{R}^d$. We omit a detailed discussion of this aspect.

In this way we obtain a triangulation $\mathcal{T}(S)$ for every subset S of the given set of hyperplanes. The analogue of (C3) is $|\mathcal{T}(S)| = O(|S|^d+1)$, which follows (assuming H in general position) because the number of simplices in each cell is proportional to the number of its vertices, and the total number of vertices is $O(|S|^d)$.

The set $I(\Delta)$ are all hyperplanes intersecting the interior of a simplex Δ, and $D(\Delta)$ consists of all the hyperplanes incident to at least one vertex of Δ. We again need to assume that our hyperplanes are in general position. Then, obviously, $|D(\Delta)| \leq d(d+1)$, and a more careful argument shows that $|D(\Delta)| \leq \frac{d(d+3)}{2}$. The important thing is that an analogue of (C0) holds, namely, that both $|D(\Delta)|$ and the number of Δ with a given $D(\Delta)$ are bounded by constants.

The condition (C1) holds trivially. The "locality" condition (C2) does need some work, although it is not too difficult, and we refer to Chazelle and Friedman [CF90] for a detailed argument.

With (C0)–(C3) in place, the whole proof proceeds exactly as in the planar case. To get the analogue of (6.4), namely $\mathbf{E}[|\mathcal{T}(S)|] = O(r^d+1)$, we need the fact that $\mathbf{E}[|S|^d] = O(r^d)$ (this is what we avoided in the proof of the higher-dimensional Clarkson's theorem on levels by passing to another way of sampling); see Exercise 2(b) or 3.

Further generalizations. An analogue of Proposition 6.5.2 can be derived from conditions (C0)–(C3) in a general abstract framework. It provides optimal $\frac{1}{r}$-cuttings not only for arrangements of hyperplanes but also in other situations, whenever one can define a suitable decomposition scheme satisfying (C0)–(C3) and bound the maximum number of cells in the decomposition (the latter is a challenging open problem for arrangements of bounded-degree algebraic surfaces). The significance of Proposition 6.5.2 reaches beyond the construction of cuttings; its variations have been used extensively, mainly in the analysis of geometric algorithms. We are going to encounter a combinatorial application in Chapter 11.

Bibliography and remarks. The proof of the cutting lemma as in this section (with a different way of sampling) is due to Chazelle and Friedman [CF90]. Analogues of Proposition 6.5.2, or more precisely the consequence stating that the expectation of the cth-degree average of the excess is bounded by a constant, were first proved and applied by Clarkson [Cla88a] (see Clarkson and Shor [CS89] for the journal version). Since then, they became one of the indispensable tools in the analysis of randomized geometric algorithms, as is illustrated by the book by Mulmuley [Mul93a], for example, as well as by many newer papers.

The bottom-vertex triangulation (also called the *canonical triangulation* in some papers) was defined in Clarkson [Cla88b].

Proposition 6.5.2 can be formulated and proved in an abstract framework, where H and *Reg* are some finite sets and $\mathcal{T}: 2^H \to 2^{Reg}$, $I: Reg \to 2^H$, and $D: Reg \to 2^H$ are mappings satisfying (C0) (with some constants), (C1), (C2), and an analogue of (C3) that bounds the expected size of $\mathcal{T}(S)$ for a random $S \subseteq H$ by a suitable function of r, typically by $O(r^k)$ for some real constant $k \geq 1$. The conclusion is $\mathbf{E}[\|\mathcal{T}(S)_{\geq t}\|] = O(2^{-t}r^k)$. Very similar abstract frameworks are discussed in Mulmuley [Mul93a] and in De Berg, Van Kreveld, Overmars, and Schwarzkopf [dBvKOS97].

The axiom (C2) can be weakened to the following:

(C2′) If $\Delta \in \mathcal{T}(S)$ and $S' \subseteq S$ satisfies $D(\Delta) \subseteq S'$, then $\Delta \in \mathcal{T}(S')$. That is, Δ cannot be destroyed by deleting elements of S unless we delete an element of $D(\Delta)$.

A typical situation where (C2′) holds although (C2) fails is that in which H is a set of lines in the plane and $\mathcal{T}(S)$ are the trapezoids in the vertical decomposition of the cell in the arrangement of S that contains some fixed point, say 0. Then Δ can be made to disappear by adding a line to S even if that line does not intersect Δ, as is illustrated below:

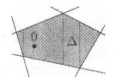 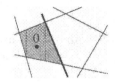

This weaker axiom was first used instead of (C2) by Chazelle, Edelsbrunner, Guibas, Sharir, and Snoeyink [CEG+93]. For a proof of a counterpart of Proposition 6.5.2 under (C2′) see Agarwal, Matoušek, and Schwarzkopf [AMS98].

Yet another proof of the cutting lemma in arbitrary dimension was invented by Chazelle [Cha93a]. An outline of the argument can also be found in Chazelle's book [Cha00c] or in the chapter by Matoušek in [SU00].

Both the proofs of the higher-dimensional cutting lemma depend crucially on the fact that the arrangement of n hyperplanes in $\mathbf{R}^d$, d fixed, can be triangulated using $O(n^d)$ simplices. As was explained in Section 6.2, the arrangement of n bounded-degree algebraic surfaces in $\mathbf{R}^d$ has $O(n^d)$ faces in total, but the faces can be arbitrarily complicated. A challenging open problem is whether each face can be further decomposed into "simple" pieces (each of them defined by a constant-bounded number of bounded-degree algebraic inequalities) such that the total number of pieces for the whole arrangement is $O(n^d)$ or not much larger. This is easy for $d = 2$ (the vertical decomposition will do), but dimension 3 is already quite challenging. Chazelle, Edelsbrunner, Guibas, and Sharir [CEGS89] found a general argument that provides an $O(n^{2d-2})$ bound in dimension d using a suitable vertical decomposition. By proving a near-optimal bound in the 3-dimensional case and using it as a basis of the induction, they obtained the bound of $O(n^{2d-3}\beta(n))$, where β is a very slowly growing function (much smaller than $\log^* n$). Recently Koltun [Kol01] established a near-tight bound in the 4-dimensional situation, which pushed the general bound to $O(n^{2d-4+\varepsilon})$ for every fixed $d \geq 4$.

This decomposition problem is the main obstacle to proving an optimal or near-optimal cutting lemma for arrangements of algebraic surfaces. For some special cases, say for an arrangement of spheres in $\mathbf{R}^d$, optimal decompositions are known and an optimal cutting lemma can be obtained. In general, if one can prove a bound of $O(n^\alpha)$ for the number of pieces in the decomposition, then the techniques of Chapter 10 yield $\frac{1}{r}$-cuttings of size $O(r^\alpha \log^\alpha r)$, and if, moreover, the locality condition (C2) can be guaranteed, then the method of the present section leads to $\frac{1}{r}$-cuttings of size $O(r^\alpha)$.

Exercises

1. Estimate the largest $k = k(n)$ such that in a row of n tosses of a fair coin we obtain k consecutive tails with probability at least $\frac{1}{2}$. In particular, using the trick with blocks in the text, check that for $k = \lfloor \frac{1}{2} \log_2 n \rfloor$, the probability of not getting all tails in any of the blocks is exponentially small (as a function of n). ②

2. Let $X = X_1 + X_2 + \cdots + X_n$, where the X_i are independent random variables, each attaining the value 1 with probability p and the value 0 with probability $1 - p$.
   (a) Calculate $\mathbf{E}[X^2]$. ②
   (b) Prove that for every integer $d \geq 1$ there exists c_d such that $\mathbf{E}[X^d] \leq (np+c_d)^d$. (You can use a Chernoff-type inequality, or prove by induction that $\mathbf{E}[(X+a)^d] \leq (np+d+a)^d$ for all nonnegative integers n, d, and a.) ③
   (c) Use (b) to prove that $\mathbf{E}[X^\alpha] \leq (np+c_\alpha)^\alpha$ also holds for nonintegral $\alpha \geq 1$. ③

3. Let $X = X_1 + X_2 + \cdots + X_n$ be as in the previous exercise. Show that $\mathbf{E}\left[\binom{X}{d}\right] = p^d \binom{n}{d}$ (where $d \geq 0$ is an integer) and conclude that $\mathbf{E}[X^d] \leq c_d(np)^d$ for $np \geq d$ and a suitable $c_d > 0$. ③

4. Let P be a d-dimensional simple convex polytope. Prove that the bottom-vertex triangulation of P has at most $C_d f_0(P)$ simplices, where C_d depends only on d and $f_0(P)$ denotes the number of vertices of P. ④

7

Lower Envelopes

This is a continuation of the chapter on arrangements. We again study the number of vertices in a certain part of the arrangement: the lower envelope. Already for segments in the plane, this problem has an unexpectedly subtle and difficult answer. The closely related combinatorial notion of *Davenport–Schinzel sequences* has proved to be a useful general tool, since the surprising phenomena encountered in the analysis of the lower envelope of segments are by no means rare in combinatorics and discrete geometry.

The chapter has two rather independent parts. After a common introduction in Section 7.1, lower envelopes in the plane are discussed in Sections 7.2 through 7.4 using Davenport–Schinzel sequences. Sections 7.5 and 7.6 gently introduce the reader to geometric methods for analyzing higher-dimensional lower envelopes, finishing with a quick overview of known results in Section 7.7.

7.1 Segments and Davenport–Schinzel Sequences

The following question is extremely natural: *What is the maximum possible combinatorial complexity of a single cell in an arrangement of n segments?* (The arrangement of segments was defined in Section 6.2.)

The complexity of a cell can be measured as the number of vertices and edges on its boundary. It is immediate that the number of edges is at most proportional to the number of vertices plus $2n$, the total number of endpoints of the segments, and so it suffices to count the vertices.

Here we mainly consider a slightly simpler question: the maximum complexity of the *lower envelope* of n segments. Informally, the lower envelope of an arrangement is the part that can be seen by an observer sitting at $(0, -\infty)$ and looking upward. In the picture below, the lower envelope of 4 segments is drawn thick:

If we think of the segments as graphs of (partially defined) functions, the lower envelope is the graph of the pointwise minimum. It consists of pieces of the segments, and we are interested in the maximum possible number of these pieces (in the drawing, we have 7 pieces). Let us denote this maximum by $\sigma(n)$.

Davenport–Schinzel sequences. A tight upper bound for $\sigma(n)$ has been obtained via a combinatorial abstraction of lower envelopes, the so-called Davenport–Schinzel sequences. These are closely related to segments, but the most natural way of introducing them is starting from curves. Let us consider a finite set of curves in the plane, such as in the following picture:

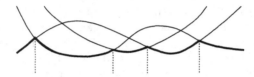

We suppose that each curve is a graph of a continuous function $\mathbf{R} \to \mathbf{R}$; in other words, each vertical line intersects it exactly once. Most significantly, we assume that every two of the curves intersect in at most s points for some constant s. This condition holds, for example, if the curves are the graphs of polynomials of degree at most s.

Let us number the curves 1 through n, and let us write down the sequence of the numbers of the curves along the lower envelope from left to right:

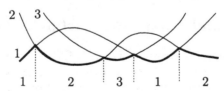

We obtain a sequence $a_1 a_2 a_3 \ldots a_\ell$ with the following properties:

(i) For all i, $a_i \in \{1, 2, \ldots, n\}$.
(ii) No two adjacent terms coincide; i.e., $a_i \neq a_{i+1}$.
(iii) There is no (not necessarily contiguous) subsequence of the form

$$\underbrace{\ldots a \ldots b \ldots a \ldots b \ldots \quad \ldots a \ldots b \ldots,}_{s + 2 \text{ letters } a \text{ and } b}$$

where $a \neq b$. In other words, there are no indices $i_1 < i_2 < i_3 < \cdots < i_{s+2}$ with $a_{i_1} \neq a_{i_2}$, $a_{i_1} = a_{i_3} = a_{i_5} = \cdots$, and $a_{i_2} = a_{i_4} = a_{i_6} = \cdots$.

Only (iii) needs a little thought: It suffices to note that between an occurrence of a curve a and an occurrence of a curve b on the lower envelope, a and b have to intersect.

Any finite sequence satisfying (i)–(iii) is called a *Davenport–Schinzel sequence of order s* over the symbols $1, 2, \ldots, n$. It is not important that the terms of the sequence are the numbers $1, 2, \ldots, n$; often it is convenient to use some other set of n distinct symbols.

Let us remark that *every* Davenport–Schinzel sequence of order s over n symbols corresponds to the lower envelope of a suitable set of n curves with at most s intersections for each pair of curves (Exercise 1). On the other hand, very little is known about the realizability of Davenport–Schinzel sequences by graphs of polynomials of degree s, say.

We will mostly consider Davenport–Schinzel sequences of order 3. This is the simplest nontrivial case and also the one closely related to lower envelopes of segments. Every two segments intersect at most once, and so it might seem that their lower envelope gives rise to a Davenport–Schinzel sequence of order 1, but this is not the case! The segments are graphs of *partially defined* functions, while the discussion above concerns graphs of functions defined on all of $\mathbf{R}$. We can convert each segment into a graph of an everywhere-defined function by appending very steep rays to both endpoints:

All the left rays are parallel, and all the right ones are parallel. Then every two of these curves have at most 3 intersections, and so if the considered segments are numbered 1 through n and we write the sequence of their numbers along the lower envelope, we get a Davenport–Schinzel sequence of order 3 (no *ababa*).

Let $\lambda_s(n)$ denote the maximum possible length of a Davenport–Schinzel sequence of order s over n symbols. Some work is needed to see that $\lambda_s(n)$ is finite for all s and n; the reader is invited to try this. The bound $\lambda_1(n) = n$ is trivial, and $\lambda_2(n) = 2n-1$ is a simple exercise. Determining the asymptotics of $\lambda_3(n)$ is a hard problem; it was posed in 1965 and solved in the mid-1980s. We will describe the solution later, but here we start more modestly: with a reasonable upper bound on $\lambda_3(n)$.

7.1.1 Proposition. *We have $\sigma(n) \leq \lambda_3(n) \leq 2n \ln n + 3n$.*

Proof. Let w be a Davenport–Schinzel sequence of order 3 over n symbols. If the length of w is ℓ, then there is a symbol a occurring at most $\frac{\ell}{n}$ times in w. Let us remove all occurrences of such a from w. The resulting sequence can contain some pairs of adjacent equal symbols. But we claim that there can be at most 2 such pairs, coming from the first and last occurrences of a.

Indeed, if some a which is neither the first nor the last a in w were surrounded by some b from both sides, we would have the situation $\dots a \dots bab \dots a \dots$ with the forbidden pattern $ababa$. So by deleting all the a and at most 2 more symbols, we obtain a Davenport–Schinzel sequence of order 3 over $n-1$ symbols.

We arrive at the recurrence

$$\lambda_3(n) \leq \frac{\lambda_3(n)}{n} + 2 + \lambda_3(n-1),$$

which can be rewritten to

$$\frac{\lambda_3(n)}{n} \leq \frac{\lambda_3(n-1)}{n-1} + \frac{2}{n-1}$$

(we saw such a recurrence in the proof of the zone theorem). Together with $\lambda_3(1) = 1$ this yields

$$\frac{\lambda_3(n)}{n} \leq 1 + 2\left(1 + \frac{1}{2} + \frac{1}{3} + \cdots + \frac{1}{n-1}\right),$$

and so $\lambda_3(n) \leq 2n \ln n + 3n$ as claimed.   □

Bibliography and remarks. A detailed account of the history of Davenport–Schinzel sequences and of the analysis of lower envelopes, with references up until 1995, can be found in the book of Sharir and Agarwal [SA95]. Somewhat more recent results are included in in their surveys [AS00b] and [AS00a]. We sketch this development and mention some newer results in the notes to Section 7.3.

Exercises

1. Let w be a Davenport–Schinzel sequence of order s over the symbols $1, 2, \dots, n$. Construct a family of planar curves $h_1, h_2, \dots, h_n$, each of them intersecting every vertical line exactly once and each two intersecting in at most s points, such that the sequence of the numbers of the curves along the lower envelope is exactly w. [2]
2. Prove that $\lambda_2(n) = 2n-1$ (the forbidden pattern is $abab$). [3]
3. Prove that for every n and s, $\lambda_s(n) \leq 1 + (s+1)\binom{n}{2}$. [3]
4. Show that the lower envelope of n rays in the plane has $O(n)$ complexity. [4]
5. (Planar zone theorem via Davenport–Schinzel sequences) Prove the zone theorem (Theorem 6.4.1) for $d = 2$ using the fact that $\lambda_2(n) = O(n)$. Consider only the part above the line g, and assign one symbol to each side of each line. [4]
6. Let $g_1, g_2, \dots, g_m \subset \mathbf{R}^2$ be graphs of piecewise linear functions $\mathbf{R} \to \mathbf{R}$ that together consist of n segments and rays. Prove that the lower envelope of $g_1, g_2, \dots, g_m$ has complexity $O(\frac{n}{m}\lambda_3(2m))$; in particular, if $m = O(1)$, then the complexity is linear. [4]

7. Let $P_1, P_2, \ldots, P_m$ be convex polygons (not necessarily disjoint!) in the plane with n vertices in total such that no vertex is common to two or more P_i and the vertices form a point set in general position. Prove that the number of lines that intersect all the P_i and are tangent to at least two of them is at most $O(\lambda_3(n))$. ③

8. (Dynamic lower envelope of lines) Let $\ell_1, \ell_2, \ldots, \ell_n$ be lines in the plane in general position (in particular, none of them is vertical). At each moment t of time, only a certain subset L_t of the lines is present: ℓ_i is inserted at time s_i and it is removed at time $t_i > s_i$. We are interested in the maximum possible total number $f(n)$ of vertices of the arrangement of the ℓ_i that appear as vertices of the lower envelope of L_t for at least one $t \in \mathbf{R}$.

   (a) Show that $f(n) = \Omega(\sigma(n))$, where $\sigma(n)$ is the maximum complexity of the lower envelope of n segments. ③

   (b) Prove that $f(n) = O(n \log n)$. (Familiarity with data structures like segment trees or interval trees may be helpful.) ⑤

   These results are from Tamir [Tam88], and improving the lower bound or the upper bound is a nice open problem.

7.2 Segments: Superlinear Complexity of the Lower Envelope

In Proposition 7.1.1 we have shown that the lower envelope of n segments has complexity at most $O(n \log n)$, but it turns out that the true complexity is still lower. With this information, the next reasonable guess would be that perhaps the complexity is linear in n. The truth is much subtler, though: On the one hand, the complexity behaves like a linear function for all practical purposes, but on the other hand, it *cannot* be bounded by any linear function: It outgrows the function $n \mapsto Cn$ for every fixed C. We present an ingenious construction witnessing this.

7.2.1 Theorem. *The function $\sigma(n)$, the maximum combinatorial complexity of the lower envelope of n segments in the plane, is superlinear. That is, for every C there exists an n_0 such that $\sigma(n_0) \geq Cn_0$. Consequently, $\lambda_3(n)$, the maximum length of a Davenport–Schinzel sequence of order 3, is superlinear, too.*

Proof. For every integers $k, m \geq 1$ we construct a set $S_k(m)$ of segments in the plane. Let $n_k(m) = |S_k(m)|$ be the number of segments and let $e_k(m)$ denote the number of arrangement vertices and segment endpoints on the lower envelope of $S_k(m)$. We prove that $e_k(m) \geq \frac{1}{2}k \cdot n_k(m)$. In particular, for $m = 1$ and $k \to \infty$, this shows that the complexity of the lower envelope is nonlinear in the number of segments.

If we really need only the case $m = 1$, then what is the parameter m good for? The answer is that we proceed by double induction, on both k and m, and in order to specify $S_k(1)$, for example, we need $S_{k-1}(2)$. Results of mathematical logic, which are beyond the scope of this book, show that double induction is in some sense unavoidable: The "usual" induction on a single variable is too crude to distinguish $\sigma(n)$ from a linear function.

The segments in $S_k(m)$ are usually not in general position, but they are aggregated in *fans* by m segments. A fan of m segments is illustrated below for $m = 4$:

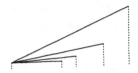

All the segments of a fan have a common left endpoint and positive slopes, and the length of the segments increases with the slope. Other than forming the fans, the segments are in general position in an obvious sense. For example, no endpoint of a segment lies inside another segment, the endpoints do not coincide unless the segments are in a common fan, and so on.

Let $f_k(m)$ denote the number of fans forming $S_k(m)$; we have $n_k(m) = m \cdot f_k(m)$.

First we describe the construction of $S_k(m)$ roughly, and later we make precise some finer aspects. As was already mentioned, we proceed by induction on k and m. One of the invariants of the construction is that the left endpoints of all the fans of $S_k(m)$ always show up on the lower envelope.

First we specify the boundary cases with $k = 1$ or $m = 1$. For $k = 1$, $S_1(m)$ is simply a single fan with m segments. For $m = 1$, $S_k(1)$ is obtained from $S_{k-1}(2)$ by the following transformation of each fan (each fan has 2 segments):

The lower segment in each fan is translated by the same tiny amount to the left.

Now we describe the construction of $S_k(m)$ for general $k, m \geq 2$. First we construct $S_k(m-1)$ inductively. We shrink this $S_k(m-1)$ both vertically and horizontally by a suitable affine transform; the vertical shrinking is much more intensive than the horizontal one, so that all segments become very short *and* almost horizontal. Let S' be the transformed $S_k(m-1)$. We will use many translated copies of S' as "microscopic" ingredients in the construction of $S_k(m)$.

The "master plan" of the construction is obtained from $S_{k-1}(M)$, where $M = f_k(m-1)$ is the number of fans in S'. Namely, we first shrink $S_{k-1}(M)$

vertically so that all segments become nearly horizontal, and then we apply the affine transform $(x, y) \mapsto (x, x + y)$ so that the slopes of all the segments are just a little over 1. Let S^* denote the resulting set.

For each fan F in the master construction S^*, we make a copy S'_F of the microscopic construction S' and place it so that its leftmost endpoint coincides with the left endpoint of F. Let the segments of F be $s_1, \ldots, s_M$, numbered by increasing slopes, and let $\ell_1, \ldots, \ell_M$ be the left endpoints of the fans in S'_F, numbered from left to right. The fan F is gigantic compared to S'_F. Now we take F apart: We translate each s_i so that its left endpoint goes to ℓ_i. The following drawing shows this schematically, since we have no chance to make a realistic drawing of $S_k(m-1)$. Only a very small part of F near its left endpoint is shown.

This construction yields $S_k(m)$. It correctly produces fans of size m, by appending one top (and long) segment to each fan in every S'_F. If S' was taken sufficiently tiny, then all the vertices of the lower envelope of S^* are preserved, as well as those in each S'_F. Crucially, we need to make sure that the above transformation of each fan F in S^* yields $M-1$ *new* vertices on the envelope, as is indicated below:

The new vertices lie on the right of S'_F but, in the scale of the master construction S^*, very close to the former left endpoint of F, and so they indeed appear on the lower envelope.

This is where we need to make the whole construction more precise, namely, to say more about the structure of the fans in $S_k(m)$. Let us call a fan r-*escalating* if the ratio of the slopes of every two successive segments in the fan is at least r. It is not difficult to check that for any given $r > 1$, the construction of $S_k(m)$ described above can be arranged so that all fans in the resulting set are r-escalating.

Then, in order to guarantee that the $M-1$ new vertices per fan arise in the general inductive step described above, we make sure that the fans in the master construction S^* are affine transforms of r-escalating fans for a suitable very large r. More precisely, let Q be a given number and let $r = r(k, Q)$ be sufficiently large and $\delta = \delta(k, Q) > 0$ sufficiently small. Let F arise from an r-escalating fan by the affine transformation described above (which makes all slopes a little bigger than 1), and assume that the shortest segment has length 1, say. Suppose that we translate the left endpoint of s_i, the segment with the ith smallest slope in F, by $\delta_1 + \delta_2 + \cdots + \delta_i$ almost horizontally to the right, where $\delta \leq \delta_i \leq Q\delta$. Then it is not difficult to see, or calculate, that the lower envelope of the translated segments of F looks combinatorially like that in the last picture and has $M-1$ new vertices. The reader who is not satisfied with this informal argument can find real and detailed calculations in the book [SA95].

We want to prove that the complexity of the lower envelope of $S_k(m)$ is at least $\frac{1}{2}km$ times the number of fans; in our notation,

$$e_k(m) \geq \tfrac{1}{2}km \cdot f_k(m).$$

This is simple to do by induction, although the numbers involved are frighteningly large. For $k = 1$, we have $f_1(m) = 1$ and $e_1(m) = m+1$, so we are fine. For $m = 1$, we obtain $f_k(1) = 2f_{k-1}(2)$ and $e_k(1) = e_{k-1}(2) + 2f_{k-1}(2) \geq \frac{1}{2}(k-1) \cdot 2 \cdot f_{k-1}(2) + 2f_{k-1}(2) = (k+1) \cdot f_{k-1}(2) > \frac{1}{2}k \cdot f_k(1)$.

In the construction of $S_k(m)$ for $k, m \geq 2$, each of the $f_{k-1}(M)$ fans of the master construction S^* produces $M = f_k(m-1)$ fans, and so

$$f_k(m) = f_{k-1}(M) \cdot M.$$

For the envelope complexity we get a contribution of $e_{k-1}(M)$ from S^*, $e_k(m-1)$ from each copy of S', and $M-1$ new vertices for each copy of S'. Putting this together and using the inductive assumption to eliminate the function e, we have

$$
\begin{aligned}
e_k(m) &\geq e_{k-1}(M) + f_{k-1}(M)\Big[e_k(m-1) + M - 1\Big] \\
&\geq f_{k-1}(M) \cdot \Big[\tfrac{1}{2}(k-1)M + \tfrac{1}{2}k(m-1)M + M - 1\Big] \\
&\geq f_{k-1}(M) \cdot \Big[\tfrac{1}{2}kM + \tfrac{1}{2}k(m-1)M\Big] \\
&= \tfrac{1}{2}km \cdot M \cdot f_{k-1}(M) = \tfrac{1}{2}km \cdot f_k(m).
\end{aligned}
$$

Theorem 7.2.1 is proved.   □

Note how the properties of the construction $S_k(m)$ contradict the intuition gained from small pictures: Most of the segments appear many times on the lower envelope, and between two successive segment endpoints on the envelope there is typically a concave arc with quite a large number of vertices.

Bibliography and remarks. An example of n segments with superlinear complexity of the lower envelope was first obtained by Wiernik and Sharir [WS88], based on an abstract combinatorial construction of Davenport–Schinzel sequences of order 3 due to Hart and Sharir [HS86]. The simpler construction shown in this section was found by Shor (in an unpublished manuscript; a detailed presentation is given in [SA95]).

Exercises

1. Construct Davenport–Schinzel sequences of order 3 of superlinear length directly. That is, rephrase the construction explained in this section in terms of Davenport–Schinzel sequences instead of segments. ③

7.3 More on Davenport–Schinzel Sequences

Here we come back to the asymptotics of the Davenport–Schinzel sequences. We have already proved that $\lambda_3(n)/n$ is unbounded. It even turns out that the construction in the proof of Theorem 7.2.1 yields an asymptotically tight lower bound for $\lambda_3(n)$, which is of order $n\alpha(n)$. Of course, we should explain what $\alpha(n)$ is.

In order to define the extremely slowly growing function α, we first introduce a hierarchy of very fast growing functions $A_1, A_2, \ldots$. We put

$$A_1(n) = 2n,$$
$$A_k(n) = A_{k-1} \circ A_{k-1} \circ \cdots \circ A_{k-1}(1) \quad (n\text{-fold composition}), \; k = 2, 3, \ldots .$$

Only the first few of these functions can be described in usual terms: We have $A_2(n) = 2^n$ and $A_3(n) = 2^{2^{\cdot^{\cdot^2}}}$ with n twos in the exponential tower. The *Ackermann function*[1] $A(n)$ is defined by diagonalizing this hierarchy:

$$A(n) = A_n(n).$$

And α is the inverse function to A:

$$\alpha(n) = \min\{k \geq 1 \colon A(k) \geq n\}.$$

Since $A(4)$ is a tower of 2's of height 2^{16}, encountering a number n with $\alpha(n) > 4$ in any physical sense is extremely unlikely.

---

[1] Several versions of the Ackermann function can be found in the literature, differing in minor details but with similar properties and orders of magnitude.

The Ackermann function was invented as an example of a function growing faster than any primitive recursive function. For people familiar with some of the usual programming languages, the following semiformal explanation can be given: No function as large as $A(n)$ can be evaluated by a program containing only FOR loops, where the number of repetitions of each loop in the program has been computed before the loop begins. For a long time, it was thought that $A(n)$ was a curiosity irrelevant to "natural" mathematical problems. Then theoretical computer scientists discovered it in the analysis of an extremely simple algorithm that manipulates rooted trees, and subsequently it was found in the backyard of elementary geometry, namely in the asymptotics of the Davenport–Schinzel sequences.

As was already remarked above, a not too difficult analysis of the construction in Theorem 7.2.1 shows that $\lambda_3(n) = \Omega(n\alpha(n))$. This is the correct order of magnitude, and we will (almost) present the matching upper bound in the next section. Even the constants in the asymptotics of $\lambda_3(n)$ are known with surprising precision. Namely, we have

$$\tfrac{1}{2}\,n\alpha(n) - 2n \leq \lambda_3(n) \leq 2n\alpha(n) + O\left(\sqrt{n\alpha(n)}\right),$$

and so the gap in the main term is only a factor of 4, in spite of the complexity of the whole problem!

Higher-order Davenport–Schinzel sequences and their generalizations. The asymptotics of the functions $\lambda_s(n)$ for fixed $s > 3$, which correspond to forbidden patterns $ababa\ldots$ with $s+2$ letters, is known quite well, although not entirely precisely. In particular, $\lambda_4(n)$ is of the (strange) order $n \cdot 2^{\alpha(n)}$, and for a general fixed s, we have

$$n \cdot 2^{p_s(\alpha(n))} \leq \lambda_s(n) \leq n \cdot 2^{q_s(\alpha(n))},$$

where $p_s(x)$ is a polynomial of degree $\lfloor \frac{s-2}{2} \rfloor$ (with a positive leading coefficient) and $q_s(x)$ is a polynomial of the same degree, for s odd multiplied by $\log x$. The proofs are similar in spirit to those shown for $s = 3$ but technically much more complicated. On the other hand, proving something like $\lambda_s(n) = O(n \log^* n)$ for every fixed s is not very difficult with the tricks from the proof of Proposition 7.4.2 below (see Exercise 7.4.1).

The Davenport–Schinzel sequences have the simple alternating forbidden pattern $ababa\ldots$. More generally, one can consider sequences with an arbitrary fixed forbidden pattern v, such as $abcdabcdabcd$, where a, b, c, d must be distinct symbols. Of course, here it is not sufficient to require that every two successive symbols in the sequence be distinct, since then the whole sequence could be $121212\ldots$ of arbitrary length. To get a meaningful problem, one can assume that if the forbidden pattern v has k distinct letters ($k = 4$ in our example), then each k consecutive letters in the considered sequence avoiding v must be distinct. Let $\mathrm{Ex}(v, n)$ denote the maximum possible length of such a sequence over n symbols. It is known that for every fixed v, we have

$$\mathrm{Ex}(v, n) \leq O\left(n \cdot 2^{\alpha(n)^c}\right)$$

for a suitable exponent $c = c(v)$. In particular, the length of such sequences is nearly linear in n. Moreover, many classes of patterns v are known with $\mathrm{Ex}(v, n) = O(n)$, although a complete characterization of such patterns is still elusive. For example, for patterns v consisting only of two letters a and b, $\mathrm{Ex}(v, n)$ is linear in n if and only if v contains no subsequence $ababa$ (not necessarily contiguous). These results have already found nice applications in combinatorial geometry and in enumerative combinatorics.

Bibliography and remarks. Davenport and Schinzel [DS65] defined the sequences now associated with their names in 1965, motivated by a geometric problem from control theory leading to lower envelopes of a collection of planar curves. They established some simple upper bounds on $\lambda_s(n)$. The next major progress was made by Szemerédi [Sze74], who proved that $\lambda_s(n) \leq C_s n \log^* n$ for a suitable C_s, where $\log^* n$ is the inverse of the tower function $A_3(n)$. Over ten more years passed until the breakthrough of Hart and Sharir [HS86], who showed that $\lambda_3(n)$ is of order $n\alpha(n)$. A recollection of Sharir about their discovery, after several months of trying to prove a linear upper bound and then learning about Szemerédi's paper, deserves to be reproduced (probably imprecisely but with Micha Sharir's kind consent): "We decided that if Szemerédi didn't manage to prove that $\lambda_3(n)$ is linear then it is probably not linear. We were aware of only one result with a nonlinear lower bound not exceeding $O(n \log^* n)$, and this was Tarjan's bound of $\Theta(n\alpha(n))$ for path compressions. In desperation, we tried to relate it to our problem, and a miracle happened: The construction Tarjan used for his lower bound could be massaged a little so as to yield a similar lower bound for $\lambda_3(n)$."

The *path compression* alluded to is an operation on a rooted tree. Let T be a tree with root r and let p be a leaf-to-root path of length at least 2 in T. The compression of p makes all the vertices on p, except for r, sons of r, while all the other father-to-son relations in T remain unchanged. Tarjan [Tar75] proved, as a part of an analysis of a simple algorithm for the so-called UNION-FIND problem, that if T is a suitably balanced rooted tree with n nodes, then the total length of all paths in any sequence of successive path compressions performed on T is no more than $O(n\alpha(n))$, and this is asymptotically tight in the worst case. Hart and Sharir put Davenport–Schinzel sequences of order 3 into correspondence with *generalized path compressions* (where only some nodes on the considered path become sons of the root, while the others retain the same father) and analyzed them in the spirit of Tarjan's proofs. Later the proofs were simplified and rephrased by Sharir to work directly with Davenport–Schinzel sequences.

The constant $\frac{1}{2}$ in the lower bound on $\lambda_3(n)$ is by Wiernik and Sharir [WS88], and the 2 in the upper bound is due to Klazar [Kla99] (he gives a self-contained proof somewhat different from that in [SA95]).

The most precise known bounds for $\lambda_s(n)$ with $s \geq 4$ were obtained by Agarwal, Sharir, and Shor [ASS89], as a slight improvement over earlier results of Sharir.

Davenport–Schinzel sequences are encountered in many geometric and nongeometric situations. Even the straightforward bound $\lambda_2(n) = 2n-1$ is often useful for simplifying proofs, and the asymptotics of the higher-order sequences allow one to prove bounds involving the function $\alpha(n)$ without too much work, although such bounds are difficult to derive from scratch. Numerous applications, mostly geometric, are listed in [SA95].

Single cell. Pollack, Sharir, and Sifrony [PSS88] proved that the complexity of a single cell in an arrangement of n segments in the plane is at most $O(n\alpha(n))$, by a reduction to Davenport–Schinzel sequences of order 3 (see Exercise 1). A similar argument shows that a single cell in an arrangement of n curves, with every two curves intersecting at most s times, has complexity $O(\lambda_{s+2}(n))$ (see [SA95]).

Generalized Davenport–Schinzel sequences were first considered by Adamec, Klazar, and Valtr [AKV92]. The near-linear upper bound $\mathrm{Ex}(v, n) = O(n \cdot 2^{\alpha(n)^c})$ mentioned in the text is from Klazar [Kla92]. The most general results about sequences u with $\mathrm{Ex}(u, n) = O(n)$ were obtained by Klazar and Valtr [KV94]. A recent survey, including applications of the generalized Davenport–Schinzel sequences, was written by Valtr [Val99a].

We mention two applications. The first one concerns Ramsey-type questions for geometric graphs (already considered in the notes to Section 4.3). We consider an n-vertex graph G drawn in the plane whose edges are straight segments, and we ask, what is the maximum possible number of edges of G so that the drawing does not contain a certain geometric configuration? Here we are interested in the following two types of configurations: *k pairwise crossing edges*

   3 pairwise crossing edges

and *k pairwise parallel edges*, where two edges are called parallel if they do not cross and their four vertices are in convex position:

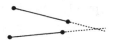

A graph with no two crossing edges is planar and thus has $O(n)$ vertices. It seems to be generally believed that forbidding k pairwise crossing edges forces $O(n)$ edges for every fixed k. This has been proved for $k = 3$ by Agarwal, Aronov, Pach, Pollack, and Sharir [AAP+97], and for all $k \geq 4$, the best known bound is $O(n \log n)$ due to Valtr (see [Val99a]). For k forbidden pairwise parallel edges, he derived an $O(n)$ bound for every fixed k using generalized Davenport–Schinzel sequences, and the $O(n \log n)$ bound for k pairwise crossing edges follows by a neat simple reduction. In this connection, let us mention a nice open question: What is the smallest $n = n(k)$ such that any straight-edge drawing of the complete graph K_n always contains k pairwise crossing edges? The best known bound is $O(k^2)$ [AEG+94], but perhaps the truth is $O(k)$ or close to it.

The second application of generalized Davenport–Schinzel sequences concerns a conjecture of Stanley and Wilf. Let σ be a fixed permutation of $\{1, 2, \ldots, k\}$. We say that a permutation π of $\{1, 2, \ldots, n\}$ *contains* σ if there are indices $i_1 < i_2 < \cdots < i_k$ such that $\sigma(u) < \sigma(v)$ if and only if $\pi(i_u) < \pi(i_v)$, $1 \leq u < v \leq k$. Let $N(\sigma, n)$ denote the number of permutations of $\{1, 2, \ldots, n\}$ that do not contain σ. The Stanley–Wilf conjecture states that for every k and σ there exists C such that $N(\sigma, n) \leq C^n$ for all n. Using generalized Davenport–Schinzel sequences, Alon and Friedgut [AF00] proved that $\log N(\sigma, n) \leq n\beta(n)$ for every fixed σ, where $\beta(n)$ denotes a very slowly growing function, and established the Stanley–Wilf conjecture for a wide class of σ (previously, much fewer cases had been known). Klazar [Kla00] observed that the Stanley–Wilf conjecture is implied by a conjecture of Füredi and Hajnal [FH92] about the maximum number of 1's in an $n \times n$ matrix of 0's and 1's that does not contain a $k \times k$ submatrix having 1's in positions specified by a given fixed $k \times k$ permutation matrix. Füredi and Hajnal conjectured that at most $O(n)$ 1's are possible. The analogous questions for other types of forbidden patterns of 1's in 0/1 matrices are also very interesting and very far from being understood; this is another direction of generalizing the Davenport–Schinzel sequences.

Exercises

1. Let C be a cell in an arrangement of n segments in the plane (assume general position if convenient).
   (a) Number the segments 1 through n and write down the sequence of the segment numbers along the boundary of C, starting from an arbitrarily chosen vertex of the boundary (decide what to do if the boundary has several connected components!). Check that there is no *ababab* subsequence, and hence that the combinatorial complexity of C is no more than $O(\lambda_4(n))$. ③

(b) Find an example where an *ababa* subsequence does appear in the sequence constructed in (a). ③

(c) Improve the argument by splitting the segments suitably, and show that the boundary of C has complexity $O(n\alpha(n))$. ⑤

2. We say that an $n \times n$ matrix A with entries 0 and 1 is good if it contains no $\begin{pmatrix} 1 & * & 1 & * \\ * & 1 & * & 1 \end{pmatrix}$; that is, if there are no indices $i_1 < i_2$ and $j_1 < j_2 < j_3 < j_4$ with $a_{i_1j_1} = a_{i_2j_2} = a_{i_1j_3} = a_{i_2j_4} = 1$.

(a) Prove that a good A has at most $\lambda_s(n) + O(n)$ ones for a suitable constant s. ③

(b) Show that one can take $s = 3$ in (a). ④

7.4 Towards the Tight Upper Bound for Segments

As we saw in Proposition 7.1.1, it is not very difficult to prove that the maximum length of a Davenport–Schinzel sequence of order 3 over n symbols satisfies $\lambda_3(n) = O(n \log n)$. Getting anywhere significantly below this bound seems much harder, and the tight bound requires double induction. But there is only one obvious parameter in the problem, namely the number n, and introducing the second variable for the induction is one of the keys to the proof.

Let $w = a_1 a_2 \ldots a_\ell$ be a sequence. A *nonrepetitive segment* in w is a contiguous subsequence $u = a_i a_{i+1} \ldots a_{i+k}$ consisting of k distinct symbols. A sequence w is *m-decomposable* if it can be partitioned into at most m nonrepetitive segments (the partition need not be unique). Here is the main definition for the inductive proof: Let $\psi(m, n)$ denote the maximum possible length of an m-decomposable Davenport–Schinzel sequence of order 3 over n symbols. First we relate $\psi(m, n)$ to $\lambda_3(n)$.

7.4.1 Lemma. *Every Davenport–Schinzel sequence of order 3 over n symbols is $2n$-decomposable, and consequently,*

$$\lambda_3(n) \leq \psi(2n, n).$$

Proof. Let w be the given Davenport–Schinzel sequence. We define a linear ordering $\preceq$ on the symbols occurring in w: We set $a \preceq b$ if the first occurrence of the symbol a in w precedes the first occurrence of the symbol b. We partition w into *maximal strictly decreasing segments* according to the ordering $\preceq$. Here is an example of such a partitioning (the sequence is chosen so that the usual ordering of the digits coincides with $\preceq$): 1|2|32|421|5|6543. Clearly, each strictly decreasing segment is a nonrepetitive segment as well, and so it suffices to show that the number of the maximal strictly decreasing segments is at most $2n$ (the tight bound is actually $2n-1$).

Let u_j and u_{j+1} be two consecutive maximal strictly decreasing segments, let a be the last symbol of u_j, let i be its position in w, and let b be the first symbol of u_{j+1} (at the $(i+1)$st position). We claim that the ith position is the last occurrence of a or the $(i+1)$st position is the first occurrence of b. This will imply that we have at most $2n$ segments u_i, because each of the n symbols has (at most) one first and one last occurrence.

Supposing that the claim is not valid, we find the forbidden subsequence $ababa$. We have $a \prec b$, for otherwise the $(i+1)$st position could be appended to u_j, contradicting the maximality. The b at position $i+1$ is not the first b, and so there is some b before the ith position. There must be another a even before that b, for otherwise we would have $b \prec a$. Finally, there is an a after the position $i+1$, and altogether we have the desired $ababa$. $\square$

Next, we derive a powerful recurrence for $\psi(m,n)$. It is perhaps best to understand the proof first, and the complicated-looking statement then becomes quite natural.

7.4.2 Proposition. *Let $m, n \geq 1$ and $p \leq m$ be integers, and let $m = m_1 + m_2 + \cdots + m_p$ be a partition of m into p nonnegative addends. Then there is a partition $n = n_1 + n_2 + \cdots + n_p + n^*$ such that*

$$\psi(m,n) \leq 4m + 4n^* + \psi(p,n^*) + \sum_{k=1}^{p} \psi(m_k, n_k).$$

Proof. Let w be an m-decomposable Davenport–Schinzel sequence of order 3 over n symbols attaining $\psi(m,n)$. Let $w = u_1 u_2 \ldots u_m$ be a partition of w into nonrepetitive segments. Let $w_1 = u_1 u_2 \ldots u_{m_1}$ consist of the first m_1 nonrepetitive segments, $w_2 = u_{m_1+1} \ldots u_{m_1+m_2}$ of the next m_2 segments, and so on until w_p. We call $w_1, w_2, \ldots, w_p$ the *parts* of w.

We divide the symbols in w into two classes: A symbol a is *local* if it occurs in (at most) one of the parts w_k, and it is *nonlocal* if it appears in at least two distinct parts. We let n^* be the number of distinct nonlocal symbols and n_k the number of distinct local symbols occurring in w_k.

If we delete all the nonlocal symbols from w_k, we obtain an m_k-decomposable sequence over n_k symbols with no $ababa$. However, this sequence can still contain consecutive repetitions of some symbols, which is forbidden for a Davenport–Schinzel sequence. So we delete all symbols in each repetition but the first one; for example, 122232244 becomes 12324. We note that consecutive repetitions can occur only at the boundaries of the nonrepetitive segments u_j, and so at most m_k-1 local symbols have been deleted from w_k. The remaining sequence is already a Davenport–Schinzel sequence, and so the total number of positions of w occupied by the local symbols is at most

$$\sum_{k=1}^{p} [m_k - 1 + \psi(m_k, n_k)] \leq m + \sum_{k=1}^{p} \psi(m_k, n_k).$$

Next, we need to deal with the nonlocal symbols. Let us say that a nonlocal symbol a is a *middle symbol* in a part w_k if it occurs both before w_k

and after w_k; otherwise, it is a *nonmiddle symbol* in w_k. We estimate the contributions of middle and nonmiddle symbols separately.

First we consider each part w_k in turn, and we delete all local symbols and all nonmiddle symbols from it. Then we look at the sequence that remains from w after these deletions, and we delete all symbols but one from each contiguous repetition. As in the case of the local symbols, we have deleted at most m middle symbols. Clearly, the resulting sequence is a Davenport–Schinzel sequence of order 3 over n^* symbols, and we claim that it is p-decomposable (this is perhaps the most surprising part of the proof). Indeed, if we consider what remained from some w_k, we see that sequence cannot contain a subsequence bab, because some a's precede and follow w_k and we would get the forbidden $ababa$. Therefore, the surviving symbols of w_k form a nonrepetitive segment. Hence the total contribution of the middle symbols to the length of w is at most $m + \psi(p, n^*)$.

The nonmiddle symbols in a given w_k can conveniently be divided into *starting* and *ending* symbols (with the obvious meaning). We concentrate on the total contribution of the starting symbols; the case of the ending symbols is symmetric. Let n_k^* be the number of distinct starting symbols in w_k; we have $\sum_{k=1}^{p} n_k^* \leq n^*$, since a symbol is starting in at most one part. Let us erase from w_k all but the starting symbols, and then we also remove all contiguous repetitions in each w_k, as in the two previous cases. The remaining starting symbols contain no subsequence $abab$, since we know that there is some a following w_k. Thus, what is left of w_k is a Davenport–Schinzel sequence of order 2 over n_k^* symbols, and as such it has length at most $2n_k^* - 1$. Therefore, the total number of starting symbols in all of w is no more than

$$\sum_{k=1}^{p} (m_k - 1 + 2n_k^* - 1) < m + 2n^*.$$

Summing up the contributions of local symbols, middle symbols, starting symbols, and ending symbols, we arrive at the bound claimed in the proposition. Here is a graphic summary of the proof:

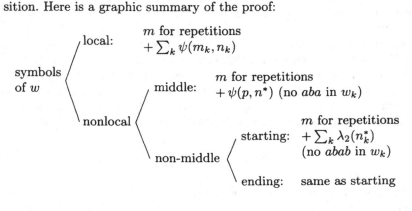

$\square$

How to prove good bounds from the recurrence. The recurrence just proved can be used to show that $\psi(m,n) = O((m+n)\alpha(m))$, and Lemma 7.4.1 then yields the desired conclusion $\lambda_3(n) = O(n\alpha(n))$. We do not give the full calculation; we only indicate how the recurrence can be used to prove better and better bounds starting from the obvious estimate $\psi(m,n) \leq mn$.

First we prove that $\psi(m,n) \leq 4m\log_2 m + 6n$, for m a power of 2. From our recurrence with $p = 2$ and $m_1 = m_2 = \frac{m}{2}$, we obtain

$$\psi(m,n) \leq 4m + 4n^* + \psi(2,n^*) + \psi(\tfrac{m}{2},n_1) + \psi(\tfrac{m}{2},n_2).$$

Proceeding by induction on $\log_2 m$ and using $\psi(2,n) = 2n$, we estimate the last expression by $4m + 4n^* + 2n^* + 2m(\log_2 m - 1) + 6n_1 + 2m(\log_2 m - 1) + 6n_2 = 4m\log_2 m + 6n$ as required.

Next, we assume that $m = A_3(r)$ (the tower function) for an integer r and prove $\psi(m,n) \leq 8rm + 10n$ by induction on r. This time we choose $p = \frac{m}{\log_2 m}$ and $m_k = \frac{m}{p} = \log_2 m = A_3(r-1)$. For estimating $\psi(p,n^*)$ we use the bound derived earlier. This gives

$$\psi(m,n) \leq 4m + 4n^* + 4p\log_2 p + 6n^* + \sum_{k=1}^{p} \psi(m_k,n_k)$$
$$\leq 4m + 4n^* + 4m + 6n^* + 8(r-1)m + 10(n-n^*) = 8rm + 10n.$$

So, by now we already know that $\lambda_3(n) = O(n\log^* n)$, where $\log^* n$ is the inverse to the tower function $A_3(n)$. This bound is as good as linear for practical purposes.

In general, one proves that for $m = A_k(r)$,

$$\psi(m,n) \leq (4k-4)rm + (4k-2)n,$$

by double induction on k and r. The inductive assumption for $k-1$ is always used to bound the term $\psi(p,n^*)$. We omit the rest of the calculation.

Bibliography and remarks. In this section we draw mostly from [SA95], with some changes in terminology.

Exercises

1. For integers $s > t \geq 1$, let $\psi_s^t(m,n)$ denote the maximum length of a Davenport–Schinzel sequence of order s (no subsequence $abab\ldots$ with $s+2$ letters) over n symbols that can be partitioned into m contiguous segments, each of them a Davenport–Schinzel sequence of order t. In particular, $\psi_s(m,n) = \psi_s^1(m,n)$ is the maximum length of a Davenport–Schinzel sequence of order s over n symbols that consists of m nonrepetitive segments.
   (a) Prove that $\lambda_s(n) \leq \psi_s^{s-1}(n,n)$. ②
   (b) Prove that

$$\psi_s^t(m,n) \leq m + \max\left\{\sum_{i=1}^{m} \lambda_t(n_i): \sum_{i=1}^{m} n_i \leq \psi_s(n,m)\right\}.$$

3

(c) Let w be a sequence witnessing $\psi_s(m,n)$ and let $m = m_1 + m_2 + \cdots + m_p$ be some partition of m. Divide w into p parts as in the proof of Proposition 7.4.2, the kth part consisting of m_k nonrepetitive segments. With the terminology and notation of that proof, check that the local symbols contribute at most $m + \sum_{k=1}^{p} \psi_s(m_k, n_k)$ to the length of w, the middle symbols at most $m + \psi_s^{s-2}(p, n^*)$, and the starting symbols no more than $m + \psi_{s-1}(m, n^*)$. 3

(d) Prove by induction that $\psi_s(n,m) \leq C_s \cdot (m+n)\log^{s-2}(m+1)$ and $\lambda_s(n) \leq C_s' n \log^{s-2}(n+1)$, for all $s \geq 2$ and suitable C_s and C_s' depending only on s (set $p = 2$ in (c)). 3

7.5 Up to Higher Dimension: Triangles in Space

As we have seen, lower envelopes in the plane can be handled by means of a simple combinatorial abstraction, the Davenport–Schinzel sequences. Unfortunately, so far, no reasonable combinatorial model has been found for higher-dimensional lower envelopes. The known upper bounds are usually much cruder than those in the plane, but their proofs are quite complex and technical. We start with almost the simplest possible case: triangles in $\mathbf{R}^3$.

Here is an example of the lower envelope of triangles viewed from below:

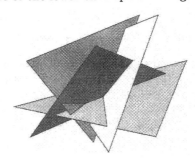

It is actually the vertical projection of the lower envelope on a horizontal plane lying below all the triangles. The projection consists of polygons, both convex and nonconvex, and the combinatorial complexity of the lower envelope is the total number of these polygons plus the number of their edges and vertices. Simple arguments, say using the Euler relation for planar graphs, show that if we do not care about constant factors, it suffices to consider the vertices of the polygons.

It turns out that the worst-case complexity of the lower envelope is of order $n^2\alpha(n)$. Here we prove a simpler, suboptimal bound:

7.5.1 Proposition. *The combinatorial complexity of the lower envelope of* n *triangles in* $\mathbf{R}^3$ *is at most* $O(n\sigma(n)\log n) = O(n^2\alpha(n)\log n)$, *where* $\sigma(n)$ *stands for the maximum complexity of the lower envelope of* n *segments in the plane.*

It is convenient, although not really essential, to work with triangles in general position. As usual, a perturbation argument shows that this is where the maximum complexity of the lower envelope is attained. The precise general position requirements can be found by inspecting the forthcoming proof, and we leave this to the reader.

Walls and boundary vertices. Let H be a set of n triangles in $\mathbf{R}^3$ in general position. We need to bound the total number of vertices in the projection of the lower envelope. The vertices are of two types: those that lie on the vertical projection of an edge of some of the triangles (*boundary vertices*), and those obtained from intersections of 3 triangles (*inner vertices*). In the above picture there are many boundary vertices but only two inner vertices. Yet the boundary vertices are rather easy to deal with, while the inner vertices present the real challenge.

We claim that the total number of boundary vertices is at most $O(n\sigma(n))$. To see this, let a be an edge of a triangle $h \in H$ and let π_a be the "vertical wall" through a, i.e., the union of all vertical lines that intersect a. Each triangle of H intersects π_a in a (possibly empty) segment. The following drawing shows the triangle h, the wall π_a, and the segments within it:

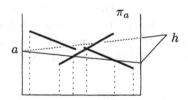

Essentially, the boundary vertices lying on the vertical projection of a correspond to breakpoints of the lower envelope of these segments within π_a. Only the segment a needs special treatment, since on the one hand, its intersections with other segments can give rise to boundary vertices, but on the other hand, it does not obscure things lying above it. To take care of this, we can consider two lower envelopes, one for the arrangement including a and another without a. So each edge a contributes at most $2\sigma(n)$ boundary vertices, and the total number of boundary vertices is $O(n\sigma(n))$.

Levels. Each inner vertex of the projected lower envelope corresponds to a vertex of the arrangement of H lying on the lower envelope, i.e., of level 0 (recall that according to our definition of arrangement, the vertices are intersections of 3 triangles). The *level* of a vertex v is defined in the usual way: It is the number of triangles of H that intersect the open ray emanating from v vertically downwards. Let $f_k(H)$ denote the number of vertices of level k,

$k = 0, 1, \ldots$. Further, let $f_k(n)$ be the maximum of $f_k(H)$ over all sets H of n triangles (in general position). So our goal is to estimate $f_0(n)$.

The first part of the proof of Proposition 7.5.1 employs a probabilistic argument, very similar to the one in the proof of the zone theorem (Theorem 6.4.1), to relate $f_0(H)$ and $f_1(H)$ to $f_0(n-1)$.

7.5.2 Lemma. *For every set H of n triangles in general position, we have*

$$\frac{n-3}{n} f_0(H) \le f_0(n-1) - \frac{1}{n} f_1(H).$$

Proof. We pick one triangle $h \in H$ at random and estimate $\mathbf{E}[f_0(H \setminus \{h\})]$, the expected number of vertices of the lower envelope after removing h. Every vertex of the lower envelope of H is determined by 3 triangles, and so its chances of surviving the removal of h are $\frac{n-3}{n}$. For a vertex v of level 1, the probability of its appearing on the lower envelope is $\frac{1}{n}$, since we must remove the single triangle lying below v. Therefore,

$$\mathbf{E}[f_0(H \setminus \{h\})] = \frac{n-3}{n} f_0(H) + \frac{1}{n} f_1(H).$$

The lemma follows by using $f_0(H \setminus \{h\}) \le f_0(n-1)$. $\qquad\square$

Before proceeding, let us inspect the inequality in the lemma just proved. Let H be a set of n triangles with $f_0(H) = f_0(n)$. If we ignored the term $\frac{1}{n} f_1(H)$, we would obtain the recurrence $\frac{n-3}{n} f_0(n) \le f_0(n-1)$. This yields only the trivial estimate $f_0(n) = O(n^3)$, which is not surprising, since we have used practically no geometric information about the triangles. In order to do better, we now want to show that $f_1(H)$ is almost as big as $f_0(H)$, in which case the term $\frac{1}{n} f_1(H)$ decreases the right-hand side significantly. Namely, we prove that

$$f_1(H) \ge f_0(H) - O(n\sigma(n)). \tag{7.1}$$

Substituting this into the inequality in Lemma 7.5.2, we arrive at

$$\frac{n-2}{n} f_0(n) \le f_0(n-1) + O(\sigma(n)).$$

We practiced this kind of recurrences in Section 6.4: The substitution $\varphi(n) = \frac{f_0(n)}{n(n-1)}$ quickly yields $f_0(n) = O(n\sigma(n) \log n)$. So in order to prove Proposition 7.5.1, it remains to derive (7.1), and this is the geometric heart of the proof.

Making someone pay for the level-0 vertices. We are going to relate the number of level-0 vertices to the number of level-1 vertices by a local charging scheme: From each vertex v of level 0, we walk around a little and find suitable vertices of level 1 to pay for v, as follows.

The level-0 vertex v is incident to 6 edges, 3 of them having level 0 and 3 level 1:

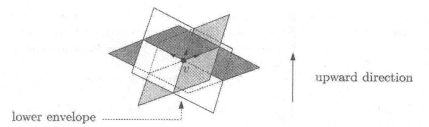

upward direction

lower envelope

The picture shows only a small square piece from each of the triangles incident to v. The lower envelope is on the bottom, and the edges of level 1 emanating from v are marked by arrows. Let e be one of the level-1 edges going from v away from the lower envelope. We follow it until one of the following events occurs:

(i) We reach the intersection v' of e with a vertical wall π_a erected from an edge a of some triangle. This v' pays 1 unit to v.

(ii) We reach the intersection v' of e with another triangle; i.e., v' is a vertex of the arrangement of H. This v' pays $\frac{1}{3}$ of a unit to v.

This is done for all 3 level-1 edges emanating from v and for all vertices v of level 0. Clearly, every v receives at least 1 unit in total. It remains to discuss what kind of vertices the v' are and to estimate the total charge paid by them.

Since there is no other vertex on e between v and v', a particular v' can be reached from at most 2 distinct v in case (i) and from at most 3 distinct v in case (ii). So a v' is charged at most 2 according to case (i) or at most 1 according to case (ii) (because of the general position of H, these cases are never combined, since no intersection of 3 triangles lies in any of the vertical walls π_a).

Next, we observe that in case (i), v' has level at most 2, and in case (ii), it has level exactly 1. This is best seen by considering the situation within the vertical plane containing the edge e. As we move along e, just after leaving v we are at level 1, with exactly one triangle h below, as is illustrated next:

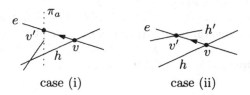

case (i)   case (ii)

The level does not change unless we enter a vertical wall π_a or another triangle $h' \in H$. If we first enter some π_a, then case (i) occurs with $v' = e \cap \pi_a$, and the level cannot change by more than 1 by entering π_a. If we first reach a triangle h', we have case (ii) with $v' = e \cap h'$, and v' has level 1.

Each v' reached in case (i) is a vertex in the arrangement of segments within one of the walls π_a, and it has level at most 2 there. It is easy to show by the technique of the proof of Clarkson's theorem on levels (Theorem 6.3.1) that the number of vertices of level at most 2 in an arrangement of n segments is $O(\sigma(n))$ (Exercise 2). Since we have $3n$ walls π_a, the total amount paid according to case (i) is $O(n\sigma(n))$.

As for case (ii), all the v' are at level 1, and each pays at most 1, so the total charge is at most $f_1(H)$.

Therefore, $f_0(H) \leq f_1(H) + O(n\sigma(n))$, which establishes (7.1) and concludes the proof of Proposition 7.5.1. $\qquad\qquad\qquad\square$

> **Bibliography and remarks.** The sharp bound of $O(n^2\alpha(n))$ for the lower envelope of n triangles in $\mathbf{R}^3$ was first proved by Pach and Sharir [PS89] using a divide-and-conquer argument. A tight bound of $O(n^{d-1}\alpha(n))$ for $(d-1)$-dimensional simplices in $\mathbf{R}^d$ was established a little later by Edelsbrunner [Ede89]. Tagansky [Tag96] found a considerably simpler argument and also proved some new results. We used his method in the proof of Proposition 7.5.1, but since we omitted a subtler analysis of the charging scheme, we obtained a suboptimal bound. To improve the bound to $O(n^2\alpha(n))$, the charging scheme is modified a little: The v' reached in case (i) pays $\frac{4}{3}$ instead of 1, and the v' reached in case (ii) pays $\frac{1}{k}$ if it was reached from $k \leq 3$ distinct v. Then it can be shown, with some work, that every vertex of the lower envelope receives a charge of at least $\frac{4}{3}$ (and not only 1); see [Tag96]. Hence $f_1(H) \geq \frac{4}{3}f_0(H) - O(n\sigma(n))$, and the resulting recurrence becomes $\frac{n-5/3}{n}f_0(n) \leq f_0(n-1) + O(\sigma(n))$. It implies $f_0(n) = O(n\sigma(n))$; proving this is somewhat complicated, since the simple substitution trick does not work here.

Exercises

1. Given a construction of a set of n segments in the plane with lower envelope of complexity $\sigma(n)$, show that the lower envelope of n triangles in $\mathbf{R}^3$ can have complexity $\Omega(n\sigma(n))$. ②

2. Show that the number of vertices of level at most k in the arrangement of n segments (in general position) in the plane is at most $O(k^2\sigma(\lfloor\frac{n}{k+1}\rfloor))$. The proof of the general case of Clarkson's theorem on levels (Theorem 6.3.1) applies almost verbatim. ①

7.6 Curves in the Plane

In the proof for triangles shown in the previous section, if we leave a vertex on the lower envelope along an edge of level 1, we cannot come back to the lower envelope before one of the events (i) or (ii) occurs. Once we start considering lower envelopes of curved surfaces, such as graphs of polynomials of degree

s for some fixed s, this is no longer true: The edge can immediately go back to another vertex on the lower envelope. Then we would be trying to charge one vertex of the lower envelope to another. This can be done, but one must define an "order" for each vertex, and charge envelope vertices of order i only to vertices of order smaller than i or to vertices of significantly higher levels.

We show this for the case of curves in the plane. This example is artificial, since using Davenport–Schinzel sequences leads to much sharper bounds. But we can thus demonstrate the ideas of the higher-dimensional proof, while avoiding many technicalities. We remark that this proof is not really an upgrade of the one for triangles: Here we aim at a much cruder bound, and so some of the subtleties in the proof for triangles can be neglected.

We consider n planar curves as discussed in Section 7.1: They are graphs of continuous functions $\mathbf{R} \to \mathbf{R}$, and every two intersect at most s times. Moreover, we assume for convenience that the curves cross at each intersection and no 3 curves have a common point.

7.6.1 Proposition. *The maximum possible number of vertices on the lower envelope of a set H of n curves as above is at most $O(n^{1+\varepsilon})$ for every fixed $\varepsilon > 0$. That is, for every s and every $\varepsilon > 0$ there exists C such that the bound is at most $Cn^{1+\varepsilon}$ for all n.*

Proof. Let v be a vertex of the arrangement of H. We say that v has *order i* if it is the ith leftmost intersection of the two curves defining it. So the order is an integer between 1 and s.

Let $f_{\leq k}^{(i)}(H)$ denote the number of vertices of order i and level at most k in the arrangement of H. Let $f_{\leq k}^{(i)}(n)$ be the maximum of this quantity over all n-element sets H of curves as in the proposition. Further, we write $f_{\leq k}(H) = \sum_{i=1}^{s} f_{\leq k}^{(i)}(H)$ for the total number of vertices of level at most k. For $k = 0$ we write just f instead of $f_{\leq 0}$ and similarly for $f^{(i)}$.

Let v be a vertex of order i on the lower envelope. We define a charging scheme; that is, we describe who is going to pay for v. We start walking from v to the left along the curve h passing through v and not being on the lower envelope on the left of v. If k_i vertices are encountered, without returning to the lower envelope or escaping to $-\infty$, then we charge each of these k_i vertices $\frac{1}{k_i}$ units. Here $k_1, k_2, \ldots, k_s$ are integer parameters whose values will be fixed later, but one can think of them as very large constants.

If we end up at $-\infty$ before encountering k_i vertices, we charge 1 to the curve h itself. Finally, if we are back at the lower envelope without having passed at least k_i vertices, then, crucially, we must have crossed the second curve h' defining the vertex v again, at a vertex v' of order $i-1$, and this v' pays 1 for v. A picture illustrates these three cases of charging:

We see that v can charge a curve or a vertex of a smaller order significantly, or it can charge many vertices of arbitrary orders, but each of them just a little.

We do this charging for all vertices v of order i on the lower envelope. A given vertex v' of the arrangement can be charged only if it has level at most k_i, and it can be charged at most twice: The vertices of the lower envelope that might possibly charge v' can be found by following the two curves passing through v' to the right. So if v' has order different from $i-1$, then it pays at most $\frac{2}{k_i}$, and if it has order $i-1$, then it can be charged 1 extra. Finally, each curve pays at most 1. Since at least 1 unit was paid for each vertex of order i on the lower envelope, we obtain

$$f^{(i)}(n) \le n + \frac{2}{k_i} f_{\le k_i}(n) + f_{\le k_i}^{(i-1)}(n). \tag{7.2}$$

Next, we want to convert this into a recurrence involving only f and $f^{(i)}$. To this end, we estimate $f_{\le k}^{(i)}$ by following the proof of Clarkson's theorem on levels almost literally (as for the case of segments in Exercise 7.5.2). We obtain

$$f_{\le k}^{(i)}(n) = O\left(k^2 f^{(i)}(\lfloor \tfrac{n}{k} \rfloor)\right).$$

By substituting this bound (and its analogue for $f_{\le k}$) into the right-hand side of (7.2), we arrive at the system of inequalities

$$f^{(i)}(n) \le n + C \cdot \left(k_i f(\lfloor \tfrac{n}{k_i} \rfloor) + k_i^2 f^{(i-1)}(\lfloor \tfrac{n}{k_i} \rfloor)\right), \quad i = 1, 2, \ldots, s, \tag{7.3}$$

where C is a suitable constant and where we put $f^{(0)} = 0$. We also have $f \le f^{(1)} + \cdots + f^{(s)}$.

It remains to derive the bound $f(n) = O(n^{1+\varepsilon})$ from this recurrence, which is not really difficult but still somewhat interesting. It is essential that $f(\lfloor \tfrac{n}{k_i} \rfloor)$ appears only with the coefficient k_i on the right-hand side, in contrast to $f^{(i-1)}(\lfloor \tfrac{n}{k_i} \rfloor)$, which has coefficient k_i^2.

Let $\varepsilon > 0$ be small but fixed. Let us see what happens if we try to prove the bounds $f^{(i)}(n) \le A_i n^{1+\varepsilon}$ and $f(n) \le A n^{1+\varepsilon}$ by induction on n using (7.3), where the A_i are suitable (large) constants and $A = \sum_{i=1}^s A_i$. The term n on the right-hand side of (7.3) is small compared to $n^{1+\varepsilon}$, and so we ignore it for the moment. We also neglect the floor functions. By substituting the inductive hypothesis $f^{(i)}(\lfloor \tfrac{n}{k_i} \rfloor) \le A_i(\tfrac{n}{k_i})^{1+\varepsilon}$ into the right-hand side of (7.3), we obtain roughly

$$n^{1+\varepsilon}(CAk_i^{-\varepsilon} + CA_{i-1}k_i^{1-\varepsilon}) \le n^{1+\varepsilon}(CAk_i^{-\varepsilon} + CA_{i-1}k_i).$$

For the induction to work, A_i must be larger than the expression in parentheses. To make A_i bigger than the second term in parentheses, we can set $A_i = 3Ck_iA_{i-1}$, say (the constant 3 is chosen to leave enough room for the other terms). Then $A_i = A_1 C_1^{i-1} k_2 k_3 \cdots k_i$, with $C_1 = 3C$. These A_i grow fast, and so $A \approx A_s$. Then the requirement that A_i be larger than the first term in parentheses yields, after a little simplification,

$$k_i^\varepsilon \geq C_1^{s-i+1} k_{i+1} k_{i+2} \cdots k_s.$$

Therefore, the k_i should decrease very fast with i. We can set $k_s = C_1^{1/\varepsilon}$ and $k_i = (C_1^{s-i+1} k_{i+1} k_{i+2} \cdots k_s)^{1/\varepsilon}$. Now setting A_1, which is still a free parameter, sufficiently (enormously) large, we can make sure that the desired bounds $f^{(i)}(n) \leq A_i n^{1+\varepsilon}$ hold at least up to $n = k_1$, so that we can really use the recurrence (7.3) in the induction with the k_i defined above. These considerations indicate that the induction works; to be completely sure, one should perform it once more in detail. But we leave this to the reader's diligence and declare Proposition 7.6.1 proved. □

Bibliography and remarks. The method shown in this section first appeared in Halperin and Sharir [HS94], who considered lower envelopes of curved objects in $\mathbf{R}^3$.

7.7 Algebraic Surface Patches

Here we state, without proofs, general bounds on the complexity of higher-dimensional lower envelopes. We also discuss a far-reaching generalization: an analogous bound for the complexity of a cell in a d-dimensional arrangement.

Roughly speaking, the lower envelope of any n "well-behaved" pieces of $(d-1)$-dimensional surfaces in $\mathbf{R}^d$ has complexity close to n^{d-1}. While for planar curves it is simple to say what "well-behaved" means, the situation is more problematic in higher dimensions. The known proofs are geometric, and listing as axioms all the geometric properties of "well-behaved pieces of surfaces" actually used in them seems too cumbersome to be useful. Thus, the most general known results, and even conjectures, are formulated for families of *algebraic surface patches*, although it is clear that the proofs apply in more general settings.

First we recall the definition of a *semialgebraic set*. This is a set in $\mathbf{R}^d$ definable by a Boolean combination of polynomial inequalities. More formally, a set $A \subseteq \mathbf{R}^d$ is called semialgebraic if there are polynomials $p_1, p_2, \ldots, p_r \in \mathbf{R}[x_1, \ldots, x_d]$ (i.e., polynomials in d variables with real coefficients) and a Boolean formula $\Phi(X_1, X_2, \ldots, X_r)$ (such as $X_1 \& (X_2 \vee X_3)$), where $X_1, \ldots, X_r$ are variables attaining values "true" or "false", such that

$$A = \left\{ x \in \mathbf{R}^d \colon \Phi\Big(p_1(x) \geq 0, p_2(x) \geq 0, \ldots, p_r(x) \geq 0\Big) \right\}.$$

Note that the formula Φ may involve negations, and so the sets $\{x \in \mathbf{R}^d \colon p_1(x) > 0\}$ and $\{x \in \mathbf{R}^d \colon p_1(x) = 0\}$ are semialgebraic, for example.

One might want to allow for quantifiers, that is, to admit sets like $\{(x_1, x_2) \in \mathbf{R}^2 : \exists y_1 \, \forall y_2 \, p(x_1, x_2, y_1, y_2) \geq 0\}$ for a 4-variate polynomial p. As is useful to know, but not very easy to prove (and we do not attempt it here), each such set is semialgebraic, too: According to a famous theorem of Tarski, it can be defined by a quantifier-free formula.

Let D be the maximum of the degrees of the polynomials $p_1, \ldots, p_r$ appearing in the definition of a semialgebraic set A. Let us call the number $\max(d, r, D)$ the *description complexity*[2] of A. The results about lower envelopes concern semialgebraic sets whose description complexity is bounded by a constant.

An *algebraic surface patch* is a special case of a semialgebraic set: It can be defined as the intersection of the zero set of some polynomial $q \in \mathbf{R}[x_1, \ldots, x_d]$ with a closed semialgebraic set B. Intuitively, $q(x) = 0$ defines a "surface" in $\mathbf{R}^d$, and B cuts off a closed patch from that surface. Note that B can be all of $\mathbf{R}^d$, and so the forthcoming results apply, among others, to graphs of polynomials or, more generally, to surfaces defined by a single polynomial equation.

Let us remark that in the papers dealing with algebraic surface patches, the definition is often more restrictive, and certainly the proofs make several extra assumptions. Most significantly, they usually suppose that the patches are smooth and they intersect transversally; that is, near each point common to the relative interior of k patches, these k patches look locally like k hyperplanes in general position, $1 \leq k \leq d$. These conditions follow from a suitable general position assumption, namely, that the coefficients of all the polynomials appearing in the descriptions of all the patches are algebraically independent numbers.[3] This can be achieved by a perturbation, but a rigorous argument, showing that a sufficiently small perturbation cannot decrease the complexity of the lower envelope too much, is not entirely easy.

The algebraic surface patches are also typically required to be x_d-monotone (every vertical line intersects them only once). This can be guaranteed by partitioning each of the original patches into smaller pieces, slicing them along the locus of points with vertical tangent hyperplanes (and eliminating the vertical pieces).

After these preliminaries, we can state the main theorem.

7.7.1 Theorem. *For every integers b and $d \geq 2$ and every $\varepsilon > 0$, there exists $C = C(d, b, \varepsilon)$ such that the following holds. Whenever $\gamma_1, \gamma_2, \ldots, \gamma_n$ are algebraic surface patches in $\mathbf{R}^d$, each of description complexity at most b, the lower envelope of the arrangement of $\gamma_1, \gamma_2, \ldots, \gamma_n$ has combinatorial complexity at most $Cn^{d-1+\varepsilon}$.*

---

[2] This terminology is not standard.

[3] Real numbers $a_1, a_2, \ldots, a_m$ are *algebraically independent* if there is no nonzero polynomial p with integer coefficients such that $p(a_1, a_2, \ldots, a_m) = 0$.

How is the combinatorial complexity of the lower envelope defined in this general case, by the way? For each γ_i, we define $M_i \subseteq \mathbf{R}^{d-1}$ as the region where γ_i is on the bottom of the arrangement; formally, M_i consists of all $(x_1, x_2, \ldots, x_{d-1}) \in \mathbf{R}^{d-1}$ such that the lowest intersection of the vertical line $\{(x_1, x_2, \ldots, x_{d-1}, t) \colon t \in \mathbf{R}\}$ with $\bigcup_{j=1}^n \gamma_j$ lies in γ_i. The arrangement of the M_i is often called the *minimization diagram* of the γ_i, and the number of its faces is the complexity of the lower envelope.

The proof of Theorem 7.7.1 is quite similar to the one shown in the preceding section. Each lower-envelope vertex is charged either to a vertex of lower order (the intersection of the same d patches but lying more to the left), or to some k_i vertices, or to a vertex within the vertical wall erected from the boundary of some patch (all the charged vertices lying at level at most k_i). The number of vertices of the last type is estimated by using the $(d-1)$-dimensional case of Theorem 7.7.1 (so the whole proof goes by induction on the dimension). To this end, one needs to show that the situation within the $(d-1)$-dimensional vertical wall, which in general is curved, can be mapped to a situation with algebraic surface patches in $\mathbf{R}^{d-1}$. Here the fact that we are dealing with semialgebraic sets is used most heavily.

Theorem 7.7.1 is a powerful result and it provides nontrivial upper bounds on the complexity of various geometric configurations. Sometimes the bound can be improved by a problem-specific proof, but the general lower-envelope result often quickly yields roughly the correct order of magnitude. For examples see Exercise 1 and [SA95] or [AS00a].

Single cell. Bounding the maximum complexity of a single cell in an arrangement is usually considerably more demanding than the lower envelope question, mainly because a cell can have a complicated topology: It can have holes, tunnels, and so on (cells in hyperplane arrangements, no more complicated than the lower envelope, are an honorable exception). The following theorem provides a bound analogous to that of Theorem 7.7.1. It was proved by similar methods but with several new ideas, especially for the topological complexity of the cell.

7.7.2 Theorem. *For every integers b and $d \geq 2$ and every $\varepsilon > 0$, there exist $C_0 = C_0(d, b)$ and $C = C(d, b, \varepsilon)$ such that the following holds. Let K be a cell in the arrangement of n algebraic surface patches in $\mathbf{R}^d$ in general position, each of description complexity at most b. Then the combinatorial complexity of K (the number of faces in its closure) is at most $Cn^{d-1+\varepsilon}$, and its topological complexity (the sum of the Betti numbers) is no more than $C_0 n^{d-1}$.*

The general position assumption can probably be removed, but I am aware of no explicit reference, except for the special case $d = 3$.

Bibliography and remarks. For a thorough discussion of semialgebraic sets and quantifier elimination we refer to books on real algebraic geometry, such as Bochnak, Coste, and Roy [BCR98].

An old conjecture of Sharir asserts that the combinatorial complexity of the lower envelope in the situation of Theorem 7.7.1 is at most $O(n^{d-2}\lambda_s(n))$ for a suitable s depending on the description complexity of the patches. The best known lower bound is $\Omega(n^{d-1}\alpha(n))$, which applies even for simplices.

The decisive advance towards proving Theorem 7.7.1 was made by Halperin and Sharir [HS94], who established the 3-dimensional case. The general case was proved, as a culmination of a long development, by Sharir [Sha94]. A discussion of the general position assumption and the perturbation argument can also be found there. Interestingly, it is not proved that the maximum complexity is attained in general position; rather, it is argued that the expected complexity after an appropriate random perturbation is always at least a fixed fraction of the original complexity minus $O(n^{d-1+\varepsilon})$.

Some applications lead to the following variation of the lower envelope problem: We have two collections $\mathcal{F}$ and $\mathcal{G}$ of algebraic surface patches in $\mathbf{R}^d$, we project the lower envelopes of both $\mathcal{F}$ and $\mathcal{G}$ into $\mathbf{R}^{d-1}$, and we are interested in the complexity of the superimposed projections (where, for $d = 3$, a vertex of the superimposed projections can arise, for example, as the intersection of an edge coming from $\mathcal{F}$ with an edge obtained from $\mathcal{G}$). In $\mathbf{R}^3$, it is known that this complexity is $O(n^{2+\varepsilon})$, where $n = |\mathcal{F}| + |\mathcal{G}|$ (Agarwal, Sharir, and Schwarzkopf [ASS96]); this is similar to the bound for the lower envelopes themselves. The problem remains open in dimensions 4 and higher.

The combinatorial complexity of a *Voronoi diagram* can also be viewed as a lower-envelope problem. Namely, let $s_1, s_2, \ldots, s_n$ be objects in $\mathbf{R}^d$ (points, lines, segments, polytopes), and let ρ be a metric on $\mathbf{R}^d$. Each s_i defines the function $f_i \colon \mathbf{R}^d \to \mathbf{R}$ by $f_i(x) = \rho(x, s_i)$, and the Voronoi diagram of the s_i is exactly the minimization diagram of the graphs of the f_i (i.e., the projection of their lower envelope). If the f_i are algebraic of bounded degree (or can be converted to such functions by a monotone transform of the range), the general lower envelope bound implies that the complexity of the Voronoi diagram in $\mathbf{R}^d$ is no more than $O(n^{d+\varepsilon})$. This result is nontrivial, but it is widely believed that it should be possible to improve it by a factor of n (and even more in some special cases). Several nice partial results are known, mostly obtained by methods similar to those for lower envelopes. Most notably, Chew, Kedem, Sharir, Tagansky, and Welzl [CKS+98] proved that if the s_i are lines in $\mathbf{R}^3$ and the metric ρ is given by a norm whose unit ball is a convex polytope with a constant-bounded number of vertices (this includes the ℓ_1 and ℓ_∞ metrics, but not the Euclidean metric), then the Voronoi diagram has complexity $O(n^2\alpha(n)\log n)$. On the other hand, Aronov [Aro00] constructed, for every $p \in [1, \infty]$, a set of n $(d-2)$-flats in $\mathbf{R}^d$ whose Voronoi diagram under the ℓ_p metric has complexity $\Omega(n^{d-1})$ (Exercise 5.7.3).

Single cell. For a single cell in the arrangement of n simplices in $\mathbf{R}^d$, Aronov and Sharir [AS94] obtained the complexity bound $O(n^{d-1}\log n)$. Halperin and Sharir [HS95] managed to prove Theorem 7.7.2 in dimension 3. The effort was crowned by Basu [Bas98], who showed by an argument inspired by Morse theory that the topological complexity of a single cell in $\mathbf{R}^d$, assuming general position, is $O(n^{d-1})$; the Halperin–Sharir technique then implies the $O(n^{d-1+\varepsilon})$ bound on the combinatorial complexity.

The research of Sharir and his colleagues in this problem (and many other problems discussed in this chapter) has been motivated by questions about automatic motion planning for a robot. For example, let us consider a square-shaped robot in the plane moving among n pairwise disjoint segment obstacles. The placement of the robot can be specified by three coordinates: the position (x, y) of the center and the angle α of rotation. Each obstacle excludes some placements of the robot. With suitable choice of coordinates, say $(x, y, \tan \frac{\alpha}{2})$, the region of excluded placements is bounded by a few algebraic surface patches. Hence all possible placements of the robot reachable from a given position by a continuous obstacle-avoiding movement correspond to a single cell in the arrangement of $O(n)$ algebraic surface patches in $\mathbf{R}^3$. Consequently, the set of reachable placements has combinatorial complexity at most $O(n^{2+\varepsilon})$. Similar reduction works for more general shapes of the robot and of the obstacles (the robot may even have movable parts), as long as the robot and each of the obstacles can be described by a bounded number of algebraic surface patches. Unfortunately, even in quite simple settings, the combinatorial complexity of the reachable region can be very large. For example, a cube robot in $\mathbf{R}^3$ has 6 degrees of freedom, and so its placements correspond to points in $\mathbf{R}^6$. Exact motion planning algorithms thus become rather impractical, and faster approximate algorithms are typically used.

The complexity of unions. This is another type of problem that often occurs in the analysis of geometric algorithms. Let $A_1, A_2, \ldots, A_n$ be sets in the plane, each of them bounded by a closed Jordan curve, and suppose that the boundaries of every A_i and A_j intersect in at most s points. For $s = 2$, the A_i are called *pseudodisks*, and the primary example is circular disks.

pseudodisks

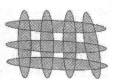
not pseudodisks

For this case Kedem, Livne, Pach, and Sharir [KLPS86] proved that the complexity of $\bigcup_{i=1}^{n} A_i$ is $O(n)$, where the complexity is measured as the sum of the complexities of the "exterior" cells of the arrangement, i.e., the cells that are not contained in any of the A_i.

For $s \geq 4$, long and skinny sets can form a grid pattern and have union complexity about n^2, but linear or near-linear bounds were proved under additional assumptions. One type of such additional assumption is metric, namely, that the objects are "fat." A rather complicated proof of Efrat and Sharir [ES00] shows that if each A_i is convex, the ratio of the circumradius and inradius is bounded by some constant K, and every two boundaries intersect at most s times, then the union complexity is at most $O(n^{1+\varepsilon})$ for any $\varepsilon > 0$, with the constant of proportionality depending on s, K, ε. Earlier, Matoušek, Pach, Sharir, Sifrony, and Welzl [MPS+94] gave a simpler and more precise bound of $O(n \log \log n)$ for fat triangles. Pach, Safruti, and Sharir [PSS01] showed that the union of n fat wedges in $\mathbf{R}^3$ (intersections of two half-spaces with angle at least some $\alpha_0 > 0$), as well as the union of n cubes in $\mathbf{R}^3$, has complexity $O(n^{2+\varepsilon})$. Various extensions of these results to nonconvex objects or to higher dimensions seem easy to conjecture but quite hard to prove.

Several results are known where one assumes that the A_i have special shapes or bounded complexity. Aronov, Sharir, and Tagansky [AST97] proved that the complexity of the union of k convex polygons in the plane with n vertices in total is $O(k^2 + n\alpha(k))$ and that the union of k convex polytopes in $\mathbf{R}^3$ with n vertices in total has complexity $O(k^3 + kn \log k)$. Boissonnat, Sharir, Tagansky, and Yvinec [BSTY98] showed that the union of n axis-parallel cubes in $\mathbf{R}^d$ has $O(n^{\lceil d/2 \rceil})$ complexity, and $O(n^{\lfloor d/2 \rfloor})$ complexity if the cubes all have the same size; both these bounds are tight.

Agarwal and Sharir [AS00c] proved that the union of n infinite cylinders of equal radius in $\mathbf{R}^3$ has complexity $O(n^{2+\varepsilon})$ (here $\Omega(n^2)$ is a lower bound), and more generally, if $A_1, \ldots, A_n$ are pairwise disjoint triangles in $\mathbf{R}^3$ and B is a ball, then $\bigcup_i (A_i + B)$ has complexity $O(n^{2+\varepsilon})$, where $A_i + B = \{a + b \colon a \in A_i, b \in B\}$ is the Minkowski sum. The proof relies on the result mentioned above about two superimposed lower envelopes.

Exercises

1. Let $p_1, \ldots, p_n$ be points in the plane. At time $t = 0$, each p_i starts moving along a straight line with a fixed velocity v_i. Use Theorem 7.7.1 to prove that the convex hull of the n moving points changes its combinatorial structure at most $O(n^{2+\varepsilon})$ times during the time interval $[0, \infty)$. ③
   The tight bound is $O(n^2)$; it was proved, together with many other related results, by Agarwal, Guibas, Herschberger, and Veach [AGHV01].

8

Intersection Patterns of Convex Sets

In Chapter 1 we covered three simple but basic theorems in the theory of convexity: Helly's, Radon's, and Carathéodory's. For each of them we present one closely related but more difficult theorem in the current chapter. These more advanced relatives are selected, among the vast number of variations on the Helly–Radon–Carathéodory theme, because of their wide applicability and also because of nice techniques and tricks appearing in their proofs.

The development started in this chapter continues in Chapters 9 and 10. One of the culminations of this route is the (p, q)-theorem of Alon and Kleitman, which we will prove in Section 10.5. The proof ingeniously combines many of the tools covered in these three chapters and illustrates their power.

Readers who do not like higher dimensions may want to consider dimensions 2 and 3 only. Even with this restriction, the results are still interesting and nontrivial.

8.1 The Fractional Helly Theorem

Helly's theorem says that if every at most $d+1$ sets of a finite family of convex sets in $\mathbf{R}^d$ intersect, then all the sets of the family intersect. What if not necessarily all, but a large fraction of $(d+1)$-tuples of sets, intersect? The following theorem states that then a large fraction of the sets must have a point in common.

8.1.1 Theorem (Fractional Helly theorem). *For every dimension $d \geq 1$ and every $\alpha > 0$ there exists a $\beta = \beta(d, \alpha) > 0$ with the following property. Let $F_1, \ldots, F_n$ be convex sets in $\mathbf{R}^d$, $n \geq d+1$, and suppose that for at least $\alpha\binom{n}{d+1}$ of the $(d+1)$-point index sets $I \subseteq \{1, 2, \ldots, n\}$, we have $\bigcap_{i \in I} F_i \neq \emptyset$. Then there exists a point contained in at least βn sets among the F_i.*

Although simple, this is a key result, and many of the subsequent developments rely on it.

The best possible value of β is $\beta = 1 - (1-\alpha)^{1/(d+1)}$. We prove the weaker estimate $\beta \geq \frac{\alpha}{d+1}$.

Proof. For a subset $I \subseteq \{1, 2, \ldots, n\}$, let us write F_I for the intersection $\bigcap_{i \in I} F_i$.

First we observe that it is enough to prove the theorem for the F_i closed and bounded (and even convex polytopes). Indeed, given some arbitrary $F_1, \ldots, F_n$, we choose a point $p_I \in F_I$ for every $(d+1)$-tuple I with $F_I \neq \emptyset$ and we define $F_i' = \mathrm{conv}\{p_I: F_I \neq \emptyset, i \in I\}$, which is a polytope contained in F_i. If the theorem holds for these F_i', then it also holds for the original F_i. In the rest of the proof we thus assume that the F_i, and hence also all the nonempty F_I, are compact.

Let $\leq_{\mathrm{lex}}$ denote the *lexicographic ordering* of the points of $\mathbf{R}^d$ by their coordinate vectors. It is easy to show that any compact subset of $\mathbf{R}^d$ has a unique lexicographically minimum point (Exercise 1). We need the following consequence of Helly's theorem.

8.1.2 Lemma. *Let $I \subseteq \{1, 2, \ldots, n\}$ be an index set with $F_I \neq \emptyset$, and let v be the (unique) lexicographically minimum point of F_I. Then there exists an at most d-element subset $J \subseteq I$ such that v is the lexicographically minimum point of F_J as well.*

In other words, the minimum of the intersection F_I is always enforced by some at most d "constraints" F_i, as is illustrated in the following drawing (note that the two constraints determining the minimum are not determined uniquely in the picture):

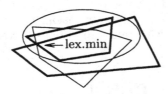

Proof. Let $C = \{x \in \mathbf{R}^d: x <_{lex} v\}$. It is easy to check that C is convex. Since v is the lexicographic minimum of F_I, we have $C \cap F_I = \emptyset$. So the family of convex sets consisting of C plus the sets F_i with $i \in I$ has an empty intersection. By Helly's theorem there are at most $d+1$ sets in this family whose intersection is empty as well. The set C must be one of them, since all the others contain v. The remaining at most d sets yield the desired index set J. $\square$

Let us remark that instead of taking the lexicographically minimum point, one can consider a point minimizing a generic linear function. That formulation is perhaps more intuitive, but it appears slightly more complicated for rigorous presentation.

We can now finish the proof of the fractional Helly theorem. For each of the $\alpha\binom{n}{d+1}$ index sets I of cardinality $d+1$ with $F_I \neq \emptyset$, we fix a d-element set $J = J(I) \subset I$ such that F_J has the same lexicographic minimum as F_I.

The theorem follows by double counting. Since the number of distinct d-tuples J is at most $\binom{n}{d}$, one of them, call it J_0, appears as $J(I)$ for at least $\alpha\binom{n}{d+1}/\binom{n}{d} = \alpha\frac{n-d}{d+1}$ distinct I. Each such I has the form $J_0 \cup \{i\}$ for some $i \in \{1, 2, \ldots, n\}$. The lexicographic minimum of F_{J_0} is contained in at least $d + \alpha\frac{n-d}{d+1} > \alpha\frac{n}{d+1}$ sets among the F_i. Hence we may set $\beta = \frac{\alpha}{d+1}$.   □

Bibliography and remarks. The fractional Helly theorem is due to Katchalski and Liu [KL79]. The quantitatively sharp version with $\beta = 1 - (1-\alpha)^{1/(d+1)}$ was proved by Kalai [Kal84] (and the main result needed for it was proved independently by Eckhoff [Eck85], too). Actually, there is an exact result: If the maximum size of an intersecting subfamily in a family of n convex sets in $\mathbf{R}^d$ is m, then the smallest possible number of intersecting $(d+1)$-tuples is attained for the family consisting of $n - m + d$ hyperplanes in general position and $m - d$ copies of $\mathbf{R}^d$. But there are many other essentially different examples attaining the same bound.

These assertions are consequences of considerably more general results about the possible intersection patterns of convex sets in $\mathbf{R}^d$. For explaining some of them it is convenient to use the language of simplicial complexes. Let $\mathcal{F} = \{F_1, F_2, \ldots, F_n\}$ be a family of convex sets in $\mathbf{R}^d$. The *nerve* $\mathcal{N}(\mathcal{F})$ of $\mathcal{F}$ is the simplicial complex with vertex set $\{1, 2, \ldots, n\}$ whose simplices are all $I \subseteq \{1, 2, \ldots, n\}$ such that $\bigcap_{i \in I} F_i \neq \emptyset$. A simplicial complex obtainable as $\mathcal{N}(\mathcal{F})$ for some family of convex sets in $\mathbf{R}^d$ is called *d-representable*. A characterization of d-representable simplicial complexes for a given d is most likely out of reach. There are several useful *necessary* conditions for d-representability. One certainly worth mentioning is *d-collapsibility*, which means that a given simplicial complex $\mathcal{K}$ can be reduced to the void complex by a sequence of *elementary d-collapsings*, where an elementary d-collapsing consists in deleting a face $S \in \mathcal{K}$ of dimension at most $d-1$ that lies in a unique maximal face of $\mathcal{K}$ and all the faces of $\mathcal{K}$ containing S. The proof of the d-collapsibility of every d-representable complex (Wegner [Weg75]) uses an idea quite similar to the proof of the fractional Helly theorem.

While no characterization of d-representable complexes is known, the possible *f-vectors* of such complexes (where f_i is the number of i-dimensional simplices, which correspond to $(i+1)$-wise intersections here) are fully characterized by a conjecture of Eckhoff, which was proved by Kalai [Kal84], [Kal86] by an impressive combination of several methods. The same characterization applies to d-collapsible complexes as well (and even to the more general *d-Leray complexes*; these are the complexes where the homology of dimension d and larger vanishes for all induced subcomplexes). We do not formulate it but mention one of its consequences, the *upper bound theorem for families of convex sets*: If $f_r(\mathcal{N}(\mathcal{F})) = 0$ for a family $\mathcal{F}$ of n convex sets in $\mathbf{R}^d$ and some r, $d \leq r \leq n$, then $f_k(\mathcal{N}(\mathcal{F})) \leq \sum_{j=0}^{d} \binom{r-d}{k-j+1}\binom{n-r+d}{j}$; equality holds, e.g., in the case mentioned above (several copies of $\mathbf{R}^d$ and hyperplanes in general position).

Exercises

1. Show that any compact set in $\mathbf{R}^d$ has a unique point with the lexicographically smallest coordinate vector. ③

2. Prove the following *colored Helly theorem*: Let $\mathcal{C}_1, \ldots, \mathcal{C}_{d+1}$ be finite families of convex sets in $\mathbf{R}^d$ such that for any choice of sets $C_1 \in \mathcal{C}_1$, $\ldots$, $C_{d+1} \in \mathcal{C}_{d+1}$, the intersection $C_1 \cap \cdots \cap C_{d+1}$ is nonempty. Then for some i, all the sets of $\mathcal{C}_i$ have a nonempty intersection. Apply a method similar to the proof of the fractional Helly theorem; i.e., consider the lexicographic minima of the intersections of suitable collections of the sets. ⑤

   The result is due to Lovász ([Lov74]; also see [Bár82]).

3. Let $F_1, F_2, \ldots, F_n$ be convex sets in $\mathbf{R}^d$. Prove that there exist convex polytopes $P_1, P_2, \ldots, P_n$ such that $\dim(\bigcap_{i \in I} F_i) = \dim(\bigcap_{i \in I} P_i)$ for every $I \subseteq \{1, 2, \ldots, n\}$ (where $\dim(\emptyset) = -1$). ②

8.2 The Colorful Carathéodory Theorem

Carathéodory's theorem asserts that if a point x is in the convex hull of a set $X \subseteq \mathbf{R}^d$, then it is in the convex hull of some at most $d+1$ points of X. Here we present a "colored version" of this statement. In the plane, it shows the following: Given a red triangle, a blue triangle, and a white triangle, each of them containing the origin, there is a vertex r of the red triangle, a vertex b of the blue triangle, and a vertex w of the white triangle such that the tricolored triangle rbw also contains the origin. (In the following pictures, the colors of points are distinguished by different shapes of the point markers.)

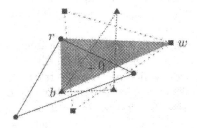

The d-dimensional statement follows.

8.2.1 Theorem (Colorful Carathéodory theorem). *Consider $d+1$ finite point sets $M_1, \ldots, M_{d+1}$ in $\mathbf{R}^d$ such that the convex hull of each M_i contains the point 0 (the origin). Then there exists a $(d+1)$-point set $S \subseteq M_1 \cup \cdots \cup M_{d+1}$ with $|M_i \cap S| = 1$ for each i and such that $0 \in \mathrm{conv}(S)$. (If we imagine that the points of M_i have "color" i, then we look for a "rainbow" $(d+1)$-point S with $0 \in \mathrm{conv}(S)$, where "rainbow" = "containing all colors.")*

Proof. Call the convex hull of a $(d+1)$-point rainbow set a *rainbow simplex*. We proceed by contradiction: We suppose that no rainbow simplex contains 0, and we choose a $(d+1)$-point rainbow set S such that the distance of $\mathrm{conv}(S)$ to 0 is the smallest possible. Let x be the point of $\mathrm{conv}(S)$ closest to 0. Consider the hyperplane h containing x and perpendicular to the segment $0x$, as in the picture:

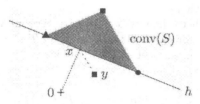

Then all of S lies in the closed half-space h^- bounded by h and not containing 0. We have $\mathrm{conv}(S) \cap h = \mathrm{conv}(S \cap h)$, and by Carathéodory's theorem, there exists an at most d-point subset $T \subseteq S \cap h$ such that $x \in \mathrm{conv}(T)$.

Let i be a color not occurring in T (i.e., $M_i \cap T = \emptyset$). If all the points of M_i lay in the half-space h^-, then 0 would not be in $\mathrm{conv}(M_i)$, which we assume. Thus, there exists a point $y \in M_i$ lying in the complement of h^- (strictly, i.e., $y \notin h$).

Let us form a new rainbow set S' from S by replacing the (unique) point of $M_i \cap S$ by y. We have $T \subset S'$, and so $x \in \mathrm{conv}(S')$. Hence the segment xy is contained in $\mathrm{conv}(S')$, and we see that $\mathrm{conv}(S')$ lies closer to 0 than $\mathrm{conv}(S)$, a contradiction. The colorful Carathéodory theorem is proved. $\square$

This proof suggests an algorithm for finding the rainbow simplex as in the theorem. Namely, start with an arbitrary rainbow simplex, and if it does not contain 0, switch one vertex as in the proof. It is not known whether the number of steps of this algorithm can be bounded by a polynomial function of the dimension and of the total number of points in the M_i. It would be very interesting to construct configurations where the number of steps is very large or to prove that it cannot be too large.

Bibliography and remarks. The colorful Carathéodory theorem is due to Bárány [Bár82]. Its algorithmic aspects were investigated by Bárány and Onn [BO97].

Exercises

1. Let S and T be $(d+1)$-point sets in $\mathbf{R}^d$, each containing 0 in the convex hull. Prove that there exists a finite sequence $S_0 = S, S_1, S_2, \ldots, S_m = T$ of $(d+1)$-point sets with $S_i \subseteq S \cup T$ and $0 \in \mathrm{conv}(S_i)$ for all i, such that S_{i+1} is obtained from S_i by deleting one point and adding another. Assume general position of $S \cup T$ if convenient. Warning: better do not try to find a $(d+1)$-term sequence. ⑤

8.3 Tverberg's Theorem

Radon's lemma states that any set of $d+2$ points in $\mathbf{R}^d$ has two disjoint subsets whose convex hulls intersect. Tverberg's theorem is a generalization of this statement, where we want not only two disjoint subsets with intersecting convex hulls but r of them.

It is not too difficult to show that if we have very many points, then such r subsets can be found. For easier formulations, let $T(d,r)$ denote the smallest integer T such that for any set A of T points in $\mathbf{R}^d$ there exist pairwise disjoint subsets $A_1, A_2, \ldots, A_r \subset A$ with $\bigcap_{i=1}^r \mathrm{conv}(A_i) \neq \emptyset$. Radon's lemma asserts that $T(d,2) = d+2$.

It is not hard to see that $T(d, r_1 r_2) \leq T(d, r_1) T(d, r_2)$ (Exercise 1). Together with Radon's lemma this observation shows that $T(d, r)$ is finite for all r, but it does not give a very good bound.

Here is another, more sophisticated, argument, leading to the (still suboptimal) bound $T(d, r) \leq n = (r-1)(d+1)^2 + 1$. Let A be an n-point set in $\mathbf{R}^d$ and let us set $s = n - (r-1)(d+1)$. A simple counting shows that every $d+1$ subsets of A of size s all have a point of A in common. Therefore, by Helly's theorem, the convex hulls of all s-tuples have a common point x (typically not in A anymore). By Carathéodory's theorem, x is contained in the convex hull of some $(d+1)$-point set $A_1 \subseteq A$. Since $A \setminus A_1$ has at least s points, x is still contained in $\mathrm{conv}(A \setminus A_1)$, and thus also in the convex hull of some $(d+1)$-point $A_2 \subseteq A \setminus A_1$, etc. We can continue in this manner and select the desired r disjoint sets $A_1, \ldots, A_r$, all of them containing x in their convex hulls.

It is not difficult to see that $T(d, r)$ cannot be smaller than $(r-1)(d+1)+1$ (Exercise 2). Tverberg's theorem asserts that this smallest conceivable value is always sufficient.

8.3.1 Theorem (Tverberg's theorem). *Let d and r be given natural numbers. For any set $A \subset \mathbf{R}^d$ of at least $(d+1)(r-1) + 1$ points there exist r pairwise disjoint subsets $A_1, A_2, \ldots, A_r \subseteq A$ such that $\bigcap_{i=1}^r \mathrm{conv}(A_i) \neq \emptyset$.*

The sets $A_1, A_2, \ldots, A_r$ as in the theorem are called a *Tverberg partition* of A (we may assume that they form a partition of A), and a point in the intersection of their convex hulls is called a *Tverberg point*. The following illustration shows what such partitions can look like for $d = 2$ and $r = 3$; both the drawings use the same 7-point set A:

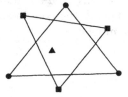 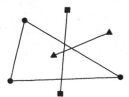

(Are these all Tverberg partitions for this set, or are there more?)

As in the colorful Carathéodory theorem, a very interesting open problem is the existence of an efficient algorithm for finding a Tverberg partition of a given set. There is a polynomial-time algorithm if the dimension is fixed, but some NP-hardness results for closely related problems indicate that if the dimension is a part of input then the problem might be algorithmically difficult.

Several proofs of Tverberg's theorem are known. The one demonstrated below is maybe not the simplest, but it shows an interesting "lifting" technique. We deduce the theorem by applying the colorful Carathéodory theorem to a suitable point configuration in a higher-dimensional space.

Proof of Tverberg's theorem. We begin with a reformulation of Tverberg's theorem that is technically easier to handle. For a set $X \subseteq \mathbf{R}^d$, the *convex cone generated by* X is defined as the set of all linear combinations of points of X with nonnegative coefficients; that is, we set

$$\mathrm{cone}(X) = \left\{ \sum_{i=1}^{n} \alpha_i x_i \colon x_1, \ldots, x_n \in X, \, \alpha_1, \ldots, \alpha_n \in \mathbf{R}, \, \alpha_i \geq 0 \right\}.$$

Geometrically, $\mathrm{cone}(X)$ is the union of all rays starting at the origin and passing through a point of $\mathrm{conv}(X)$. The following statement is equivalent to Tverberg's theorem:

8.3.2 Proposition (Tverberg's theorem: cone version). *Let A be a set of $(d+1)(r-1) + 1$ points in $\mathbf{R}^{d+1}$ such that $0 \notin \mathrm{conv}(A)$. Then there exist r pairwise disjoint subsets $A_1, A_2, \ldots, A_r \subseteq A$ such that $\bigcap_{i=1}^{r} \mathrm{cone}(A_i) \neq \{0\}$.*

Let us verify that this proposition implies Tverberg's theorem. Embed $\mathbf{R}^d$ into $\mathbf{R}^{d+1}$ as the hyperplane $x_{d+1} = 1$ (as in Section 1.1). A set $A \subset \mathbf{R}^d$ thus becomes a subset of $\mathbf{R}^{d+1}$; moreover, its convex hull lies in the $x_{d+1} = 1$ hyperplane, and thus it does not contain 0. By Proposition 8.3.2, the set A can be partitioned into groups $A_1, \ldots, A_r$ with $\bigcap_{i=1}^{r} \mathrm{cone}(A_i) \neq \{0\}$. The intersection of these cones thus contains a ray originating at 0. It is easily checked that such a ray intersects the hyperplane $x_{d+1} = 1$ and that the intersection point is a Tverberg point for A. Hence it suffices to prove Proposition 8.3.2.

Proof of Proposition 8.3.2. Let us put $N = (d+1)(r-1)$; thus, A has $N+1$ points. First we define linear maps $\varphi_j \colon \mathbf{R}^{d+1} \to \mathbf{R}^N$, $j = 1, 2, \ldots, r$. We group the coordinates in the image space $\mathbf{R}^N$ into $r-1$ blocks by $d+1$ coordinates each. For $j = 1, 2, \ldots, r-1$, $\varphi_j(x)$ is the vector having the coordinates of x in the jth block and zeros in the other blocks; symbolically,

$$\varphi_j(x) = \big(\,\underbrace{0\,|\,0\,|\,\cdots\,|\,0}_{(j-1)\times}\,|\,x\,|\,0\,|\,\cdots\,|\,0\,\big).$$

The last mapping, φ_r, has $-x$ in each block: $\varphi_r(x) = (-x\,|\,-x\,|\,\cdots\,|\,-x\,)$.

These maps have the following property: For any r vectors $u_1, \ldots, u_r \in \mathbf{R}^{d+1}$,

$$\sum_{j=1}^{r} \varphi_j(u_j) = 0 \ \text{ holds if and only if } \ u_1 = u_2 = \cdots = u_r. \tag{8.1}$$

Indeed, this can be easily seen by expressing

$$\sum_{j=1}^{r} \varphi_j(u_j) = (u_1 - u_r\,|\,u_2 - u_r\,|\,\cdots\,|\,u_{r-1} - u_r)\,.$$

Next, let $A = \{a_1, \ldots, a_{N+1}\} \subset \mathbf{R}^{d+1}$ be a set with $0 \notin \mathrm{conv}(A)$. We consider the set $M = \varphi_1(A) \cup \varphi_2(A) \cup \cdots \cup \varphi_r(A)$ in $\mathbf{R}^N$ consisting of r copies of A. The first $r-1$ copies are placed into mutually orthogonal coordinate subspaces of $\mathbf{R}^N$. The last copy of each a_i sums up to 0 with the other $r-1$ copies of a_i. Then we color the points of M by $N+1$ colors; all copies of the same a_i get the color i. In other words, we set $M_i = \{\varphi_1(a_i), \varphi_2(a_i), \ldots, \varphi_r(a_i)\}$. As we have noted, the points in each M_i sum up to 0, which means that $0 \in \mathrm{conv}(M_i)$, and thus the assumptions of the colorful Carathéodory theorem hold for $M_1, \ldots, M_{N+1}$.

Let $S \subseteq M$ be a rainbow set (containing one point of each M_i) with $0 \in \mathrm{conv}(S)$. For each i, let $f(i)$ be the index of the point of M_i contained in S; that is, we have $S = \{\varphi_{f(1)}(a_1), \varphi_{f(2)}(a_2), \ldots, \varphi_{f(N+1)}(a_{N+1})\}$. Then $0 \in \mathrm{conv}(S)$ means that

$$\sum_{i=1}^{N+1} \alpha_i \varphi_{f(i)}(a_i) = 0$$

for some nonnegative real numbers $\alpha_1, \ldots, \alpha_{N+1}$ summing to 1. Let I_j be the set of indices i with $f(i) = j$, and set $A_j = \{a_i \colon i \in I_j\}$. The above sum can be rearranged:

$$\sum_{i=1}^{N+1} \alpha_i \varphi_{f(i)}(a_i) = \sum_{j=1}^{r} \sum_{i \in I_j} \alpha_i \varphi_j(a_i) = \sum_{j=1}^{r} \varphi_j\left(\sum_{i \in I_j} \alpha_i a_i\right)$$

(the last equality follows from the linearity of each φ_j). Write $u_j = \sum_{i \in I_j} \alpha_i a_i$. This is a linear combination of points of A_j with nonnegative coefficients, and hence $u_j \in \mathrm{cone}(A_j)$. Above we have derived $\sum_{j=1}^{r} \varphi_j(u_j) = 0$, and so by (8.1) we get $u_1 = u_2 = \cdots = u_r$. Hence the common value of all the u_j belongs to $\bigcap_{j=1}^{r} \mathrm{cone}(A_j)$.

It remains to check that $u_j \neq 0$. Since we assume $0 \notin \mathrm{conv}(A)$, the only nonnegative linear combination of points of A equal to 0 is the trivial one, with all coefficients 0. On the other hand, since not all the α_i are 0, at least one u_j is expressed as a nontrivial linear combination of points of A. This proves Proposition 8.3.2 and Tverberg's theorem as well. $\qquad\square$

The colored Tverberg theorem. If we have 9 points in the plane, 3 of them red, 3 blue, and 3 white, it turns out that we can always partition them into 3 triples in such a way that each triple has one red, one blue, and one white point, and the 3 triangles determined by the triples have a nonempty intersection.

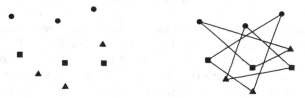

The colored Tverberg theorem is a generalization of this statement for arbitrary d and r. We will need it in Section 9.2, for a result about many simplices with a common point. In that application, the colored version is essential (and Tverberg's theorem alone is not sufficient).

8.3.3 Theorem (Colored Tverberg theorem). *For any integers $r, d \geq 2$ there exists an integer t such that given any $t(d+1)$-point set $Y \subset \mathbf{R}^d$ partitioned into $d+1$ color classes $Y_1, \ldots, Y_{d+1}$ with t points each, there exist r pairwise disjoint sets $A_1, \ldots, A_r$ such that each A_i contains exactly one point of each Y_j, $j = 1, 2, \ldots, d+1$ (that is, the A_i are rainbow), and $\bigcap_{i=1}^{r} \mathrm{conv}(A_i) \neq \emptyset$.*

Let $T_{\mathrm{col}}(d, r)$ denote the smallest t for which the conclusion of the theorem holds. It is known that $T_{\mathrm{col}}(2, r) = r$ for all r. It is possible that $T_{\mathrm{col}}(d, r) = r$ for all d and r, but only weaker bounds have been proved. The strongest known result guarantees that $T_{\mathrm{col}}(d, r) \leq 2r - 1$ whenever r is a prime power.

Recall that in Tverberg's theorem, if we need only the existence of $T(d, r)$, rather than the precise value, several simple arguments are available. In contrast, for the colored version, even if we want only the existence of $T_{\mathrm{col}}(d, r)$, there is essentially only one type of proof, which is not easy and which uses topological methods. Since such methods are not considered in this book, we have to omit a proof of the colored Tverberg theorem.

Bibliography and remarks. Tverberg's theorem was conjectured by Birch and proved by Tverberg (really!) [Tve66]. His original proof is technically complicated, but the idea is simple: Start with some point configuration for which the theorem is valid and convert it to a given configuration by moving one point at a time. During the movement, the current partition may stop working at some point, and it must be shown that it can be replaced by another suitable partition by a local change.

Later on, Tverberg found a simpler proof [Tve81]. For the proof presented in the text above, the main idea is due to Sarkaria [Sar92], and our presentation is based on a simplification by Onn (see [BO97]). Another proof, also due to Tverberg and inspired by the proof of the colorful Carathéodory theorem, was published in a paper by Tverberg and Vrećica [TV93]. Here is an outline.

Let $\pi = (A_1, A_2, \ldots, A_r)$ be a partition of $(d+1)(r-1)+1$ given points into r disjoint nonempty subsets. Consider a ball intersecting all the sets $\mathrm{conv}(A_j)$, $j = 1, 2, \ldots, r$, whose radius $\rho = \rho(\pi)$ is the smallest possible. By a suitable general position assumption, it can be assured that the smallest ball is always unique for any partition. (Alternatively, among all balls of the smallest possible radius, one can take the one with the lexicographically smallest center, which again guarantees uniqueness.) If $\rho(\pi) = 0$, then π is a Tverberg partition. Supposing that $\rho(\pi) > 0$, it can be shown that π can be locally changed (by reassigning one point from one class to another) to another partition π' with $\rho(\pi') < \rho(\pi)$. Another proof, based on a similar idea, was found by Roudneff [Rou01a]. Instead of $\rho(\pi)$, he considers $w(\pi) = \min_{x \in \mathbf{R}^d} w(\pi, x)$, where $w(\pi, x) = \sum_{i=1}^{r} \mathrm{dist}(x, \mathrm{conv}(A_i))^2$. He actually proves a "cone version" of Tverberg's theorem (but different from our cone version and stronger).

Several extensions of Tverberg's theorem are known or conjectured. Here we mention only two conjectures related to the dimension of the set of Tverberg points. For $X \subset \mathbf{R}^d$, let $T_r(X)$ denote the set of all Tverberg points for r-partitions of A (the points of $T_r(X)$ are usually called r-*divisible*). Reay [Rea68] conjectured that if X is in general position and has k more points than is generally necessary for the existence of a Tverberg r-partition, i.e., $|X| = (d+1)(r-1) + 1 + k$, then $\dim T_r(X) \geq k$. This holds under various strong general position assumptions, and special cases for small k have also been established (see Roudneff [Rou01a], [Rou01b]). Kalai asked the following sophisticated question in 1974: Does $\sum_{r=1}^{|X|} \dim T_r(X) \geq 0$ hold for every finite $X \subset \mathbf{R}^d$? Here $\dim \emptyset = -1$, and so the nonexistence of Tverberg r-partitions for large r must be compensated by sufficiently large dimensions of $T_r(X)$ for small r. Together with other interesting aspects of Tverberg's theorem, this is briefly discussed in Kalai's lively survey [Kal01]. There he also notes that edge 3-colorability of a 3-regular graph can be reformulated as the existence of a Tverberg 3-partition of a suitable high-dimensional point set. This implies that deciding whether $T_3(X) = \emptyset$ for a $(2d+3)$-point $X \subset \mathbf{R}^d$ is NP-complete.

It is interesting to note that Tverberg's theorem implies the center-point theorem (Theorem 1.4.2). More generally, if x is an r-divisible point of a finite $X \subset \mathbf{R}^d$, then each closed half-space containing x contains at least r points of X (at least one from each of the r parts); in particular, if $|X| = n$ and $r = \lceil \frac{n}{d+1} \rceil$, we get that every r-divisible point is a centerpoint. On the other hand, as an example of Avis [Avi93] in $\mathbf{R}^3$ shows, a point x such that each closed half-space h containing x satisfies $|h \cap X| \geq r$ need not be r-divisible in general; these two properties are equivalent only in the plane.

A conjecture of Sierksma asserts that the number of Tverberg partitions for a set of $(r-1)(d+1)+1$ points in $\mathbf{R}^d$ in general position is at least $((r-1)!)^d$. A lower bound of $\frac{1}{(r-1)!} \left(\frac{r}{2} \right)^{(r-1)(d+1)/2}$, provided that $r \geq 3$ is a prime number, was proved by Vučić and Živaljević [VŽ93] by an ingenious topological argument.

The colored Tverberg theorem was conjectured by Bárány, Füredi, and Lovász [BFL90], who also proved the planar case. The general case was established by Živaljević and Vrećica [ŽV92]; simplified proofs were given later by Björner, Lovász, Živaljević, and Vrećica [BLŽV94] and by Matoušek [Mat96a] (using a method of Sarkaria). As was mentioned in the text, all these proofs are topological. They show that $T_{\mathrm{col}}(d,r) \leq 2r-1$ for r a prime. Recently, this was extended to all prime powers r by Živaljević [Živ98] (a similar approach in a different problem was used earlier by Özaydin, by Sarkaria, and by Volovikov). Bárány and Larman [BL92] proved that $T(2,r) = r$ for all r.

We outline a beautiful topological proof, due to Lovász (reproduced in [BL92]), showing that $T_{\mathrm{col}}(d,2) = 2$ for all d. Let X be the surface of the $(d+1)$-dimensional crosspolytope. We recall that the crosspolytope is the convex hull of $V = \{e_1, -e_1, e_2, -e_2, \ldots, e_{d+1}, -e_{d+1}\}$, where $e_1, e_2, \ldots, e_{d+1}$ is the standard orthonormal basis in $\mathbf{R}^{d+1}$. Note that X consists of 2^{d+1} simplices of dimension d, each of them the convex hull of $d+1$ points of V. Let $Y_i = \{u_i, v_i\} \subset \mathbf{R}^d$, $i = 1, 2, \ldots, d+1$, be the given two-point color classes. Define the mapping $f \colon V \to \mathbf{R}^d$ by setting $f(e_i) = u_i$, $f(-e_i) = v_i$. This mapping has a unique extension $\bar{f} \colon X \to \mathbf{R}^d$ such that $\bar{f}$ is affine on each of the d-dimensional simplices mentioned above. This $\bar{f}$ is a continuous mapping of $X \to \mathbf{R}^d$. Since X is homeomorphic to the d-dimensional sphere S^d, the Borsuk–Ulam theorem guarantees that there is an $x \in X$ such that $\bar{f}(x) = \bar{f}(-x)$. If $V_1 \subset V$ is the vertex set of a d-dimensional simplex containing x, then $V_1 \cap (-V_1) = \emptyset$, $-x \in \mathrm{conv}(-V_1)$, and as is easy to check, $S_1 = f(V_1)$ and $S_2 = f(-V_1)$ are vertex sets of intersecting rainbow simplices ($\bar{f}(x) = \bar{f}(-x)$ is a common point).

Exercises

1. Prove (directly, without using Tverberg's theorem) that for any integers $d, r_1, r_2 \geq 2$, we have $T(d, r_1 r_2) \leq T(d, r_1) T(d, r_2)$. ②

2. For each $r \geq 2$ and $d \geq 2$, find $(d+1)(r-1)$ points in $\mathbf{R}^d$ with no Tverberg r-partition. ③

3. Prove that Tverberg's theorem implies Proposition 8.3.2. Why is the assumption $0 \notin \mathrm{conv}(A)$ necessary in Proposition 8.3.2? ①

4. (a) Derive the following Radon-type theorem (use Radon's lemma): For every $d \geq 1$ there exists $\ell = \ell(d)$ such that every ℓ points in $\mathbf{R}^d$ in general position can be partitioned into two disjoint subsets A, B such that not only $\mathrm{conv}(A) \cap \mathrm{conv}(B) \neq \emptyset$, but this property is preserved by deleting any single point; that is, $\mathrm{conv}(A \setminus \{a\}) \cap \mathrm{conv}(B) \neq \emptyset$ for each $a \in A$ and $\mathrm{conv}(A) \cap \mathrm{conv}(B \setminus \{b\}) \neq \emptyset$ for each $b \in B$. ③
   (b) Show that $\ell(2) \geq 7$. ②

   Remark. The best known value of $\ell(d)$ is $2d+3$; this was established by Larman [Lar72], and his proof is difficult. The original question is, What is the largest $n = n(k)$ such that every n points in $\mathbf{R}^k$ in general position can be brought to a convex position by some projective transform? Both formulations are related via the Gale transform.

5. Show that for any $d, r \geq 1$ there is an $(N+1)$-point set in $\mathbf{R}^d$ in general position, $N = (d+1)(r-1)$, having no more than $((r-1)!)^d$ Tverberg partitions. ③

6. Why does Tverberg's theorem imply the centerpoint theorem (Theorem 1.4.2)? ①

9

Geometric Selection Theorems

As in Chapter 3, the common theme of this chapter is geometric Ramsey theory. Given n points, or other geometric objects, where n is large, we want to select a not too small subset forming a configuration that is "regular" in some sense.

As was the case for the Erdős–Szekeres theorem, it is not difficult to prove the existence of a "regular" configuration via Ramsey's theorem in some of the subsequent results, but the size of that configuration is very small. The proofs we are going to present give much better bounds. In many cases we obtain "positive-fraction theorems": The regular configuration has size at least cn, where n is the number of the given objects and c is a positive constant independent of n.

In the proofs we encounter important purely combinatorial results: a weak version of the Szemerédi regularity lemma and a theorem of Erdős and Simonovits on the number of complete k-partite subhypergraphs in dense k-uniform hypergraphs. We also apply tools from Chapter 8, such as Tverberg's theorem.

9.1 A Point in Many Simplices: The First Selection Lemma

Consider n points in the plane in general position, and draw all the $\binom{n}{3}$ triangles with vertices at the given points. Then there exists a point of the plane common to at least $\frac{2}{9}\binom{n}{3}$ of these triangles. Here $\frac{2}{9}$ is the optimal constant; the proof below, which establishes a similar statement in arbitrary dimension, gives a considerably smaller constant.

For easier formulations we introduce the following terminology: If $X \subset \mathbf{R}^d$ is a finite set, an X-*simplex* is the convex hull of some $(d+1)$-tuple of points of X. We make the convention that X-simplices are in bijective correspondence with their vertex sets. This means that two X-simplices determined by two distinct $(d+1)$-point subsets of X are considered different even if they coincide as subsets of $\mathbf{R}^d$. Thus, the X-simplices form a multiset in general. This concerns only sets X in degenerate positions; if X is in general position, then distinct $(d+1)$-point sets have distinct convex hulls.

9.1.1 Theorem (First selection lemma). *Let X be an n-point set in $\mathbf{R}^d$. Then there exists a point $a \in \mathbf{R}^d$ (not necessarily belonging to X) contained in at least $c_d \binom{n}{d+1}$ X-simplices, where $c_d > 0$ is a constant depending only on the dimension d.*

The best possible value of c_d is not known, except for the planar case. The first proof below shows that for n very large, we may take $c_d \approx (d+1)^{-(d+1)}$.

The first proof: from Tverberg and colorful Carathéodory. We may suppose that n is sufficiently large ($n \geq n_0$ for a given constant n_0), for otherwise, we can set c_d to be sufficiently small and choose a point contained in a single X-simplex.

Put $r = \lceil n/(d+1) \rceil$. By Tverberg's theorem (Theorem 8.3.1), there exist r pairwise disjoint sets $M_1, \ldots, M_r \subseteq X$ whose convex hulls all have a point in common; call this point a. (A typical M_i has $d+1$ points, but some of them may be smaller.)

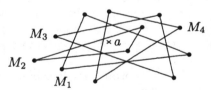

We want show that the point a is contained in many X-simplices (so far we have $\text{const} \cdot n$ and we need $\text{const} \cdot n^{d+1}$).

Let $J = \{j_0, \ldots, j_d\} \subseteq \{1, 2, \ldots, r\}$ be a set of $d+1$ indices. We apply the colorful Carathéodory's theorem (Theorem 8.2.1) for the $(d+1)$ "color" sets $M_{j_0}, \ldots, M_{j_d}$, which all contain a in their convex hull. This yields a rainbow X-simplex S_J containing a and having one vertex from each of the M_{j_i}, as illustrated below:

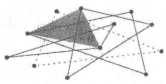

If $J' \neq J$ are two $(d+1)$-tuples of indices, then $S_J \neq S_{J'}$. Hence the number of X-simplices containing the point a is at least

$$\binom{r}{d+1} = \binom{\lceil n/(d+1)\rceil}{d+1} \geq \frac{1}{(d+1)^{d+1}} \frac{n(n-(d+1))\cdots(n-d(d+1))}{(d+1)!}.$$

For n sufficiently large, say $n \geq 2d(d+1)$, this is at least $(d+1)^{-(d+1)}2^{-d}\binom{n}{d+1}$.
$\square$

The second proof: from fractional Helly. Let $\mathcal{F}$ denote the family of all X-simplices. Put $N = |\mathcal{F}| = \binom{n}{d+1}$. We want to apply the fractional Helly theorem (Theorem 8.1.1) to $\mathcal{F}$. Call a $(d+1)$-tuple of sets of $\mathcal{F}$ *good* if its $d+1$ sets have a common point. To prove the first selection lemma, it suffices to show that there are at least $\alpha\binom{N}{d+1}$ good $(d+1)$-tuples for some $\alpha > 0$ independent of n, since then the fractional Helly theorem provides a point common to at least βN members of $\mathcal{F}$.

Set $t = (d+1)^2$ and consider a t-point set $Y \subset X$. Using Tverberg's theorem, we find that Y can be partitioned into $d+1$ pairwise disjoint sets, of size $d+1$ each, whose convex hulls have a common point. (Tverberg's theorem does not guarantee that the parts have size $d+1$, but if they don't, we can move points from the larger parts to the smaller ones, using Carathéodory's theorem.) Therefore, each t-point $Y \subset X$ provides at least one good $(d+1)$-tuple of members of $\mathcal{F}$. Moreover, the members of this good $(d+1)$-tuple are pairwise vertex-disjoint, and therefore the $(d+1)$-tuple uniquely determines Y. It follows that the number of good $(d+1)$-tuples is at least $\binom{n}{t} = \Omega(n^{(d+1)^2}) \geq \alpha\binom{N}{d+1}$.
$\square$

In the first proof we have used Tverberg's theorem for a large point set, while in the second proof we applied it only to configurations of bounded size. For the latter application, if we do not care about the constant of proportionality in the first selection lemma, a weaker version of Tverberg's theorem suffices, namely the finiteness of $T(d, d+1)$, which can be proved by quite simple arguments, as we have seen.

The relation of Tverberg's theorem to the first selection lemma in the second proof somewhat resembles the derivation of macroscopic properties in physics (pressure, temperature, etc.) from microscopic properties (laws of motion of molecules, say). From the information about small (microscopic) configurations we obtained a global (macroscopic) result, saying that a significant portion of the X-simplices have a common point.

A point in the interior of many X-simplices. In applications of the first selection lemma (or its relatives) we often need to know that there is a point contained in the *interior* of many of the X-simplices. To assert anything like that, we have to assume some kind of nondegenerate position of X. The following lemma helps in most cases.

9.1.2 Lemma. *Let $X \subset \mathbf{R}^d$ be a set of $n \geq d+1$ points in general position, meaning that no $d+1$ points of X lie on a common hyperplane, and let $\mathcal{H}$ be the set of the $\binom{n}{d}$ hyperplanes determined by the points of X. Then no point*

$a \in \mathbf{R}^d$ is contained in more than dn^{d-1} hyperplanes of $\mathcal{H}$. Consequently, at most $O(n^d)$ X-simplices have a on their boundary.

Proof. For each d-tuple S whose hyperplane contains a, we choose an inclusion-minimal set $K(S) \subseteq S$ whose affine hull contains a. We claim that if $|K(S_1)| = |K(S_2)| = k$, then either $K(S_1) = K(S_2)$ or $K(S_1)$ and $K(S_2)$ share at most $k-2$ points.

Indeed, if $K(S_1) = \{x_1, \ldots, x_{k-1}, x_k\}$ and $K(S_2) = \{x_1, \ldots, x_{k-1}, y_k\}$, $x_k \neq y_k$, then the affine hulls of $K(S_1)$ and $K(S_2)$ are distinct, for otherwise, we would have $k+1$ points in a common $(k-1)$-flat, contradicting the general position of X. But then the affine hulls intersect in the $(k-2)$-flat generated by $x_1, \ldots, x_{k-1}$ and containing a, and $K(S_1)$ and $K(S_2)$ are not inclusion-minimal.

Therefore, the first $k-1$ points of $K(S)$ determine the last one uniquely, and the number of distinct sets of the form $K(S)$ of cardinality k is at most n^{k-1}. The number of hyperplanes determined by X and containing a given k-point set $K \subset X$ is at most n^{d-k}, and the lemma follows by summing over k.   □

Bibliography and remarks. The planar version of the first selection lemma, with the best possible constant $\frac{2}{9}$, was proved by Boros and Füredi [BF84]. A generalization to an arbitrary dimension, with the first of the two proofs given above, was found by Bárány [Bár82]. The idea of the proof of Lemma 9.1.2 was communicated to me by János Pach.

Boros and Füredi [BF84] actually showed that any centerpoint of X works; that is, it is contained in at least $\frac{2}{9}\binom{n}{3}$ X-triangles. Wagner and Welzl (private communication) observed that a centerpoint works in every fixed dimension, being common to at least $c_d\binom{n}{d+1}$ X-simplices. This follows from known results on the face numbers of convex polytopes using the Gale transform, and it provides yet another proof of the first selection lemma, yielding a slightly better value of the constant c_d than that provided by Bárány's proof. Moreover, for a centrally symmetric point set X this method implies that the origin is contained in the largest possible number of X-simplices.

As for lower bounds, it is known that no n-point $X \subset \mathbf{R}^d$ in general position has a point common to more than $\frac{1}{2^d}\binom{n}{d+1}$ X-simplices [Bár82]. It seems that suitable sets might provide stronger lower bounds, but no results in this direction are known.

9.2 The Second Selection Lemma

In this section we continue using the term X-simplex in the sense of Section 9.1; that is, an X-simplex is the convex hull of a $(d+1)$-point subset of X. In that section we saw that if X is a set in $\mathbf{R}^d$ and we consider *all* the X-simplices, then at least a fixed fraction of them have a point in common.

What if we do not have all, but many X-simplices, some α-fraction of all? It turns out that still many of them must have a point in common, as stated in the second selection lemma below.

9.2.1 Theorem (Second selection lemma). *Let X be an n-point set in $\mathbf{R}^d$ and let $\mathcal{F}$ be a family of $\alpha\binom{n}{d+1}$ X-simplices, where $\alpha \in (0,1]$ is a parameter. Then there exists a point contained in at least*

$$c\alpha^{s_d}\binom{n}{d+1}$$

X-simplices of $\mathcal{F}$, where $c = c(d) > 0$ and s_d are constants.

This result is already interesting for α fixed. But for the application that motivated the discovery of the second selection lemma, namely, trying to bound the number of k-sets (see Chapter 11), the dependence of the bound on α is important, and it would be nice to determine the best possible values of the exponent s_d.

For $d = 1$ it is not too difficult to obtain an asymptotically sharp bound (see Exercise 1). For $d = 2$ the best known bound (probably still not sharp) is as follows: If $|\mathcal{F}| = n^{3-\nu}$, then there is a point contained in at least $\Omega(n^{3-3\nu}/\log^5 n)$ X-triangles of $\mathcal{F}$. In the parameterization as in Theorem 9.2.1, this means that s_2 can be taken arbitrarily close to 3, provided that α is sufficiently small, say $\alpha \leq n^{-\delta}$ for some $\delta > 0$. For higher dimensions, the best known proof gives $s_d \approx (4d+1)^{d+1}$.

Hypergraphs. It is convenient to formulate some of the subsequent considerations in the language of hypergraphs. Hypergraphs are a generalization of graphs where edges can have more than 2 points (from another point of view, a hypergraph is synonymous with a set system). A *hypergraph* is a pair $H = (V, E)$, where V is the vertex set and $E \subseteq 2^V$ is a system of subsets of V, the edge set. A *k-uniform hypergraph* has all edges of size k (so a graph is a 2-uniform hypergraph). A *k-partite hypergraph* is one where the vertex set can be partitioned into k subsets $V_1, V_2, \ldots, V_k$, the *classes*, so that each edge contains at most one point from each V_i. The notions of *subhypergraph* and *isomorphism* are defined analogously to these for graphs. A subhypergraph is obtained by deleting some vertices and some edges (all edges containing the deleted vertices, but possibly more). An isomorphism is a bijection of the vertex sets that maps edges to edges in both directions (a renaming of the vertices).

Proof of the second selection lemma. The proof is somewhat similar to the second proof of the first selection lemma (Theorem 9.1.1). We again use the fractional Helly theorem. We need to show that many $(d+1)$-tuples of X-simplices of $\mathcal{F}$ are good (have nonempty intersections).

We can view $\mathcal{F}$ as a $(d+1)$-uniform hypergraph. That is, we regard X as the vertex set and each X-simplex corresponds to an edge, i.e., a subset of X of size $d+1$. This hypergraph captures the "combinatorial type" of the family $\mathcal{F}$, and a specific placement of the points of X in $\mathbf{R}^d$ then gives a concrete "geometric realization" of $\mathcal{F}$.

First, let us concentrate on the simpler task of exhibiting at least one good $(d+1)$-tuple; even this seems quite nontrivial. Why cannot we proceed as in the second proof of the first selection lemma? Let us give a concrete example with $d = 2$. Following that proof, we would consider 9 points in $\mathbf{R}^2$, and Tverberg's theorem would provide a partition into triples with intersecting convex hulls:

But it can easily happen that one of these triples, say $\{a, b, c\}$, is not an edge of our hypergraph. Tverberg's theorem gives us no additional information on which triples appear in the partition, and so this argument would guarantee a good triple only if *all* the triples on the considered 9 points were contained in $\mathcal{F}$. Unfortunately, a 3-uniform hypergraph on n vertices can contain more than half of all possible $\binom{n}{3}$ triples without containing all triples on some 9 points (even on 4 points). This is a "higher-dimensional" version of the fact that the complete bipartite graph on $\frac{n}{2} + \frac{n}{2}$ vertices has about $\frac{1}{4} n^2$ edges without containing a triangle.

Hypergraphs with many edges need not contain complete hypergraphs, but they have to contain complete multipartite hypergraphs. For example, a graph on n vertices with significantly more than $n^{3/2}$ edges contains $K_{2,2}$, the complete bipartite graph on $2 + 2$ vertices (see Section 4.5). Concerning hypergraphs, let $K^{d+1}(t)$ denote the complete $(d+1)$-partite $(d+1)$-uniform hypergraph with t vertices in each of its $d+1$ vertex classes. The illustration shows a $K^3(4)$; only three edges are drawn as a sample, although of course, all triples connecting vertices at different levels are present.

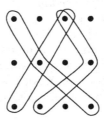

If t is a constant and we have a $(d+1)$-uniform hypergraph on n vertices with sufficiently many edges, then it has to contain a copy of $K^{d+1}(t)$ as a subhypergraph. We do not formulate this result precisely, since we will need a stronger one later.

In geometric language, given a family $\mathcal{F}$ of sufficiently many X-simplices, we can color some t points of X red, some other t points blue,..., t points by color $(d+1)$ in such a way that all the rainbow X-simplices on the $(d+1)t$ colored points are present in $\mathcal{F}$. And in such a situation, if t is a sufficiently large constant, the colored Tverberg theorem (Theorem 8.3.3) with $r = d+1$ claims that we can find a $(d+1)$-tuple of vertex-disjoint rainbow X-simplices whose convex hulls intersect, and so there is a good $(d+1)$-tuple! In fact, these are the considerations that led to the formulation of the colored Tverberg theorem.

For the fractional Helly theorem, we need not only one but many good $(d+1)$-tuples. We use an appropriate stronger hypergraph result, saying that if a hypergraph has enough edges, then it contains many copies of $K^{d+1}(t)$:

9.2.2 Theorem (The Erdős–Simonovits theorem). *Let d and t be positive integers. Let $\mathcal{H}$ be a $(d+1)$-uniform hypergraph on n vertices and with $\alpha\binom{n}{d+1}$ edges, where $\alpha \geq Cn^{-1/t^d}$ for a certain sufficiently large constant C. Then $\mathcal{H}$ contains at least*

$$c\alpha^{t^{d+1}} n^{(d+1)t}$$

copies of $K^{d+1}(t)$, where $c = c(d,t) > 0$ is a constant.

For completeness, a proof is given at the end of this section.

Note that in particular, the theorem implies that a $(d+1)$-uniform hypergraph having at least a constant fraction of all possible edges contains at least a constant fraction of all possible copies of $K^{d+1}(t)$.

We can now finish the proof of the second selection lemma by double counting. The given family $\mathcal{F}$, viewed as a $(d+1)$-uniform hypergraph, has $\alpha\binom{n}{d+1}$ edges, and thus it contains at least $c\alpha^{t^{d+1}} n^{(d+1)t}$ copies of $K^{d+1}(t)$ by Theorem 9.2.2. As was explained above, each such copy contributes at least one good $(d+1)$-tuple of vertex-disjoint X-simplices of $\mathcal{F}$. On the other hand, $d+1$ vertex-disjoint X-simplices have together $(d+1)^2$ vertices, and hence their vertex set can be extended to a vertex set of some $K^{d+1}(t)$ (which has $t(d+1)$ vertices) in at most $n^{t(d+1)-(d+1)^2} = n^{(t-d-1)(d+1)}$ ways. This is the maximum number of copies of $K^{d+1}(t)$ that can give rise to the same good $(d+1)$-tuple. Hence there are at least $c\alpha^{t^{d+1}} n^{(d+1)^2}$ good $(d+1)$-tuples of X-simplices of $\mathcal{F}$. By the fractional Helly theorem, at least $c'\alpha^{t^{d+1}} n^{d+1}$ X-simplices of $\mathcal{F}$ share a common point, with $c' = c'(d) > 0$. This proves the second selection lemma, with the exponent $s_d \leq (4d+1)^{d+1}$. □

Proof of the Erdős–Simonovits theorem (Theorem 9.2.2). By induction on k, we are going to show that a k-uniform hypergraph on n vertices and with m edges contains at least $f_k(n,m)$ copies of $K^k(t)$, where

$$f_k(n,m) = c_k n^{tk} \left(\frac{m}{n^k}\right)^{t^k} - C_k n^{t(k-1)},$$

with $c_k > 0$ and C_k suitable constants depending on k and also on t (t is not shown in the notation, since it remains fixed). This claim with $k = d+1$ implies the Erdős–Simonovits theorem.

For $k = 1$, the claim holds.

So let $k > 1$ and let $\mathcal{H}$ be k-uniform with vertex set V, $|V| = n$, and edge set E, $|E| = m$. For a vertex $v \in V$, define a $(k-1)$-uniform hypergraph $\mathcal{H}_v$ on V, whose edges are all edges of $\mathcal{H}$ that contain v, but with v deleted; that is, $\mathcal{H}_v = (V, \{e \setminus \{v\}: e \in E, v \in e\})$. Further, let $\mathcal{H}'$ be the $(k-1)$-uniform hypergraph whose edge set is the union of the edge sets of all the $\mathcal{H}_v$.

Let $\mathcal{K}$ denote the set of all copies of the complete $(k-1)$-partite hypergraph $K^{k-1}(t)$ in $\mathcal{H}'$. The key notion in the proof is that of an *extending vertex* for a copy $K \in \mathcal{K}$: A vertex $v \in V$ is extending for a $K \in \mathcal{K}$ if K is contained in $\mathcal{H}_v$, or in other words, if for each edge e of K, $e \dot\cup \{v\}$ is an edge in $\mathcal{H}$. The picture below shows a $K^2(2)$ and an extending vertex for it (in a 3-regular hypergraph).

The idea is to count the number of all pairs (K, v), where $K \in \mathcal{K}$ and v is an extending vertex of K, in two ways.

On the one hand, if a fixed copy $K \in \mathcal{K}$ has q_K extending vertices, then it contributes $\binom{q_K}{t}$ distinct copies of $K^k(t)$ in $\mathcal{H}$. We note that one copy of $K^k(t)$ comes from at most $O(1)$ distinct $K \in \mathcal{K}$ in this way, and therefore it suffices to bound $\sum_{K \in \mathcal{K}} \binom{q_K}{t}$ from below.

On the other hand, for a fixed vertex v, the hypergraph $\mathcal{H}_v$ contains at least $f_{k-1}(n, m_v)$ copies $K \in \mathcal{K}$ by the inductive assumption, where m_v is the number of edges of $\mathcal{H}_v$. Hence

$$\sum_{K \in \mathcal{K}} q_K \geq \sum_{v \in V} f_{k-1}(n, m_v).$$

Using $\sum_{v \in V} m_v = km$, the convexity of f_{k-1} in the second variable, and Jensen's inequality (see page xvi), we obtain

$$\sum_{K \in \mathcal{K}} q_K \geq n\, f_{k-1}(n, km/n). \tag{9.1}$$

To conclude the proof, we define a convex function extending the binomial coefficient $\binom{x}{t}$ to the domain $\mathbf{R}$:

$$g(x) = \begin{cases} 0 & \text{for } x \leq t-1, \\ \frac{x(x-1)\cdots(x-t+1)}{t!} & \text{for } x > t-1. \end{cases}$$

We want to bound $\sum_{K \in \mathcal{K}} g(q_K)$ from below, and we have the bound (9.1) for $\sum_{K \in \mathcal{K}} q_K$. Using the bound $|\mathcal{K}| \leq n^{t(k-1)}$ (clear, since $K^{k-1}(t)$ has $t(k-1)$ vertices) and Jensen's inequality, we derive that the number of copies of $K^k(t)$ in $\mathcal{H}$ is at least

$$cn^{t(k-1)} g\left(\frac{n f_{k-1}(n, km/n)}{n^{t(k-1)}}\right).$$

A calculation finishes the induction step; we omit the details. $\square$

> **Bibliography and remarks.** The second selection lemma was conjectured, and proved in the planar case, by Bárány, Füredi, and Lovász [BFL90]. The missing part for higher dimensions was the colored Tverberg theorem (discussed in Section 8.3). A proof for the planar case by a different technique, with considerably better quantitative bounds than can be obtained by the method shown above, was given by Aronov, Chazelle, Edelsbrunner, Guibas, Sharir, and Wenger [ACE+91] (the bounds were mentioned in the text). The full proof of the second selection lemma for arbitrary dimension appears in Alon, Bárány, Füredi, and Kleitman [ABFK92].
>
> Several other "selection lemmas," sometimes involving geometric objects other than simplices, were proved by Chazelle, Edelsbrunner, Guibas, Herschberger, Seidel, and Sharir [CEG+94].
>
> Theorem 9.2.2 is from Erdős and Simonovits [ES83].

Exercises

1. (a) Prove a one-dimensional selection lemma: Given an n-point set $X \subset \mathbf{R}$ and a family $\mathcal{F}$ of $\alpha\binom{n}{2}$ X-intervals, there exists a point common to $\Omega(\alpha^2\binom{n}{2})$ intervals of $\mathcal{F}$. What is the best value of the constant of proportionality you can get? ③
   (b) Show that this result is sharp (up to the value of the multiplicative constant) in the full range of α. ②

2. (a) Show that the exponent s_2 in the second selection lemma in the plane cannot be smaller than 2. ②
   (b) Show that $s_3 \geq 2$. ④ Can you also show that $s_d \geq 2$?
   (c) Show that the proof method via the fractional Helly theorem cannot give a better value of s_2 than 3 in Theorem 9.2.1. That is, construct an n-point set and $\alpha\binom{n}{3}$ triangles on it in such a way that no more than $O(\alpha^5 n^9)$ triples of these triangles have a point in common. ②

9.3 Order Types and the Same-Type Lemma

The order type of a set. There are infinitely many 4-point sets in the plane in general position, but there are only two "combinatorially distinct" types of such sets:

and

What is an appropriate equivalence relation that would capture the intuitive notion of two finite point sets in $\mathbf{R}^d$ being "combinatorially the same"? We have already encountered one suitable notion of combinatorial isomorphism in Section 5.6. Here we describe an equivalent but perhaps more intuitive approach based on the *order type* of a configuration. First we explain this notion for planar configurations in general position, where it is quite simple. Let $\boldsymbol{p} = (p_1, p_2, \ldots, p_n)$ and $\boldsymbol{q} = (q_1, q_2, \ldots, q_n)$ be two sequences of points in $\mathbf{R}^2$, both in general position (no 2 points coincide and no 3 are collinear). Then $\boldsymbol{p}$ and $\boldsymbol{q}$ have the same order type if for any indices $i < j < k$ we turn in the same direction (right or left) when going from p_i to p_k via p_j and when going from q_i to q_k via q_j:

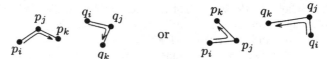

We say that both the triples (p_i, p_j, p_k) and (q_i, q_j, q_k) have the same orientation.

If the point sequences $\boldsymbol{p}$ and $\boldsymbol{q}$ are in $\mathbf{R}^d$, we require that every $(d+1)$-element subsequence of $\boldsymbol{p}$ have the same orientation as the corresponding subsequence of $\boldsymbol{q}$. The notion of orientation is best explained for d-tuples of vectors in $\mathbf{R}^d$. If $v_1, \ldots, v_d$ are vectors in $\mathbf{R}^d$, there is a unique linear mapping sending the vector e_i of the standard basis of $\mathbf{R}^d$ to v_i, $i = 1, 2, \ldots, d$. The matrix A of this mapping has the vectors $v_1, \ldots, v_d$ as the columns. The orientation of $(v_1, \ldots, v_d)$ is defined as the sign of $\det(A)$; so it can be $+1$ (positive orientation), -1 (negative orientation), or 0 (the vectors are linearly dependent and lie in a $(d-1)$-dimensional linear subspace). For a $(d+1)$-tuple of points $(p_1, p_2, \ldots, p_{d+1})$, we define the orientation to be the orientation of the d vectors $p_2 - p_1, p_3 - p_1, \ldots, p_{d+1} - p_1$. Geometrically, the orientation of a 4-tuple (p_1, p_2, p_3, p_4) tells us on which side of the plane $p_1 p_2 p_3$ the point p_4 lies (if p_1, p_2, p_3, p_4 are affinely independent).

Returning to the order type, let $\boldsymbol{p} = (p_1, p_2, \ldots, p_n)$ be a point sequence in $\mathbf{R}^d$. The *order type* of $\boldsymbol{p}$ (also called the *chirotope* of $\boldsymbol{p}$) is defined as the mapping assigning to each $(d+1)$-tuple $(i_1, i_2, \ldots, i_{d+1})$ of indices, $1 \le i_1 < i_2 < \cdots < i_{d+1} \le n$, the orientation of the $(d+1)$-tuple $(p_{i_1}, p_{i_2}, \ldots, p_{i_{d+1}})$. Thus, the order type of $\boldsymbol{p}$ can be described by a sequence of $+1$'s, -1's, and 0's with $\binom{n}{d+1}$ terms.

The order type makes good sense only for point sequences in $\mathbf{R}^d$ containing some $d+1$ affinely independent points. Then one can read off various properties of the sequence from its order type, such as general position, convex position, and so on; see Exercise 1.

In this section we prove a powerful Ramsey-type result concerning order types, called the same-type lemma.

Same-type transversals. Let $(Y_1, Y_2, \ldots, Y_m)$ be an m-tuple of finite sets in $\mathbf{R}^d$. By a *transversal* of this m-tuple we mean any m-tuple $(y_1, y_2, \ldots, y_m)$ such that $y_i \in Y_i$ for all i. We say that $(Y_1, Y_2, \ldots, Y_m)$ *has same-type transversals* if all of its transversals have the same order type. Here is an example of 4 planar sets with same-type transversals:

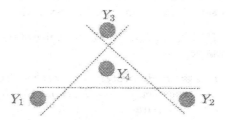

If $(X_1, X_2, \ldots, X_m)$ are very large finite sets such that $X_1 \dot\cup \cdots \dot\cup X_m$ is in general position,[1] we can find not too small subsets $Y_1 \subseteq X_1, \ldots,$ $Y_m \subseteq X_m$ such that $(Y_1, \ldots, Y_m)$ has same-type transversals. To see this, color each transversal of $(X_1, X_2, \ldots, X_m)$ by its order type. Since the number of possible order types of an m-point set in general position cannot exceed $r = 2^{\binom{m}{d+1}}$, we have a coloring of the edges of the complete m-partite hypergraph on $(X_1, \ldots, X_m)$ by r colors. By the Erdős–Simonovits theorem (Theorem 9.2.2), there are sets $Y_i \subseteq X_i$, not too small, such that all edges induced by $Y_1 \cup \cdots \cup Y_m$ have the same color, i.e., $(Y_1, \ldots, Y_m)$ has same-type transversals.

As is the case for many other geometric applications of Ramsey-type theorems, this result can be quantitatively improved tremendously by a geometric argument: For m and d fixed, the size of the sets Y_i can be made a constant fraction of $|X_i|$.

9.3.1 Theorem (Same-type lemma). *For any integers $d, m \geq 1$, there exists $c = c(d, m) > 0$ such that the following holds. Let $X_1, X_2, \ldots, X_m$ be finite sets in $\mathbf{R}^d$ such that $X_1 \dot\cup \cdots \dot\cup X_m$ is in general position. Then there are $Y_1 \subseteq X_1, \ldots, Y_m \subseteq X_m$ such that the m-tuple $(Y_1, Y_2, \ldots, Y_m)$ has same-type transversals and $|Y_i| \geq c|X_i|$ for all $i = 1, 2, \ldots, m$.*

Proof. First we observe that it is sufficient to prove the same-type lemma for $m = d+1$. For larger m, we begin with $(X_1, X_2, \ldots, X_m)$ as the current m-tuple of sets. Then we go through all $(d+1)$-tuples $(i_1, i_2, \ldots, i_{d+1})$ of indices, and if $(Z_1, \ldots, Z_m)$ is the current m-tuple of sets, we apply the same-type lemma to the $(d+1)$-tuple $(Z_{i_1}, \ldots, Z_{i_{d+1}})$. These sets are replaced by smaller

---

[1] This is a shorthand for saying that $X_i \cap X_j = \emptyset$ for all $i \neq j$ and $X_1 \cup \cdots \cup X_m$ is in general position.

sets $(Z'_{i_1}, \ldots, Z'_{i_{d+1}})$ such that this $(d+1)$-tuple has same-type transversals. After this step is executed for all $(d+1)$-tuples of indices, the resulting current m-tuple of sets has same-type transversals.

This method gives the rather small lower bound

$$c(d, m) \geq c(d, d+1)^{\binom{m-1}{d}}.$$

To handle the crucial case $m = d+1$, we will use the following criterion for a $(d+1)$-tuple of sets having same-type transversals.

9.3.2 Lemma. *Let $C_1, C_2, \ldots, C_{d+1} \subseteq \mathbf{R}^d$ be convex sets. The following two conditions are equivalent:*

(i) *There is no hyperplane simultaneously intersecting all of $C_1, C_2, \ldots, C_{d+1}$.*
(ii) *For each nonempty index set $I \subset \{1, 2, \ldots, d+1\}$, the sets $\bigcup_{i \in I} C_i$ and $\bigcup_{j \notin I} C_j$ can be strictly separated by a hyperplane.*

Moreover, if $X_1, X_2, \ldots, X_{d+1} \subset \mathbf{R}^d$ are finite sets such that the sets $C_i = \mathrm{conv}(X_i)$ have property (i) *(and* (ii)*), then $(X_1, \ldots, X_{d+1})$ has same-type transversals.*

In particular, planar convex sets C_1, C_2, C_3 have no line transversal if and only if each of them can be separated by a line from the other two. The proof of this neat result is left to Exercise 3. We will not need the assertion that (i) implies (ii).

Same-type lemma for $d+1$ sets. To prove the same-type lemma for the case $m = d+1$, it now suffices to choose the sets $Y_i \subseteq X_i$ in such a way that their convex hulls are separated in the sense of (ii) in Lemma 9.3.2. This can be done by an iterative application of the ham-sandwich theorem (Theorem 1.4.3).

Suppose that for some nonempty index set $I \subset \{1, 2, \ldots, d+1\}$, the sets $\mathrm{conv}\left(\bigcup_{i \in I} X_i\right)$ and $\mathrm{conv}\left(\bigcup_{j \notin I} X_j\right)$ cannot be separated by a hyperplane. For notational convenience, we assume that $d+1 \in I$. Let h be a hyperplane simultaneously bisecting $X_1, X_2, \ldots, X_d$, whose existence is guaranteed by the ham-sandwich theorem. Let γ be a closed half-space bounded by h and containing at least half of the points of X_{d+1}. For all $i \in I$, including $i = d+1$, we discard the points of X_i not lying in γ, and for $j \notin I$ we throw away the points of X_j that lie in the interior of γ (note that points on h are never discarded); see Figure 9.1.

We claim that union of the resulting sets with indices in I is now strictly separated from the union of the remaining sets. If h contains no points of the sets, then it is a separating hyperplane. Otherwise, let the points contained in h be $a_1, \ldots, a_t$; we have $t \leq d$ by the general position assumption. For each a_j, choose a point a'_j very near to a_j. If a_j lies in some X_i with $i \in I$, then a'_j is chosen in the complement of γ, and otherwise, it is chosen in the interior of γ. We let h' be a hyperplane passing through $a'_1, \ldots, a'_t$ and lying

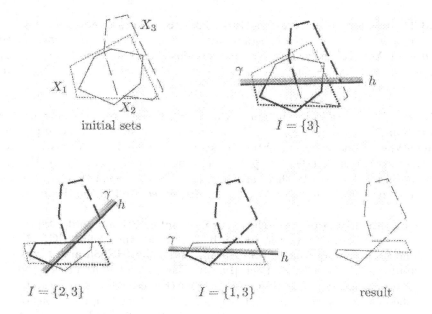

Figure 9.1. Proof of the same-type lemma for $d = 2$, $m = 3$.

very close to h. Then h' is the desired separating hyperplane, provided that the a'_j are sufficiently close to the corresponding a_j, as in the picture below:

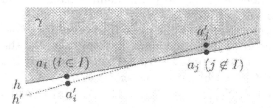

Thus, we have "killed" the index set I, at the price of halving the sizes of the current sets; more precisely, the size of a set X_i is reduced from $|X_i|$ to $\lceil |X_i|/2 \rceil$ (or larger). We can continue with the other index sets in the same manner. After no more than 2^{d-1} halvings, we obtain sets satisfying the separation condition and thus having same-type transversals. The same-type lemma is proved. The lower bound for $c(d, d+1)$ is doubly exponential, roughly 2^{-2^d}. $\qquad\square$

A simple application. We recall that by the Erdős–Szekeres theorem, for any natural number k there is a natural number $n = n(k)$ such that any n-point set in the plane in general position contains a subset of k points in convex position (forming the vertices of a convex k-gon). The same-type lemma immediately gives the following result:

9.3.3 Theorem (Positive-fraction Erdős–Szekeres theorem). *For every integer $k \geq 4$ there is a constant $c_k > 0$ such that every sufficiently large finite set $X \subset \mathbf{R}^2$ in general position contains k disjoint subsets $Y_1, \ldots, Y_k$, of size at least $c_k|X|$ each, such that each transversal of $(Y_1, \ldots, Y_k)$ is in convex position.*

Proof. Let $n = n(k)$ be the number as in the Erdős–Szekeres theorem. We partition X into n sets $X_1, \ldots, X_n$ of almost equal sizes, and we apply the same-type lemma to them, obtaining sets $Y_1, \ldots, Y_n$, $Y_i \subseteq X_i$, with same-type transversals. Let $(y_1, \ldots, y_n)$ be a transversal of $(Y_1, \ldots, Y_n)$. By the Erdős–Szekeres theorem, there are $i_1 < i_2 < \cdots < i_k$ such that $y_{i_1}, \ldots, y_{i_k}$ are in convex position. Then $Y_{i_1}, \ldots, Y_{i_k}$ are as required in the theorem. $\square$

Bibliography and remarks. For more information on order types, the reader can consult the survey by Goodman and Pollack [GP93]. The same-type lemma is from Bárány and Valtr [BV98], and a very similar idea was used by Pach [Pac98]. Bárány and Valtr proved the positive-fraction Erdős–Szekeres theorem (the case $k = 4$ was established earlier by Nielsen), and they gave several more applications of the same-type lemma, such as a positive-fraction Radon lemma and a positive-fraction Tverberg theorem.

Another, simple proof of the positive-fraction Erdős–Szekeres theorem was found by Pach and Solymosi [PS98b]; see Exercise 4 for an outline.

The equivalence of (i) and (ii) in Lemma 9.3.2 is from Goodman, Pollack, and Wenger [GPW96].

A nice strengthening of the same-type lemma was proved by Pór [Pór02]: Instead of just selecting a Y_i from each X_i, the X_i can be completely partitioned into such Y_i. That is, for every d and m there exists $n = n(d, m)$ such that whenever $X_1, X_2, \ldots, X_m \subset \mathbf{R}^d$ are finite sets with $|X_1| = |X_2| = \cdots = |X_m|$ and with $\bigcup X_i$ in general position, there are partitions $X_i = Y_{i1} \dot\cup Y_{i2} \dot\cup \cdots \dot\cup Y_{in}$, $i = 1, 2, \ldots, m$, such that for each $j = 1, 2, \ldots, n$, the sets $Y_{1j}, Y_{2,j}, \ldots, Y_{mj}$ have the same size and same-type transversals. Schematically:

(the sets in each column have same-type transversals). For the proof, one first observes that it suffices to prove the existence of $n(d, d+1)$; the larger m follow as in the proof of the same-type lemma, by refining the partitions for every $(d+1)$-tuple of the indices i. The key step is showing $n(d, d+1) \leq 2n(d-1, d+1)$. The X_i are projected on a generic hyperplane h and the appropriate partitions are found for the projections by induction. Let $X_i' \subset h$ be the projection of X_i, let $Y_1', \ldots, Y_{(d+1)}'$ be one of the "columns" in the partitions of the X_i' (we omit the index j for simpler notation), let $k = |Y_i'|$, and let $Y_i \subseteq X_i$ be the preimage of Y_i'. As far as separation by hyperplanes is concerned, the Y_i' behave like $d+1$ points in general position in $\mathbf{R}^{d-1}$, and so there is only one inseparable (Radon) partition (see Exercise 1.3.9), i.e., an $I \subset \{1, 2, \ldots, d+1\}$ (unique up to complementation) such that $\bigcup_{i \in I} Y_i'$ cannot be separated from $\bigcup_{i \notin I} Y_i'$. By an argument resembling proofs of the ham-sandwich theorem, it can be shown that there is a half-space γ in $\mathbf{R}^d$ and a number k_1 such that $|\gamma \cap Y_i| = k_1$ for $i \in I$ and $|\gamma \cap Y_i| = k - k_1$ for $i \notin I$. Letting $Z_i = Y_i \cap \gamma$ for $i \in I$ and $Z_i = Y_i \setminus \gamma$ for $i \notin I$ and $T_i = Y_i \setminus Z_i$, one obtains that $(Z_1, \ldots, Z_{d+1})$ satisfy condition (ii) in Lemma 9.3.2, and so they have same-type transversals, and similarly for the T_i. A 2-dimensional picture illustrates the construction:

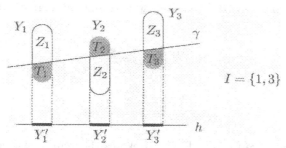

The problem of estimating $n(d, m)$ (the proof produces a doubly exponential bound) is interesting even for $d = 1$, and there Pór showed, by ingenious arguments, that $n(1, m) = \Theta(m^2)$.

Exercises

1. Let $\boldsymbol{p} = (p_1, p_2, \ldots, p_n)$ be a sequence of points in $\mathbf{R}^d$ containing $d+1$ affinely independent points. Explain how we can decide the following questions, knowing the order type of $\boldsymbol{p}$ and nothing else about it:
   (a) Is it true that for every k points among the p_i, $k = 2, 3, \ldots, d+1$, the affine hull has the maximum dimension $k-1$? ①
   (b) Does p_{d+2} lie in $\mathrm{conv}(\{p_1, \ldots, p_{d+1}\})$? ③
   (c) Are the points $p_1, \ldots, p_n$ convex independent (i.e., is each of them a vertex of their convex hull)? ①
2. Let $\boldsymbol{p} = (p_1, p_2, \ldots, p_n)$ be a sequence of points in $\mathbf{R}^d$ whose affine hull is the whole of $\mathbf{R}^d$. Explain how we can determine the order type of $\boldsymbol{p}$, up to a global change of all signs, from the knowledge of $\mathrm{sgn}(\mathsf{AffVal}(\boldsymbol{p}))$ (the signs of affine functions on the p_i; see Section 5.6). ②

(Conversely, $\text{sgn}(\text{AffVal}(\boldsymbol{p}))$ can be reconstructed from the order type, but the proof is more complicated; see, e.g., [BVS^{+}99].)

3. (a) Prove that in the setting of Lemma 9.3.2, if the convex hulls of the X_i have property (i), then $(X_1, \ldots, X_{d+1})$ has same-type transversals. Proceed by contradiction. [3]

(b) Prove that property (ii) (separation) implies property (i) (no hyperplane transversal). Proceed by contradiction and use Radon's lemma. [3]

(c) Prove that (i) implies (ii). [4]

4. Let $k \geq 3$ be a fixed integer.

(a) Show that for n sufficiently large, any n-point set X in general position in the plane contains at least cn^{2k} convex independent subsets of size $2k$, for a suitable $c = c(k) > 0$. [3]

(b) Let $S = \{p_1, p_2, \ldots, p_{2k}\}$ be a convex independent subset of X, where the points are numbered along the circumference of the convex hull in a clockwise order, say. The *holder* of S is the set $H(S) = \{p_1, p_3, \ldots, p_{2k-1}\}$. Show that there is a set H that is the holder of at least $\Omega(n^k)$ sets S. [1]

(c) Derive that each of the indicated triangular regions of such an H contain $\Omega(n)$ points of X:

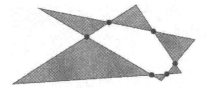

Infer the positive-fraction Erdős–Szekeres theorem in the plane. [2]

(d) Show that the positive-fraction Erdős–Szekeres theorem in higher dimensions is implied by the planar version. [1]

5. (A Ramsey-type theorem for segments)

(a) Let L be a set of n lines and P a set of n points in the plane, both in general position and with no point of P lying on any line of L. Prove that we can select subsets $L' \subseteq L$, $|L'| \geq \alpha n$, and $P' \subseteq P$, $|P'| \geq \alpha n$, such that P' lies in a single cell of the arrangement of L' (where $\alpha > 0$ is a suitable absolute constant). You can use the same-type lemma for $m = 3$ (or an elementary argument). [4]

(b) Given a set S of n segments and a set L of n lines in the plane, both in general position and with no endpoint of a segment lying on any of the lines, show that there exist $S' \subseteq S$ and $L' \subseteq L$, $|S'|, |L'| \geq \beta n$, with a suitable constant $\beta > 0$, such that either each segment of S' intersects each line of L' or all segments of S' are disjoint from all lines of L'. [3]

(c) Given a set R of n red segments and a set B of n blue segments in the plane, with $R \dot{\cup} B$ in general position, prove that there are subsets $R' \subseteq R$, $|R'| \geq \gamma n$, and $B' \subseteq B$, $|B'| \geq \gamma n$, such that either each segment

of R' intersects each segment of B' or each segment of R' is disjoint from each segment of B' ($\gamma > 0$ is another absolute constant). [3]
The result in (c) is due to Pach and Solymosi [PS01].

9.4 A Hypergraph Regularity Lemma

Here we consider a fine tool from the theory of hypergraphs, which we will need for yet another version of the selection lemma in the subsequent section. It is a result inspired by the famous *Szemerédi regularity lemma* for graphs. Very roughly speaking, the Szemerédi regularity lemma says that for given $\varepsilon > 0$, the vertex set of any sufficiently large graph G can be partitioned into some number, not too small and not too large, of parts in such a way that the bipartite graphs between "most" pairs of the parts look like random bipartite graphs, up to an "error" bounded by ε. An exact formulation is rather complicated and is given in the notes below. The result discussed here is a hypergraph analogue of a weak version of the Szemerédi regularity lemma. It is easier to prove than the Szemerédi regularity lemma.

Let $\mathcal{H} = (X, E)$ be a k-partite hypergraph whose vertex set is the union of k pairwise disjoint n-element sets $X_1, X_2, \ldots, X_k$, and whose edges are k-tuples containing precisely one element from each X_i. For subsets $Y_i \subseteq X_i$, $i = 1, 2, \ldots, k$, let $e(Y_1, \ldots, Y_k)$ denote the number of edges of $\mathcal{H}$ contained in $Y_1 \cup \cdots \cup Y_k$. In this notation, the total number of edges of $\mathcal{H}$ is equal to $e(X_1, \ldots, X_k)$. Further, let

$$\rho(Y_1, \ldots, Y_k) = \frac{e(Y_1, \ldots, Y_k)}{|Y_1| \cdot |Y_2| \cdots |Y_k|}$$

denote the *density* of the subhypergraph induced by the Y_i.

9.4.1 Theorem (Weak regularity lemma for hypergraphs). *Let $\mathcal{H}$ be a k-partite hypergraph as above, and suppose that $\rho(\mathcal{H}) \geq \beta$ for some $\beta > 0$. Let $0 < \varepsilon < \frac{1}{2}$. Suppose that n is sufficiently large in terms of k, β, and ε.*

Then there exist subsets $Y_i \subseteq X_i$ of equal size $|Y_i| = s \geq \beta^{1/\varepsilon^k} n$, $i = 1, 2, \ldots, k$, such that

(i) *(High density) $\rho(Y_1, \ldots, Y_k) \geq \beta$, and*
(ii) *(Edges on all large subsets) $e(Z_1, \ldots, Z_k) > 0$ for any $Z_i \subseteq Y_i$ with $|Z_i| \geq \varepsilon s$, $i = 1, 2, \ldots, k$.*

The following scheme illustrates the situation (but of course, the vertices of the Y_i and Z_i need not be contiguous).

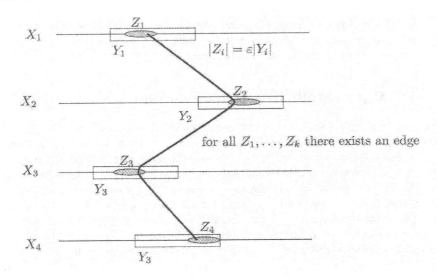

Proof. Intuitively, the sets Y_i should be selected in such a way that the subhypergraph induced by them is as dense as possible. We then want to show that if there were $Z_1, \ldots, Z_k$ of size at least εs with no edges on them, we could replace the Y_i by sets with a still larger density. But if we looked at the usual density $\rho(Y_1, \ldots, Y_k)$, we would typically get too small sets Y_i. The trick is to look at a modified density parameter that slightly favors larger sets. Thus, we define the *magical density* $\mu(Y_1, \ldots, Y_k)$ by

$$\mu(Y_1, \ldots, Y_k) = \frac{e(Y_1, \ldots, Y_k)}{(|Y_1| \cdot |Y_2| \cdots |Y_k|)^{1-\varepsilon^k/k}}.$$

We choose $Y_1, \ldots, Y_k$, $Y_i \subseteq X_i$, as sets of *equal size* that have the maximum possible magical density $\mu(Y_1, \ldots, Y_k)$. We denote the common size $|Y_1| = \cdots = |Y_k|$ by s.

First we derive the condition (i) in the theorem for this choice of the Y_i. We have

$$\frac{e(Y_1, \ldots, Y_k)}{s^{k-\varepsilon^k}} = \mu(Y_1, \ldots, Y_k) \geq \mu(X_1, \ldots, X_k) = \beta n^{\varepsilon^k} \geq \beta s^{\varepsilon^k},$$

and so $e(Y_1, \ldots, Y_k) \geq \beta s^k$, which verifies (i). Since obviously $e(Y_1, \ldots, Y_k) \leq s^k$, we have $\mu(Y_1, \ldots, Y_k) \leq s^{\varepsilon^k}$. Combining with $\mu(Y_1, \ldots, Y_k) \geq \beta n^{\varepsilon^k}$ derived above, we also obtain that $s \geq \beta^{1/\varepsilon^k} n$.

It remains to prove (ii). Since εs is a large number by the assumptions, rounding it up to an integer does not matter in the subsequent calculations (as can be checked by a simple but somewhat tedious analysis). In order to simplify matters, we will thus assume that εs is an integer, and we let $Z_1 \subseteq Y_1, \ldots, Z_k \subseteq Y_k$ be εs-element sets. We want to prove $e(Z_1, \ldots, Z_k) > 0$. We have

$$e(Z_1, \ldots, Z_k) = e(Y_1, \ldots, Y_k) \tag{9.2}$$
$$- e(Y_1 \setminus Z_1, Y_2, Y_3, \ldots, Y_k)$$
$$- e(Z_1, Y_2 \setminus Z_2, Y_3, \ldots, Y_k)$$
$$- e(Z_1, Z_2, Y_3 \setminus Z_3, \ldots, Y_k)$$
$$\vdots$$
$$- e(Z_1, Z_2, Z_3, \ldots, Y_k \setminus Z_k).$$

We want to show that the negative terms are not too large, using the assumption that the magical density of $Y_1, \ldots, Y_k$ is maximum. The problem is that $Y_1, \ldots, Y_k$ maximize the magical density only among the sets of equal size, while we have sets of different sizes in the terms. To get back to sets of equal size, we use the following observation. If, say, R_1 is a randomly chosen subset of Y_1 of some given size r, we have

$$\mathbf{E}[\rho(R_1, Y_2, \ldots, Y_k)] = \rho(Y_1, \ldots, Y_k),$$

where $\mathbf{E}[\,\cdot\,]$ denotes the expectation with respect to the random choice of an r-element $R_1 \subseteq Y_1$. This preservation of density by choosing a random subset is quite intuitive, and it is not difficult to verify it by counting (Exercise 1). For estimating the term $e(Y_1 \setminus Z_1, Y_2, \ldots, Y_k)$, we use random subsets $R_2, \ldots, R_k$ of size $(1-\varepsilon)s$ of $Y_2, \ldots, Y_k$, respectively. Thus,

$$e(Y_1 \setminus Z_1, Y_2, \ldots, Y_k) = (1-\varepsilon)s^k \mathbf{E}[\rho(Y_1 \setminus Z_1, R_2, \ldots, R_k)].$$

Now for any choice of $R_2, \ldots, R_k$, we have

$$\rho(Y_1 \setminus Z_1, R_2, \ldots, R_k) = ((1-\varepsilon)s)^{-\varepsilon^k} \mu(Y_1 \setminus Z_1, R_2, \ldots, R_k)$$
$$\leq ((1-\varepsilon)s)^{-\varepsilon^k} \mu(Y_1, Y_2, \ldots, Y_k)$$
$$= (1-\varepsilon)^{-\varepsilon^k} \rho(Y_1, \ldots, Y_k).$$

Therefore,

$$e(Y_1 \setminus Z_1, Y_2, \ldots, Y_k) \leq (1-\varepsilon)^{1-\varepsilon^k} e(Y_1, \ldots, Y_k) \leq (1-\varepsilon)e(Y_1, \ldots, Y_k).$$

To estimate the term $e(Z_1, Z_2, \ldots, Z_{i-1}, Y_i \setminus Z_i, Y_{i+1}, \ldots, Y_k)$, we use random subsets $R_i \subset Y_i \setminus Z_i$ and $R_{i+1} \subset Y_{i+1}, \ldots, R_k \subset Y_k$, this time all of size εs. A similar calculation as before yields

$$e(Z_1, Z_2, \ldots, Z_{i-1}, Y_i \setminus Z_i, Y_{i+1}, \ldots, Y_k) \leq \varepsilon^{i-1-\varepsilon^k}(1-\varepsilon)e(Y_1, \ldots, Y_k).$$

(This estimate is also valid for $i = 1$, but it is worse than the one derived above and it would not suffice in the subsequent calculation.) From (9.2) we obtain that $e(Z_1, \ldots, Z_k)$ is at least $e(Y_1, \ldots, Y_k)$ multiplied by the factor

$$1 - (1 - \varepsilon) - (1 - \varepsilon)\varepsilon^{-\varepsilon^k} \sum_{i=2}^{k} \varepsilon^{i-1} = \varepsilon - \varepsilon^{1-\varepsilon^k}\left(1 - \varepsilon^{k-1}\right)$$

$$= \varepsilon\left(1 + \varepsilon^{-\varepsilon^k}(\varepsilon^{k-1} - 1)\right)$$

$$= \varepsilon\left(1 + e^{\varepsilon^k \ln(1/\varepsilon)}(\varepsilon^{k-1} - 1)\right)$$

$$\geq \varepsilon\left(1 + (1 + \varepsilon^k \ln\tfrac{1}{\varepsilon})(\varepsilon^{k-1} - 1)\right)$$

$$= \varepsilon^{k+1}\left(\tfrac{1}{\varepsilon} - \ln\tfrac{1}{\varepsilon} + \varepsilon^k \ln\tfrac{1}{\varepsilon}\right)$$

$$\geq \varepsilon^{k+1}\left(\tfrac{1}{\varepsilon} - \ln\tfrac{1}{\varepsilon}\right)$$

$$> 0.$$

Theorem 9.4.1 is proved.   ☐

Bibliography and remarks. The Szemerédi regularity lemma is from [Sze78], and in its full glory it goes as follows: *For every $\varepsilon > 0$ and for every k_0, there exist K and n_0 such that every graph G on $n \geq n_0$ vertices has a partition $(V_0, V_1, \ldots, V_k)$ of the vertex set into $k+1$ parts, $k_0 \leq k \leq K$, where $|V_0| \leq \varepsilon n$, $|V_1| = |V_2| = \cdots = |V_k| = m$, and all but at most εk^2 of the $\binom{k}{2}$ pairs $\{V_i, V_j\}$ are ε-regular, which means that for every $A \subseteq V_i$ and $B \subseteq V_j$ with $|A|, |B| \geq \varepsilon m$ we have $|\rho(A, B) - \rho(V_i, V_j)| \leq \varepsilon$.* Understanding the idea of the proof is easier than understanding the statement. The regularity lemma is an extremely powerful tool in modern combinatorics. A survey of applications and variations can be found in Komlós and Simonovits [KS96].

Our presentation of Theorem 9.4.1 essentially follows Pach [Pac98], whose treatment is an adaptation of an approach of Komlós and Sós.

One can formulate various hypergraph analogues of the Szemerédi regularity lemma in its full strength. For instance, for a 3-uniform hypergraph, one can define a triple V_1, V_2, V_2 of disjoint subsets of vertices to be ε-regular if $|\rho(A_1, A_2, A_3) - \rho(V_1, V_2, V_3)| \leq \varepsilon$ for every $A_i \subseteq V_i$ with $|A_i| \geq \varepsilon|V_i|$, and formulate a statement about a partition of the vertex set of every 3-regular hypergraph in which almost all triples of classes are ε-regular. Such a result indeed holds, but this formulation has significant shortcomings. For example, the Szemerédi regularity lemma allows approximate *counting* of small subgraphs in the given graph (see Exercise 3 for a simple example), which is the key to many applications, but the notion of ε-regularity for triple systems just given does not work in this way (Exercise 4). A technically quite complicated but powerful regularity lemma for 3-regular hypergraphs that does admit counting of small subhypergraphs was proved by Frankl and Rödl [FR01]. The first insight is that for triple systems, one should not partition only vertices but also *pairs* of vertices.

Let us mention a related innocent-looking problem of geometric flavor. For a point $c \in S = \{1, 2, \ldots, n\}^d$, we define a *jack* with center

c as the set of all points of S that differ from c in at most 1 coordinate. The problem, formulated by Székely, asks for the maximum possible cardinality of a system of jacks in S such that no two jacks share a line (i.e., every two centers differ in at least 2 coordinates) and no point is covered by d jacks. It is easily seen that no more than n^{d-1} jacks can be taken, and the problem is to prove an $o(n^{d-1})$ bound for every fixed d. The results of Frankl and Rödl [FR01] imply this bound for $d = 4$, and recently Rödl and Skokan announced a positive solution for $d = 5$ as well; these results are based on sophisticated hypergraph regularity lemmas. A positive answer would imply the famous theorem of Szemerédi on arithmetic progressions (see, e.g., Gowers [Gow98] for recent work and references) and would probably provide a "purely combinatorial" proof.

Exercises

1. Verify the equality $\mathbf{E}[\rho(R_1, Y_2, \ldots, Y_k)] = \rho(Y_1, \ldots, Y_k)$, where the expectation is with respect to a random choice of an r-element $R_1 \subseteq Y_1$. Also derive the other similar equalities used in the proof in the text. ②

2. (Density Ramsey-type result for segments)
   (a) Let $c > 0$ be a given positive constant. Using Exercise 9.3.5(c) and the weak regularity lemma, prove that there exists $\beta = \beta(c) > 0$ such that whenever R and B are sets of segments in the plane with $R \dot\cup B$ in general position and such that the number of pairs (r, b) with $r \in R$, $b \in B$, and $r \cap b \neq \emptyset$ is at least cn^2, then there are subsets $R' \subseteq R$ and $B' \subseteq B$ such that $|R'| \geq \beta n$, $|B'| \geq \beta n$, and each $r \in R'$ intersects each $b \in B'$. ③
   (b) Prove the analogue of (a) for noncrossing pairs. Assuming at least cn^2 pairs (r, b) with $r \cap b = \emptyset$, select R' and B' of size βn such that $r \cap b = \emptyset$ for each $r \in R'$ and $b \in B'$. ①
   These results are from Pach and Solymosi [PS01].

3. (a) Let $G = (V, E)$ be a graph, and let V be partitioned into classes V_1, V_2, V_3 of size m each. Suppose that there are no edges with both vertices in the same V_i, that $|\rho(V_i, V_j) - \frac{1}{2}| \leq \varepsilon$ for all $i < j$, and that each pair (V_i, V_j) is ε-regular (this means that $|\rho(A, B) - \rho(V_i, V_j)| \leq \varepsilon$ for any $A \subseteq V_i$ and $B \subseteq V_j$ with $|A|, |B| \geq \varepsilon m$). Prove that the number of triangles in G is $(\frac{1}{8} + o(1))m^3$, where the $o(1)$ notation refers to $\varepsilon \to 0$ (while m is considered arbitrary but sufficiently large in terms of ε). ③
   (b) Generalize (a) to counting the number of copies of K_4, where G has 4 classes $V_1, \ldots, V_4$ of equal size (if all the densities are about $\frac{1}{2}$, then the number should be $(2^{-6} + o(1))m^4$). ③

4. For every $\varepsilon > 0$ and for arbitrarily large m, construct a 3-uniform 4-partite hypergraph with vertex classes $V_1, \ldots, V_4$, each of size m, that contains no $K_4^{(3)}$ (the system of all triples on 4 vertices), but where $|\rho(V_i, V_j, V_k) - \frac{1}{2}| \leq \varepsilon$ for all $i < j < k$ and each triple (V_i, V_j, V_k) is

ε-regular. The latter condition means $|\rho(A_i, A_j, A_k) - \rho(V_i, V_j, V_k)| \le \varepsilon$ for every $A_i \subseteq V_i$, $A_j \subseteq V_j$, $A_k \subseteq V_k$ of size at least εm. $\boxed{5}$

9.5 A Positive-Fraction Selection Lemma

Here we discuss a stronger version of the first selection lemma (Theorem 9.1.1). Recall that for any n-point set $X \subset \mathbf{R}^d$, the first selection lemma provides a "heavily covered" point, that is, a point contained in at least a fixed fraction of the $\binom{n}{d+1}$ simplices with vertices in points of X. The theorem below shows that we can even get a large collection of simplices with a quite special structure. For example, in the plane, given n red points, n white points, and n blue points, we can select $\frac{n}{12}$ red, $\frac{n}{12}$ white, and $\frac{n}{12}$ blue points in such a way that *all* the red–white–blue triangles for the resulting sets have a point in common. Here is the d-dimensional generalization.

9.5.1 Theorem (Positive-fraction selection lemma). *For all natural numbers d, there exists $c = c(d) > 0$ with the following property. Let $X_1, X_2, \ldots, X_{d+1} \subset \mathbf{R}^d$ be finite sets of equal size, with $X_1 \dot\cup X_2 \dot\cup \cdots \dot\cup X_{d+1}$ in general position. Then there is a point $a \in \mathbf{R}^d$ and subsets $Z_1 \subseteq X_1, \ldots, Z_{d+1} \subseteq X_{d+1}$, with $|Z_i| \ge c|X_i|$, such that the convex hull of every transversal of $(Z_1, \ldots, Z_{d+1})$ contains a.*

As was remarked above, for $d = 2$, one can take $c = \frac{1}{12}$. There is an elementary and not too difficult proof (which the reader is invited to discover). In higher dimensions, the only known proof uses the weak regularity lemma for hypergraphs.

Proof. Let $X = X_1 \cup \cdots \cup X_{d+1}$. We may suppose that all the X_i are large (for otherwise, one-point Z_i will do). Let $\mathcal{F}_0$ be the set of all "rainbow" X-simplices, i.e., of all transversals of $(X_1, \ldots, X_{d+1})$, where the transversals are formally considered as sets for the moment. The size of $\mathcal{F}_0$ is, for d fixed, at least a constant fraction of $\binom{|X|}{d+1}$ (here we use the assumptions that the X_i are of equal size). Therefore, by the second selection lemma (Theorem 9.2.1), there is a subset $\mathcal{F}_1 \subseteq \mathcal{F}_0$ of at least βn^{d+1} X-simplices containing a common point a, where $\beta = \beta(d) > 0$. (Note that we do not need the full power of the second selection lemma here, since we deal with the complete $(d+1)$-partite hypergraph.)

For the subsequent argument we need the common point a to lie in the *interior* of many of the X-simplices. One way of ensuring this would be to assume a suitable strongly general position of X and use a perturbation argument for arbitrary X. Another, perhaps simpler, way is to apply Lemma 9.1.2, which guarantees that a lies on the boundary of at most $O(n^d)$ of the X-simplices of $\mathcal{F}_1$. So we let $\mathcal{F}_2 \subseteq \mathcal{F}_1$ be the X-simplices containing a in the interior, and for a sufficiently large n we still have $|\mathcal{F}_2| \ge \beta' n^{d+1}$.

Next, we consider the $(d+1)$-partite hypergraph $\mathcal{H}$ with vertex set X and edge set $\mathcal{F}_2$. We let $\varepsilon = c(d, d+2)$, where $c(d, m)$ is as in the same-type lemma, and we apply the weak regularity lemma (Theorem 9.4.1) to $\mathcal{H}$. This yields sets $Y_1 \subseteq X_1, \ldots, Y_{d+1} \subseteq X_{d+1}$, whose size is at least a fixed fraction of the size of the X_i, and such that any subsets $Z_1 \subseteq Y_1, \ldots, Z_{d+1} \subseteq Y_{d+1}$ of size at least $\varepsilon |Y_i|$ induce an edge; this means that there is a rainbow X-simplex with vertices in the Z_i and containing the point a.

The argument is finished by applying the same-type lemma with the $d+2$ sets $Y_1, Y_2, \ldots, Y_{d+1}$ and $Y_{d+2} = \{a\}$. We obtain sets $Z_1 \subseteq Y_1, \ldots, Z_{d+1} \subseteq Y_{d+1}$ and $Z_{d+2} = \{a\}$ with same-type transversals, and with $|Z_i| \geq \varepsilon |Y_i|$ for $i = 1, 2, \ldots, d+1$. (Indeed, the same-type lemma guarantees that at least one point is selected even from an 1-point set.) Now either all transversals of $(Z_1, \ldots, Z_{d+1})$ contain the point a in their convex hull or none does (use Exercise 9.3.1(d)). But the latter possibility is excluded by the choice of the Y_i (by the weak regularity lemma). The positive-fraction selection lemma is proved.   □

It is amazing how many quite heavy tools are used in this proof. It would be nice to find a more direct argument.

Bibliography and remarks. The planar case of Theorem 9.5.1 was proved by Bárany, Füredi, and Lovász [BFL90] (with $c(2) \geq \frac{1}{12}$), and the result for arbitrary dimension is due to Pach [Pac98].

10

Transversals and Epsilon Nets

Here we are going to consider problems of the following type: We have a family $\mathcal{F}$ of geometric shapes satisfying certain conditions, and we would like to conclude that $\mathcal{F}$ can be "pierced" by not too many points, meaning that we can choose a bounded number of points such that each set of $\mathcal{F}$ contains at least one of them. Such questions are sometimes called *Gallai-type problems*, because of the following nice problem raised by Gallai: Let $\mathcal{F}$ be a finite family of closed disks in the plane such that every two disks in $\mathcal{F}$ intersect. What is the smallest number of points needed to pierce $\mathcal{F}$? For this problem, the exact answer is known: 4 points always suffice and are sometimes necessary.

We will not cover this particular (quite difficult) result; rather, we consider general methods for proving that the number of piercing points can be bounded. These methods yield numerous results where no other proofs are available. On the other hand, the resulting estimates are usually quite large, and in some simpler cases (such as Gallai's problem mentioned above), specialized geometric arguments provide much better bounds.

Some of the tools introduced in this chapter are widely applicable and sometimes more significant than the particular geometric results. Such important tools include the transversal and matching numbers of set systems, their fractional versions (connected via the duality of linear programming), the Vapnik–Chervonenkis dimension and ways of estimating it, and epsilon nets.

10.1 General Preliminaries: Transversals and Matchings

Let $\mathcal{F}$ be a system of sets on a ground set X; both $\mathcal{F}$ and X may generally be infinite. A subset $T \subseteq X$ is called a *transversal* of $\mathcal{F}$ if it intersects all the sets of $\mathcal{F}$.

The *transversal number* of $\mathcal{F}$, denoted by $\tau(\mathcal{F})$, is the smallest possible cardinality of a transversal of $\mathcal{F}$.

Many combinatorial and geometric problems, some them considered in this chapter, can be rephrased as questions about the transversal number of suitable set systems.

Another important parameter of a set system $\mathcal{F}$ is the *packing number* (or *matching number*) of $\mathcal{F}$, usually denoted by $\nu(\mathcal{F})$. This is the maximum cardinality of a system of pairwise disjoint sets in $\mathcal{F}$:

$$\nu(\mathcal{F}) = \sup\{|\mathcal{M}|: \mathcal{M} \subseteq \mathcal{F}, \ M_1 \cap M_2 = \emptyset \text{ for all } M_1, M_2 \in \mathcal{M}, \ M_1 \neq M_2\}.$$

A subsystem $\mathcal{M} \subseteq \mathcal{F}$ of pairwise disjoint sets is called a *packing* (or a *matching*; this refers to graph-theoretic matching, which is a system of pairwise disjoint edges).

Any transversal is at least as large as any packing, and so always

$$\nu(\mathcal{F}) \leq \tau(\mathcal{F}).$$

In the reverse direction, very little can be said in general, since $\tau(\mathcal{F})$ can be arbitrarily large even if $\nu(\mathcal{F}) = 1$. As a simple geometric example, we can take the plane as the ground set X and let the sets of $\mathcal{F}$ be n lines in general position. Then $\nu(\mathcal{F}) = 1$, since every two lines intersect, but $\tau(\mathcal{F}) \geq \frac{1}{2} n$, because no point is contained in more than two of the lines.

Fractional packing and transversal numbers. Now we introduce another parameter of a set system, which always lies between ν and τ and which has proved extremely useful in arguments estimating τ or ν. First we restrict ourselves to set systems on *finite* ground sets.

Let $\mathcal{F}$ be a system of subsets of a finite set X. A *fractional transversal* for $\mathcal{F}$ is a function $\varphi: X \to [0, 1]$ such that for each $S \in \mathcal{F}$, we have $\sum_{x \in S} \varphi(x) \geq 1$. The size of a fractional transversal φ is $\sum_{x \in X} \varphi(x)$, and the *fractional transversal number* $\tau^*(\mathcal{F})$ is the infimum of the sizes of fractional transversals. So in a fractional transversal, we can take one-third of one point, one-fifth of another, etc., but we must put total weight of at least one full point into every set.

Similarly, a *fractional packing* for $\mathcal{F}$ is a function $\psi\colon \mathcal{F} \to [0,1]$ such that for each $x \in X$, we have $\sum_{S\in\mathcal{F}\colon x\in S} \psi(S) \leq 1$. So sets receive weights and the total weight of sets containing any given point must not exceed 1. The *size* of a fractional packing ψ is $\sum_{S\in\mathcal{F}} \psi(S)$, and the *fractional packing number* $\nu^*(\mathcal{F})$ is the supremum of the sizes of all fractional packings for $\mathcal{F}$.

It is instructive to consider the "triangle" system of 3 sets on 3 points,

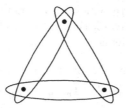

and check that $\nu = 1$, $\tau = 2$, and $\nu^* = \tau^* = \frac{3}{2}$.

Any packing $\mathcal{M}$ yields a fractional packing (by assigning weight 1 to the sets in $\mathcal{M}$ and 0 to others), and so $\nu \leq \nu^*$. Similarly, $\tau^* \leq \tau$.

We promised one parameter but introduced two: τ^* and ν^*. But they happen to be the same.

10.1.1 Theorem. *For every set system $\mathcal{F}$ on a finite ground set, we have $\nu^*(\mathcal{F}) = \tau^*(\mathcal{F})$. Moreover, the common value is a rational number, and there exist an optimal fractional transversal and an optimal fractional packing attaining only rational values.*

This is not a trivial result; the proof is a nice application of the duality of linear programming. Here is the version of the linear programming duality we need.

10.1.2 Proposition (Duality of linear programming). *Let A be an $m \times n$ real matrix, $b \in \mathbf{R}^m$ a (column) vector, and $c \in \mathbf{R}^n$ a (column) vector. Let*
$$P = \{x \in \mathbf{R}^n\colon x \geq 0,\ Ax \geq b\}$$
and
$$D = \{y \in \mathbf{R}^m\colon y \geq 0,\ y^T A \leq c^T\}$$
(the inequalities between vectors should hold in every component). If both $P \neq \emptyset$ and $D \neq \emptyset$, then
$$\min\{c^T x\colon x \in P\} = \max\{y^T b\colon y \in D\};$$
in particular, both the minimum and the maximum are well-defined and attained.

This result can be quickly proved by piecing together a larger matrix from A, b, and c and applying a suitable version of the Farkas lemma (Lemma 1.2.5) to it (Exercise 6). It can also be derived directly from the separation theorem.

Let us remark that there are several versions of the linear programming duality (differing, for example, in including or omitting the requirement $x \geq 0$, or replacing $Ax \geq b$ by $Ax = b$, or exchanging minima and maxima), and they are easy to mix up.

Proof of Theorem 10.1.1. Set $n = |X|$ and $m = |\mathcal{F}|$, and let A be the $m \times n$ incidence matrix of the set system $\mathcal{F}$: Rows correspond to sets, columns to points, and the entry corresponding to a point p and a set S is 1 if $p \in S$ and 0 if $p \notin S$. It is easy to check that $\nu^*(\mathcal{F})$ and $\tau^*(\mathcal{F})$ are solutions to the following optimization problems:

$$\tau^*(\mathcal{F}) = \min\left\{\mathbf{1}_n^T x\colon x \geq 0,\ Ax \geq \mathbf{1}_m\right\},$$
$$\nu^*(\mathcal{F}) = \max\left\{y^T \mathbf{1}_m\colon y \geq 0,\ y^T A \leq \mathbf{1}_n^T\right\},$$

where $\mathbf{1}_n \in \mathbf{R}^n$ denotes the (column) vector of all 1's of length n. Indeed, the vectors $x \in \mathbf{R}^n$ satisfying $x \geq 0$ and $Ax \geq \mathbf{1}_m$ correspond precisely to the fractional transversals of $\mathcal{F}$, and similarly, the $y \in \mathbf{R}^n$ with $y \geq 0$ and $y^T A \leq \mathbf{1}_n^T$ correspond to the fractional packings. There is at least one fractional transversal, e.g., $x = \mathbf{1}_n$, and at least one fractional packing, namely, $y = 0$, and so Proposition 10.1.2 applies and shows that $\nu^*(\mathcal{F}) = \tau^*(\mathcal{F})$.

At the same time, $\tau^*(F)$ is the minimum of the linear function $x \mapsto \mathbf{1}_n^T x$ over a polyhedron, and such a minimum, since it is finite, is attained at a vertex. The inequalities describing the polyhedron have rational coefficients, and so all vertices are rational points. □

Remark about infinite set systems. Set systems encountered in geometry are usually infinite. In almost all the considerations concerning transversals, the problem can be reduced to a problem about finite sets, usually by a simple ad hoc argument. Nevertheless, we include here a few remarks that can aid a simple consistent treatment of the infinite case. However, they will not be used in the sequel in any essential way.

There is no problem with the definitions of ν and τ in the infinite case, but one has to be a little careful with the definition of ν^* and τ^* to preserve the equality $\nu^* = \tau^*$. Everything is still fine if we have finitely many sets on an infinite ground set: The infinite ground set can be factored into finitely many equivalence classes, where two points are equivalent if they belong to the same subcollection of the sets. One can choose one point in each equivalence class and work with a finite system.

For infinitely many sets, some sort of compactness condition is certainly needed. For example, the system of intervals $\{[i, \infty)\colon i = 1, 2, \ldots\}$ has, according to any reasonable definition, $\nu^* = 1$ but $\tau^* = \infty$.

If we let $\mathcal{F}$ be a family of closed sets in a compact metric space X (compact Hausdorff space actually suffices), we can define $\nu^*(\mathcal{F})$ as $\sup_\psi \sum_{S \in \mathcal{F}} \psi(S)$, where the supremum is over all $\psi\colon \mathcal{F} \to [0, 1]$ attaining only finitely many nonzero values and such that $\sum_{S \in \mathcal{F}\colon x \in S} \psi(S) \leq 1$ for each $x \in X$.

For the definition of τ^*, the first attempt might be to consider all functions $\varphi\colon X \to [0,1]$ attaining only finitely many nonzero values and summing up to at least 1 over every set. But this does not work very well: For example, if we let $\mathcal{F}$ be the system of all compact subsets of $[0,1]$ of Lebesgue measure $\frac{1}{2}$, say, then $\nu^* \leq 2$ but τ^* would be infinite, since any finite subset is avoided by some member of $\mathcal{F}$. It is better to define a *fractional transversal* of $\mathcal{F}$ as a Borel measure μ on X such that $\mu(S) \geq 1$ for all $S \in \mathcal{F}$, and $\tau^*(\mathcal{F})$ as the infimum of $\mu(X)$ over all such μ. With this definition, the validity of the first part Theorem 10.1.1 is preserved; i.e., $\nu^*(\mathcal{F}) = \tau^*(\mathcal{F})$ for all systems $\mathcal{F}$ of closed sets in a compact X. The proof uses a little of functional analysis, and we omit it; it can be found in [KM97a]. The rationality of ν^* and τ^* no longer holds in the infinite case.

Bibliography and remarks. Gallai's problem about pairwise intersecting disks mentioned at the beginning of this chapter was first solved by Danzer in 1956, but he hasn't published the solution. For another solution and a historical account see Danzer [Dan86].

Attempting to summarize the contemporary knowledge about the transversal number and the packing number in combinatorics would mean taking a much larger bite than can be swallowed, so we restrict ourselves to a few sketchy remarks. An excellent source for many combinatorial results is Lovász's problem collection [Lov93].

A quite old result relating ν and τ is the famous König's edge-covering theorem from 1912, asserting that $\nu(\mathcal{F}) = \tau(\mathcal{F})$ if $\mathcal{F}$ is the system of edges of a bipartite graph (this is also easily seen to be equivalent to Hall's marriage theorem, proved by Frobenius in 1917; see Lovász and Plummer [LP86] for the history). On the other hand, an appropriate generalization to systems of triples, namely, $\tau \leq 2\nu$ for any tripartite 3-uniform hypergraph, is a celebrated recent result of Aharoni [Aha01] (based on Aharoni and Haxell [AH00]), while the generalization $\tau \leq (k-1)\nu$ for k-partite k-uniform hypergraphs, known as Ryser's conjecture, remains unproved for $k \geq 4$.

While computing ν or τ for a given $\mathcal{F}$ is well known to be NP-hard, τ^* can be computed in time polynomial in $|X| + |\mathcal{F}|$ by linear programming (this is another reason for the usefulness of the fractional parameter). The problem of approximating τ is practically very important and has received considerable attention. More often it is considered in the dual form, as the *set cover problem*: Given $\mathcal{F}$ with $\bigcup \mathcal{F} = X$, find the smallest subcollection $\mathcal{F}' \subseteq \mathcal{F}$ that still covers X. The size of such $\mathcal{F}'$ is the transversal number of the set system *dual* to $(X, \mathcal{F})$, where each set $S \in \mathcal{F}$ is assigned a point y_S and each point $x \in X$ gives rise to the set $\{y_S\colon x \in S\}$.

For the set cover problem, it was shown by Chvátal and independently by Lovász that the greedy algorithm (always take a set covering the maximum possible number of yet uncovered points) achieves a solution whose size is no more than $(1 + \ln |X|)$ times larger than the optimal one.[1] Lovász actually observed that the proof implies, for any finite set system $\mathcal{F}$,

$$\tau(\mathcal{F}) \leq \tau^*(\mathcal{F}) \cdot (1 + \ln \Delta(\mathcal{F})),$$

where $\Delta(\mathcal{F})$ is the maximum degree of $\mathcal{F}$, i.e., the maximum number of sets with a common point (Exercise 4). The weaker bound with $\Delta(\mathcal{F})$ replaced by $|\mathcal{F}|$ is easy to prove by probabilistic argument (Exercise 3). It shows that in order to have a large gap between τ^* and τ, the set system must have very many sets.

Exercises

1. (a) Find examples of set systems with τ^* bounded by a constant and τ arbitrarily large. ☐1
   (b) Find examples of set systems with ν bounded by a constant and ν^* arbitrarily large. ☐1
2. Let $\mathcal{F}$ be a system of finitely many closed intervals on the real line. Prove that $\nu(\mathcal{F}) = \tau(\mathcal{F})$. ☐3
3. Prove that
$$\tau(\mathcal{F}) \leq \tau^*(\mathcal{F}) \cdot \ln(|\mathcal{F}|+1)$$
   for all (finite) set systems $\mathcal{F}$. Choose a transversal as a random sample. ☐3
4. (Analysis of the greedy algorithm for transversal) Let $\mathcal{F}$ be a finite set system. We choose points $x_1, x_2, \ldots, x_t$ of a transversal one by one: x_i is taken as a point contained in the maximum possible number of uncovered sets (i.e., sets of $\mathcal{F}$ containing none of $x_1, \ldots, x_{i-1}$).
   (a) Prove that the size t of the resulting transversal satisfies
$$t \leq \frac{1}{1 \cdot 2} \nu_1(\mathcal{F}) + \frac{1}{2 \cdot 3} \nu_2(\mathcal{F}) + \cdots + \frac{1}{(d-1)d} \nu_{d-1}(\mathcal{F}) + \frac{1}{d} \nu_d(\mathcal{F}),$$
   where $d = \Delta(\mathcal{F})$ is the maximum degree of $\mathcal{F}$ and $\nu_k(\mathcal{F})$ is the maximum size of a *simple k-packing* in $\mathcal{F}$. A subsystem $\mathcal{M} \subseteq \mathcal{F}$ is a simple k-packing if $\Delta(\mathcal{M}) \leq k$ (so $\nu_1(\mathcal{F}) = \nu(\mathcal{F})$). ☐5
   (b) Conclude that $\tau(\mathcal{F}) \leq t \leq \tau^*(\mathcal{F}) \cdot \sum_{k=1}^{d} \frac{1}{k}$. ☐2
5. König's edge-covering theorem asserts that if E is the set of edges of a bipartite graph, then $\nu(E) = \tau(E)$. Hall's marriage theorem states that if G is a bipartite graph with color classes A and B such that every subset $S \subseteq A$ has at least $|S|$ neighbors in B, then there is a matching in G containing all vertices of A.

---

[1] As a part of a very exciting development in complexity theory, it was recently proved that no polynomial-time algorithm can do better in general unless P = NP; see, e.g., [Hoc96] for proofs and references.

(a) Derive König's edge-covering theorem from Hall's marriage theorem. ☐2

(b) Derive Hall's marriage theorem from König's edge-covering theorem. ☐2

6. Let A, b, c, P, and D be as in Proposition 10.1.2.

(a) Check that $c^T x \geq y^T b$ for all $x \in P$ and all $y \in D$. ☐1

(b) Prove that if $P \neq \emptyset$ and $D \neq \emptyset$, then the system $Ax \leq b$, $y^T A \geq c$, $c^T x \geq y^T b$ has a nonnegative solution x, y (which implies Proposition 10.1.2). Apply the version of the Farkas lemma as in Exercise 1.2.7(b). ☐4

10.2 Epsilon Nets and VC-Dimension

Large sets should be easier to hit by a transversal than small ones. The notion of ε-net and the related theory elaborate on this intuition. We begin with a special case, where the ground set is finite and the size of a set is simply measured as the cardinality.

10.2.1 Definition (Epsilon net, a special case). *Let $(X, \mathcal{F})$ be a set system with X finite and let $\varepsilon \in [0, 1]$ be a real number. A set $N \subseteq X$ (not necessarily one of the sets of $\mathcal{F}$) is called an ε-net for $(X, \mathcal{F})$ if $N \cap S \neq \emptyset$ for all $S \in \mathcal{F}$ with $|S| \geq \varepsilon |X|$.*

So an ε-net is a transversal for all sets larger than $\varepsilon |X|$. Sometimes it is convenient to write $\frac{1}{r}$ instead of ε, with $r \geq 1$ a real parameter. A beautiful result (Theorem 10.2.4 below) describes a simple combinatorial condition on the structure of $\mathcal{F}$ that guarantees the existence of $\frac{1}{r}$-nets of size only $O(r \log r)$ for all $r \geq 2$.

If we want to deal with infinite sets, measuring the size as the number of points is no longer appropriate. For example, a "large" subset of the unit square could naturally be defined as one with large Lebesgue measure. So in general we consider an arbitrary probability measure μ on the ground set. In concrete situations we will most often encounter μ concentrated on finitely many points. This means that there is a finite set $Y \subseteq X$ and a positive function $w \colon Y \to (0, 1]$ with $\sum_{y \in Y} w(y) = 1$, and μ is given by $\mu(A) = \sum_{y \in A \cap Y} w(y)$. In particular, if the weights of all points $y \in Y$ are the same, i.e., $\frac{1}{|Y|}$, we speak of the *uniform measure on Y*. Another common example of μ is a suitable multiple of the Lebesgue measure restricted to some geometric figure.

10.2.2 Definition (Epsilon net). *Let X be a set, let μ be a probability measure on X, let $\mathcal{F}$ be a system of μ-measurable subsets of X, and let $\varepsilon \in [0, 1]$ be a real number. A subset $N \subseteq X$ is called an ε-net for $(X, \mathcal{F})$ with respect to μ if $N \cap S \neq \emptyset$ for all $S \in \mathcal{F}$ with $\mu(S) \geq \varepsilon$.*

VC-dimension. In order to describe the result promised above, about existence of small ε-nets, we need to introduce a parameter of a set system called the *Vapnik–Chervonenkis dimension*, or *VC-dimension* for short. Its applications are much wider than for the existence of ε-nets.

Let $\mathcal{F}$ be a set system on X and let $Y \subseteq X$. We define the *restriction of $\mathcal{F}$ on Y* (also called the *trace* of $\mathcal{F}$ on Y) as

$$\mathcal{F}|_Y = \{S \cap Y \colon S \in \mathcal{F}\}.$$

It may happen that several distinct sets in $\mathcal{F}$ have the same intersection with Y; in such a case, the intersection is still present only once in $\mathcal{F}|_Y$.

10.2.3 Definition (VC-dimension). *Let $\mathcal{F}$ be a set system on a set X. Let us say that a subset $A \subseteq X$ is shattered by $\mathcal{F}$ if each of the subsets of A can be obtained as the intersection of some $S \in \mathcal{F}$ with A, i.e., if $\mathcal{F}|_A = 2^A$. We define the VC-dimension of $\mathcal{F}$, denoted by $\dim(\mathcal{F})$, as the supremum of the sizes of all finite shattered subsets of X. If arbitrarily large subsets can be shattered, the VC-dimension is ∞.*

Let us consider two examples. First, let $\mathcal{H}$ be the system of all closed half-planes in the plane. We claim that $\dim(\mathcal{H}) = 3$. If we have 3 points in general position, each of their subsets can be cut off by a half-plane, and so such a 3-point set is shattered. Next, let us check that no 4-point set can be shattered. Up to possible degeneracies, there are only two essentially different positions of 4 points in the plane:

In both these cases, if the black points are contained in a half-plane, then a white point also lies in that half-plane, and so the 4 points are not shattered. This is a rather ad hoc argument, and later we will introduce tools for bounding the VC-dimension in geometric situations. We will see that bounded VC-dimension is rather common for families of simple geometric objects in Euclidean spaces.

A rather different example is the system $\mathcal{K}_2$ of all convex sets in the plane. Here the VC-dimension is infinite, since any finite convex independent set A is shattered: Each $B \subseteq A$ can be expressed as the intersection of A with a convex set, namely, $B = A \cap \text{conv}(B)$.

We can now formulate the promised result about small ε-nets.

10.2.4 Theorem (Epsilon net theorem). *If X is a set with a probability measure μ, $\mathcal{F}$ is a system of μ-measurable subsets of X with $\dim(\mathcal{F}) \leq d$, $d \geq 2$, and $r \geq 2$ is a parameter, then there exists a $\frac{1}{r}$-net for $(X, \mathcal{F})$ with respect to μ of size at most $Cdr\ln r$, where C is an absolute constant.*

The proof below gives the estimate $C \leq 20$, but a more accurate calculation shows that C can be taken arbitrarily close to 1 for sufficiently large r. More precisely, for any $d \geq 2$ there exists an $r_0 > 1$ such that for all $r > r_0$, each set system of VC-dimension d admits a $\frac{1}{r}$-net of size at most $dr\ln r$. Moreover, this bound is tight in the worst case up to smaller-order terms.

For the proof (and also later on) we need a fundamental lemma bounding the number of distinct sets in a system of given VC-dimension. First we define the *shatter function* of a set system $\mathcal{F}$ by

$$\pi_{\mathcal{F}}(m) = \max_{Y \subseteq X,\ |Y|=m} |\mathcal{F}|_Y| .$$

In words, $\pi_{\mathcal{F}}(m)$ is the maximum possible number of distinct intersections of the sets of $\mathcal{F}$ with an m-point subset of X.

10.2.5 Lemma (Shatter function lemma). *For any set system $\mathcal{F}$ of VC-dimension at most d, we have $\pi_{\mathcal{F}}(m) \leq \Phi_d(m)$ for all m, where $\Phi_d(m) = \binom{m}{0} + \binom{m}{1} + \cdots + \binom{m}{d}$.*

Thus, the shatter function for any set system is either 2^m for all m (the case of infinite VC-dimension) or it is bounded by a fixed polynomial.

For d fixed and $m \to \infty$, $\Phi_d(m)$ can be simply estimated by $O(m^d)$. For more precise calculations, where we are interested in the dependence on d, we can use the estimate $\Phi_d(m) \leq \left(\frac{em}{d}\right)^d$, where e is the basis of natural logarithms. This is valid for all $m, d \geq 1$.

Proof of Lemma 10.2.5. Since VC-dimension does not increase by passing to a subsystem, it suffices to show that any set system of VC-dimension at most d on an n-point set has no more than $\Phi_d(n)$ sets. We proceed by induction on d, and for a fixed d we use induction on n.

Consider a set system $(X, \mathcal{F})$ with $|X| = n$ and $\dim(\mathcal{F}) = d$, and fix some $x \in X$. In the induction step, we would like to remove x and pass to the set system $\mathcal{F}_1 = \mathcal{F}|_{X \setminus \{x\}}$ on $n-1$ points. This $\mathcal{F}_1$ has VC-dimension at most d, and hence $|\mathcal{F}_1| \leq \Phi_d(n-1)$ by the inductive hypothesis. How many more sets can $\mathcal{F}$ have compared to $\mathcal{F}_1$? The only way that the number of sets decreases by removing x is when two sets $S, S' \in \mathcal{F}$ give rise to the same set in $\mathcal{F}_1$, which means that $S' = S \cup \{x\}$, $x \notin S$, or the other way round. This suggests that we define an auxiliary set system $\mathcal{F}_2$ consisting of all sets in $\mathcal{F}_1$ that correspond to such pairs $S, S' \in \mathcal{F}$: $\mathcal{F}_2 = \{S \in \mathcal{F} : x \notin S, S \cup \{x\} \in \mathcal{F}\}$.

By the above discussion, we have $|\mathcal{F}| = |\mathcal{F}_1| + |\mathcal{F}_2|$. Crucially, we observe that $\dim(\mathcal{F}_2) \leq d-1$, since if $A \subseteq X \setminus \{x\}$ is shattered by $\mathcal{F}_2$, then $A \cup \{x\}$ is shattered by $\mathcal{F}$. Therefore, $|\mathcal{F}_2| \leq \Phi_{d-1}(n-1)$. The resulting recurrence has already been solved in the first proof of Proposition 6.1.1. $\qquad \square$

The rest of the proof of the epsilon net theorem is a clever probabilistic argument; one might be tempted to believe that it works by some magic. First we need a technical lemma concerning the binomial distribution.

10.2.6 Lemma. *Let* $X = X_1 + X_2 + \cdots + X_n$, *where the* X_i *are independent random variables,* X_i *attaining the value* 1 *with probability* p *and the value* 0 *with probability* $1-p$. *Then* $\mathrm{Prob}\left[X \geq \frac{1}{2}np\right] \geq \frac{1}{2}$, *provided that* $np \geq 8$.

Proof. This is a routine consequence of Chernoff-type tail estimates for the binomial distribution, and in fact, considerably stronger estimates hold. The simple result we need can be quickly derived from Chebyshev's inequality for X, stating that $\mathrm{Prob}[|X - \mathbf{E}[X]| \geq t] \leq \mathrm{Var}[X]/t^2$, $t \geq 0$. Here $\mathbf{E}[X] = np$ and $\mathrm{Var}[X] = \sum_{i=1}^n \mathrm{Var}[X_i] \leq np$. So

$$\mathrm{Prob}\left[X < \tfrac{1}{2}np\right] \leq \mathrm{Prob}\left[|X - \mathbf{E}[X]| \geq \tfrac{1}{2}np\right] \leq \frac{4}{np} \leq \tfrac{1}{2}.$$

$\qquad \square$

Proof of the epsilon net theorem. Let us put $s = Cdr \ln r$ (assuming without harm that it is an integer), and let N be a random sample picked by s independent random draws, where each element is drawn from X according to the probability distribution μ. (So the same element can be drawn several times; this does not really matter much, and this way of random sampling is chosen to make calculations simpler.) The goal is to show that N is a $\frac{1}{r}$-net with a positive probability.

To simplify formulations, let us assume that all $S \in \mathcal{F}$ satisfy $\mu(S) \geq \frac{1}{r}$; this is no loss of generality, since the smaller sets do not play any role. The probability that the random sample N misses any given set $S \in \mathcal{F}$ is at most $(1 - \frac{1}{r})^s \leq e^{-s/r}$, and so if s were at least $r \ln(|\mathcal{F}|+1)$, say, the conclusion would follow immediately. But r is typically much smaller than $|\mathcal{F}|$ (it can be a constant, say), and so we need to do something more sophisticated.

Let E_0 be the event that the random sample N fails to be a $\frac{1}{r}$-net, i.e., misses some $S \in \mathcal{F}$. We bound $\mathrm{Prob}[E_0]$ from above using the following thought experiment.

By s more independent random draws we pick another random sample M.[2] We put $k = \frac{s}{2r}$, again assuming that it is an integer, and we let E_1 be the following event:

There exists an $S \in \mathcal{F}$ with $N \cap S = \emptyset$ and $|M \cap S| \geq k$.

---

[2] This double sampling resembles the proof of Proposition 6.5.2, and indeed these proofs have a lot in common, although they work in different settings.

Here an explanation concerning repeated elements is needed. Formally, we regard N and M as *sequences* of elements of X, with possible repetitions, so $N = (x_1, x_2, \ldots, x_s)$, $M = (y_1, y_2, \ldots, y_s)$. The notation $|M \cap S|$ then really means $|\{i \in 1, 2, \ldots, s \colon y_i \in S\}|$, and so an element repeated in M and lying in S is counted the appropriate number of times.

Clearly, $\mathrm{Prob}[E_1] \leq \mathrm{Prob}[E_0]$, since E_1 requires E_0 plus something more. We are going to show that $\mathrm{Prob}[E_1] \geq \frac{1}{2}\mathrm{Prob}[E_0]$. Let us investigate the conditional probability $\mathrm{Prob}[E_1 \mid N]$, that is, the probability of E_1 when N is fixed and M is random. If N is a $\frac{1}{r}$-net, then E_1 cannot occur, and $\mathrm{Prob}[E_0 \mid N] = \mathrm{Prob}[E_1 \mid N] = 0$.

So suppose that there exists an $S \in \mathcal{F}$ with $N \cap S = \emptyset$. There may be many such S, but let us fix one of them and denote it by S_N. We have $\mathrm{Prob}[E_1 \mid N] \geq \mathrm{Prob}[|M \cap S_N| \geq k]$. The quantity $|M \cap S_N|$ behaves like the random variable X in Lemma 10.2.6 with $n = s$ and $p = \frac{1}{r}$, and so $\mathrm{Prob}[|M \cap S_N| \geq k] \geq \frac{1}{2}$. Hence $\mathrm{Prob}[E_0 \mid N] \leq 2\,\mathrm{Prob}[E_1 \mid N]$ for all N, and thus $\mathrm{Prob}[E_0] \leq 2\,\mathrm{Prob}[E_1]$.

Next, we are going to bound $\mathrm{Prob}[E_1]$ differently. Instead of choosing N and M at random directly as above, we first make a sequence $A = (z_1, z_2, \ldots, z_{2s})$ of $2s$ independent random draws from X. Then, in the second stage, we randomly choose s positions in A and put the elements at these positions into N, and the remaining elements into M (so there are $\binom{2s}{s}$ possibilities for A fixed). The resulting distribution of N and M is the same as above. We now prove that for every fixed A, the conditional probability $\mathrm{Prob}[E_1 \mid A]$ is small. This implies that $\mathrm{Prob}[E_1]$ is small, and therefore $\mathrm{Prob}[E_0]$ is small as well.

So let A be fixed. First let $S \in \mathcal{F}$ be a fixed set and consider the conditional probability $P_S = \mathrm{Prob}[N \cap S = \emptyset, |M \cap S| \geq k \mid A]$. If $|A \cap S| < k$, then $P_S = 0$. Otherwise, we bound $P_S \leq \mathrm{Prob}[N \cap S = \emptyset \mid A]$. The latter is the probability that a random sample of s positions out of $2s$ in A avoids the at least k positions occupied by elements of S. This is at most

$$\frac{\binom{2s-k}{s}}{\binom{2s}{s}} \leq \left(1 - \frac{k}{2s}\right)^s \leq e^{-(k/2s)s} = e^{-k/2} = e^{-(Cd\ln r)/4} = r^{-Cd/4}.$$

This was an estimate of P_S for a fixed $S \in \mathcal{F}$. Now, finally, we use the assumption about the VC-dimension of $\mathcal{F}$, via the shatter function lemma: The sets of $\mathcal{F}$ have at most $\Phi_d(2s)$ distinct intersections with A. Since the event "$N \cap S = \emptyset$ and $|M \cap S| \geq k$" depends only on $A \cap S$, it suffices to consider at most $\Phi_d(2s)$ distinct sets S, and so for every fixed A,

$$\mathrm{Prob}[E_1 \mid A] \leq \Phi_d(2s) \cdot r^{-Cd/4} \leq \left(\frac{2es}{d}\right)^d r^{-Cd/4} = \left(2er\ln r \cdot r^{-C/4}\right)^d < \frac{1}{2}$$

if $d, r \geq 2$ and C is sufficiently large. So $\mathrm{Prob}[E_0] \leq 2\,\mathrm{Prob}[E_1] < 1$, which proves Theorem 10.2.4. $\qquad\square$

The epsilon net theorem implies that for set systems of small VC-dimension, the gap between the fractional transversal number and the transversal number cannot be too large.

10.2.7 Corollary. *Let $\mathcal{F}$ be a finite set system on a ground set X with $\dim(\mathcal{F}) \le d$. Then we have*

$$\tau(\mathcal{F}) \le Cd\tau^*(\mathcal{F}) \ln \tau^*(\mathcal{F}),$$

where C is as in the epsilon net theorem.

Proof. Let $r = \tau^*(\mathcal{F})$. Since $\mathcal{F}$ is finite, we may assume that an optimal fractional transversal $\varphi \colon X \to [0,1]$ is concentrated on a finite set Y. This φ, after rescaling, defines a probability measure μ on X, by letting $\mu(\{y\}) = \frac{1}{r}\varphi(y)$, $y \in Y$. Each $S \in \mathcal{F}$ has $\mu(S) \ge \frac{1}{r}$ by the definition of fractional transversal, and so a $\frac{1}{r}$-net for $(X, \mathcal{F})$ with respect to μ is a transversal. By the epsilon net theorem, there exists a transversal of size at most $Cdr \ln r$. $\square$

We mention a concrete application of the corollary in the next section, where we collect examples of set systems of bounded VC-dimension.

Bibliography and remarks. The notion of VC-dimension originated in statistics. It was introduced by Vapnik and Chervonenkis [VC71]. Under different names, it has also appeared in other papers (Sauer [Sau72] and Shelah [She72]), but the work [VC71] was probably the most influential for subsequent developments. The name VC-dimension and some other, by now more or less standard, terminology were introduced by Haussler and Welzl [HW87]. VC-dimension and the related theory play an important role in several mathematical fields, such as statistics (the theory of empirical processes), computational learning theory, computational geometry, discrete geometry, combinatorics of hypergraphs, and discrepancy theory.

The shatter function lemma was independently discovered in the three already mentioned papers [VC71], [Sau72], [She72].

The shatter function, together with the dual shatter function (defined as the shatter function of the dual set system) was introduced and applied by Welzl [Wel88]. Implicitly, these notions were used much earlier, and they appear in the literature under various names, such as *growth functions*.

The notion of ε-net and the epsilon net theorem (with X finite and μ uniform) are due to Haussler and Welzl [HW87]. Their proof is essentially the one shown in the text, and it closely follows an earlier proof by Vapnik and Chervonenkis [VC71] concerning the related notion of ε-*approximations*. In the same setting as in the definition of

ε-nets, a set $A \subseteq X$ is an ε-approximation for $(X, \mathcal{F})$ with respect to μ if for all $S \in \mathcal{F}$,

$$\left| \mu(S) - \frac{|A \cap S|}{|A|} \right| < \varepsilon.$$

So while an ε-net intersects each large set at least once, an ε-approximation provides a "proportional representation" up to the error of ε. Vapnik and Chervonenkis [VC71] proved the existence of $\frac{1}{r}$-approximations of size $O(dr^2 \log r)$ for all set system of VC-dimension d.

Komlós, Pach, and Wöginger [KPW92] improved the dependence on d in the Haussler–Welzl bound on the size of ε-nets. The improvement is achieved by choosing the second sample M of size t somewhat larger than s and doing the calculations more carefully. They also proved an almost matching lower bound using suitable random set systems. The proofs can be found in [PA95] as well.

The proof in the Vapnik–Chervonenkis style, while short and clever, does not seem to convey very well the reasons for the existence of small ε-nets. Somewhat longer but more intuitive proofs have been found in the investigation of deterministic algorithms for constructing ε-approximations and ε-nets; one such proof is given in [Mat99a], for instance.

Exercises

1. Show that for any integer d there exists a convex set C in the plane such that the family of all isometric copies of C has VC-dimension at least d. [4]
2. Show that the shatter function lemma is tight. That is, for all d and n construct a system of VC-dimension d on n points with $\Phi_d(n)$ sets. [2]

10.3 Bounding the VC-Dimension and Applications

The VC-dimension can be determined without great difficulty in several simple cases, such as for half-spaces or balls in $\mathbf{R}^d$, but for only slightly more complicated families its computation becomes challenging. On the other hand, a few simple steps explained below show that the VC-dimension is bounded for any family whose sets can be defined by a formula consisting of polynomial equations and inequalities combined by Boolean connectives (conjunctions, disjunctions, etc.) and involving a bounded number of real parameters. This includes families like all ellipsoids in $\mathbf{R}^d$, all boxes in $\mathbf{R}^d$, arbitrary intersections of pairs of circular disks in the plane, and so on. On the other hand, arbitrary convex polygons are not covered (since a general convex polygon cannot be described by a bounded number of real parameters) and indeed, this family has infinite VC-dimension.

We begin by determining the VC-dimension for half-spaces.

10.3.1 Lemma. *The VC-dimension of the system of all (closed) half-spaces in $\mathbf{R}^d$ equals $d+1$.*

Proof. Obviously, any set of $d+1$ affinely independent points can be shattered. On the other hand, no $d+2$ points can be shattered by Radon's lemma. $\square$

Next, we turn to the family $\mathcal{P}_{d,D}$ of all sets in $\mathbf{R}^d$ definable by a single polynomial inequality of degree at most D.

10.3.2 Proposition. *Let $\mathbf{R}[x_1, x_2, \ldots, x_d]_{\leq D}$ denote the set of all real polynomials in d variables of degree at most D, and let*

$$\mathcal{P}_{d,D} = \Big\{ \{x \in \mathbf{R}^d \colon p(x) \geq 0\} \colon p \in \mathbf{R}[x_1, x_2, \ldots, x_d]_{\leq D} \Big\}.$$

Then $\dim(\mathcal{P}_{d,D}) \leq \binom{d+D}{d}$.

Proof. The following simple but powerful trick is known as the *Veronese mapping* in algebraic geometry (or as *linearization*; it is also related to the reduction of Voronoi diagrams to convex polytopes in Section 5.7). Let M be the set of all possible nonconstant monomials of degree at most D in $x_1, \ldots, x_d$. For example, for $D = d = 2$, we have $M = \{x_1, x_2, x_1 x_2, x_1^2, x_2^2\}$. Let $m = |M|$ and let the coordinates in $\mathbf{R}^m$ be indexed by the monomials in M. Define the map $\varphi \colon \mathbf{R}^d \to \mathbf{R}^m$ by $\varphi(x)_\mu = \mu(x)$, where the monomial μ serves as a formal symbol (index) on the left-hand side, while on the right-hand side we have the number obtained by evaluating μ at the point $x \in \mathbf{R}^d$. For example, for $d = D = 2$, the map is

$$\varphi \colon (x_1, x_2) \in \mathbf{R}^2 \mapsto (x_1, x_2, x_1 x_2, x_1^2, x_2^2) \in \mathbf{R}^5.$$

We claim that if $A \subset \mathbf{R}^d$ is shattered by $\mathcal{P}_{d,D}$, then $\varphi(A)$ is shattered by half-spaces in $\mathbf{R}^m$. To see this, let $B \subseteq A$, and let $p \in \mathcal{P}_{d,D}$ be a polynomial that is nonnegative at the points of B and negative at $A \setminus B$. We let a_μ be the coefficient of μ in p and a_0 the constant term of p, and we define the half-space $h_p \in \mathbf{R}^m$ as $\{y \in \mathbf{R}^m \colon a_0 + \sum_{\mu \in M} a_\mu y_\mu \geq 0\}$. For example, if $p(x_1, x_2) = 7 + 3x_2 - x_1 x_2 + x_1^2 \in \mathcal{P}_{2,2}$, the corresponding half-space is $h_p = \{y \in \mathbf{R}^5 \colon 7 + 3y_2 - y_3 + y_4 \geq 0\}$. Then we get $h_p \cap \varphi(A) = \varphi(B)$. Since, finally, φ is injective, we obtain a set of size $|A|$ in $\mathbf{R}^m$ shattered by half-spaces. By Lemma 10.3.1, we have $\dim(\mathcal{P}_{d,D}) \leq |M| + 1 = \binom{D+d}{d}$. $\square$

Geometrically, the Veronese map embeds $\mathbf{R}^d$ into $\mathbf{R}^m$ as a curved manifold in such a way that any subset of $\mathbf{R}^d$ definable by a single polynomial inequality of degree at most D can be cut off by a half-space in $\mathbf{R}^m$. Except for few simple cases, this is hard to visualize, but the formulas work in a really simple way.

By Proposition 10.3.2, any subfamily of some $\mathcal{P}_{d,D}$ has bounded VC-dimension; this applies, e.g., to balls in $\mathbf{R}^d$ ($D = 2$) and ellipsoids in $\mathbf{R}^d$ ($D = 2$ as well). For concrete families, the bound from Proposition 10.3.2 is often very weak. First, if we deal only with special polynomials involving fewer than $\binom{D+d}{d}$ monomials, then we can use an embedding into $\mathbf{R}^m$ with a smaller m. We also do not have to use only coordinates corresponding to monomials in the embedding. For example, for the family of all balls in $\mathbf{R}^d$, a suitable embedding is $\varphi\colon \mathbf{R}^d \to \mathbf{R}^{d+1}$ given by $(x_1, \ldots, x_d) \mapsto (x_1, x_2, \ldots, x_1^2 + x_2^2 + \cdots + x_d^2)$. It is closely related to the "lifting" transforming Voronoi diagrams in $\mathbf{R}^d$ to convex polytopes in $\mathbf{R}^{d+1}$ discussed in Section 5.7. Estimates for the VC-dimension can also be obtained from Theorem 6.2.1 about the number of sign patterns of polynomials or from similar results.

Combinations of polynomial inequalities. Families like all rectangular boxes in $\mathbf{R}^d$ or lunes (differences of two disks in the plane) can be handled using the following result.

10.3.3 Proposition. *Let $F(X_1, X_2, \ldots, X_k)$ be a fixed set-theoretic expression (using the operations of union, intersection, and difference) with variables $X_1, \ldots X_k$ standing for sets; for instance,*

$$F(X_1, X_2, X_3) = (X_1 \cup X_2 \cup X_3) \setminus (X_1 \cap X_2 \cap X_3).$$

Let $\mathcal{S}$ be a set system on a ground set X with $\dim(\mathcal{S}) = d < \infty$. Let

$$\mathcal{T} = \{F(S_1, \ldots, S_k)\colon S_1, \ldots, S_k \in \mathcal{S}\}.$$

Then $\dim(\mathcal{T}) = O(kd \ln k)$.

Proof. The trick is to look at the shatter functions. Let $A \subseteq X$ be an m-point set. It is easy to verify by induction on the structure of F that for any $S_1, S_2, \ldots, S_k$, we have $F(S_1, \ldots, S_k) \cap A = F(S_1 \cap A, \ldots, S_k \cap A)$. In particular, $F(S_1, \ldots, S_k) \cap A$ depends only on the intersections of the S_i with A. Therefore, $\pi_{\mathcal{T}}(m) \leq \pi_{\mathcal{S}}(m)^k$. By the shatter function lemma, we have $\pi_{\mathcal{S}}(m) \leq \Phi_d(m)$. If A is shattered by $\mathcal{T}$, then $\pi_{\mathcal{T}}(m) = 2^m$. From this we have the inequality $2^m \leq \Phi_d(m)^k$. Calculation using the estimate $\Phi_d(m) \leq (\frac{em}{d})^d$ leads to the claimed bound. $\qquad\square$

Propositions 10.3.3 and 10.3.2 together show that families of geometric shapes definable by formulas of bounded size involving polynomial equations and inequalities have bounded VC-dimension. (In the terminology introduced in Section 7.7, families of semialgebraic sets of bounded description complexity have bounded VC-dimension.) In the subsequent example we will encounter a family of quite different nature with bounded VC-dimension. First we present a general observation.

VC-dimension of the dual set system. Let $(X, \mathcal{F})$ be a set system. The *dual set system* to $(X, \mathcal{F})$ is defined as follows: The ground set is $Y =$

$\{y_S\colon S \in \mathcal{F}\}$, where the y_S are pairwise distinct points, and for each $x \in X$ we have the set $\{y_S\colon S \in \mathcal{F},\ x \in S\}$ (the same set may be obtained for several different x, but this does not matter for the VC-dimension).

10.3.4 Lemma. *Let $(X, \mathcal{F})$ be a set system and let $(Y, \mathcal{G})$ be the dual set system. Then $\dim(\mathcal{G}) < 2^{\dim(\mathcal{F})+1}$.*

Proof. We show that if $\dim(\mathcal{G}) \geq 2^d$, then $\dim(\mathcal{F}) \geq d$. Let A be the incidence matrix of $(X, \mathcal{F})$, with columns corresponding to points of X and rows corresponding to sets of $\mathcal{F}$. Then the transposed matrix A^T is the incidence matrix of $(Y, \mathcal{G})$. If Y contains a shattered set of size 2^d, then A has a $2^d \times 2^{2^d}$ submatrix M with all the possible 0/1 vectors of length 2^d as columns. We claim that M contains as a submatrix the $2^d \times d$ matrix M_1 with all possible 0/1 vectors of length d as rows. This is simply because the d columns of M_1 are pairwise distinct and they all occur as columns of M. This M_1 corresponds to a shattered subset of size d in $(X, \mathcal{F})$. Here is an example for $d = 2$:

$$M = \begin{pmatrix} 0 & 0 & 0 & \mathbf{0} & 0 & \mathbf{0} & 0 & 0 & 1 & 1 & 1 & 1 & 1 & 1 & 1 & 1 \\ 0 & 0 & 0 & \mathbf{0} & 1 & \mathbf{1} & 1 & 1 & 0 & 0 & 0 & 0 & 1 & 1 & 1 & 1 \\ 0 & 0 & 1 & \mathbf{1} & 0 & \mathbf{0} & 1 & 1 & 0 & 0 & 1 & 1 & 0 & 0 & 1 & 1 \\ 0 & 1 & 0 & \mathbf{1} & 0 & \mathbf{1} & 0 & 1 & 0 & 1 & 0 & 1 & 0 & 1 & 0 & 1 \end{pmatrix};$$

the submatrix M_1 is marked bold. $\square$

An art gallery problem. An *art gallery*, for the purposes of this section, is a compact set X in the plane, such as the one drawn in the following picture:

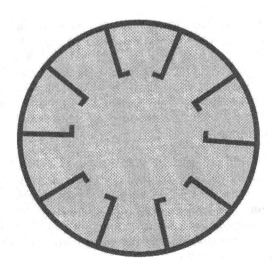

The set X is the lightly shaded area, while the black regions are walls that are not part of X. We want to choose a small set $G \subset X$ of guards that together can see all points of X, where a point $x \in X$ *sees* a point $y \in X$ if the segment xy is fully contained in X. The *visibility region* $V(x)$ of a point $x \in X$ is the set of all points $y \in X$ seen by x, as is illustrated below:

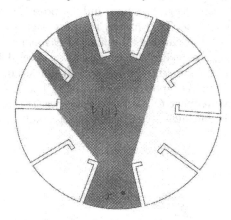

It is easy to construct galleries that require arbitrarily many guards; it suffices to include many small niches so that each of them needs an individual guard. To forbid this cheap way of making a gallery difficult to guard, we consider only galleries where each point can be seen from a reasonably large part of the gallery. That is, we suppose that the gallery X has Lebesgue measure 1 and that $\mu(V(x)) \geq \varepsilon$ for every $x \in X$, where $\varepsilon > 0$ is a parameter (say $\frac{1}{10}$) and μ is the Lebesgue measure restricted to X. Can every such gallery be guarded by a number of guards that depends only on ε?

The answer to this question is still no, although an example is not entirely easy to construct. The problem is with galleries with many "holes," i.e., many connected components of the complement (corresponding to pillars in a real-world gallery, say). But if we forbid holes, then the answer becomes yes.

10.3.5 Theorem. *Let X be a simply connected art gallery (i.e., with $\mathbf{R}^2 \setminus X$ connected) of Lebesgue measure 1, and let $r \geq 2$ be a real number such that $\mu(V(x)) \geq \frac{1}{r}$ for all $x \in X$. Then X can be guarded by at most $Cr \log r$ points, where C is a suitable absolute constant.*

Proof. The bound $O(r \log r)$ for the number of guards is obtained from the epsilon net theorem (Theorem 10.2.4). Namely, we introduce the set system $\mathcal{V} = \{V(x)\colon x \in X\}$, and note that G is a set guarding all of X if and only if it is a transversal of $\mathcal{V}$. Further, an ε-net for $(X, \mathcal{V})$ with respect to μ is a transversal of $\mathcal{V}$, since by the assumption, $\mu(V) \geq \varepsilon = \frac{1}{r}$ for each $V \in \mathcal{V}$. So the theorem will be proved if we can show that $\dim(\mathcal{V})$ is bounded by some constant (independent of X).

Tools like Proposition 10.3.2 and Proposition 10.3.3 seem to be of little use, since the visibility regions can be arbitrarily complicated. We thus need a different strategy, one that can make use of the simple connectedness. We proceed by contradiction: Assuming the existence of an extremely large set $A \subset X$ shattered by $\mathcal{V}$, we find, by a sequence of Ramsey-type steps, a configuration forcing a hole in X.

Let d be a sufficiently large number, and suppose that there is a d-point set $A \subset X$ shattered by $\mathcal{V}$. This means that for each subset $B \subseteq A$ there exists a point $\sigma_B \in X$ that can see all points of B but no point of $A \setminus B$. We put $\Sigma = \{\sigma_B : B \subseteq A\}$. In such a situation, we say that A is *shattered by* Σ.

Starting with A and Σ, we find a smaller shattered set in a special position. We draw a line through each pair of points of A. The arrangement of these at most $\binom{d}{2}$ lines has at most $O(d^4)$ faces (vertices, edges, and open convex polygons), so there is one such face F_0 containing a subset $\Sigma' \subseteq \Sigma$ of at least $2^d/O(d^4)$ points of Σ.

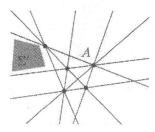

These points correspond to subsets of A, and so they define a set system $\mathcal{V}_1$ on A. If $d_1 = \dim(\mathcal{V}_1)$ were bounded by a constant independent of d, then the number of sets in $\mathcal{V}_1$ would grow at most polynomially with d (by Lemma 10.2.5). But we know that it grows exponentially, and so $d_1 \to \infty$ as $d \to \infty$. Thus, we may assume that some subset $A_1 \subseteq A$ is shattered by a subset $\Sigma_1 \subseteq \Sigma'$, with $d_1 = |A_1|$ large, and the whole of Σ_1 lies in a single face of the arrangement of the lines determined by points of A_1.

Next, we would like to ensure a similar condition in the reverse direction, that is, all the points being shattered lying in a single cell of the arrangement of the lines determined by the shattering points.

A simple, although wasteful, way is to apply Lemma 10.3.4 about the dimension of the dual set system. This means that we can select sets $A_2 \subseteq \Sigma_1$ and $\Sigma_2 \subseteq A_1$ such that A_2 is shattered by Σ_2 and $d_2 = |A_2|$ is still large (about $\log_2 d_1$).

Now we can repeat the procedure from the first step of the proof, this time selecting a set $A_3 \subseteq A_2$ of size d_3 (still sufficiently large) and $\Sigma_3 \subseteq \Sigma_2$ such that A_3 is shattered by Σ_3 and all of Σ_3 lies in a single face of the arrangement of the lines determined by the pairs of points of A_3. This face must be 2-dimensional, since if it were an edge, all the points of A_3 and Σ_3 would be collinear, which is impossible.

We thus have all points of A_3 within a single 2-face of the arrangement of the lines determined by Σ_3 and vice versa. In other words, no line determined by two points of A_3 intersects $\text{conv}(\Sigma_3)$, and no line determined by two points of Σ_3 intersects $\text{conv}(A_3)$. In particular, $\text{conv}(A_3) \cap \text{conv}(\Sigma_3) = \emptyset$. It follows that each point of Σ_3 sees all points of A_3 within an angle smaller than π and in the same clockwise angular order; let $\leq_A$ be this linear order of the points of A_3. Similarly, we have a common counterclockwise angular order $\leq_\Sigma$ of points of Σ_3 around any point of A_3.

Suppose that the initial d was so large that $d_3 = |A_3| = 5$. For each $a \in A_3$, we consider the point $\sigma(a) \in \Sigma_3$ that sees all points of A_3 but a. Let these 5 points form a set $\Sigma_4 \subset \Sigma_3$. We have a situation indicated below, where dashed connecting segments correspond to *invisibility* and they form a matching between A_3 and Σ_4.

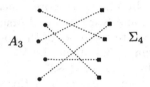

Since we have 5 points on each side, we may choose an $a \in A_3$ such that a is neither the first nor the last point of A_3 in $\leq_A$, and at the same time $\sigma = \sigma(a) \in \Sigma_4$ is not the first or last point in $\leq_\Sigma$. Then we have the following situation (full segments indicate visibility, and the dashed segment means invisibility):

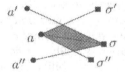

The segments $a\sigma'$ and $a'\sigma$ both lie above the line $a\sigma$, and they intersect as indicated (a' cannot line in the triangle $a\sigma\sigma'$, because the line aa' would go between σ and σ', and neither can the segment $\sigma a'$ be outside that triangle, because then the line $\sigma\sigma'$ would separate a from a'). Similarly, the segments $a\sigma''$ and $a''\sigma$ intersect as shown. The four segments $a\sigma'$, $a'\sigma$, $a\sigma''$, and $a''\sigma$ are contained in X, and since X is simply connected, the shaded quadrilateral bounded by them must be a part of X. Hence a and σ can see each other. This contradiction proves Theorem 10.3.5.   □

The bound on the VC-dimension obtained from this proof is rather large: about 10^{12}. By a more careful analysis, avoiding the use of Lemma 10.3.4 on the dual VC-dimension where one loses the most, the bound has been improved to 23. Determining the exact VC-dimension in the worst case might be quite challenging. The art gallery drawn in the initial picture is not chosen only because of the author's liking for several baroque buildings with pentagonal symmetry, but also because it is an example where $\mathcal{V}$ has VC-dimension at least 5 (Exercise 2). A more complicated example gives VC-dimension 6, and this is the current best lower bound.

Bibliography and remarks. As was remarked in the text, for bounding the VC-dimension of set systems defined by polynomial inequalities, we can use the linearization method (as in the proof of Proposition 10.3.2) or results like Theorem 6.2.1 on the number of sign patterns. The latter can often provide asymptotically sharp bounds on the shatter functions (which are usually the more important quantitative parameters in applications); for linearizations, this happens only in quite simple cases.

There are fairly general results bounding the VC-dimension for families of sets defined by functions more general than polynomials; see, e.g., Wilkie [Wil99] and Karpinski and Macintyre [KM97b].

Considerations similar to the proof of Proposition 10.3.3 appear in Dudley [Dud78]. Lemma 10.3.4 about the VC-dimension of the dual set system was noted by Assouad [Ass83].

The art gallery problem considered in this section was raised by Kavraki, Latombe, Motwani, and Raghavan [KLMR98] in connection with automatic motion planning for robots. Theorem 10.3.5, with the proof shown, is from Kalai and Matoušek [KM97a]. That paper also proves that for galleries with h holes, the number of guards can be bounded by a function of ε and h, and provides an example showing that one may need at least $\Omega(\log h)$ guards in the worst case for a suitable fixed ε. Valtr [Val98] greatly improved the quantitative bounds, obtaining the lower bound of 6 and upper bound of 23 for $\dim(\mathcal{V})$ for simply connected galleries, as well as a bound of $O(\log_2 h)$ for galleries with h holes. In another paper [Val99b], he constructed contractible 3-dimensional galleries where the visibility region of each point occupies almost half of the total volume of the gallery but the number of guards is unbounded, which shows that Theorem 10.3.5 has no straightforward analogue in dimension 3 and higher. Here is another result from [KM97a]: If a planar gallery X is such that among every k points of X there are 3 that can be guarded by a single guard, then all of X can be guarded by $O(k^3 \log k)$ guards. Let us stress that our example was included mainly as an illustration to VC-dimension, rather than as a typical specimen of the extensive subject of studying guards in art galleries from the mathematical point of view. This field has a large number results, some of them very nice; see, e.g., the handbook chapter [Urr00] for a survey.

Exercises

1. (a) Determine the VC-dimension of the set system consisting of all triangles in the plane. ③
   (b) What is the VC-dimension of the system of all convex k-gons in the plane, for a given integer k? ②

2. Show that $\dim(\mathcal{V}) \geq 5$ for the art gallery shown above Theorem 10.3.5. ②

Can you construct an example with VC-dimension 6, or even higher?

3. Show that the unit square cannot be expressed as $\{(x,y) \in \mathbf{R}^2 : p(x,y) \geq 0\}$ for any polynomial $p(x,y)$. ③

4. (a) Let H be a finite set of lines in the plane. For a triangle T, let H_T be the set of lines of H intersecting the interior of T, and let $\mathcal{T} \subseteq 2^H$ be the system of the sets H_T for all triangles T. Show that the VC-dimension of $\mathcal{T}$ is bounded by a constant. ②

(b) Using (a) and the epsilon net theorem, prove the suboptimal cutting lemma (Lemma 6.5.1): For every finite set H of lines in the plane and for every r, $1 < r < |H|$, there exists a $\frac{1}{r}$-cutting for L consisting of $O(r^2 \log^2 r)$ generalized triangles. Use the proof in Section 4.6 as an inspiration. ③

(c) Generalize (a) and (b) to obtain a cutting lemma for circles with the same bound $O(r^2 \log^2 r)$ (see Exercise 4.6.3). ②

5. Let $d \geq 1$ be an integer, let $U = \{1, 2, \ldots, d\}$ and $V = 2^U$. Let the *shattering graph* SG_d have vertex set $U \cup V$ and edge set $\{\{a, A\} : a \in U, A \in V, a \in A\}$. Prove that if H is a bipartite graph with classes R and S, $|R| = r$ and $|S| = s$, such that $r + \log_2 s \leq d$, then there is an r-element subset $R_1 \subseteq U$ and an s-element subset $S_1 \subseteq V$ such that the subgraph induced in SG_d by $R_1 \cup S_1$ is isomorphic to H. Thus, the shattering graph is "universal": It contains all sufficiently small bipartite subgraphs. ③

6. For a graph G, let $\mathcal{N}(G) = \{N_G(v) : v \in V(G)\}$ be the system of vertex neighborhoods (where $N_G(v) = \{u \in V(G) : \{u, v\} \in E(G)\}$).

(a) Prove that there is a constant d_0 such that $\dim(\mathcal{N}(G)) \leq d_0$ for all planar G. ③

(b) Show that for every C there exists $d = d(C)$ such that if G is a graph in which every subgraph on n vertices has at most Cn edges, for all $n \geq 1$, then $\dim(\mathcal{N}(G)) \leq d$. (This implies (a) and, more generally, shows that bounded genus of G implies bounded $\dim(\mathcal{N}(G))$.) ③

(c) Show that for every k there exists $d = d(k)$ such that if $\dim(\mathcal{N}(G)) \geq d$, then G contains a subdivision of the complete graph K_k as a subgraph. (This gives an alternative proof that if $\dim(\mathcal{N}(G))$ is large, then the genus of G is large, too.) ③

10.4 Weak Epsilon Nets for Convex Sets

Weak ε-nets. Let $\mathcal{H}$ be the system of all closed half-planes in the plane, and let μ be the planar Lebesgue measure restricted to a (closed) disk D of unit area. What should the smallest possible ε-net for $(\mathbf{R}^2, \mathcal{H})$ with respect to μ look like? A natural idea would be to place the points of the ε-net equidistantly around the perimeter of the disk:

Is this the best way? No; according to Definition 10.2.2, three points placed as in the picture below form a valid ε-net for every $\varepsilon \geq 0$, since any half-plane cutting into D necessarily contains at least one of them!

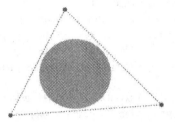

One may feel that this is a cheating. The problem is that the points of this ε-net are far away from where the measure is concentrated. For some applications of ε-nets this is not permissible, and for this reason, ε-nets of this kind are usually called *weak ε-nets* in the literature, while a "real" ε-net in the above example would be required to have all of its points inside the disk D.

For ε-nets obtained using the epsilon net theorem (Theorem 10.2.4), this presents no real problem, since we can always restrict the considered set system to the subset where we want our ε-net to lie. In the above example we would simply require an ε-net for the set system $(D, \mathcal{H}|_D)$. The restriction to a subset does not increase the VC-dimension.

On the other hand, there are set systems of infinite VC-dimension, and there we cannot require small ε-nets to exist for every restriction of the ground set. Indeed, if $(X, \mathcal{F})$ has infinite VC-dimension, then by definition, there is an arbitrarily large $A \subseteq X$ that is shattered by $\mathcal{F}$, meaning that $\mathcal{F}|_A = 2^A$. And the complete set system $(A, 2^A)$ certainly does not admit small ε-nets: Any $\frac{1}{2}$-net, say, for $(A, 2^A)$ with respect to the uniform measure on A must have at least $\frac{1}{2}|A|$ elements! In this sense, the epsilon net theorem is an "if and only if" result: A set system $(X, \mathcal{F})$ *and all of its restrictions to smaller ground sets* admit ε-nets of size depending only on ε if and only if $\dim(\mathcal{F})$ is finite.

As was mentioned after the definition of VC-dimension, the (important) system $\mathcal{K}_2$ of convex sets in the plane has infinite VC-dimension. Therefore, the epsilon net theorem is not applicable, and we know that restrictions of $\mathcal{K}_2$ to some bad ground sets (convex independent sets, in this case) provide arbitrarily large complete set systems. But yet it turns out that not too large (weak) ε-nets exist if the ground set is taken to be the whole plane (or, actually, it can be restricted to any convex set). These are much less understood than the ε-nets in the case of finite VC-dimensions, and many interesting questions remain open.

As has been done in the literature, we will restrict ourselves to measures concentrated on finite point sets, and first we will talk about uniform measures. To be on the safe side, let us restate the definition for this particular case, keeping the traditional terminology of "weak ε-nets."

10.4.1 Definition (Weak epsilon net for convex sets). *Let X be a finite point set in $\mathbf{R}^d$ and $\varepsilon > 0$ a real number. A set $N \subseteq \mathbf{R}^d$ is called a weak ε-net for convex sets with respect to X if every convex set containing at least $\varepsilon|X|$ points of X contains a point of N.*

In the rest of this section we consider exclusively ε-nets with respect to convex sets, and so instead of "weak ε-net for convex sets with respect to X" we simply say "weak ε-net for X."

10.4.2 Theorem (Weak epsilon net theorem). *For every $d \geq 1$, $\varepsilon > 0$, and finite $X \subset \mathbf{R}^d$, there exists a weak ε-net for X of size at most $f(d, \varepsilon)$, where $f(d, \varepsilon)$ depends on d and ε but not on X.*

The best known bounds are $f(2, \frac{1}{r}) = O(r^2)$ in the plane and $f(d, \frac{1}{r}) = O(r^d(\log r)^{b(d)})$ for every fixed d, with a suitable constant $b(d) > 0$. The proof shown below gives $f(d, \frac{1}{r}) = O(r^{d+1})$. On the other hand, no lower bound superlinear in r is known (for fixed d).

Proof. The proof is simple once we have the first selection lemma (Theorem 9.1.1) at our disposal.

Let an $X \subset \mathbf{R}^d$ be an n-point set. The required weak ε-net N is constructed by a greedy algorithm. Set $N_0 = \emptyset$. If N_i has already been constructed, we look whether there is a convex set C containing at least εn points of X and no point of N_i. If not, N_i is a weak ε-net by definition. If yes, we set $X_i = X \cap C$, and we apply the first selection lemma to X_i. This gives us a point a_i contained in at least $c_d\binom{|X_i|}{d+1} = \Omega(\varepsilon^{d+1}n^{d+1})$ X_i-simplices. We set $N_{i+1} = N_i \cup \{a_i\}$ and continue with the next step of the algorithm.

Altogether there are $\binom{n}{d+1}$ X-simplices. In each step of the algorithm, at least $\Omega(\varepsilon^{d+1}n^{d+1})$ of them are "killed," meaning that they were not intersected by N_i but are intersected by N_{i+1}. Hence the algorithm takes at most $O(\varepsilon^{-(d+1)})$ steps. $\qquad\square$

In a forthcoming application, we also need weak ε-nets for convex sets with respect to a nonuniform measure (but still concentrated on finitely many points).

10.4.3 Corollary. *Let μ be a probability measure concentrated on finitely many points in $\mathbf{R}^d$. Then weak ε-nets for convex sets with respect to μ exist, of size bounded by a function of d and ε.*

Sketch of proof. By taking ε a little smaller, we can make the point weights rational. Then the problem is reduced to the weak epsilon net theorem with X a multiset. One can check that all ingredients of the proof go through in this case, too. □

10.4.4 Corollary. *For every finite system $\mathcal{F}$ of convex sets in $\mathbf{R}^d$, we have $\tau(\mathcal{F}) \le f(d, 1/\tau^*(\mathcal{F}))$, where $f(d, \varepsilon)$ is as in the weak epsilon net theorem.*

The proof of the analogous consequence of the epsilon net theorem, Corollary 10.2.7, can be copied almost verbatim.

Bibliography and remarks. Weak ε-nets were introduced by Haussler and Welzl [HW87]. The existence of weak ε-nets for convex sets was proved by Alon, Bárány, Füredi, and Kleitman [ABFK92] by the method shown in the text but with a slight quantitative improvement, achieved by using the second selection lemma (Theorem 9.2.1) instead of the first selection lemma.

The estimates for $f(d, \frac{1}{r})$ mentioned after Theorem 10.4.2 have the following sources: The bound $O(r^2)$ in the plane is from [ABFK92] (see Exercise 1), and the best general bound in $\mathbf{R}^d$, close to $O(r^d)$, is due to Chazelle, Edelsbrunner, Grini, Guibas, Sharir, and Welzl [CEG+95]. It seems that these bounds are quite far from the truth. Intuitively, one of the "worst" cases for constructing a weak ε-net should be a convex independent set X. For such sets in the plane, though, near-linear bounds have been obtained by Chazelle et al. [CEG+95]; they are presented in Exercises 2 and 3 below. The original proof of the result in Exercise 3 was formulated using hyperbolic geometry. A simple lower bound for the size of weak ε-nets was noted in [Mat01]; it concerns the dependence on d for ε fixed and shows that $f(d, \frac{1}{50}) = \Omega\left(e^{\sqrt{d/2}}\right)$ as $d \to \infty$.

Exercises

1. Complete the following sketch of an alternative proof of the weak epsilon net theorem.

   (a) Let X be an n-point set in the plane (assume general position if convenient). Let h be a vertical line with half of the points of X on each side, and let X_1, X_2 be these halves. Let M be the set of all intersections of segments of the form $x_1 x_2$ with h, where $x_1 \in X_1$ and $x_2 \in X_2$. Let N_0 be a weak ε'-net for M (this is a one-dimensional situation!). Recursively construct weak ε''-nets N_1, N_2 for X_1 and X_2, respectively, and set $N = N_0 \cup N_1 \cup N_2$. Show that with a suitable choice of ε' and ε'', N is a weak ε-net for X of size $O(\varepsilon^{-2})$. ③

   (b) Generalize the proof from (a) to $\mathbf{R}^d$ (use induction on d). Estimate the exponent of ε in the resulting bound on the size of the constructed weak ε-net. ③

2. The aim of this exercise is to show that if X is a finite set in the plane in convex position, then for any $\varepsilon > 0$ there exists a weak ε-net for X of size nearly linear in $\frac{1}{\varepsilon}$.

(a) Let an n-point convex independent set $X \subset \mathbf{R}^2$ be given and let $\ell \le n$ be a parameter. Choose points $p_0, p_1, \ldots, p_{\ell-1}$ of X, appearing in this order around the circumference of $\mathrm{conv}(X)$, in such a way that the set X_i of points of X lying (strictly) between p_{i-1} and p_i has at most n/ℓ points for each i. Construct a weak ε'-net N_i for each X_i (recursively) with $\varepsilon' = \ell\varepsilon/3$, and let M be the set containing the intersection of the segment $p_0 p_{j-1}$ with $p_j p_i$, for all pairs i, j, $1 \le i < j-1 \le \ell-2$. Show that the set $N = \{p_0, \ldots, p_{\ell-1}\} \cup N_1 \cup \cdots \cup N_\ell \cup M$ is a weak ε-net for X. ☐3

(b) If $f(\varepsilon)$ denotes the minimum necessary size of a weak ε-net for a finite convex independent point set in the plane, derive a recurrence for $f(\varepsilon)$ using (a) with a suitably chosen ℓ, and prove the bound for $f(\varepsilon) = O\left(\frac{1}{\varepsilon}\left(\log \frac{1}{\varepsilon}\right)^c\right)$. What is the smallest c you can get? ☐3

3. In this exercise we want to show that if X is the vertex set of a regular convex n-gon in the plane, then there exists a weak ε-net for X of size $O(\frac{1}{\varepsilon})$.

Suppose X lies on the unit circle u centered at 0. For an arc length $\alpha \le \pi$ radians, let $r(\alpha)$ be the radius of the circle centered at 0 and touching a chord of u connecting two points on u at arc distance α. For $i = 0, 1, 2, \ldots$, let N_i be a set of $\lfloor \frac{100}{\varepsilon(1.01)^i} \rfloor$ points placed at regular intervals on the circle of radius $r(\varepsilon(1.01)^i/10)$ centered at 0 (we take only those i for which this is well-defined). Show that $0 \cup \bigcup_i N_i$ is a weak ε-net of size $O(\frac{1}{\varepsilon})$ for X (the constants 1.01, etc., are rather arbitrary and can be greatly improved). ☐3

10.5 The Hadwiger–Debrunner (p, q)-Problem

Let $\mathcal{F}$ be a finite family of convex sets in the plane. By Helly's theorem, if every 3 sets from $\mathcal{F}$ intersect, then all sets of $\mathcal{F}$ intersect (unless $\mathcal{F}$ has 2 sets, that is). What if we know only that out of every 4 sets of $\mathcal{F}$, there are some 3 that intersect? Let us say that $\mathcal{F}$ satisfies the $(4,3)$-condition. In such a case, $\mathcal{F}$ may consist, for instance, of $n-1$ sets sharing a common point and one extra set lying somewhere far away from the others. So we cannot hope for a nonempty intersection of all sets. But can all the sets of $\mathcal{F}$ be pierced by a bounded number of points? That is, does there exist a constant C such that for any family $\mathcal{F}$ of convex sets in $\mathbf{R}^2$ satisfying the $(4,3)$-condition there are at most C points such that each set of $\mathcal{F}$ contains at least one of them?

This is the simplest nontrivial case of the so-called (p,q)-problem raised by Hadwiger and Debrunner and solved, many years later, by Alon and Kleitman.

10.5.1 Theorem (The (p, q)-theorem). *Let p, q, d be integers with $p \geq q \geq d+1$. Then there exists a number $\mathrm{HD}_d(p, q)$ such that the following is true: Let $\mathcal{F}$ be a finite family of convex sets in $\mathbf{R}^d$ satisfying the (p, q)-condition; that is, among any p sets of $\mathcal{F}$ there are q sets with a common point. Then $\mathcal{F}$ has a transversal consisting of at most $\mathrm{HD}_d(p, q)$ points.*

Clearly, the condition $q \geq d+1$ is necessary, since n hyperplanes in general position in $\mathbf{R}^d$ satisfy the (d, d)-condition but cannot be pierced by any bounded number of points independent of n.

It has been known for a long time that if $p(d-1) < (q-1)d$, then $\mathrm{HD}_d(p, q)$ exists and equals $p-q+1$ (Exercise 2). This is the only nontrivial case where exact values, or even good estimates, of $\mathrm{HD}_d(p, q)$ are known.

The reader might (rightly) wonder how one can get interesting examples of families satisfying the $(4, 3)$-condition, say. A large collection of examples can be obtained as follows: Choose a probability measure μ in the plane $(\mu(\mathbf{R}^2) = 1)$, and let $\mathcal{F}$ consist of all convex sets S with $\mu(S) > 0.5$. The $(4, 3)$-condition holds, because 4 sets together have measure larger than 2, and so some point has to be covered at least 3 times. The proof below shows that *every* family $\mathcal{F}$ of planar convex sets fulfilling the $(4, 3)$-condition somewhat resembles this example; namely, that there is a probability measure μ such that $\mu(S) > c$ for all $S \in \mathcal{F}$, with some small positive constant $c > 0$ (independent of $\mathcal{F}$). Note that the existence of such μ implies the $(p, 3)$ condition for a sufficiently large $p = p(c)$.

The Alon–Kleitman proof combines an amazing number of tools. The whole structure of the proof, starting from basic results like Helly's theorem, is outlined in Figure 10.1. The emphasis is on simplicity of the derivation rather than on the best quantitative bounds (so, for example, Tverberg's theorem is not required in full strength). The most prominent role is played by the fractional Helly theorem and by weak ε-nets for convex sets. An unsatisfactory feature of this method is that the resulting estimates for $\mathrm{HD}_d(p, q)$ are enormously large, while the truth is probably much smaller.

Since we have prepared all of the tools and notions in advance, the proof is now short. We do not attempt to optimize the constant resulting from the proof, and so we may as well assume that $q = d+1$.

By Corollary 10.4.4, we know that τ is bounded by a function of τ^* for any finite system of convex sets in $\mathbf{R}^d$. So it remains to show that if $\mathcal{F}$ satisfies the $(p, d+1)$-condition, then $\tau^*(\mathcal{F}) = \nu^*(\mathcal{F})$ is bounded.

10.5.2 Lemma (Bounded ν^*). *Let $\mathcal{F}$ be a finite family of convex sets in $\mathbf{R}^d$ satisfying the $(p, d+1)$-condition. Then $\nu^*(\mathcal{F}) \leq C$, where C depends on p and d but not on $\mathcal{F}$.*

Proof. The first observation is that if $\mathcal{F}$ satisfies the $(p, d+1)$-condition, then many $(d+1)$-tuples of sets of $\mathcal{F}$ intersect. This can be seen by double counting. Every p-tuple of sets of $\mathcal{F}$ contains (at least) one intersecting $(d+1)$-tuple,

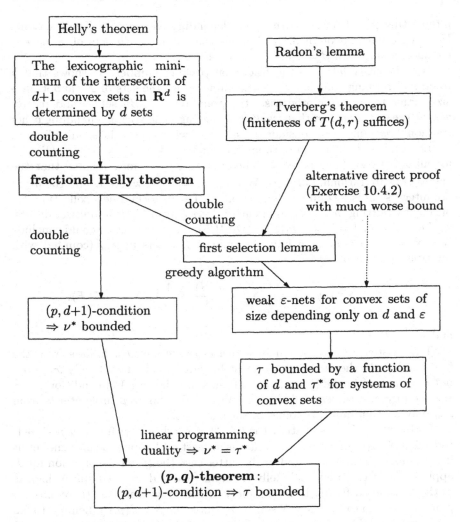

Figure 10.1. Main steps in the proof of the (p,q)-theorem.

and a single $(d+1)$-tuple is contained in $\binom{n-d+1}{p-d+1}$ p-tuples (where $n = |\mathcal{F}|$). Therefore, there are at least

$$\frac{\binom{n}{p}}{\binom{n-d+1}{p-d+1}} \geq \alpha \binom{n}{d+1}$$

intersecting $(d+1)$-tuples, with $\alpha > 0$ depending on p, d only. The fractional Helly theorem (Theorem 8.1.1) implies that at least βn sets of $\mathcal{F}$ have a common point, with $\beta = \beta(d, \alpha) > 0$ a constant.[3]

How is this related to the fractional packing number? It shows that a fractional packing that has the same value on all the sets of $\mathcal{F}$ cannot have size larger than $\frac{1}{\beta}$, for otherwise, the point lying in βn sets would receive weight greater than 1 in that fractional packing. The trick for handling other fractional packings is to consider the sets in $\mathcal{F}$ with appropriate multiplicities.

Let $\psi \colon \mathcal{F} \to [0, 1]$ be an optimal fractional packing ($\sum_{S \in \mathcal{F}: x \in S} \psi(S) \leq 1$ for all x). As we have noted in Theorem 10.1.1, we may assume that the values of ψ are rational numbers. Write $\psi(S) = \frac{m(S)}{D}$, where D and the $m(S)$ are integers (D is a common denominator). Let us form a new collection $\mathcal{F}_m$ of sets, by putting $m(S)$ copies of each S into $\mathcal{F}_m$; so $\mathcal{F}_m$ is a multiset of sets.

Let $N = |\mathcal{F}_m| = \sum_{S \in \mathcal{F}} m(S) = D \cdot \nu^*(\mathcal{F})$. Suppose that we could conclude the existence of a point a lying in at least βN sets of $\mathcal{F}_m$ (counted with multiplicity). Then

$$1 \geq \sum_{S \in \mathcal{F}: a \in S} \psi(S) = \sum_{S \in \mathcal{F}: a \in S} \frac{m(S)}{D} = \frac{1}{D} \cdot \beta N = \beta \nu^*(\mathcal{F}),$$

and so $\nu^*(\mathcal{F}) \leq \frac{1}{\beta}$.

The existence of a point a in at least βN sets of $\mathcal{F}_m$ follows from the fractional Helly theorem, but we must be careful: The new family $\mathcal{F}_m$ does not have to satisfy the $(p, d+1)$-condition, since the $(p, d+1)$-condition for $\mathcal{F}$ speaks only of p-tuples of *distinct* sets from $\mathcal{F}$, while a p-tuple of sets from $\mathcal{F}_m$ may contain multiple copies of the same set.

Fortunately, $\mathcal{F}_m$ does satisfy the $(p', d+1)$-condition with $p' = d(p-1)+1$. Indeed, a p'-tuple of sets of $\mathcal{F}_m$ contains at least $d+1$ copies of the same set or it contains p distinct sets, and in the latter case the $(p, d+1)$-condition for $\mathcal{F}$ applies. Using the fractional Helly theorem (which does not require the sets in the considered family to be distinct) as before, we see that there exists a point a common to at least βN sets of $\mathcal{F}_m$ for some $\beta = \beta(p, d)$. Lemma 10.5.2 is proved, and this also concludes the proof of the (p, q)-theorem.   □

Bibliography and remarks. The (p, q)-problem was posed by Hadwiger and Debrunner in 1957, who also solved the special case in Exercise 2 below. The solution described in this section follows Alon and Kleitman [AK92].

Much better quantitative bounds on $\mathrm{HD}_d(p, q)$ were obtained by Kleitman, Gyárfás, and Tóth [KGT01] for the smallest nontrivial values of p, q, d: $3 \leq \mathrm{HD}_2(4, 3) \leq 13$.

---

[3] By removing these βn sets and iterating, we would get that $\mathcal{F}$ can be pierced by $O(\log n)$ points. The main point of the (p, q)-theorem is to get rid of this $\log n$ factor.

Exercises

1. For which values of p and r does the following hold? Let $\mathcal{F}$ be a finite family of convex sets in $\mathbf{R}^d$, and suppose that any subfamily consisting of at most p sets can be pierced by at most r points. Then $\mathcal{F}$ can be pierced by at most C points, for some $C = C_d(p,r)$. ②

2. Let $p \geq q \geq d+1$ and $p(d-1) < (q-1)d$. Prove that $\mathrm{HD}_d(p,q) \leq p-q+1$. You may want to start with the case of $\mathrm{HD}_2(5,4)$. ④

3. Let $X \subset \mathbf{R}^2$ be a $(4k+1)$-point set, and let $\mathcal{F} = \{\mathrm{conv}(Y) : Y \subset X, |Y| = 2k+1\}$.
   (a) Verify that $\mathcal{F}$ has the $(4,3)$-property, and show that if X is in convex position, then $\tau(\mathcal{F}) \geq 3$. ③
   (b) Show that $\tau(\mathcal{F}) \leq 5$ (for any X). ⑤

These results are due to Alon and Rosenfeld (private communication).

10.6 A (p,q)-Theorem for Hyperplane Transversals

The technique of the proof of the (p,q)-theorem is quite general and allows one to prove (p,q)-theorems for various families. That is, if we have some basic family $\mathcal{B}$ of sets, such as the family $\mathcal{K}$ of all convex sets in Theorem 10.5.1, a (p,q)-theorem for $\mathcal{B}$ means that if $\mathcal{F} \subseteq \mathcal{B}$ satisfies the (p,q)-condition, then $\tau(\mathcal{F})$ is bounded by a function of p and q (depending on $\mathcal{B}$ but not on the choice of $\mathcal{F}$).

To apply the technique in such a situation, we first need to bound $\nu^*(\mathcal{F})$ using the (p,q)-condition. To this end, it suffices to derive a fractional Helly-type theorem for $\mathcal{B}$. Next, we need to bound $\tau(\mathcal{F})$ as a function of $\tau^*(\mathcal{F})$. If the VC-dimension of $\mathcal{F}$ is bounded, this is just Corollary 10.2.7, and otherwise, we need to prove a "weak ε-net theorem" for $\mathcal{F}$. Here we present one sophisticated illustration.

10.6.1 Theorem (A (p,q)-theorem for hyperplane transversals).
Let $p \geq d+1$ and let $\mathcal{F}$ be a finite family of convex sets in $\mathbf{R}^d$ such that among every p members of $\mathcal{F}$, there exist $d+1$ that have a common hyperplane transversal (i.e., there is a hyperplane intersecting all of them). Then there are at most $C = C(p,d)$ hyperplanes whose union intersects all members of $\mathcal{F}$.

Note that here the piercing is not by points but by hyperplanes. Let $\tau_{\mathrm{hyp}}(\mathcal{F})$, $\tau^*_{\mathrm{hyp}}(\mathcal{F})$, and $\nu^*_{\mathrm{hyp}}(\mathcal{F})$ be the notions corresponding to the transversal number, fractional transversal number, and fractional packing number in this setting.[4] We prove only the planar case, since some of the required auxiliary results become more complicated in higher dimensions.

---

[4] We could reformulate everything in terms of piercing by points if we wished to do so, by assigning to every $S \in \mathcal{F}$ the set T_S of all hyperplanes intersecting S. Then, e.g., $\tau_{\mathrm{hyp}}(\mathcal{F}) = \tau(\{T_S : S \in \mathcal{F}\})$.

To prove Theorem 10.6.1 for $d = 2$, we first want to derive a fractional Helly theorem.

10.6.2 Lemma (Fractional Helly for line transversals). *If $\mathcal{F}$ is a family of n convex sets in the plane such that at least $\alpha\binom{n}{3}$ triples have line transversals, then at least βn of the sets have a line transversal, $\beta = \beta(\alpha) > 0$.*

Proof. Let $\mathcal{F}$ be a family as in the lemma. We distinguish two cases depending on the number of pairs of sets in $\mathcal{F}$ that intersect.

First, suppose that at least $\frac{\alpha}{7}\binom{n}{2}$ pairs $\{S, S'\} \in \binom{\mathcal{F}}{2}$ satisfy $S \cap S' \neq \emptyset$. Project all sets of $\mathcal{F}$ vertically on the x-axis. The projections form a family of intervals with at least $\frac{\alpha}{7}\binom{n}{2}$ intersecting pairs, and so by the one-dimensional fractional Helly theorem, at least $\beta'n$ of these have a common point x. The vertical line through x intersects $\beta'n$ sets of $\mathcal{F}$.

Next, it remains to deal with the case of at most $\frac{\alpha}{7}\binom{n}{2}$ intersecting pairs in $\mathcal{F}$. Call a triple $\{S_1, S_2, S_3\}$ *good* if it has a line transversal and its three members are pairwise disjoint. Since each intersecting pair gives rise to at most n triples whose members are not pairwise disjoint, there are at most $n \cdot \frac{\alpha}{7}\binom{n}{2} \leq \frac{\alpha}{2}\binom{n}{3}$ nondisjoint triples, and so at least $\frac{\alpha}{2}\binom{n}{3}$ good triples remain.

Let $\{S_1, S_2, S_3\}$ be a good triple; we claim that its sets have a line transversal that is a common tangent to (at least) two of them. To see this, start with an arbitrary line transversal, translate it until it becomes tangent to one of the S_i, and then rotate it while keeping tangent to S_i until it becomes tangent to an S_j, $i \neq j$.

Let L denote the set of all lines that are common tangents to at least two *disjoint* members of $\mathcal{F}$. Since two disjoint convex sets in the plane have exactly 4 common tangents, $|L| \leq 4\binom{n}{2}$.

First, to see the idea, let us make the simplifying assumption that no 3 sets of $\mathcal{F}$ have a common tangent. Then each line $\ell \in L$ has a unique *defining pair* of disjoint sets for which it is a common tangent. As we have seen, for each good triple $\{S_1, S_2, S_3\}$ there is a line $\ell \in L$ such that two sets of the triple are the defining pair of ℓ and the third is intersected by ℓ. Now, since we have $\frac{\alpha}{2}\binom{n}{3}$ good triples and $|L| \leq 4\binom{n}{2}$, there is an $\ell_0 \in L$ playing this role for at least δn of the good triples, $\delta > 0$. Each of these δn triples contains the defining pair of ℓ_0 plus some other set, so altogether ℓ_0 intersects at least δn sets. (Note the similarity to the proof of the fractional Helly theorem.)

Now we need to relax the simplifying assumption. Instead of working with lines, we work with pairs $(\ell, \{S, S'\})$, where $S, S' \in \mathcal{F}$ are disjoint and ℓ is one of their common tangents, and we let L be the set of all such pairs. We still have $|L| \leq 4\binom{n}{2}$, and each good triple $\{S_1, S_2, S_3\}$ gives rise to at least one $(\ell, \{S, S'\}) \in L$, where $\{S, S'\} \subset \{S_1, S_2, S_3\}$. The rest of the argument is as before.   □

The interesting feature is that while this fractional Helly theorem is valid, there is no Helly theorem for line transversals! That is, for all n one can find families of n disjoint planar convex sets (even segments) such that any $n-1$ have a line transversal but there is no line transversal for all of them (Exercise 5.1.9).

Lemma 10.6.2 implies, exactly as in the proof of Lemma 10.5.2, that ν^*_{hyp} is bounded for any family satisfying the $(p, d+1)$-condition. It remains to prove a weak ε-net result.

10.6.3 Lemma. *Let L be a finite set (or multiset) of lines in the plane and let $r \geq 1$ be given. Then there exists a set N of $O(r^2)$ lines (a weak ε-net) such that whenever $S \subseteq \mathbf{R}^2$ is an (arcwise) connected set intersecting more than $\frac{|L|}{r}$ lines of L, then it intersects a line of N.*

Proof. Recall from Section 4.5 that a $\frac{1}{r}$-cutting for a set L of lines is a collection $\{\Delta_1, \ldots, \Delta_t\}$ of generalized triangles covering the plane such that the interior of each Δ_i is intersected by at most $\frac{|L|}{r}$ lines of L. The cutting lemma (Lemma 4.5.3) guarantees the existence of a $\frac{1}{r}$-cutting of size $O(r^2)$.

The cutting lemma does not directly cover multisets of lines. Nevertheless, with some care one can check that the perturbation argument works for multisets of lines as well.

Thus, let $\{\Delta_1, \ldots, \Delta_t\}$ be a $\frac{1}{r}$-cutting for the considered L, $t = O(r^2)$. The weak ε-net N is obtained by extending each side of each Δ_i into a line. Indeed, if an arcwise connected set S intersects more than $\frac{|L|}{r}$ lines of L, then it cannot be contained in the interior of a single Δ_i, and consequently, it intersects a line of N.   □

Conclusion of the proof of Theorem 10.6.1. Lemma 10.6.3 is now used exactly as the ε-nets results were used before, to show that $\tau_{\text{hyp}}(\mathcal{F}) = O(\tau^*_{\text{hyp}}(\mathcal{F})^2)$ in this case. This proves the planar version of Theorem 10.6.1.   □

Bibliography and remarks. Theorem 10.6.1 was proved by Alon and Kalai [AK95], as well as the results indicated in Exercises 3 and 4 below. It is related to the following conjecture of Grünbaum and Motzkin: *Let $\mathcal{F}$ be a family of sets in $\mathbf{R}^d$ such that the intersection of any at most k sets of $\mathcal{F}$ is a disjoint union of at most k closed convex sets. Then the Helly number of $\mathcal{F}$ is at most $k(d+1)$.* So here, in contrast to Exercise 4, the Helly number is determined exactly. I mention this mainly because of a neat proof by Amenta [Ame96] using a technique originally developed for algorithmic purposes.

It is not completely honest to say that there is no Helly theorem for line (and hyperplane) transversals, since there are very nice theorems of this sort, but the assumptions must be strengthened. For example, *Hadwiger's transversal theorem* asserts that if $\mathcal{F}$ is a finite family of disjoint convex sets in the plane with a linear ordering $\leq$ such that every 3 members of $\mathcal{F}$ can be intersected by a directed line in the order given by $\leq$, then $\mathcal{F}$ has a line transversal. This has been generalized to hyperplane transversals in $\mathbf{R}^d$, and many related results are known; see, e.g., the survey Goodman, Pollack, and Wenger [GPW93].

The application of the Alon–Kleitman technique for transversals of d-intervals in Exercise 2 below is due to Alon [Alo98]. Earlier, a similar result with the slightly stronger bound $\tau \leq (d^2 - d)\nu$ was proved by Kaiser [Kai97] by a topological method, following an initial breakthrough by Tardos [Tar95], who dealt with the case $d = 2$. By the Alon–Kleitman method, Alon [Alo] proved analogous bounds for families whose sets are subgraphs with at most d components of a given tree, or, more generally, subgraphs with at most d components of a graph G of bounded tree-width. In a sense, the latter is an "if and only if" result, since for every k there exists $w(k)$ such that every graph of tree-width $w(k)$ contains a collection of subtrees with $\nu = 1$ and $\tau \geq k$.

Alon, Kalai, Matoušek, and Meshulam [AKMM01] investigated generalizations of the Alon–Kleitman technique in the setting of abstract set systems. They showed that $(p, d+1)$-theorems for all p follow from a suitable fractional Helly property concerning $(d+1)$-tuples, and further that a set system whose nerve is d-Leray (see the notes to Section 8.1) has the appropriate fractional Helly property and consequently satisfies $(p, d+1)$-theorems.

Exercises

1. (a) Prove that if $\mathcal{F}$ is a finite family of circular disks in the plane such that every two members of $\mathcal{F}$ intersect, then $\tau(\mathcal{F})$ is bounded by a constant (this is a very weak version of Gallai's problem mentioned at the beginning of this chapter). ③

   (b) Show that for every $p \geq 2$ there is an n_0 such that if a family of n_0 disks in the plane satisfies the $(p, 2)$-condition, then there is a point common to at least 3 disks of the family. ④

   (c) Prove a $(p, 2)$-theorem for disks in the plane (or for balls in $\mathbf{R}^d$). ③

2. A *d-interval* is a set $J \subseteq \mathbf{R}$ of the form $J = I_1 \cup I_2 \cup \cdots \cup I_d$, where the $I_j \subset \mathbf{R}$ are closed intervals on the real line. (In the literature this is customarily called a *homogeneous d-interval*.)

   (a) Let $\mathcal{F}$ be a finite family of d-intervals with $\nu(\mathcal{F}) = k$. The family may contain multiple copies of the same d-interval. Show that there is a

$\beta = \beta(d,k) > 0$ such that for any such $\mathcal{F}$, there is a point contained in at least $\beta \cdot |\mathcal{F}|$ members of $\mathcal{F}$. ③ Can you prove this with $\beta = \frac{1}{2dk}$? ⑤

(b) Prove that $\tau(\mathcal{F}) \leq d\tau^*(\mathcal{F})$ for any finite family of d-intervals. ③

(c) Show that $\tau(\mathcal{F}) \leq 2d^2\nu(\mathcal{F})$ for any finite family of d-intervals, or at least that τ is bounded by a function of d and ν. ③

3. Let $\mathcal{K}_d^k$ denote the family of all unions of at most k convex sets in $\mathbf{R}^d$ (so the d-intervals from Exercise 2 are in $\mathcal{K}_1^d$). Prove a $(p,d+1)$-theorem for this family by the Alon–Kleitman technique: Whenever a finite family $\mathcal{F} \subset \mathcal{K}_d^k$ satisfies the $(p,d+1)$-condition, $\tau(\mathcal{F}) \leq f(p,d,k)$ for some function f. ④

4. (a) Show that the family $\mathcal{K}_2^2$ as in Exercise 3 has no finite Helly number. That is, for every h there exists a subfamily $\mathcal{F} \subset \mathcal{K}_2^2$ of $h+1$ sets in which every h members intersect but $\bigcap \mathcal{F} = \emptyset$. ④

(b) Use the result of Exercise 3 to derive that for every $k, d \geq 1$, there exists an h with the following property. Let $\mathcal{F} \subset \mathcal{K}_d^k$ be a finite family such that the intersection of any subfamily of $\mathcal{F}$ lies in $\mathcal{K}_d^k$ (i.e., is a union of at most k convex sets). Suppose that every at most h members of $\mathcal{F}$ have a common point. Then all the sets of $\mathcal{F}$ have a common point. (This is expressed by saying that the family $\mathcal{K}_d^k$ has *Helly order* at most h.) ⑤

11

Attempts to Count k-Sets

Consider an n-point set $X \subset \mathbf{R}^d$, and fix an integer k. Call a k-point subset $S \subseteq X$ a *k-set of* X if there exists an open half-space γ such that $S = X \cap \gamma$; that is, S can be "cut off" by a hyperplane. In this chapter we want to estimate the maximum possible number of k-sets of an n-point set in $\mathbf{R}^d$, as a function of n and k.

This question is known as the *k-set problem*, and it seems to be extremely challenging. Only partial results have been found so far, and there is a substantial gap between the upper and lower bounds even for the number of planar k-sets, in spite of considerable efforts by many researchers. So this chapter presents work in progress, much more so than the other parts of this book. I believe that the k-set problem deserves to be such an exception, since it has stimulated several interesting directions of research, and the partial results have elegant proofs.

11.1 Definitions and First Estimates

For technical reasons, we are going to investigate a quantity slightly different from the number of k-sets, which turns out to be asymptotically equivalent, however.

First we consider a planar set $X \subset \mathbf{R}^2$ in general position. A *k-facet* of X is a directed segment xy, $x, y \in X$, such that exactly k points of X lie (strictly) to the left of the directed line determined by x and y.

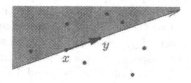

a 4-facet

Similarly, for $X \subset \mathbf{R}^d$, a *k-facet* is an oriented $(d-1)$-dimensional simplex with vertices $x_1, x_2, \ldots, x_d \in X$ such that the hyperplane h determined by $x_1, x_2, \ldots, x_d$ has exactly k points of X (strictly) on its positive side. (The orientation of the simplex means that one of the half-spaces determined by h is designated as positive and the other one as negative.)

Let us stress that we consider k-facets *only* for sets X in general position (no $d+1$ points on a common hyperplane). In such a case, the 0-facets are precisely the facets of the convex hull of X, and this motivates the name k-facet (so k-facets are *not* k-dimensional!).

A special case of k-facets are the *halving facets*. These exist only if $n - d$ is even, and they are the $\frac{n-d}{2}$-facets; i.e., they have exactly the same number of points on both sides of their hyperplane. Each halving facet appears as an $\frac{n-d}{2}$-facet with both orientations, and so halving facets can be considered unoriented. In the plane, instead of k-facets and halving facets, one often speaks of *k-edges* and *halving edges*. The drawing shows a planar point set with the halving edges:

We let $\mathrm{KFAC}(X, k)$ denote the number of k-facets of X, and $\mathrm{KFAC}_d(n, k)$ is the maximum of $\mathrm{KFAC}(X, k)$ over all n-point sets $X \subset \mathbf{R}^d$ in general position.

Levels, k-sets, and k-facets. The maximum possible number of k-sets is attained for point sets in general position: Each k-set is defined by an open half-space, and so a sufficiently small perturbation of X loses no k-sets (while it may create some new ones).

Next, we want to show that for sets in general position, the number of k-facets and the number of k-sets are closely related (although the exact relations are not simple). The best way seems to be to view both notions in the dual setting.

Let $X \subset \mathbf{R}^d$ be a finite set in general position. Let $H = \{\mathcal{D}(x): x \in X\}$ be the collection of hyperplanes dual to the points of X, where $\mathcal{D}$ is the duality "with the origin at $x_d = -\infty$" as defined in Section 5.1.

We may assume that each k-set S of X is cut off by a nonvertical hyperplane h_S that does not pass through any point of X. If S lies *below* h_S, then the dual point $y_S = \mathcal{D}(h_S)$ is a point lying on no hyperplane of H and having exactly k hyperplanes of H below it. So y_S lies in the interior of a cell at level k of the arrangement of H. Similarly, if S lies *above* h_S, then y_S is in a cell at level $n-k$. Moreover, if y_{S_1} and y_{S_2} lie in the same cell, then $S_1 = S_2$, and so k-sets exactly correspond to cells of level k and $n-k$.

Similarly, we find that the k-facets of X correspond to vertices of the arrangement of H of levels k or $n-k-d$ (we need to subtract d because of the d hyperplanes passing through the vertex that are not counted in its level).

The arrangement of H has at most $O(n^{d-1})$ unbounded cells (Exercise 6.1.2). Therefore, all but at most $O(n^{d-1})$ cells of level k have a topmost vertex, and the level of such a vertex is between $k-d+1$ and k. On the other hand, every vertex is the topmost vertex of at most one cell of level k. A similar relation exists between cells of level $n-k$ and vertices of level $n-k-d$. Therefore, the number of k-sets of X is at most $O(n^{d-1}) + \sum_{j=0}^{d-1} \text{KFAC}(X, k-j)$. Conversely, $\text{KFAC}(X, k)$ can be bounded in terms of the number of k-sets; this we leave to Exercise 2. From now on, we thus consider only estimating $\text{KFAC}_d(n, k)$.

Viewing $\text{KFAC}_d(n, k)$ in terms of the k-level in a hyperplane arrangement, we obtain some immediate bounds from the results of Section 6.3. The k-level has certainly no more vertices than all the levels 0 through k together, and hence

$$\text{KFAC}_d(n, k) = O\left(n^{\lfloor d/2 \rfloor}(k+1)^{\lceil d/2 \rceil}\right)$$

by Theorem 6.3.1. On the other hand, the arrangements showing that Theorem 6.3.1 is tight (constructed using cyclic polytopes) prove that for $k \leq n/2$, we have

$$\text{KFAC}_d(n, k) = \Omega\left(n^{\lfloor d/2 \rfloor}(k+1)^{\lceil d/2 \rceil - 1}\right);$$

this determines $\text{KFAC}_d(n, k)$ up to a factor of k.

The levels 0 through n together have $O(n^d)$ vertices, and so for any particular arrangement of n hyperplanes, if k is chosen at random, the expected k-level complexity is $O(n^{d-1})$. This means that a level with a substantially higher complexity has to be exceptional, much bigger than most other levels. It seems hard to imagine how this could happen. Indeed, it is widely believed that $\text{KFAC}_d(n, k)$ is never much larger than n^{d-1}. On the other hand, levels with somewhat larger complexity *can* appear, as we will see in Section 11.2.

Halving facets versus k-facets. In the rest of this chapter we will mainly consider bounds on the halving facets; that is, we will prove estimates for the function

$$\text{HFAC}_d(n) = \tfrac{1}{2} \text{KFAC}_d(n, \tfrac{n-d}{2}), \quad n-d \text{ even.}$$

It is easy to see that for all k, we have $\text{KFAC}_d(n, k) \leq 2 \cdot \text{HFAC}_d(2n+d)$ (Exercise 1). Thus, for proving asymptotic bounds on $\max_{0 \leq k \leq n-d} \text{KFAC}_d(n, k)$, it suffices to estimate the number of halving facets. It turns out that even a stronger result is true: The following theorem shows that upper bounds on $\text{HFAC}_d(n)$ automatically provide upper bounds on $\text{KFAC}_d(n, k)$ *sensitive to k*.

11.1.1 Theorem. *Suppose that for some d and for all n, $\text{HFAC}_d(n)$ can be bounded by $O(n^{d-c_d})$, for some constant $c_d > 0$. Then we have, for all $k \leq \frac{n-d}{2}$,*

$$\text{KFAC}_d(n, k) = O\left(n^{\lfloor d/2 \rfloor}(k+1)^{\lceil d/2 \rceil - c_d}\right).$$

Proof. We use the method of the probabilistic proof of the cutting lemma from Section 6.5 with only small modifications; we assume familiarity with that proof. We work in the dual setting, and so we need to bound the number of vertices of level k in the arrangement of a set H of n hyperplanes in general position. Since for k bounded by a constant, the complexity of the k-level is asymptotically determined by Clarkson's theorem on levels (Theorem 6.3.1), we can assume $2 \leq k \leq \frac{n}{2}$.

We set $r = \frac{n}{k}$ and $p = \frac{r}{n} = \frac{1}{k}$, and we let $S \subseteq H$ be a random sample obtained by independent Bernoulli trials with success probability p. This time we let $\mathcal{T}(S)$ denote the bottom-vertex triangulation *of the bottom unbounded cell* of the arrangement of S (actually, in this case it seems simpler to use the top-vertex triangulation instead of the bottom-vertex one); the rest of the arrangement is ignored. (For $d = 2$, we can take the vertical decomposition instead.) Here is a schematic illustration for the planar case:

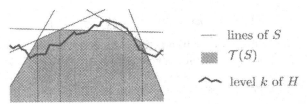

        ---- lines of S

        ▓▓▓ $\mathcal{T}(S)$

        ∿∿∿   level k of H

The conditions (C0)–(C2) as in Section 6.5 are satisfied for this $\mathcal{T}(S)$ (in (C0) we have constants depending on d, of course), and as for (C3), we have $|\mathcal{T}(S)| = O(|S|^{\lfloor d/2 \rfloor} + 1)$ for all $S \subseteq H$ by the asymptotic upper bound theorem (Theorem 5.5.2) and by the properties of the bottom-vertex triangulation. Thus, the analogy of Proposition 6.5.2 can be derived: For every $t \geq 0$, the expected number of simplices with excess at least t in $\mathcal{T}(S)$ is bounded as follows:

$$\mathbf{E}[|\mathcal{T}(S)_{\geq t}|] = O\left(2^{-t} r^{\lfloor d/2 \rfloor}\right) = O\left(2^{-t}(\tfrac{n}{k})^{\lfloor d/2 \rfloor}\right). \qquad (11.1)$$

Let V_k denote the set of the vertices of level k in the arrangement of H, whose size we want to estimate, and let $V_k(S)$ be the vertices in V_k that have level 0 with respect to the arrangement of S; i.e., they are covered by a simplex of $\mathcal{T}(S)$.

First we claim that, typically, a significant fraction of the vertices of V_k appears in $V_k(S)$, namely, $\mathbf{E}[|V_k(S)|] \geq \frac{1}{4}|V_k|$. For every $v \in V_k$, the probability that $v \in V_k(S)$, i.e., that none of the at most k hyperplanes below v goes into S, is at least $(1-p)^k = (1 - \frac{1}{k})^k \geq \frac{1}{4}$, and the claim follows.

It remains to bound $\mathbf{E}[|V_k(S)|]$ from above. Let $\Delta \in \mathcal{T}(S)$ be a simplex and let $\tilde{H}_\Delta$ be the set of all hyperplanes of H intersecting Δ. Not all of these hyperplanes have to intersect the interior of Δ (and thus be counted in the excess of Δ), but since H is in general position, there are at most a constant number of such exceptional hyperplanes. We note that all the vertices in

$V_k(S) \cap \Delta$ have the same level in the arrangement of $\tilde{H}_\Delta$ (it is k minus the number of hyperplanes below Δ). By the assumption in the theorem, we thus have $|V_k(S) \cap \Delta| = O(|\tilde{H}_\Delta|^{d-c_d}) = O((t_\Delta \frac{n}{r})^{d-c_d}) = O((t_\Delta k)^{d-c_d})$, where t_Δ is the excess of Δ. Therefore,

$$\mathbf{E}[|V_k(S)|] \leq O(k^{d-c_d}) \cdot \sum_{\Delta \in \mathcal{T}(S)} t_\Delta^{d-c_d}.$$

Using (11.1), the sum is bounded by $O((\frac{n}{k})^{\lfloor d/2 \rfloor})$; this is as in Section 6.5. We have shown that

$$|V_k| \leq 4\mathbf{E}[|V_k(S)|] = O\left(n^{\lfloor d/2 \rfloor} k^{\lceil d/2 \rceil - c_d}\right),$$

and Theorem 11.1.1 is proved. □

Bibliography and remarks. We summarize the bibliography of k-sets here, and in the subsequent sections we only mention the origins of the particular results described there. In the following we always assume $k \geq 1$, which allows us to write k instead of $k+1$ in the bounds.

The first paper concerning k-sets is by Lovász [Lov71], who proved an $O(n^{3/2})$ bound for the number of halving edges. Straus (unpublished) showed an $\Omega(n \log n)$ lower bound. This appeared, together with the bound $O(n\sqrt{k})$ for planar k-sets, in Erdős, Lovász, Simmons, and Straus [ELSS73]. The latter bound was independently found by Edelsbrunner and Welzl [EW85]. It seems to be the natural bound to come up with if one starts thinking about planar k-sets; there are numerous variations of the proof (see Agarwal, Aronov, Chan, and Sharir [AACS98]), and breaking this barrier took quite a long time. The first progress was made by Pach, Steiger, and Szemerédi [PSS92], who improved the upper bound by the tiny factor of $\log^* k$. A significant breakthrough, and the current best planar upper bound of $O(nk^{1/3})$, was achieved by Dey [Dey98]. A simpler version of his proof, involving new insights, was provided by Andrzejak, Aronov, Har-Peled, Seidel, and Welzl [AAHP+98].

An improvement over the $\Omega(n \log k)$ lower bound [ELSS73] was obtained by Tóth [Tót01b], namely, $\mathrm{KFAC}_2(n, k) \geq n \exp(c\sqrt{\log k})$ for a constant $c > 0$ (a similar bound was found by Klawe, Paterson, and Pippenger in the 1980s in an unpublished manuscript, but only for the number of vertices of level k in an arrangement of n *pseudolines* in the plane).

The first nontrivial bound on k-sets in higher dimension was proved by Bárány, Füredi, and Lovász [BFL90]. They showed that $\mathrm{HFAC}_3(n) = O(n^{2.998})$. Their method includes the main ingredients of most of the subsequent improvements; in particular, they proved a planar version of the second selection lemma (Theorem 9.2.1) and conjectured the colored Tverberg theorem (see the notes to Sections 8.3

and 9.2). Aronov, Chazelle, Edelsbrunner, Guibas, Sharir, and Wenger [ACE+91] improved the bound for the planar second selection lemma (with a new proof) and showed that $\mathrm{HFAC}_3(n) = O(n^{8/3} \log^{5/3} n)$. A nontrivial upper bound for every fixed dimension d, $\mathrm{HFAC}_d(n) = O(n^{d-c_d})$ for a suitable $c_d > 0$, was obtained by Alon, Bárány, Füredi, and Kleitman [ABFK92], following the method of [BFL90] and using the recently established colored Tverberg theorem. Dey and Edelsbrunner [DE94] proved a slightly better 3-dimensional bound $\mathrm{HFAC}_3(n) = O(n^{8/3})$ by a direct and simple 3-dimensional argument avoiding the use of a planar selection lemma (see Exercise 11.3.8). A new significant improvement to $\mathrm{HFAC}_3(n) = O(n^{2.5})$ was achieved by Sharir, Smorodinsky, and Tardos [SST01]; their argument is sketched in the notes to Section 11.4.

Theorem 11.1.1 is due to Agarwal et al. [AACS98]. Their proof uses a way of random sampling different from ours, but the idea is the same.

Another interesting result on planar k-sets, due to Welzl [Wel86], is $\sum_{k \in K} \mathrm{KFAC}(X, k) = O\left(n\sqrt{\sum_{k \in K} k}\right)$ for every n-point set $X \subset \mathbf{R}^2$ and every index set $K \subseteq \{1, 2, \dots, \lfloor n/2 \rfloor\}$ (see Exercise 11.3.2). Using identities derived by Andrzejak et al. [AAHP+98] (based on Dey's method), the bound can be improved to $O\left(n\left(|K| \cdot \sum_{k \in K} k\right)^{1/3}\right)$; this was communicated to me by Emo Welzl.

Edelsbrunner, Valtr, and Welzl [EVW97] showed that "dense" sets X, i.e., n-point $X \subset \mathbf{R}^d$ such that the ratio of the maximum to minimum interpoint distance is $O(n^{1/d})$, cannot asymptotically maximize the number of k-sets. For example, in the plane, they proved that a bound of $\mathrm{HFAC}_2(n) = O(n^{1+\alpha})$ for arbitrary sets implies that any n-point *dense* set has at most $O(n^{1+\alpha/2})$ halving edges. Alt, Felsner, Hurtado, and Noy [AFH+00] showed that if $X \subset \mathbf{R}^2$ is a set contained in a union of C convex curves, then $\mathrm{KFAC}(X, k) = O(n)$ for all k, with the constant of proportionality depending on C.

Several upper bounds concern the maximum combinatorial complexity of level k for objects other than hyperplanes. For segments in the plane, the estimate obtained by combining a result of Dey [Dey98] with the general tools in Agarwal et al. [AACS98] is $O(nk^{1/3}\alpha(\frac{n}{k}))$. Their method yields the same result for the level k in an arrangement of n *extendible pseudosegments* (defined in Exercise 6.2.5). For arbitrary pseudosegments, the result of Chan mentioned in that exercise (n pseudosegments can be cut into $O(n \log n)$ extendible pseudosegments) gives the slightly worse bound $O(nk^{1/3}\alpha(\frac{n}{k}) \log^{2/3}(k+1))$.

The study of levels in arrangements of curves with more than one pairwise intersection was initiated by Tamaki and Tokuyama [TT98], who considered a family of n parabolas in $\mathbf{R}^2$ (here is a neat motivation: Given n points in the plane, each of them moving along a straight line with constant velocity, how many times can the pair of points with median distance change?). They showed that n parabolas can be cut into $O(n^{5/3})$ pieces in total so that the resulting collection of curves is a family of pseudosegments (see Exercise 6). This idea of cutting curves into pseudosegments proved to be of great importance for other problems as well; see the notes to Section 4.5. Tamaki and Tokuyama obtained the bound of $O(n^{2-1/12})$ for the maximum complexity of the k-level for n parabolas. Using the tools from [AACS98] and a cutting into extendible pseudosegments, Chan [Cha00a] improved this bound to $O(nk^{1-2/9} \log^{2/3}(k+1))$.

All these results can be transferred without much difficulty from parabolas to *pseudocircles*, which are closed planar Jordan curves, every two intersecting at most twice. Aronov and Sharir [AS01a] proved that if the curves are *circles*, then even cutting into $O(n^{3/2+\varepsilon})$ pseudosegments is possible (the best known lower bound is $\Omega(n^{4/3})$; see Exercise 5). This upper bound was extended by Nevo, Pach, Pinchasi, and Sharir [NPPS01] to certain families of pseudocircles: The pseudocircles in the family should be selected from a 3-parametric family of real algebraic curves and satisfy an additional condition; for example, it suffices that their interiors can be pierced by $O(1)$ points (also see Alon, Last, Pinchasi, and Sharir [ALPS01] for related things).

Tamaki and Tokuyama constructed a family of n curves with at most 3 pairwise intersections that cannot be cut into fewer than $\Omega(n^2)$ pseudosegments, demonstrating that their approach cannot yield nontrivial bounds for the complexity of levels for such general curves (Exercise 5). However, for graphs of polynomials of degree at most s, Chan [Cha00a] obtained a cutting into roughly $O(n^{2-1/3^{s-1}})$ pseudosegments and consequently a nontrivial upper bound for levels. His bound was improved by Nevo et al. [NPPS01].

As for higher-dimensional results, Katoh and Tokuyama [KT99] proved the bound $O(n^2 k^{2/3})$ for the complexity of the k-level for n triangles in $\mathbf{R}^3$.

Bounds on k-sets have surprising applications. For example, Dey's results for planar k-sets mentioned above imply that if G is a graph with n vertices and m edges and each edge has weight that is a linear function of time, then the minimum spanning tree of G changes at most $O(mn^{1/3})$ times; see Eppstein [Epp98]. The number of k-sets of the infinite set $(\mathbf{Z}_0^+)^d$ (lattice points in the nonnegative orthant) appears in computational algebra in connection with Gröbner bases of certain ideals. The bounds of $O((k \log k)^{d-1})$ and $\Omega(k^{d-1} \log k)$ for every fixed d, as well as references, can be found in Wagner [Wag01].

Exercises

1. Verify that for all k and all dimensions d, $\mathrm{KFAC}_d(n,k) \leq 2 \cdot \mathrm{HFAC}_d(2n+d)$. ②

2. Show that every vertex in an arrangement of hyperplanes in general position is the topmost vertex of exactly one cell. For $X \subset \mathbf{R}^d$ finite and in general position, bound $\mathrm{KFAC}(X,k)$ using the numbers of j-sets of X, $k \leq j \leq k+d-1$. ③

3. Suppose that we have a construction that provides an n-point set in the plane with at least $f(n)$ halving edges for all even n. Show that this implies $\mathrm{KFAC}_2(n,k) = \Omega(\lfloor n/2k \rfloor f(2k))$ for all $k \leq \frac{n}{2}$. ③

4. Suppose that for all even n, we can construct a planar n-point set with at least $f(n)$ halving edges. Show that one can construct n-point sets with $\Omega(nf(n))$ halving facets in $\mathbf{R}^3$ (for infinitely many n, say). ④ Can you extend the construction to $\mathbf{R}^d$, obtaining $\Omega(n^{d-2}f(n))$ halving facets?

5. (Lower bounds for cutting curves into pseudosegments) In this exercise, Γ is a family of n curves in the plane, such as those considered in connection with Davenport–Schinzel sequences: Each curve intersects every vertical line exactly once, every two curves intersect at most s times, and no 3 have a common point.

   (a) Construct such a family Γ with $s = 2$ (a family of *pseudoparabolas*) whose arrangement has $\Omega(n^{4/3})$ *empty lenses*, where an empty lens is a bounded cell of the arrangement of Γ bounded by two of the curves. (The number of empty lenses is obviously a lower bound for the number of cuts required to turn Γ into a family of pseudosegments.) ③

   (b) Construct a family Γ with $s = 3$ and with $\Omega(n^2)$ empty lenses. ②

6. (Cutting pseudoparabolas into pseudosegments) Let Γ be a family of n pseudoparabolas in the plane as in Exercise 5(a). For every two curves $\gamma, \gamma' \in \Gamma$ with exactly two intersection points, the *lens* defined by γ and γ' consists of the portions of γ and γ' between their two intersection points, as indicated in the picture:

   (a) Let Λ be a family of *pairwise nonoverlapping* lenses in the arrangement of Γ, where two lenses are nonoverlapping if they do not share any edge of the arrangement (but they may intersect, or one may be enclosed in the other). The goal is to bound the maximum size of Λ. We define a bipartite graph G with $V(G) = \Gamma \times \{0,1\}$ and with $E(G)$ consisting of all edges $\{(\gamma,0),(\gamma',1)\}$ such that there is a lens in Λ whose lower portion comes from γ and upper portion from γ'. Prove that G contains no $K_{3,4}$ and hence $|\Lambda| = O(n^{5/3})$. Supposing that $K_{3,4}$ were present, corresponding to "lower" curves $\gamma_1, \gamma_2, \gamma_3$ and "upper" curves $\gamma'_1, \ldots, \gamma'_4$, consider the upper envelope U of $\gamma_1, \gamma_2, \gamma_3$ and the lower envelope L of $\gamma_1', \ldots, \gamma_4'$. (A more careful argument shows that even $K_{3,3}$ is excluded.) [4]

(b) Show that the graph G in (a) can contain a $K_{2,r}$ for arbitrarily large r. [1]

(c) Given Γ, define the *lens set system* $(X, \mathcal{L})$ with X consisting of all bounded edges of the arrangement of Γ and the sets of $\mathcal{L}$ corresponding to lenses (each lens contributes the set of arrangement edges contained in its two arcs). Check that $\tau(\mathcal{L})$ is the smallest number of cuts needed to convert Γ into a collection of pseudosegments, and that the result of (a) implies $\nu(\mathcal{L}) = O(n^{5/3})$. [1]

(d) Using the method of the proof of Clarkson's theorem on levels and the inequality in Exercise 10.1.4(a), prove that $\tau(\mathcal{L}) = O(n^{5/3})$. [5]

7. (The k-set polytope) Let $X \subset \mathbf{R}^d$ be an n-point set in general position and let $k \in \{1, 2, \ldots, n-1\}$. The *k-set polytope* $Q_k(X)$ is the convex hull of the set

$$\left\{ \sum_{x \in S} x : S \subset X, |S| = k \right\}$$

in $\mathbf{R}^d$. Prove that the vertices of $Q_k(X)$ correspond bijectively to the k-sets of X. [4]

The k-set polytope was introduced by Edelsbrunner, Valtr, and Welzl [EVW97]. It can be used for algorithmic enumeration of k-sets, for example by the reverse search method mentioned in the notes to Section 5.5.

11.2 Sets with Many Halving Edges

Here we are going to construct n-point planar sets with a superlinear number of halving edges. It seems more intuitive to present the constructions in the dual setting, that is, to construct arrangements of n lines with many vertices of level $\frac{n-2}{2}$.

A simpler construction. We begin with a construction providing $\Omega(n \log n)$ vertices of the middle level.

By induction on m, we construct a set L_m of 2^m lines in general position with at least $f_m = (m+1)2^{m-2}$ vertices of the middle level (i.e., level $2^{m-1}-1$). We note that each line of L_m contains at least one of the middle-level vertices.

For $m = 1$ we take two nonvertical intersecting lines.

Let $m \geq 1$ and suppose that an L_m satisfying the above conditions has already been constructed. First, we select a subset $M \subset L_m$ of 2^{m-1} lines, and to each line of $\ell \in M$ we assign a vertex $v(\ell)$ of the middle level lying on ℓ, in such a way that $v(\ell) \neq v(\ell')$ for $\ell \neq \ell'$. The selection can be done greedily: We choose a line into M, take a vertex of the middle level on it, and exclude the other line passing through that vertex from further consideration.

Next, we replace each line of L_m by a pair of lines, both almost parallel to the original line. For a line $\ell \in M$, we let the two lines replacing ℓ intersect at $v(\ell)$. Each of the remaining lines is replaced by two almost parallel lines whose intersection is not near to any vertex of the arrangement of L_m. This yields the set L_{m+1}.

As the following picture shows, a middle-level vertex of the form $v(\ell)$ yields 3 vertices of the new middle level (level $2^m - 1$ in the arrangement of L_{m+1}):

Each of the other middle-level vertices yields 2 vertices of the new middle level:

Hence the number of middle-level vertices for L_{m+1} is at least $2f_m + 2^{m-1} = 2\left[(m+1)2^{m-2}\right] + 2^{m-1} = f_{m+1}$. $\square$

A better construction. This construction is more complicated, but it shows the lower bound

$$n \cdot e^{\Omega\left(\sqrt{\log n}\right)}$$

for the number of vertices of the middle level (and thus for the number of halving edges). This bound is smaller than $n^{1+\delta}$ for every $\delta > 0$ but much larger than $n(\log n)^c$ for any constant c.

For simplicity, we will deal only with values of n of a special form, thus providing a lower bound for infinitely many n. Simple considerations show that $\mathrm{HFAC}_2(n)$ is nondecreasing, and this gives a bound for all n.

The construction is again inductive. We first explain the idea, and then we describe it more formally.

In the first step, we let L_0 consist of two intersecting nonvertical lines. Suppose that after m steps, a set of lines L_m in general position has already been constructed, with many vertices of the middle level. First we replace every line $\ell \in L_m$ by a_m parallel lines; let us call these lines the *bundle of ℓ*. So if v is a vertex of the middle level of L_m, we get a_m vertices of the middle level near v after the replacement.

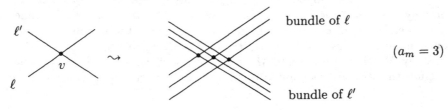

$(a_m = 3)$

Then we add two new lines λ_v and μ_v as indicated in the next picture, and we obtain $2a_m$ vertices of the middle level:

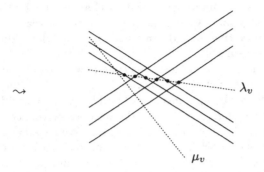

If $n_m = |L_m|$ and f_m is the number of vertices of the middle level in L_m, the construction gives roughly $n_{m+1} \approx a_m n_m + 2f_m$ and $f_{m+1} \approx 2a_m f_m$. This recurrence is good: With a suitable choice of the multiplicities a_m, it leads to the claimed bound. But the construction as presented so far is not at all guaranteed to work, because the new lines λ_v and μ_v might mess up the levels of the other vertices. We must make some extra provisions to get this under control.

First of all, we want the auxiliary lines λ_v and μ_v to be nearly parallel to the old line ℓ' in the picture. This is achieved by letting the vertical spacing of the a_m lines in the bundle of ℓ' be much smaller than the spacing in the bundle of ℓ:

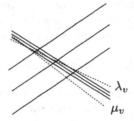

Namely, if the lines of L_m are $\ell_1, \ell_2, \ldots, \ell_{n_m}$, then the vertical spacing in the bundle of ℓ_i is set to ε^i, where $\varepsilon > 0$ is a suitable very small number.

Let ℓ_i be a line of L_m, and let d_i denote the number of indices $j < i$ such that ℓ_j intersects ℓ_i in a vertex of the middle level. In the new arrangement of L_{m+1} we obtain a_m lines of the bundle of ℓ_i and $2d_i$ lines of the form λ_v and μ_v, which are almost parallel to ℓ_i, and d_i of them go above the bundle and d_i below. Thus, for points not very close to ℓ_i, the effect is as if ℓ_i were replicated $(a_m + 2d_i)$ times. This is still not good; we would need that all lines have the same multiplicities. So we let D be the maximum of the d_i, and for each i, we add $D - d_i$ more lines parallel to ℓ_i below ℓ_i and $D - d_i$ parallel lines above it.

How do we control D? We do not know how many middle-level vertices can appear on the lines of L_{m+1}; some vertices are necessarily there by the construction, but some might arise "just by chance," say by the interaction of the various auxiliary lines λ_v and μ_v, which we do not really want to analyze. So we take a conservative attitude and deal only with the middle-level vertices we know about for sure.

Here is the whole construction, this time how it really goes. Suppose that we have already constructed a set $L_m = \{\ell_1, \ldots, \ell_{n_m}\}$ of lines in general position (which includes being nonvertical) and a set V_m of middle-level vertices in the arrangement of L_m, such that the number of vertices of V_m lying on ℓ_i is no more than D_m, for all $i = 1, 2, \ldots, n_m$. We let $\varepsilon = \varepsilon_m$ be sufficiently small, and we replace each ℓ_i by a_m parallel lines with vertical spacing ε^i. Then for each $v \in V_m$, we add the two lines λ_v and μ_v as explained above, and finally we add, for each i, the $2(D_m - d_i)$ lines parallel to ℓ_i, half above and half below the bundle, where d_i is the number of vertices of V_m lying on ℓ_i.

Since L_{m+1} is supposed to be in general position, we should not forget to apply a very small perturbation to L_{m+1} after completing the step just described.

For each old vertex $v \in V_m$, we now really get the $2a_m$ new middle-level vertices near v as was indicated in the drawing above, and we put these into V_{m+1}. So we have

$$n_{m+1} = (a_m + 2D_m)n_m, \qquad f_{m+1} = |V_{m+1}| = 2a_m f_m.$$

What about D_{m+1}, the maximum number of points of V_{m+1} lying on a single line? Each line in the bundle of ℓ_i has exactly d_i vertices of V_{m+1}. The lines λ_v get $2a_m$ vertices of V_{m+1}, and the remaining auxiliary lines get none. So

$$D_{m+1} = \max(D_m, 2a_m).$$

It remains to define the a_m, which are free parameters of the construction. A good choice is to let $a_m = 4D_m$. Then we have $D_0 = 1$, $D_m = 8^m$, and $a_m = 4 \cdot 8^m$. From the recurrences above, we further calculate

$$n_m = 2 \cdot 6^m \cdot 8^{1+2+\cdots+(m-1)}, \qquad f_m = 8^m \cdot 8^{1+2+\cdots+(m-1)}.$$

So $\log n_m$ is $O(m^2)$, while $\log(f_m/n_m) = \log\left(\frac{1}{2}(\frac{8}{6})^m\right) = \Omega(m)$. We indeed have $f_m \geq n_m \cdot e^{\Omega\left(\sqrt{\log n_m}\right)}$ as promised. □

Bibliography and remarks. The first construction is from Erdős et al. [ELSS73] and the second one from Tóth [Tót01b]. In the original papers, they are phrased in the primal setting.

11.3 The Lovász Lemma and Upper Bounds in All Dimensions

In this section we prove a basic property of the halving facets, usually called the Lovász lemma. It implies nontrivial upper bounds on the number of halving facets, by a simple self-contained argument in the planar case and by the second selection lemma (Theorem 9.2.1) in an arbitrary dimension. We prove a slightly more precise version of the Lovász lemma than is needed here, since we will use it in a subsequent section. On the other hand, we consider only halving facets, although similar results can be obtained for k-facets as well. Sticking to halving facets simplifies matters a little, since for other k-facets one has to be careful about the orientations.

Let $X \subset \mathbf{R}^d$ be an n-point set in general position with $n - d$ even. Let T be a $(d-1)$-point subset of X and let

$$V_T = \{x \in X \setminus T \colon T \cup \{x\} \text{ is a halving facet of } X\}.$$

In the plane, T has a single point and V_T are the other endpoints of the halving edges emanating from it. In 3 dimensions, $\mathrm{conv}(T)$ is a segment, and a typical picture might look as follows:

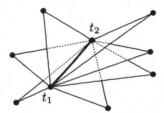

where $T = \{t_1, t_2\}$ and the triangles are halving facets.

Let h be a hyperplane containing T and no point of $X \setminus T$. Since $|X \setminus T|$ is odd, one of the open half-spaces determined by h, the *larger half-space*, contains more points of X than the other, the *smaller half-space*.

11.3.1 Lemma (Halving-facet interleaving lemma). *Every hyperplane h as above "almost halves" the halving facets containing T. More precisely, if r is the number of points of V_T in the smaller half-space of h, then the larger half-space contains exactly $r+1$ points of V_T.*

Proof. To get a better picture, we project T and V_T to a 2-dimensional plane ρ orthogonal to T. (For dimension 2, no projection is necessary, of course.) Let the projection of T, which is a single point, be denoted by t and the projection of V_T by V_T'. Note that the points of V_T project to distinct points. The halving facets containing T project to segments emanating from t. The hyperplane h is projected to a line h', which we draw vertically in the following indication of the situation in the plane ρ:

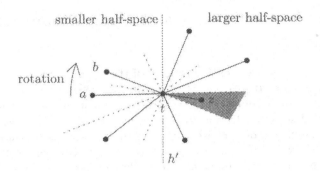

We claim that for any two angularly consecutive segments, such as at and bt, the angle opposite the angle atb contains a point of V_T' (such as z). Indeed, the hyperplane passing through t and a has exactly $\frac{n-d}{2}$ points of X in both of its open half-spaces. If we start rotating it around T towards b, the point a enters one of the open half-spaces (in the picture, the one below the rotating hyperplane). But just before we reach b, that half-space again has $\frac{n-d}{2}$ points. Hence there was a moment when the number of points in this half-space went from $\frac{n-d}{2}+1$ to $\frac{n-d}{2}$, and this must have been a moment of reaching a suitable z.

This means that for every two consecutive points of V_T', there is at least one point of V_T' in the corresponding opposite wedge. There is actually exactly one, for if there were two, their opposite wedge would have to contain another point. Therefore, the numbers of points of V_T in the half-spaces determined by h differ exactly by 1.

To finish the proof of the lemma, it remains to observe that if we start rotating the hyperplane h around T in either direction, the first point of V_T encountered must be in the larger half-space. So the larger half-space has one more point of V_T than the smaller half-space. (Recall that the larger half-space is defined with respect to X, and so we did not just parrot the definition here.)   □

11.3.2 Corollary (Lovász lemma). *Let $X \subset \mathbf{R}^d$ be an n-point set in general position, and let ℓ be a line that is not parallel to any of the halving facets of X. Then ℓ intersects the relative interior of at most $O(n^{d-1})$ halving facets of X.*

Proof. We can move ℓ a little so that it intersects the relative interiors of the same halving facets as before but intersects no boundary of a halving facet. Next, we start translating ℓ in a suitably chosen direction. (In the plane there are just two directions, and both of them will do.) The direction is selected so that we never cross any $(d-3)$-dimensional flat determined by the points of X. To this end, we need to find a two-dimensional plane passing through ℓ and avoiding finitely many $(d-3)$-dimensional flats in $\mathbf{R}^d$, none of them intersecting ℓ; this is always possible.

As we translate the line ℓ, the number of halving facets currently inter-sected by ℓ may change only as ℓ crosses the boundary of a halving facet F, i.e., a $(d-2)$-dimensional face of F. By the halving-facet interleaving lemma, by crossing one such face T, the number of intersected halving facets changes by 1. After moving far enough, the translated line ℓ intersects no halving facet at all. On its way, it crossed no more than $O(n^{d-1})$ boundaries, since there are only $O(n^{d-1})$ simplices of dimension $d-2$ with vertices at X. This proves the corollary. $\qquad\square$

11.3.3 Theorem. *For each $d \geq 2$, the maximum number of halving facets satisfies*
$$\mathrm{HFAC}_d(n) = O\big(n^{d-1/s_{d-1}}\big),$$
where s_{d-1} is an exponent for which the statement of the second selection lemma (Theorem 9.2.1) holds in dimension $d-1$. In particular, in the plane we obtain $\mathrm{HFAC}_2(n) = O(n^{3/2})$.

For higher dimensions, this result shows that $\mathrm{HFAC}_d(n)$ is asymptotically somewhat smaller than n^d, but the proof method is inadequate for proving bounds close to n^{d-1}.

Theorem 11.3.3 is proved from Corollary 11.3.2 using the second selection lemma. Let us first give a streamlined proof for the planar case, although later on we will prove a considerably better planar bound.

Proof of Theorem 11.3.3 for $d = 2$. Let us project the points of X ver-tically on the x-axis, obtaining a set Y. The projections of the halving edges of X define a system of intervals with endpoints in Y. By Corollary 11.3.2, any point is contained in the interior of at most $O(n)$ of these intervals, for otherwise, a vertical line through that point would intersect too many halving edges.

Mark every qth point of Y (with q a parameter to be set suitably later). Divide the intervals into two classes: those containing some marked point in their interior and those lying in a gap between two marked points. The number of intervals of the first class is at most $O(n)$ per marked point, i.e., at most $O(n^2/q)$ in total. The number of intervals of the second class is no more than $\binom{q+1}{2}$ per gap, i.e., at most $(\frac{n}{q}+1)\binom{q+1}{2}$ in total. Balancing both bounds by setting $q = \lceil\sqrt{n}\,\rceil$, we get that the total number of halving edges is $O(n^{3/2})$ as claimed. $\qquad\square$

Note that we have implicitly applied and proved a one-dimensional second selection lemma (Exercise 9.2.1).

Proof of Theorem 11.3.3. We consider an n-point $X \subset \mathbf{R}^d$. We project X vertically into the coordinate hyperplane $x_d = 0$, obtaining a point set Y, which we regard as lying in $\mathbf{R}^{d-1}$. If the coordinate system is chosen suitably, Y is in general position.

Each halving facet of X projects to a $(d-1)$-dimensional Y-simplex in $\mathbf{R}^{d-1}$; let $\mathcal{F}$ be the family of these Y-simplices. If we write $|\mathcal{F}| = \alpha\binom{n}{d}$, then by the second selection lemma, there exists a point a contained in at least $c\alpha^{s_{d-1}}\binom{n}{d}$ simplices of $\mathcal{F}$. Only at most $O(n^{d-1})$ of these contain a in their boundary, by Lemma 9.1.2, and the remaining ones have a in the interior. By the Lovász lemma (Corollary 11.3.2) applied on the vertical line in $\mathbf{R}^d$ passing through the point a, we thus get $c\alpha^{s_{d-1}}\binom{n}{d} = O(n^{d-1})$. We calculate that $|\mathcal{F}| = \alpha\binom{n}{d} = O(n^{d-1/s_{d-1}})$ as claimed.   □

Bibliography and remarks.   The planar version of the Lovász lemma (Corollary 11.3.2) originated in Lovász [Lov71]; the proof implicitly contains the halving-facet interleaving lemma. A higher-dimensional version of the Lovász lemma appeared in Bárány, Füredi, and Lovász [BFL90].

Welzl [Wel01] proved an exact version of the Lovász lemma, as is outlined in Exercises 5 and 6 below. This question is equivalent to the upper bound theorem for convex polytopes, via the Gale transform. The connection of k-facets and h-vectors of convex polytopes was noted earlier by several authors (Lee [Lee91], Clarkson [Cla93], and Mulmuley [Mul93b]), sometimes in a slightly different but essentially equivalent form. Using this correspondence and the generalized lower bound theorem mentioned in Section 5.5, Welzl also proved that the maximum total number of j-facets with $j \leq k$ for an n-point set in $\mathbf{R}^3$ (or, equivalently, the maximum possible number of vertices of level at most k in an arrangement of n planes in general position in $\mathbf{R}^3$) is attained for a set in convex position, from which the exact maximum can be calculated. It also implies that in $\mathbf{R}^3$, a set in convex position minimizes the number of halving facets (triangles).

An interesting connection of this result to another problem was discovered by Sharir and Welzl [SW01]. They quickly derived the following theorem, which was previously established by Pach and Pinchasi [PP01] by a difficult elementary proof: If $R, B \subset \mathbf{R}^2$ are n-point sets ("red" and "blue") with $R \dot\cup B$ in general position, then there are at least n *balanced lines*, where a line ℓ is balanced if $|R \cap \ell| = |B \cap \ell| = 1$ and on both sides of ℓ the number of red points equals the number of blue points (for odd n, the existence of at least one balanced line follows from the ham-sandwich theorem). A proof based on Welzl's result in $\mathbf{R}^3$ mentioned above is outlined in Exercise 4. Let us remark that conversely, the Pach–Pinchasi theorem implies the generalized lower bound theorem for $(d+4)$-vertex polytopes in $\mathbf{R}^d$.

Exercises

1. (a) Prove the following version the Lovász lemma in the planar case: *For a set $X \subset \mathbf{R}^2$ in general position, every vertical line ℓ intersects the interiors of at most $k+1$ of the k-edges.* ☐4

(b) Using (a), prove the bound $\text{KFAC}_2(n,k) = O(n\sqrt{k+1})$ (without appealing to Theorem 11.1.1). ③

2. Let $K \subseteq \{1, 2, \ldots, \lfloor n/2 \rfloor\}$. Using Exercise 1, prove that for any n-point set $X \subset \mathbf{R}^2$ in general position, the total number of k-edges with $k \in K$ (or equivalently, the total number of vertices of levels $k \in K$ in an arrangement of n lines) is at most $O\left(n\sqrt{\sum_{k \in K} k}\right)$. (Note that this is better than applying the bound $\text{KFAC}_2(n,k) = O(n\sqrt{k})$ for each $k \in K$ separately.) ③

3. (Exact planar Lovász lemma) Let $X \subset \mathbf{R}^2$ be a $2n$-point set in general position, and let ℓ be a vertical line having k points of X on the left and $2n-k$ points on the right. Prove that ℓ crosses exactly $\min(k, 2n-k)$ halving edges of X. ②

4. Let X be a set of $2n+1$ points in $\mathbf{R}^3$ in general position, and let $p_1, p_2, \ldots, p_{2n+1}$ be the points of X listed by increasing height (z-coordinate).
   (a) Using Exercise 3, check that if p_{k+1} is a vertex of $\text{conv}(X)$, then there are exactly $\min(k, 2n-k)$ halving triangles having p_{k+1} as the middle-height vertex (that is, the triangle is $p_i p_{k+1} p_j$ with $i < k+1 < j$). ③
   (b) Prove that every $(2n+1)$-point convex independent set $X \subset \mathbf{R}^3$ in general position has at least n^2 halving triangles. ②
   (c) Assuming that each $(2n+1)$-point set in $\mathbf{R}^3$ in general position has at least n^2 halving triangles (which follows from (b) and the result mentioned in the notes above about the number of halving triangles being minimized by a set in convex position), infer that if $X = \{p_1, \ldots, p_{2n+1}\} \subset \mathbf{R}^3$ is in general position, then for every k, there are always *at least* $\min(k, 2n-k)$ halving triangles having p_{k+1} as the middle-height vertex (even if p_{k+1} is not extremal in X). ③
   (d) Derive from (c) the result about balanced lines mentioned in the notes to this section: If $R, B \subset \mathbf{R}^2$ are n-point sets (red and blue points), with $R \cup B$ in general position, then there are at least n balanced lines ℓ (with $|R \cap \ell| = |B \cap \ell| = 1$ and such that on both sides of ℓ the number of red points equals the number of blue points). Embed $\mathbf{R}^2$ as the $z = 1$ plane in $\mathbf{R}^3$ and use a central projection on the unit sphere in $\mathbf{R}^3$ centered at 0. ③
   See [SW01] for solutions and related results.

5. (Exact Lovász lemma) Let $X \subset \mathbf{R}^d$ be an n-point set in general position and let ℓ be a directed line disjoint from the convex hulls of all $(d-1)$-point subsets of X. We think of ℓ as being vertical and directed upwards. We say that ℓ *enters* a j-facet F if it passes through F from the positive side (the one with j points) to the negative side. Let $\bar{h}_j = \bar{h}_j(\ell, X)$ denote the number of j-facets entered by ℓ, $j = 0, 1, \ldots, n-d$. Further, let $s_k(\ell, X)$ be the number of $(d+k)$-element subsets $S \subseteq X$ such that $\ell \cap \text{conv}(S) \neq \emptyset$.
   (a) Prove that for every X and ℓ as above, $s_k = \sum_{j=k}^{n-d} \binom{j}{k} \bar{h}_j$. ④

(b) Use (a) to show that $\bar{h}_0, \ldots, \bar{h}_{n-d}$ are uniquely determined by $s_0, s_1, \ldots, s_{n-d}$. [2]

(c) Infer from (b) that if X' is a set in general position obtained from X by translating each point in a direction parallel to ℓ, then $\bar{h}_j(\ell, X) = \bar{h}_j(\ell, X')$ for all j. Derive $\bar{h}_j = \bar{h}_{n-d-j}$. [2]

(d) Prove that for every $x \in X$ and all j, we have $\bar{h}_j(\ell, X \setminus \{x\}) \leq \bar{h}_j(\ell, X)$. [3]

(e) Choose $x \in X$ uniformly at random. Check that $\mathbf{E}\big[\bar{h}_j(\ell, X \setminus \{x\})\big] = \frac{n-d-j}{n}\bar{h}_j + \frac{j+1}{n}\bar{h}_{j+1}$. [2]

(f) From (d) and (e), derive $\bar{h}_{j+1} \leq \frac{j+d}{j+1}\bar{h}_j$, and conclude the exact Lovász lemma:

$$\bar{h}_j \leq \min\left\{\binom{j+d-1}{d-1}, \binom{n-j-1}{d-1}\right\}.$$

[2]

6. (The upper bound theorem and k-facets) Let $\boldsymbol{a} = (a_1, a_2, \ldots, a_n)$ be a sequence of $n \geq d+1$ convex independent points in $\mathbf{R}^d$ in general position, and let P be the d-dimensional simplicial convex polytope with vertex set $\{a_1, \ldots, a_n\}$. Let $\bar{\boldsymbol{g}} = (\bar{g}_1, \ldots, \bar{g}_n)$ be the Gale transform of $\boldsymbol{a}$, $\bar{g}_1, \ldots, \bar{g}_n \in \mathbf{R}^{n-d-1}$, and let b_i be a point in $\mathbf{R}^{n-d}$ obtained from g_i by appending a number t_i as the last coordinate, where the t_i are chosen so that $X = \{b_1, \ldots, b_n\}$ is in general position.

(a) Let ℓ be the x_{n-d}-axis in $\mathbf{R}^{n-d}$ oriented upwards, and let $s_k = s_k(\ell, X)$ and $\bar{h}_j = \bar{h}_j(\ell, X)$ be as in Exercise 5. Show that $f_k(P) = s_{d-k-1}(\ell, X)$, $k = 0, 1, \ldots, d-1$. [3]

(b) Derive that $h_j(P) = \bar{h}_j(\ell, X)$, $j = 0, 1, \ldots, d$, where h_j is as at the end of Section 5.5, and thus (f) of the preceding exercise implies the upper bound theorem in the formulation with the h-vector (5.3). [2]

If (a) and (b) are applied to the cyclic polytopes, we get equality in the bound for $\bar{h}_j$ in Exercise 5(f). In fact, the reverse passage (from an $X \subset \mathbf{R}^{n-d}$ in general position to a simplicial polytope in $\mathbf{R}^d$) is possible as well (see [Wel01]), and so the exact Lovász lemma can also be derived from the upper bound theorem.

7. This exercise shows limits for what can be proved about k-sets using Corollary 11.3.2 alone.

(a) Construct an n-point set $X \subset \mathbf{R}^2$ and a collection of $\Omega(n^{3/2})$ segments with endpoints in X such that no line intersects more than $O(n)$ of these segments. [2]

(b) Construct an n-point set in $\mathbf{R}^3$ and a collection of $\Omega(n^{5/2})$ triangles with vertices at these points such that no line intersects more than $O(n^2)$ triangles. [4]

8. (The Dey–Edelsbrunner proof of $\mathrm{HFAC}_3(n) = O(n^{8/3})$) Let X be an n-point set in $\mathbf{R}^3$ in general position (make a suitable general position assumption), and let $\mathcal{T}$ be a collection of t triangles with vertices at points of X. By a *crossing* we mean a pair (T, e), where $T \in \mathcal{T}$ is a triangle

and e is an edge of another triangle from $\mathcal{T}$, such that e intersects the interior of T in a single point (in particular, e is vertex-disjoint from T).
(a) Show that if $t \geq Cn^2$ for a suitable constant C, then two triangles sharing exactly one vertex intersect in a segment, and conclude that at least one crossing exists. [4]
(b) Show that at least $t - Cn^2$ crossings exist. [2]
(c) Show that for $t \geq C'n^2$, with $C' > C$ being a sufficiently large constant, at least $\Omega(t^3/n^4)$ crossings exist. Infer that there is an edge crossing $\Omega(t^3/n^6)$ triangles. (Proceed as in the proof of the crossing number theorem.) [3]
(d) Use Corollary 11.3.2 to conclude that $\mathrm{HFAC}_3(n) = O(n^{8/3})$. [1]

11.4 A Better Upper Bound in the Plane

Here we prove an improved bound on the number of halving edges in the plane.

11.4.1 Theorem. *The maximum possible number of halving edges of an n-point set in the plane is at most $O(n^{4/3})$.*

Let X be an n-point set in the plane in general position, and let us draw all the halving edges as segments. In this way we get a drawing of a graph (the *graph of halving edges*) in the plane. Let $\deg(x)$ denote the degree of x in this graph, i.e., the number of halving edges incident to x, and let $\mathrm{cr}(X)$ denote the number of pairs of the halving edges that cross. In the following example we have $\mathrm{cr}(X) = 2$, and the degrees are $(1, 1, 1, 1, 1, 3)$.

Theorem 11.4.1 follows from the crossing number theorem (Theorem 4.3.1) and the following remarkable identity.

11.4.2 Theorem. *For each n-point set X in the plane in general position, where n is even, we have*

$$\mathrm{cr}(X) + \sum_{x \in X} \binom{\frac{1}{2}(\deg(x) + 1)}{2} = \binom{n/2}{2}. \tag{11.2}$$

Proof of Theorem 11.4.1. Theorem 11.4.2 implies, in particular, that $\mathrm{cr}(X) = O(n^2)$. The crossing number theorem shows that $\mathrm{cr}(X) = \Omega(t^3/n^2) - O(n)$, where t is the number of halving edges, and this implies $t = O(n^{4/3})$.
$\square$

Proof of Theorem 11.4.2.   First we note that by the halving-facet in-terleaving lemma, $\deg(x)$ is odd for every $x \in X$, and so the expression $\frac{1}{2}(\deg(x)+1)$ in the identity (11.2) is always an integer.

For the following arguments, we formally regard the set X as a sequence $(x_1, x_2, \ldots, x_n)$. From Section 9.3 we recall the notion of *orientation* of a triple (x_i, x_j, x_k): Assuming $i < j < k$, the orientation is positive if we make a right turn when going from x_i to x_k via x_j, and it is negative if we make a left turn. The *order type* of X describes the orientations of all the triples (x_i, x_j, x_k), $1 \le i < j < k \le n$. We observe that the order type uniquely determines the halving edge graph: Whether $\{x_i, x_j\}$ is a halving edge or not can be deduced from the orientations of the triples involving x_i and x_j. Similarly, the orientations of all triples determine whether two halving edges cross.

The theorem is proved by a *continuous motion argument*. We start with the given point sequence X, and we move its points continuously until we reach some suitable configuration X_0 for which the identity (11.2) holds. For example, X_0 can consist of n points in convex position, where we have $\frac{n}{2}$ halving edges and every two of them cross.

The continuous motion transforming X into X_0 is such that the current sequence remains in general position, except for finitely many moments when exactly one triple (x_i, x_j, x_k) changes its orientation. The points x_i, x_j, x_k thus become collinear at such a moment, but we assume that they always remain distinct, and we also assume that no other collinearities occur at that moment. Let us call such a moment a *mutation at* $\{x_i, x_j, x_k\}$.

We will investigate the changes of the graph of halving edges during the motion, and we will show that mutations leave the left-hand side of the identity (11.2) unchanged.

Both the graph and the crossings of its edges remain combinatorially unchanged between the mutations. Moreover, some thought reveals that by a mutation at $\{x, y, z\}$, only the halving edges with both endpoints among x, y, z and their crossings with other edges can be affected; all the other halving edges and crossings remain unchanged.

Let us first assume that $\{x, y\}$ is a halving edge before the mutation at $\{x, y, z\}$ and that z lies on the segment xy at the moment of collinearity:

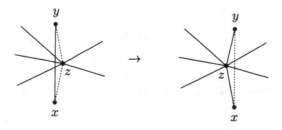

Figure 11.1. Welzl's Little Devils.

We note that $\{x, z\}$ and $\{y, z\}$ cannot be halving edges before the mutation. After the mutation, $\{x, y\}$ ceases to be halving, while $\{x, z\}$ and $\{y, z\}$ become halving.

Let $\deg(z) = 2r+1$ (before the mutation) and let h be the line passing through z and parallel to xy. The larger side of h, i.e., the one with more points of X, is the one containing x and y, and by the halving-facet interleaving lemma, $r+1$ of the halving edges emanating from z go into the larger side of h and thus cross xy. So the following changes in degrees and crossings are caused by the mutation:

- $\deg(z)$, which was $2r+1$, increases by 2, and
- $\mathrm{cr}(X)$ decreases by $r+1$.

It is easy to check that the left-hand side of the identity (11.2) remains the same after this change.

What other mutations are possible? One is the mutation inverse to the one discussed above, with z moving in the reverse direction. We show that there are no other types of mutations affecting the graph of halving edges. Indeed, for any mutation, the notation can be chosen so that z crosses over the segment xy. Just before the mutation or just after it, it is not possible for $\{x, z\}$ to be a halving edge and $\{y, z\}$ not. The last remaining possibility is a mutation with no halving edge on $\{x, y, z\}$, which leaves the graph unchanged. Theorem 11.4.2 is proved. □

Tight bounds for small n. Using the identity (11.2) and the fact that all vertices of the graph of halving edges must have odd degrees, one can determine the exact maximum number of halving edges for small point configurations (Exercise 1). Figure 11.1 shows examples of configurations with the maximum possible number of halving edges for $n = 8$, 10, and 12. These small examples seem to be misleading in various respects: For example, we know that the number maximum of halving edges is superlinear, and so the graph of halving edges cannot be planar for large n, and yet all the small pictures are planar.

Bibliography and remarks. Theorem 11.4.1 was first proved by Dey [Dey98], who discovered the surprising role of the crossings of the halving edges. His proof works partially in the dual setting, and it relies on a technique of decomposing the k-level in an arrangement into convex chains discussed in Agarwal et al. [AACS98]. The identity (11.2), with the resulting considerable simplification of Dey's proof, were found by Andrzejak et al. [AAHP+98]. They also computed the exact maximum number of halving edges up to $n = 12$ and proved results about k-facets and k-sets in dimension 3.

Improved upper bound for k-sets in $\mathbf{R}^3$. We outline the argument of Sharir et al. [SST01] proving that an n-point set $X \subset \mathbf{R}^3$ in general position has at most $O(n^{2.5})$ halving triangles. Let $\mathcal{T}$ be the set of halving triangles and let $t = |\mathcal{T}|$. We will count the number N of *crossing pairs* of triangles in $\mathcal{T}$ in two ways, where a crossing pair looks like this:

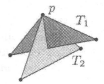

The triangles share one vertex p, and the edge of T_1 opposite to p intersects the interior of T_2.

The Lovász lemma (Corollary 11.3.2: no line intersects more than $O(n^2)$ halving triangles) implies $N = O(n^4)$. To see this, we first consider pairs (ℓ, T), where ℓ is a line spanned by two points $p, q \in X$, $t \in \mathcal{T}$, and ℓ intersects the interior of T. Each of the $\binom{n}{2}$ lines ℓ contributes at most $O(n^2)$ pairs, and each pair (ℓ, T) yields at most 3 crossing pairs of triangles, one for each vertex of T.

Now we are going to show that $N = \Omega(t^2/n) - O(tn)$, which together with $N = O(n^4)$ implies $t = O(n^{2.5})$. Let ρ be a horizontal plane lying below all of X. For a set $A \subset \mathbf{R}^3$, let A^* denote the central projection of A from p into ρ. To bound N from below, we consider each $p \in X$ in turn, and we associate to it a graph G_p drawn in ρ. Let γ_p be the open half-space below the horizontal plane through p. The vertex set of the geometric graph G_p is $V_p = (X \cap \gamma_p)^*$. Let $\mathcal{H}_p \subseteq \mathcal{T}$ be the set of the halving triangles having p as the highest vertex, and let $\mathcal{M}_p \subseteq \mathcal{T}$ be the triangles with p as the middle-height vertex. Each $T \in \mathcal{H}_p$ contributes an edge of G_p, namely, the segment T^*:

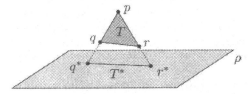

Each $T \in \mathcal{M}_p$ gives rise to an unbounded ray in G_p, namely, $(T \cap \gamma_p)^*$:

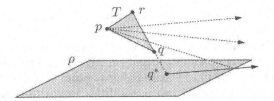

Formally, we can interpret such a ray as an edge connecting the vertex $q^* \in V_p$ to a special vertex at infinity.

Let $m_p = |\mathcal{H}_p| + |\mathcal{M}_p|$ be the total number of edges of G_p, including the rays, and let $r_p = |\mathcal{M}_p|$ be the number of rays. Write x_p for the number of edge crossings in the drawing of G_p. We have

$$\sum_{p \in X} m_p = 2t \quad \text{and} \quad \sum_{p \in X} r_p = t,$$

because each $T \in \mathcal{T}$ contributes to one $\mathcal{H}_p$ and one $\mathcal{M}_p$. We note that $N \geq \sum_{p \in X} x_p$, since an edge crossing in G_p corresponds to a crossing pair of triangles with a common vertex p.

A lower bound for x_p is obtained using a decomposition of G_p into *convex chains*, which is an idea from Agarwal et al. [AACS98] (used in Dey's original proof of the $O(n^{4/3})$ bound for planar halving edges). We fix a vertical direction in ρ so that no edges of G_p are vertical. Each convex chain is a contiguous sequence of (bounded or unbounded) edges of G_p that together form a graph of a convex function defined on an interval. Each edge lies in exactly one convex chain. Let e be an edge of G_p whose right end is a (finite) vertex v. We specify how the convex chain containing e continues to the right of v: It follows an edge e' going from v to the right and turning upwards with respect to v but as little as possible.

If there is no e' like this, then the considered chain ends at v:

By the halving-facet interleaving lemma, the fan of edges emanating from v has an "antipodal" structure: For every two angularly consecutive edges, the opposite wedge contains exactly one edge. This implies that e is uniquely determined by e', and so we have a well-defined decomposition of the edges of G_p into convex chains. Moreover, exactly one convex chain begins or ends at each vertex. Thus, the number c_p of chains equals $\frac{1}{2}(n_p + r_p)$.

A lower bound for the number of edge crossings x_p is the number of pairs $\{C_1, C_2\}$ of chains such that an edge of C_1 crosses an edge of C_2. The trick is to estimate the number of pairs $\{C_1, C_2\}$ that do *not* cross in this way. There are two possibilities for such pairs: C_1 and C_2 can be disjoint or they can cross at a vertex:

The number of pairs $\{C_1, C_2\}$ crossing at a vertex is at most $m_p n_p$, because the edge e_1 of C_1 entering the crossing determines both C_1 and the crossing vertex, and C_2 can be specified by choosing one of the at most n_p edges incident to that vertex. Finally, suppose that C_1 and C_2 are disjoint and C_2 is above C_1. If we fix an edge e_1 of C_1, then C_2 is determined by the vertex where the line parallel to e_1 translated upwards first hits C_2:

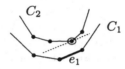

We obtain $x_p \geq \binom{c_p}{2} - 2m_p n_p$, and a calculation leads to $N \geq \sum x_p = \Omega(t^2/n) - O(nt)$. This concludes the proof of the $O(n^{2.5})$ bound for halving facets in $\mathbf{R}^3$.

Having already introduced the decomposition of the graph of halving edges into convex chains as above, one can give an extremely simple alternative proof of Theorem 11.4.1. Namely, the graph of halving edges is decomposed into at most n convex chains and, similarly, into at most n concave chains. Any convex chain intersects any concave chain in at most 2 points, and it follows that the number of edge crossings in the graph of halving edges is $O(n^2)$. The application of the crossing number theorem finishes the proof.

Exercises

1. (a) Find the maximum possible number of halving edges for $n = 4$ and $n = 6$, and construct the corresponding configurations. [3]

    (b) Check that the three graphs in Figure 11.1 are graphs of halving edges of the depicted point sets. [1]

    (c) Show that the configurations in Figure 11.1 maximize the number of halving edges. [4]

12

Two Applications of High-Dimensional Polytopes

From this chapter on, our journey through discrete geometry leads us to the high-dimensional world. Up until now, although we have often been considering geometric objects in arbitrary dimension, we could mostly rely on the intuition from the familiar dimensions 2 and 3. In the present chapter we can still use dimensions 2 and 3 to picture examples, but these tend to be rather trivial. For instance, in the first section we are going to prove things about graphs via convex polytopes, and for an n-vertex graph we need to consider an n-dimensional polytope. It is clear that graphs with 2 or 3 vertices cannot serve as very illuminating examples. In order to underline this shift to high dimensions, from now on we mostly denote the dimension by n instead of d as before, in agreement with the habits prevailing in the literature on high-dimensional topics.

In the first and third sections we touch upon *polyhedral combinatorics*. Let E be a finite set, for example the edge set of a graph G, and let $\mathcal{F}$ be some interesting system of subsets of E, such as the set of all matchings in G or the set of all Hamiltonian circuits of G. In polyhedral combinatorics one usually considers the convex hull of the characteristic vectors of the sets of $\mathcal{F}$; the characteristic vectors are points of $\{0,1\}^E \subset \mathbf{R}^E$. For the two examples above, we thus obtain the *matching polytope* of G and the *traveling salesman polytope* of G. The basic problem of polyhedral combinatorics is to find, for a given $\mathcal{F}$, inequalities describing the facets of the resulting polytope. Sometimes one succeeds in describing all facets, as is the case for the matching polytope. This may give insights into the combinatorial structure of $\mathcal{F}$, and often it has algorithmic consequences. If we know the facets and they have a sufficiently nice structure, we can optimize any linear function over the polytope in polynomial time. This means that given some real weights of the elements of E, we can find in polynomial time a maximum-weight set in $\mathcal{F}$

(e.g., a maximum-weight matching). In other cases, such as for the traveling salesman polytope, describing all facets is beyond reach. The knowledge of some facets may still yield interesting consequences, and on the practical side, it can provide a good approximation algorithm for the maximum-weight set. Indeed, the largest traveling salesman problems solved in practice, with thousands of vertices, have been attacked by these methods.

We do not treat polyhedral combinatorics in any systematic manner; rather we focus on two gems (partially) belonging to this area. The first one is the celebrated weak perfect graph conjecture, stating that the complement of any perfect graph is perfect, which is proved by combining combinatorial and polyhedral arguments. The second one is an algorithmically motivated problem of sorting with partial information, discussed in Section 12.3. We associate a polytope with every finite partially ordered set, and we reduce the question to slicing the polytope into two parts of roughly equal volume by a hyperplane. A key role in this proof is played by the Brunn–Minkowski inequality. This fundamental geometric inequality is explained and proved in Section 12.2.

12.1 The Weak Perfect Graph Conjecture

First we recall a few notions from graph theory. Let $G = (V, E)$ be a finite undirected graph on n vertices. By $\overline{G}$ we denote the *complement* of G, that is, the graph $(V, \binom{V}{2} \setminus E)$. An *induced subgraph* of G is any graph that can be obtained from G by deleting some vertices and all edges incident to the deleted vertices (but an edge must not be deleted if both of its vertices remain in the graph). Let $\omega(G)$ denote the *clique number* of G, which is the maximum size of a complete subgraph of G, and let $\alpha(G) = \omega(\overline{G})$ be the *independence number* of G. Explicitly, $\alpha(G)$ is the maximum size of an *independent set* in G, where a set $S \subseteq V(G)$ is independent if the subgraph induced by S in G has no edges. The *chromatic number* of G is the smallest number of independent sets covering all vertices of G, and it is denoted by $\chi(G)$.

Both the problems of finding $\omega(G)$ and finding $\chi(G)$ are computationally hard. It is NP-complete to decide whether $\omega(G) \geq k$, where k is a part of the input, and it is NP-complete to decide whether $\chi(G) = 3$. Even approximating $\chi(G)$ or $\omega(G)$ is hard. So classes of graphs where the clique number and/or the chromatic number are computationally tractable are of great interest.

Perfect graphs are one of the most important such classes, and they include many other classes found earlier. A graph $G = (V, E)$ is called *perfect* if $\omega(G') = \chi(G')$ for every induced subgraph G' of G (including $G' = G$).

For every graph G we have $\chi(G) \geq \omega(G)$, so a high clique number is a "reason" for a high chromatic number. But in general it is not the only possible reason, since there are graphs with $\omega(G) = 2$ but $\chi(G)$ arbitrarily large.

Perfect graphs are those whose chromatic number is exclusively controlled by the cliques, and this is true for G and also for all of its subgraphs.

For perfect graphs, the clique number, and hence also the chromatic number, can be computed in polynomial time by a sophisticated algorithm (related to semidefinite programming briefly discussed in Section 15.5). It is not known how hard it is to decide perfectness of a given graph. No polynomial-time algorithm has been found, but neither has any hardness result (such as coNP-hardness) been proved. But for graphs arising in many applications we know in advance that they are perfect.

Typical nonperfect graphs are the odd cycles C_{2k+1} of length 5 and larger, since $\omega(C_{2k+1}) = 2$ for $k \geq 2$, while $\chi(C_{2k+1}) = 3$.

The following two conjectures were formulated by Berge at early stages of research on perfect graphs. Here is the stronger one:

Strong perfect graph conjecture. *A graph G is perfect if and only if neither G nor its complement contain an odd cycle of length 5 or larger as an induced subgraph.*

This is still open, in spite of considerable effort. The second conjecture is this:

Weak perfect graph conjecture. *A graph is perfect if and only if its complement is perfect.*

This was proved in 1972. We reproduce a proof using convex polytopes.

12.1.1 Definition. *Let $G = (V, E)$ be a graph on n vertices. We assign a convex polytope $P(G) \subset \mathbf{R}^n$ to G. Let the coordinates in $\mathbf{R}^n$ be indexed by the vertices of G; i.e., if $V = \{v_1, \ldots, v_n\}$, then the points of $P(G)$ are of the form $x = (x_{v_1}, \ldots, x_{v_n})$. For an $x \in \mathbf{R}^n$ and a subset $U \subseteq V$, we put $x(U) = \sum_{v \in U} x_v$.*
The polytope $P(G)$ is defined by the following inequalities:

(i) $x_v \geq 0$ *for each vertex $v \in V$, and*
(ii) $x(K) \leq 1$ *for each clique (complete subgraph) K in the graph G.*

Observations.

- $P(G) \subseteq [0, 1]^n$. The inequality $x_v \leq 1$ is obtained from (ii) by choosing $K = \{v\}$.
- The characteristic vector of each independent set lies in $P(G)$.
- If a vector $x \in P(G)$ is integral (i.e., it is a 0/1 vector), then it is the characteristic vector of an independent set.

Before we start proving the weak perfect graph conjecture, let us introduce some more notation. Let $w: V \to \{0, 1, 2, \ldots\}$ be a function assigning nonnegative integer weights to the vertices of G. We define the *weighted clique number* $\omega(G, w)$ as the maximum possible weight of a clique, where the weight of a clique is the sum of the weights of its vertices. We also define the *weighted chromatic number* $\chi(G, w)$ as the minimum number of independent sets such that each vertex $v \in V$ is covered by $w(v)$ of them.

Now we can formulate the main theorem.

12.1.2 Theorem. *The following conditions are equivalent for a graph G:*

(i) G *is perfect.*

(ii) $\omega(G, w) = \chi(G, w)$ *for any nonnegative integral weight function w.*

(iii) *All vertices of the polytope $P(G)$ are integral (and thus correspond to the independent sets in G).*

(iv) *The graph $\overline{G}$ is perfect.*

Proof of (i) $\Rightarrow$ **(ii).** This part is purely graph-theoretic. For every weight function $w: V \to \{0, 1, 2, \ldots\}$, we need to exhibit a covering of V by independent sets witnessing $\chi(G, w) = \omega(G, w)$. If w attains only values 0 and 1, then we can use (i) directly, since selecting an induced subgraph of G is the same as specifying a 0/1 weight function on the vertices.

For other values of w we proceed by induction on $w(V)$. Let w be given and let v_0 be a vertex with $w(v_0) > 1$. We define a new weight function w':

$$w'(v) = \begin{cases} w(v) - 1 & \text{for } v = v_0, \\ w(v) & \text{for } v \neq v_0. \end{cases}$$

Since $w'(V) < w(V)$, by the inductive hypothesis we assume that we have independent sets $I_1, I_2, \ldots, I_N$ covering each v exactly $w'(v)$ times, where $N = \omega(G, w')$. If $\omega(G, w) > N$, then we can obtain the appropriate covering for w by adding the independent set $\{v_0\}$, so let us suppose $\omega(G, w) = N$.

Let the notation be chosen so that $v_0 \in I_1$. We define another weight function w'':

$$w''(v) = \begin{cases} w(v) - 1 & \text{for } v \in I_1, \\ w(v) & \text{for } v \notin I_1. \end{cases}$$

We claim that $\omega(G, w'') < N$. If not, then there exists a clique K with $w''(K) = N = \omega(G, w')$. By the choice of the I_i, we have $N \leq w'(K) = \sum_{i=1}^{N} |I_i \cap K|$. Since a clique intersects an independent set in at most one vertex, K has to intersect each I_i. In particular, it intersects I_1, and so $w(K) > w''(K) = N$, contradicting $\omega(G, w) = N$.

We thus have $\omega(G, w'') < N$. By the inductive hypothesis, we can produce a covering by independent sets showing that $\chi(G, w'') < N$. By adding I_1 to it we obtain a covering witnessing $\chi(G, w) = N$.

Proof of (ii) $\Rightarrow$ **(iii).** Let $x = (x_{v_1}, \ldots, x_{v_n})$ be a vertex of the convex polytope $P(G)$. Since all the inequalities defining $P(G)$ have rational coefficients, x has rational coordinates, and we can find a natural number q such that $w = qx$ is an integral vector. We interpret the coordinates of w as weights of the vertices of G. Let K be a clique with weight $N = \omega(G, w)$. One of the inequalities defining $P(G)$ is $x(K) \leq 1$, and hence $N = w(K) \leq q$.

By (ii) we have $\chi(G,w) = \omega(G,w) \leq q$, and so there are independent sets $I_1, \ldots, I_q$ (some of them may be empty) covering each vertex $v \in V$ precisely w_v times. Let c_i be the characteristic vector of I_i; then this property of the sets I_i can be written as $x = \sum_{i=1}^{q} \frac{1}{q} c_i$. Thus x is a convex combination of the c_i, and since it is a vertex of $P(G)$, it must be equal to some c_i, which is a characteristic vector of an independent set in G.

Proof of (iii) $\Rightarrow$ (iv). It suffices to prove $\chi(\overline{G}) = \omega(\overline{G})$ for every G satisfying (iii), since (iii) is preserved by passing to an induced subgraph (right?).

We prove that a graph G fulfilling (iii) has a clique K intersecting all independent sets of the maximum size $\alpha(G)$. Then the graph $G \setminus K$ has independence number $\alpha(G) - 1$, and by repeating the same procedure we can cover G by $\alpha(G)$ cliques.

To find the required K, let us consider all the independent sets of size $\alpha = \alpha(G)$ in G and let $M \subseteq P(G)$ be the convex hull of their characteristic vectors. We note that M lies in the hyperplane $h = \{x \colon x(V) = \alpha\}$. This h defines a (proper) face of $P(G)$, for otherwise, we would have vertices of $P(G)$ on both sides of h, and in particular, there would be a vertex z with $z(V) > \alpha$. This is impossible, since by (iii), z would correspond to an independent set bigger than α.

Each facet of $P(G)$ corresponds to an equality in some of the inequalities defining $P(G)$. The equality can be either of the form $x_v = 0$ or of the form $x(K) = 1$. The face $F = P(G) \cap h$ is the intersection of some of the facets. Not all of these facets can be of the type $x_v = 0$, since then their intersection would contain 0, while $0 \notin h$. Hence all $x \in M$ satisfy $x(K) = 1$ for a certain clique K, and this means that $K \cap I \neq \emptyset$ for each independent set I of size α.

Proof of (iv) $\Rightarrow$ (i). This is the implication (i) $\Rightarrow$ (iv) for the graph $\overline{G}$. $\square$

Bibliography and remarks. Perfect graphs were introduced by Berge [Ber61],[Ber62], who also formulated the two perfect graph conjectures. The weak perfect graph conjecture was first proved (combinatorially) by Lovász [Lov72]. The proof shown in this section follows Grötschel, Lovász, and Schrijver [GLS88], whose account is based on the ideas of [Lov72] and of Fulkerson [Ful70].

Grötschel et al. [GLS88] denote the polytope $P(G)$ by QSTAB(G) and call it the *clique-constrained stable set polytope* (another name in the literature is the *fractional stable set polytope*). Here *stable set* is another common name for an independent set, and the *stable set polytope* STAB$(G) \subset \mathbf{R}^{|E|}$ is the convex hull of the characteristic vectors of all independent sets in G. As we have seen, STAB$(G) =$ QSTAB(G) if and only if G is a perfect graph. Polynomial-time algorithms for perfect graphs are based on beautiful geometric ideas (related to the famous Lovász ϑ-function), and they are presented in [GLS88] or in Lovász [Lov] (as well as in many other sources).

Polyhedral combinatorics was initiated mainly by the results of Edmonds [Edm65]. For a graph $G = (V, E)$, let $M(G)$ denote the *matching polytope* of G, that is, the convex hull of the characteristic vectors of the matchings in a graph G. According to Edmonds' matching polytope theorem, $M(G)$ is described by the following inequalities: $x_e \geq 0$ for all $e \in E$, $\sum_{e \in E: v \in e} x_e \leq 1$ for all $v \in V$, and $\sum_{e \in E: e \subseteq S} x_e \leq \frac{1}{2}(|S|-1)$ for all $S \subseteq V$ of odd cardinality. For bipartite G, the constraints of the last type are not necessary (this is an older result of Birkhoff).

A modern textbook on combinatorial optimization, with an introduction to polyhedral combinatorics, is Cook, Cunningham, Pulleyblank, and Schrijver [CCPS98]. It also contains references to theoretical and practical studies of the traveling salesman problem by polyhedral methods.

A key step in many results of polyhedral combinatorics is proving that a certain system of inequalities defines an *integral polytope*, i.e., one with all vertices in $\mathbf{Z}^n$. Let us mention just one important related concept: the *total unimodularity*. An $m \times n$ matrix A is totally unimodular if every square submatrix of A has determinant 0, 1, or -1. Total unimodularity can be tested in polynomial time (using a deep characterization theorem of Seymour). All polyhedra defined by totally unimodular matrices are integral, in the sense formulated in Exercise 6. For other aspects of integral polytopes (sometimes also called *lattice polytopes*) see, e.g., Barvinok [Bar97] (and Section 2.2).

Exercises

1. What are the integral vertices of the polytope $P(C_5)$? Find some nonintegral vertex (and prove that it is really a vertex!). ③

2. Prove that for every graph G and every clique K in G, the inequality $x(K) \leq 1$ defines a *facet* of the polytope $P(G)$. In other words, there is an $x \in P(G)$ for which $x(K) = 1$ is the only inequality among those defining $P(G)$ that is satisfied with equality. ②

3. (On König's edge-covering theorem) Explain why bipartite graphs are perfect, and why the perfectness of the complements of bipartite graphs is equivalent to König's edge-covering theorem asserting that the maximum number of vertex-disjoint edges in a bipartite graph equals the minimum number of vertices needed to intersect all edges (also see Exercise 10.1.5). ②

4. (Comparability graphs and Dilworth's theorem) For a finite partially ordered set $(X, \leq)$ (see Section 12.3 for the definition), let $G = (X, E)$ be the graph with $E = \{\{u, v\} \in \binom{X}{2}: u < v$ or $v < u\}$; that is, edges correspond to pairs of comparable elements. Any graph isomorphic to such a G is called a *comparability graph*. We also need the notions of a

chain (a subset of X linearly ordered by $\leq$) and an *antichain* (a subset of X with no two elements comparable under $\leq$).

(a) Prove that any finite $(X, \leq)$ is the union of at most c antichains, where c is the length of the longest chain, and check that this implies the perfectness of comparability graphs. ④

(b) Derive from (a) the *Erdős–Szekeres lemma*: If $a_1, a_2, \ldots, a_n$ are arbitrary real numbers, then there exist indices $i_1 < i_2 < \cdots < i_k$ with $k^2 \geq n$ and such that the subsequence $a_{i_1}, a_{i_2}, \ldots, a_{i_k}$ is monotone (nondecreasing or decreasing). ③

(c) Check that the perfectness of the complements of comparability graphs is equivalent to the following theorem of Dilworth [Dil50]: Any finite $(X, \leq)$ is the union of at most a chains, where a is the maximum number of elements of an antichain. ①

5. (Hoffman's characterization of polytope integrality) Let P be a (bounded) convex polytope in $\mathbf{R}^n$ such that for every $a \in \mathbf{Z}^n$, the minimum of the function $x \mapsto \langle a, x \rangle$ over all $x \in P$ is an integer. Prove that all vertices of P are integral (i.e., they belong to $\mathbf{Z}^n$). ⑤

6. (Kruskal–Hoffman theorem)
   (a) Show that if A is a nonsingular $n \times n$ totally unimodular matrix (all square submatrices have determinant 0 or ± 1), then the mapping $x \mapsto Ax$ maps $\mathbf{Z}^n$ bijectively onto $\mathbf{Z}^n$. ③

   (b) Show that if A is an $m \times n$ totally unimodular matrix and b is an m-dimensional integer vector such that the system $Ax = b$ has a real solution x, then $Ax = b$ has an integral solution as well. ③

   (c) Let A be an $m \times n$ totally unimodular matrix and let $u, v \in \mathbf{Z}^n$ and $w, z \in \mathbf{Z}^m$ be integer vectors. Show that all vertices of the convex polyhedron given by the inequalities $u \leq x \leq v$ and $w \leq Ax \leq z$ are integral. ①

7. (Helly-type theorem for lattice points in convex sets)
   (a) Let A be a set of $2^d + 1$ points in $\mathbf{Z}^d$. Prove that there are $a, b \in A$ with $\frac{1}{2}(a + b) \in \mathbf{Z}^d$. ③

   (b) Let $\gamma_1, \ldots, \gamma_n$ be closed half-spaces in $\mathbf{R}^d$, $n \geq 2^d + 1$, and suppose that the intersection of every 2^d of them contains a lattice point (a point of $\mathbf{Z}^d$). Prove that there exists a lattice point common to all the γ_i. ④

   (c) Prove that the number 2^d in (b) is the best possible, i.e., there are 2^d half-spaces such that every $2^d - 1$ of them have a common lattice point but there is no lattice point common to all of them. ②

   (d) Extend the Helly-type theorem in (b) to arbitrary convex sets instead of half-spaces. ③

   The result in (d) was proved by Doignon [Doi73]; his proof starts with (a) and proceeds on the level of abstract convexity (while the proof suggested in (b) is more geometric).

12.2 The Brunn–Minkowski Inequality

Let us consider a 3-dimensional convex loaf of bread and slice it by three parallel planar cuts.

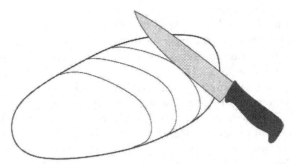

As we will derive below, the middle cut cannot have area smaller than both of the other two cuts. Let us choose the coordinate system so that the cuts are perpendicular to the x_1-axis and denote by $v(t)$ the area of the cut by the plane $x_1 = t$. Then the claim can be stated as follows: For any $t_1 < t < t_2$ we have $v(t) \geq \min(v(t_1), v(t_2))$. Thus, there is some t_0 such that the function $t \mapsto v(t)$ is nondecreasing on $(-\infty, t_0]$ and nonincreasing on $[t_0, \infty)$. Such a function is called *unimodal*. A similar result is true for any convex body C in $\mathbf{R}^{n+1}$ if $v(t)$ denotes the n-dimensional volume of the intersection of C with the hyperplane $\{x_1 = t\}$.

How can one prove such a statement? In the planar case, with $n = 1$, it is easy to see that $v(t)$ is a concave function on the interval obtained by projecting C on the x_1-axis.

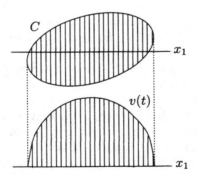

This might tempt one to think that $v(t)$ is concave on the appropriate interval in higher dimension, too, but this is false in general! (See Exercise 1.) There is concavity in the game, but the right function to look at in $\mathbf{R}^{n+1}$ is $v(t)^{1/n}$. Perhaps a little more intuitively, we can define $r(t)$ as the radius of the n-dimensional ball whose volume equals $v(t)$. We have $r(t) = R_n v(t)^{1/n}$, where

R_n is the radius of a unit-volume ball in $\mathbf{R}^n$; let us call $r(t)$ the *equivalent radius* of C at t.

12.2.1 Theorem (Brunn's inequality for slice volumes). *Let $C \subset \mathbf{R}^{n+1}$ be a compact convex body and let the interval $[t_{\min}, t_{\max}]$ be the projection of C on the x_1-axis. Then the equivalent radius function $r(t)$ (or, equivalently, the function $v(t)^{1/n}$) is concave on $[t_{min}, t_{max}]$. Consequently, for any $t_1 < t < t_2$ we have $v(t) \geq \min(v(t_1), v(t_2))$.*

Brunn's inequality is a consequence of the following more general and more widely applicable statement dealing with two arbitrary compact sets.

12.2.2 Theorem (Brunn–Minkowski inequality). *Let A and B be nonempty compact sets in $\mathbf{R}^n$. Then*

$$\operatorname{vol}(A + B)^{1/n} \geq \operatorname{vol}(A)^{1/n} + \operatorname{vol}(B)^{1/n}.$$

Here $A + B = \{a + b\colon a \in A, b \in B\}$ denotes the *Minkowski sum* of A and B. If A' is a translated copy of A, and B' a translated copy of B, then $A' + B'$ is a translated copy of $A + B$. So the position of $A + B$ with respect to A and B depends on the choice of coordinate system, but the shape of $A + B$ does not. One way of interpreting the Minkowski sum is as follows: Keep A fixed, pick a point $b_0 \in B$, and translate B into all possible positions for which b_0 lies in A. Then $A + B$ is the union of all such translates. Here is a planar example:

Sometimes it is also useful to express the Minkowski sum $A + B$ as a projection of the Cartesian product $A \times B \subset \mathbf{R}^{2n}$ by the mapping $(x, y) \mapsto x + y$, $x, y \in \mathbf{R}^n$.

Proof of Brunn's inequality for slice volumes from the Brunn–Minkowski inequality. First we consider "convex combinations" of sets $A, B \subset \mathbf{R}^n$ of the form $(1-t)A + tB$, where $t \in [0, 1]$ and where tA stands for $\{ta\colon a \in A\}$. As t goes from 0 to 1, $(1-t)A + tB$ changes shape continuously from A to B.

Now, if A and B are both convex and we place them into $\mathbf{R}^{n+1}$ so that A lies in the hyperplane $\{x_1 = 0\}$ and B in the hyperplane $\{x_1 = 1\}$, it is not difficult to check that $(1-t)A + tB$ is the slice of the convex body $\operatorname{conv}(A \cup B)$ by the hyperplane $\{x_1 = t\}$; see Exercise 2:

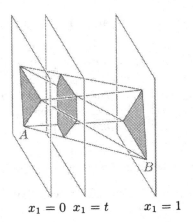

$$x_1 = 0 \quad x_1 = t \qquad x_1 = 1$$

Let us consider the situation as in Brunn's inequality, where $C \subset \mathbf{R}^{n+1}$ is a convex body. Let A and B be the slices of C by the hyperplanes $\{x_1 = t_1\}$ and $\{x = t_2\}$, respectively, where $t_1 < t_2$ are such that $A, B \neq \emptyset$. For convenient notation, we change the coordinate system so that $t_1 = 0$ and $t_2 = 1$. To prove the concavity of the function $v(t)^{1/n}$ in Brunn's inequality, we need to show that for all $t \in (0, 1)$,

$$(1-t)\operatorname{vol}(A)^{1/n} + t\operatorname{vol}(B)^{1/n} \leq \operatorname{vol}(M)^{1/n}, \qquad (12.1)$$

where M is the slice of C by the hyperplane $h_t = \{x_1 = t\}$. Let $C' = \operatorname{conv}(A \cup B)$ and $M' = C' \cap h_t$. We have $C' \subseteq C$ and $M' \subseteq M$. By the remark above, $M' = (1-t)A + tB$, and so the Brunn–Minkowski inequality applied to the sets $(1-t)A$ and tB yields

$$\begin{aligned} \operatorname{vol}(M)^{1/n} &\geq \operatorname{vol}(M')^{1/n} = \operatorname{vol}((1-t)A + tB)^{1/n} \\ &\geq \operatorname{vol}((1-t)A)^{1/n} + \operatorname{vol}(tB)^{1/n} \\ &= (1-t)\operatorname{vol}(A)^{1/n} + t\operatorname{vol}(B)^{1/n}. \end{aligned}$$

This verifies (12.1).   □

Proof of the Brunn–Minkowski inequality. The idea of this proof is simple but perhaps surprising in this context. Call a set $A \subseteq \mathbf{R}^d$ a *brick set* if it is a union of finitely many closed axis-parallel boxes with disjoint interiors. First we show that it suffices to prove the inequality for brick sets (which is easy but a little technical), and then for brick sets the proof goes by induction on the number of bricks.

12.2.3 Lemma. *If the Brunn–Minkowski inequality holds for all nonempty brick sets $A', B' \subset \mathbf{R}^n$, then it is valid for all nonempty compact sets $A, B \subset \mathbf{R}^n$ as well.*

Proof. We use a basic fact from measure theory, namely, that if $X_1 \supseteq X_2 \supseteq X_3 \supseteq \cdots$ is a sequence of measurable sets in $\mathbf{R}^n$ such that $X = \bigcap_{i=1}^{\infty} X_i$, then the numbers $\operatorname{vol}(X_i)$ converge to $\operatorname{vol}(X)$.

Let $A, B \subset \mathbf{R}^n$ be nonempty and compact. For $k = 1, 2, \ldots$, consider the closed axis-parallel cubes with side length 2^{-k} centered at the points of the scaled grid $2^{-k}\mathbf{Z}^n$ (these cubes cover $\mathbf{R}^n$ and have disjoint interiors). Let A_k be the union of all such cubes intersecting the set A, and similarly for B_k.

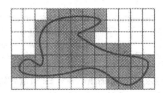

We have $A_1 \supseteq A_2 \supseteq \cdots$ and $\bigcap_k A_k = A$ (since any point not belonging to A has a positive distance from it, and the distance of any point of A_k from A is at most $2^{-k}\sqrt{n}$). Therefore, $\mathrm{vol}(A_k) \to \mathrm{vol}(A)$ and $\mathrm{vol}(B_k) \to \mathrm{vol}(B)$.

We claim that $A + B \supseteq \bigcap_k (A_k + B_k)$. To see this, let $x \in A_k + B_k$ for all k. We pick $y_k \in A_k$ and $z_k \in B_k$ with $x = y_k + z_k$, and by passing to convergent subsequences we may assume that $y_k \to y \in A$ and $z_k \to z \in B$. Then we obtain $x = y + z \in A + B$. Thus $\lim_{k\to\infty}\mathrm{vol}(A_k + B_k) \leq \mathrm{vol}(A + B)$. By the Brunn–Minkowski inequality for the brick sets A_k, B_k, we have $\mathrm{vol}(A + B)^{1/n} \geq \lim_{k\to\infty}\mathrm{vol}(A_k + B_k)^{1/n} \geq \lim_{k\to\infty}(\mathrm{vol}(A_k)^{1/n} + \mathrm{vol}(B_k)^{1/n}) = \mathrm{vol}(A)^{1/n} + \mathrm{vol}(B)^{1/n}$.   □

Proof of the Brunn–Minkowski inequality for brick sets. Let A and B be brick sets consisting of k bricks in total. If $k = 2$, then both A and B, and $A + B$ too, are bricks. Then if $x_1, \ldots, x_n$ are the sides of A and $y_1, \ldots, y_n$ are the sides of B, it suffices to establish the inequality $\left(\prod_{i=1}^n x_i\right)^{1/n} + \left(\prod_{i=1}^n y_i\right)^{1/n} \leq \left(\prod_{i=1}^n (x_i + y_i)\right)^{1/n}$; we leave this to Exercise 3.

Now let $k > 2$ and suppose that the Brunn–Minkowski inequality holds for all pairs A, B of brick sets together consisting of fewer than k bricks. Let A and B together have k bricks, and let the notation be chosen so that A has at least two bricks. Then it is easily seen that there exists a hyperplane h parallel to some of the coordinate hyperplanes and with at least one full brick of A on one side and at least one full brick of A on the other side (Exercise 4). By a suitable choice of the coordinate system, we may assume that h is the hyperplane $\{x_1 = 0\}$.

Let A' be the part of A on one side of h and A'' the part on the other side. More precisely, A' is the closure of $A \cap h^{\oplus}$, where $h^{\oplus}$ is the open half-space $\{x_1 > 0\}$, and similarly, A'' is the closure of $A \cap h^{\ominus}$. Hence both A' and A'' have at least one brick fewer than A.

Next, we translate the set B in the x_1-direction in such a way that the hyperplane h divides its volume in the same ratio as A is divided (translation does not influence the validity of the Brunn–Minkowski inequality). Let B' and B'' be the respective parts of B.

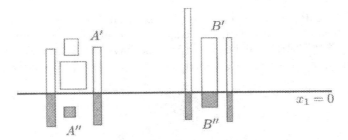

Putting $\rho = \operatorname{vol}(A')/\operatorname{vol}(A)$, we also have $\rho = \operatorname{vol}(B')/\operatorname{vol}(B)$. (If $\operatorname{vol}(A) = 0$ or $\operatorname{vol}(B) = 0$, then the Brunn–Minkowski inequality is obvious.)

The sets A' and B' together have fewer than k bricks, so we can use the inductive assumption for them, and similarly for A'', B''.

The set $A' + B'$ is contained in the closed half-space $\{x_1 \geq 0\}$, and $A'' + B''$ lies in the opposite closed half-space $\{x_1 \leq 0\}$. Therefore, crucially, $\operatorname{vol}(A + B) \geq \operatorname{vol}(A' + B') + \operatorname{vol}(A'' + B'')$. We calculate

$$
\begin{aligned}
\operatorname{vol}(A + B) &\geq \operatorname{vol}(A' + B') + \operatorname{vol}(A'' + B'') \\
\text{(induction)} &\geq \Big[\operatorname{vol}(A')^{1/n} + \operatorname{vol}(B')^{1/n}\Big]^n + \Big[\operatorname{vol}(A'')^{1/n} + \operatorname{vol}(B'')^{1/n}\Big]^n \\
&= \Big[\rho^{1/n}\operatorname{vol}(A)^{1/n} + \rho^{1/n}\operatorname{vol}(B)^{1/n}\Big]^n \\
&\quad + \Big[(1-\rho)^{1/n}\operatorname{vol}(A)^{1/n} + (1-\rho)^{1/n}\operatorname{vol}(B)^{1/n}\Big]^n \\
&= \Big[\operatorname{vol}(A)^{1/n} + \operatorname{vol}(B)^{1/n}\Big]^n.
\end{aligned}
$$

This concludes the proof of the Brunn–Minkowski inequality. ☐

Bibliography and remarks. Brunn's inequality for slice volumes appears in Brunn's dissertation from 1887 and in his *Habilitationsschrift* from 1889. Minkowski's formulation of Theorem 12.2.2 (proved for convex sets) was published in the 1910 edition of his book [Min96]. A proof for arbitrary compact sets was given by Lusternik in 1935; see, e.g., the Sangwine-Yager [SY93] for references.

The proof of the Brunn–Minkowski inequality presented here follows Appendix III in Milman and Schechtman [MS86]. Several other proofs are known. A modern one, explained in Ball [Bal97], derives a more general inequality dealing with functions. Namely, if $t \in (0,1)$ and f, g, and h are nonnegative measurable functions $\mathbf{R}^n \to \mathbf{R}$ such that $h\big((1-t)x + ty\big) \geq f(x)^{1-t}g(y)^t$ for all $x, y \in \mathbf{R}^n$, then $\int_{\mathbf{R}^n} h \geq \big(\int_{\mathbf{R}^n} f\big)^{1-t}\big(\int_{\mathbf{R}^n} g\big)^t$ (the *Prékopa–Leindler inequality*). By letting f, g, and h be the characteristic functions of A, B, and $A+B$, respectively, we obtain $\operatorname{vol}\big((1-t)A + tB\big) \geq \operatorname{vol}(A)^{1-t}\operatorname{vol}(B)^t$. This is an alternative form of the Brunn–Minkowski inequality, from which the version in Theorem 12.2.2 follows quickly (see Exercise 5). Advantageously, the dimension does not appear in the Prékopa–Leindler inequality, and it is simple to derive the general case from the 1-dimensional case by induction; see Exercise 7. This passage to a dimension-free form of the inequality, which can be proved from the 1-dimensional case by a simple product argument, is typical in the modern theory of geometric inequalities (a similar phenomenon for measure concentration inequalities is mentioned in the notes to Section 14.2).

The Brunn–Minkowski inequality is just the first step in a sophisticated theory; see Schneider [Sch93] or Sangwine-Yager [SY93]. Among the most prominent notions are the *mixed volumes*. As was discovered by Minkowski, if $K_1, \ldots, K_r \subset \mathbf{R}^n$ are convex bodies and $\lambda_1, \lambda_2, \ldots, \lambda_r$ are nonnegative real parameters, then $\mathrm{vol}(\lambda_1 K_1 + \lambda_2 K_2 + \cdots + \lambda_r K_r)$ is a homogeneous symmetric polynomial of degree n. For $1 \leq i_1 \leq i_2 \leq \cdots \leq i_n \leq r$, the coefficient of $\lambda_{i_1} \lambda_{i_2} \cdots \lambda_{i_n}$ is denoted by $V(K_{i_1}, K_{i_2}, \ldots, K_{i_n})$ and called the *mixed volume* of $K_{i_1}, K_{i_2}, \ldots, K_{i_n}$. A powerful generalization of the Brunn–Minkowski inequality, the *Alexandrov–Fenchel inequality*, states that for any convex $A, B, K_3, K_2, \ldots, K_n \subset \mathbf{R}^n$, we have

$$V(A, B, K_2, \ldots, K_n)^2 \geq V(A, A, K_3, \ldots, K_n) \cdot V(B, B, K_3, \ldots, K_n).$$

Exercises

1. Let A be a single point and B the n-dimensional unit cube. What is the function $v(t) = \mathrm{vol}\big((1-t)A + tB\big)$? Show that $v(t)^\beta$ is not concave on $[0, 1]$ for any $\beta > \frac{1}{n}$. ☐1

2. Let $A, B \subseteq \mathbf{R}^n$ be convex sets. Show that the sets $\mathrm{conv}\big(({\{0\}} \times A) \cup ({\{1\}} \times B)\big)$ and $\bigcup_{t \in [0,1]} \big[{\{t\}} \times ((1-t)A + tB)\big]$ (in $\mathbf{R}^{n+1}$) are equal. ☐2

3. Prove that

$$\left(\prod_{i=1}^{n} x_i\right)^{1/n} + \left(\prod_{i=1}^{n} y_i\right)^{1/n} \leq \left(\prod_{i=1}^{n} (x_i + y_i)\right)^{1/n}$$

   for arbitrary positive reals x_i, y_i. ☐3

4. Show that for any brick set A with at least two bricks, there exists a hyperplane h parallel to one of the coordinate hyperplanes that has at least one full brick of A on each side. ☐2

5. (Dimension-free form of Brunn–Minkowski) Consider the following two statements:
   (i) Theorem 12.2.2, i.e., $\mathrm{vol}(A + B)^{1/n} \geq \mathrm{vol}(A)^{1/n} + \mathrm{vol}(B)^{1/n}$ for every nonempty compact $A, B \subset \mathbf{R}^n$.
   (ii) For all compact $C, D \subset \mathbf{R}^n$ and all $t \in (0, 1)$, $\mathrm{vol}\big((1-t)C + tD\big) \geq \mathrm{vol}(C)^{1-t} \mathrm{vol}(D)^t$.
   (a) Derive (ii) from (i); prove and use the inequality $(1-t)x + ty \geq x^{1-t}y^t$ (x, y positive reals, $t \in (0, 1)$). ☐3

(b) Prove (i) from (ii). ③

6. Give a short proof of the 1-dimensional Brunn–Minkowski inequality: $\text{vol}(A + B) \geq \text{vol}(A) + \text{vol}(B)$ for any nonempty measurable $A, B \subset \mathbf{R}$. ②

7. (Brunn–Minkowski via Prékopa–Leindler) The goal is to establish statement (ii) in Exercise 5.

(a) Let $f, g, h \colon \mathbf{R} \to \mathbf{R}$ be bounded nonnegative measurable functions such that $h\big((1-t)x+ty\big) \geq f(x)^{1-t}g(y)^t$ for all $x, y \in \mathbf{R}$ and all $t \in (0,1)$. Use the one-dimensional Brunn–Minkowski inequality (Exercise 6) to prove $\int h \geq (1-t) \left(\int f\right) + t \left(\int g\right)$ (all integrals over $\mathbf{R}$); by the inequality in Exercise 5(a), the latter expression is at least $\left(\int f\right)^{1-t} \left(\int g\right)^t$. First show that we may assume $\sup f = \sup g = 1$. ④

(b) Prove statement (ii) in Exercise 5 by induction on the dimension, using (a) in the induction step. ③

12.3 Sorting Partially Ordered Sets

Here we present an amazing application of polyhedral combinatorics and of the Brunn–Minkowski inequality in a problem in theoretical computer science: sorting of partially ordered sets. We recall that a *partially ordered set*, or *poset* for short, is a pair $(X, \preceq)$, where X is a set and $\preceq$ is a binary relation on X (called an *ordering*) satisfying three axioms: reflexivity ($x \preceq x$ for all x), transitivity ($x \preceq y$ and $y \preceq z$ implies $x \preceq z$), and weak antisymmetry (if $x \preceq y$ and $y \preceq x$, then $x = y$). The ordering $\preceq$ is *linear* if every two elements of $x, y \in X$ are comparable; that is, $x \preceq y$ or $y \preceq x$.

Let X be a given finite set with some linear ordering $\leq$. For example, the elements of X could be identical-looking golden coins ordered by their weights (assuming that no two weights exactly coincide). We want to sort X according to $\leq$; that is, to list the elements of X in increasing order. We can get information about $\leq$ by *pairwise comparisons*: We can choose two elements $a, b \in X$ and ask an oracle whether $a \leq b$ or $a \geq b$. In our example, we have precise scales such that only one coin fits on each scale, which allows us to make pairwise comparisons. Our sorting procedure may be adaptive: The elements to be compared next may be selected depending on the outcome of previous comparisons. We want to make as few comparisons as possible.

In the usual sorting problem we begin with no information about the ordering $\leq$ whatsoever. As is well known, $\Theta(n \log n)$ comparisons are sufficient and also necessary in the worst case. Here we consider a different setting, when we start with some information already given. Namely, we obtain (explicitly) some partial ordering $\preceq$ on X, and we are guaranteed that $x \preceq y$ implies $x \leq y$; that is, $\leq$ is a *linear extension* of $\preceq$. In the example with coins, some weighings have already been made for us before we start. How many comparisons do we need to sort?

Let $E(\preceq)$denote the set of all linear extensions of a partial ordering $\preceq$ and let $e(\preceq) = |E(\preceq)|$ be the number of linear extensions. To sort means to select one among the $e(\preceq)$ possible linear extensions. Since a comparison of distinct elements a and b can have two outcomes, we need at least $\log_2 e(\preceq)$ comparisons in the worst case to distinguish the appropriate linear extension. Is this lower bound always asymptotically tight? Can one always sort using $O(\log_2 e(\preceq))$ comparisons, for any $\preceq$? An affirmative answer is implied by the following theorem:

12.3.1 Theorem (Efficient comparison theorem). *Let $(X, \preceq)$ be a poset, and suppose that $\preceq$ is not linear. Then there exist elements $a, b \in X$ such that*

$$\delta \leq \frac{e(\preceq + (a,b))}{e(\preceq)} \leq 1 - \delta,$$

where $\delta > 0$ is an absolute constant and $\preceq + (a, b)$ stands for the transitive closure of the relation $\preceq \cup \{(a,b)\}$, that is, the partial ordering we obtain from $\preceq$ if we are told that a precedes b.

How do we use this for sorting $\preceq$? For the first comparison, we choose the two elements a, b as in the theorem. Depending on the outcome of this comparison, we pass either to the partial ordering $\preceq +(a, b)$ or to $\preceq +(b, a)$. In both cases, the number of linear extensions has been reduced by the factor $1-\delta$: For $a \leq b$ this is clear by the theorem, and for $a \geq b$ this follows from the equality $e(\preceq + (a, b)) + e(\preceq + (b, a)) = e(\preceq)$. Hence, proceeding by induction, we can sort any partial ordering $\preceq$ using at most $\lceil \log_{1/(1-\delta)} e(\preceq)\rceil$ comparisons.

The conjectured "right" value of δ in Theorem 12.3.1 is $\frac{1}{3} \approx 0.33$; obviously, one cannot do any better for the poset

(meaning that (a, b) is the only pair of distinct elements in the relation $\preceq$). The proof below gives $\delta = \frac{1}{2e} \approx 0.184$, and more complicated proofs yield better values, although $\frac{1}{3}$ seems still elusive.

Order polytopes. We assign certain convex polytopes to partial orderings.

12.3.2 Definition (Order polytope). *Let $(X, \preceq)$ be an n-element poset. Let the coordinates in $\mathbf{R}^n$ be indexed by the elements of X. We define a polytope $P(\preceq)$, the order polytope of $\preceq$, as the set of all $x \in [0, 1]^n$ satisfying the following inequalities:*

$$x_a \leq x_b \quad \text{for every } a, b \in X \text{ with } a \preceq b.$$

Here is an alternative description of the order polytope:

12.3.3 Observation. *The vertices of the order polytope $P(\preceq)$ are precisely the characteristic vectors of all up-sets in $(X, \preceq)$, where an up-set is a subset $U \subseteq X$ such that if $a \in U$ and $a \preceq b$, then $b \in U$.*

Proof. It is easy to see that the characteristic vector of an up-set is in $P(\preceq)$, and that any 0/1 vector in $P(\preceq)$ determines an up-set. It remains to check that all vertices of $P(\preceq)$ are integral. Any vertex is the intersection of some n facet hyperplanes. Since all potential facet hyperplanes have the form $x_a = x_b$, or $x_a = 0$, or $x_a = 1$, the integrality is obvious.   □

12.3.4 Observation. *Let X be an n-element set.*

(i) *If $\leq$ is a linear ordering on X, then $P(\leq)$ is a simplex of volume $1/n!$.*

(ii) *For any partial ordering $\preceq$ on X, the simplices of the form $P(\leq)$, where $\leq$ is a linear extension of $\preceq$, cover $P(\preceq)$ and have disjoint interiors. Hence $\mathrm{vol}(P(\preceq)) = \frac{1}{n!}e(\preceq)$.*

Here is the order polytope of a 3-element poset:

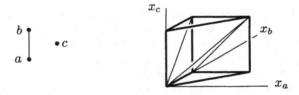

It is subdivided into 3 tetrahedra corresponding to linear extensions.

Proof of Observation 12.3.4. In (i), consider the ordering $1 \leq 2 \leq \cdots \leq n$. The characteristic vectors of up-sets have the form $(0, 0, \ldots, 0, 1, 1, \ldots, 1)$. There are $n+1$ of them, and they are affinely independent, so $P(\leq)$ is a simplex. Other linear orderings differ by a permutation of coordinates, so we get congruent simplices. The volume could be calculated directly, but it follows easily from considerations below.

As for (ii), any point $(x_1, \ldots, x_n) \in P(\preceq)$ with distinct coordinates determines a unique linear extension of $\preceq$, namely the one given by the natural ordering of its coordinates as real numbers. Conversely, for any linear extension $\leq \in E(\preceq)$, we have $P(\leq) \subseteq P(\preceq)$ by definition. Hence the congruent simplices corresponding to linear extensions subdivide $P(\preceq)$.

To see that the simplices have volume $1/n!$, take the discrete ordering (no two distinct elements are comparable) for $\preceq$. The order polytope is the unit cube $[0, 1]^n$, and it is subdivided into $n!$ congruent simplices corresponding to the $n!$ possible linear orderings.   □

Height and center of gravity. Let X be a finite set and $\leq$ a linear ordering on it. For $a \in X$, we define the *height of a* in $\leq$, denoted by $h_{\leq}(a)$, as $|\{x \in X : x \leq a\}|$. For a poset $(X, \preceq)$, the height of an element is defined as the average height over all linear extensions:

$$h_{\preceq}(a) = \frac{1}{e(\preceq)} \sum_{\leq \in E(\preceq)} h_{\leq}(a).$$

If $\preceq$ is clear from context, we omit it in the subscript and we write just $h(a)$.

The "good" elements a, b in the efficient comparison theorem can be selected using the height. Namely, we show that any two distinct a, b with $|h(a) - h(b)| < 1$ will do. (It is simple to check that if $\preceq$ is not a linear ordering, then such a and b always exist; see Exercise 1.)

We now relate the height to the order polytope.

12.3.5 Lemma. *For any n-element poset $(X, \preceq)$, the center of gravity of the order polytope $P(\preceq)$ is $c = (c_a \colon a \in X)$, where $c_a = \frac{1}{n+1} h_{\preceq}(a)$.*

Proof. The center of gravity of $P(\preceq)$ is the arithmetic average of centers of gravity of the simplices $P(\leq)$ with $\leq \in E(\preceq)$. Hence it suffices to prove the lemma for a linear ordering $\leq$. By permuting coordinates, it suffices to calculate that for the simplex with vertices of the form $(0, \ldots, 0, 1, \ldots, 1)$, the center of gravity is $\frac{1}{n+1}(1, 2, \ldots, n)$. This is left as Exercise 2. $\quad\square$

Proof of the efficient comparison theorem. Given the poset $(X, \preceq)$, we consider two elements $a, b \in X$ with $|h(a) - h(b)| < 1$. We want to show that the number of linear extensions of both $\preceq + (a, b)$ and $\preceq + (b, a)$ is at least a constant fraction of $e(\preceq)$. Consider the order polytopes $P = P(\preceq)$, $P_{\leq} = P(\preceq + (a, b))$, and $P_{\geq} = P(\preceq + (b, a))$. Geometrically, P is sliced into $P_{\leq}$ and $P_{\geq}$ by the hyperplane $h = \{x \in \mathbf{R}^n \colon x_a = x_b\}$.

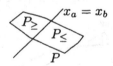

By Observation 12.3.4(ii), it suffices to show that the volumes of both $P_{\leq}$ and $P_{\geq}$ are at least a constant fraction of $\mathrm{vol}(P)$.

For convenience, let us introduce a new coordinate system in $\mathbf{R}^n$, where the first coordinate y_1 is $x_b - x_a$ and the others complete it to an orthonormal coordinate system $(y_1, \ldots, y_n)$. Hence h is the hyperplane $y_1 = 0$. Let $c(P)$ denote the center of gravity of P, and let $c_1 = c_1(P)$ be its y_1-coordinate.

What geometric information do we have about P? It is a convex body with the following properties:

- The projection of P onto the y_1-axis is the interval $[-1, 1]$. This is because there is an up-set of $\preceq$ containing a and not b, and also an up-set containing b but not a, and thus P has a vertex with $x_a = 1$, $x_b = 0$ and a vertex with $x_a = 0$, $x_b = 1$.
- We have $-\frac{1}{n+1} < c_1 < \frac{1}{n+1}$, since $c_1 = \frac{1}{n+1}(h(a) - h(b))$ and $|h(a) - h(b)| < 1$.

The proof of Theorem 12.3.1 is finished by showing that any compact convex body $P \subset \mathbf{R}^n$ with these two properties satisfies

$$\mathrm{vol}(P_\leq) \geq \frac{1}{2e}\,\mathrm{vol}(P) \quad \text{and} \quad \mathrm{vol}(P_\geq) \geq \frac{1}{2e}\,\mathrm{vol}(P),$$

where $P_\leq$ is the part of P in the half-space $\{y_1 \leq 0\}$ and $P_\geq$ is the other part.

For $t \in [-1, 1]$, let P_t be the $(n-1)$-dimensional slice of P by the hyperplane $\{y_1 = t\}$, and let $r(t)$ be the equivalent radius of P_t, i.e., the radius of an $(n-1)$-dimensional ball of volume $\mathrm{vol}_{n-1}(P_t)$. By Brunn's inequality for slice volumes (Theorem 12.2.1), $r(t)$ is concave on $[-1, 1]$.

The y_1-coordinate of the center of gravity of P can be expressed as

$$c_1(P) = \frac{1}{\mathrm{vol}(P)} \int_{-1}^{1} t\,\mathrm{vol}_{n-1}(P_t)\,\mathrm{d}t$$

(imagine P composed of thin plates perpendicular to the y_1-axis). Hence c_1 is fully determined by the function $r(t)$. In other words, the shapes of the slices of P do not really matter; only their volumes do, and so we may imagine that P is a rotational body whose slice P_t is an $(n-1)$-dimensional ball of radius $r(t)$ centered at $(t, 0, \ldots, 0)$.

We want to show that if $c_1(P) \geq -\frac{1}{n+1}$, then $\mathrm{vol}(P_\geq) \geq \frac{1}{2e}\,\mathrm{vol}(P)$. The inequality for $\mathrm{vol}(P_\leq)$ follows by symmetry. The key step is to pass to another, especially simple, rotational convex body K. The slice K_t of K has radius $\kappa(t)$; the functions $\kappa(t)$ and $r(t)$ are schematically plotted below:

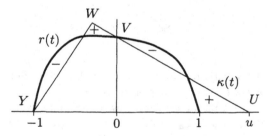

The graph of the function $\kappa(t)$ consists of two linear segments, and so K is a double cone. First we construct the function $\kappa(t)$ for t positive. Here the graph is a segment starting at the point $V = (0, r(0))$ and ending at the point $U = (u, 0)$. The number u is chosen so that $\mathrm{vol}(K_\geq) = \mathrm{vol}(P_\geq)$. Since $r(t)$ is concave and $\kappa(t)$ is linear on $[0, u]$, we have $u \geq 1$. Moreover, as t grows from 0 to 1, we first have $r(t) \geq \kappa(t)$, and then from some point on $r(t) \leq \kappa(t)$. This ensures that the center of gravity of $K_\geq$ is to the right of the center of gravity of $P_\geq$ (we can imagine that $P_\geq$ is transformed into $K_\geq$ by peeling off some mass in the region labeled "$-$" and moving it right, to the region labeled "$+$").

Next, we define $\kappa(t)$ for $t < 0$. We extend the segment UV to the left until the (unique) point W such that when YWV is the graph of $\kappa(t)$ for negative t, we have $\operatorname{vol}(K_\leq) = \operatorname{vol}(P_\leq)$. As t goes from 0 down to -1, $\kappa(t)$ is first above $r(t)$ and then below it. This is because at V, the segment WU decreases more steeply than the function $r(t)$. Therefore, we also have $c_1(K_\leq) \geq c_1(P_\leq)$, and hence $c_1(K) \geq c_1(P) \geq -\frac{1}{n+1}$. So, as was noted above, it remains to show that $\operatorname{vol}(K_\geq) \geq \frac{1}{2e} \operatorname{vol}(K)$, which is a more or less routine calculation.

We fix the notation as in the following picture:

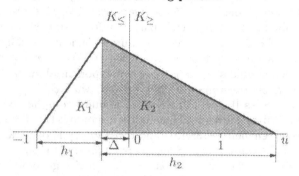

We note that $c_1(K)$ is a weighted average of $c_1(K_1)$ and $c_1(K_2)$; the weights are the volumes of K_1 and K_2 whose ratio is $h_1 : h_2$. The center of gravity of an n-dimensional cone is at $\frac{1}{n+1}$ of its height, and hence $c_1(K_1) = -\frac{h_1}{n+1} - \Delta$ and $c_1(K_2) = \frac{h_2}{n+1} - \Delta$. Therefore,

$$c_1(K) = \frac{h_1\left(-\frac{h_1}{n+1}\right) + h_2\left(\frac{h_2}{n+1}\right)}{h_1 + h_2} - \Delta = \frac{h_2 - h_1}{n+1} - \Delta.$$

We have $\Delta = 1 - h_1$, and so from the condition $c_1(K) \geq -\frac{1}{n+1}$ we obtain $h_2 + nh_1 \geq n$. We substitute $h_1 = u - h_2 + 1$ and rearrange, which yields

$$\frac{u}{h_2} \geq 1 - \frac{1}{n}. \tag{12.2}$$

We are interested in bounding $\operatorname{vol}(K_\geq)$ from below. The cone $K_\geq$ is similar to K_2, with ratio u/h_2. So

$$\operatorname{vol}(K_\geq) = \left(\frac{u}{h_2}\right)^n \operatorname{vol}(K_2) = \left(\frac{u}{h_2}\right)^n \frac{h_2}{h_1 + h_2} \operatorname{vol}(K)$$

$$= \frac{u}{u+1}\left(\frac{u}{h_2}\right)^{n-1} \operatorname{vol}(K).$$

Now we substitute for u/h_2 from (12.2), obtaining

$$\operatorname{vol}(K_\geq) \geq \frac{u}{u+1}\left(1 - \frac{1}{n}\right)^{n-1} \operatorname{vol}(K).$$

Finally, $\frac{u}{u+1} \geq \frac{1}{2}$ (as $u \geq 1$) and $(1 - \frac{1}{n})^{n-1} > e^{-1}$ for all n, so $\mathrm{vol}(K_\geq) \geq \frac{1}{2e} \mathrm{vol}(K)$ follows. $\square$

Bibliography and remarks. The statement of the efficient comparison theorem with $\delta = \frac{1}{3}$, known as the "$\frac{1}{3}$-$\frac{2}{3}$ conjecture," was conjectured by Kislitsyn [Kis68] and, later but independently, by Fredman (unpublished) and by Linial [Lin84]. In this strongest possible form it remains a challenging open problem in the theory of partially ordered sets (see Trotter [Tro92], [Tro95] for overviews of this interesting area).

The problem of sorting with partial information was considered by Fredman [Fre76], who proved that any n-element partially ordered set $(X, \preceq)$ can be sorted by at most $\log_2(e(\preceq)) + 2n$ comparisons. This is optimal unless $e(\preceq)$ is only subexponential in n. The efficient comparison theorem was first proved, with $\delta = \frac{3}{11} \approx 0.2727$, by Kahn and Saks [KS84]. Their proof is quite complicated, and instead of the Brunn–Minkowski inequality it employs the more powerful Aleksandrov–Fenchel inequality. The constant $\frac{3}{11}$ is optimal for their approach, in the sense that if a and b are elements of a poset such that $|h(a) - h(b)| < 1$, then the comparison of a and b generally need not reduce the number of linear extensions by any better ratio.

The simpler proof presented in this section is due to Kahn and Linial [KL91], and a similar one, with a slightly worse δ, was found by Karzanov and Khachiyan; see [Kha89]. The method is inspired by proofs of a result about splitting a convex body by a hyperplane passing exactly through the center of gravity (Exercise 3), proved by Grünbaum [Grü60] (see [KL91] for more remarks on the history). Observation 12.3.4, on which all the proofs of Theorem 12.3.1 are based, is from Linial [Lin84].

The current best value of $\delta = (5 - \sqrt{5})/10 \approx 0.2764$ was achieved by Brightwell, Felsner, and Trotter [BFT95]. They extend the Kahn–Saks method, and instead of two elements a and b with $|h(a) - h(b)| < 1$, they consider three elements a, b, c with $h(a) \leq h(b) \leq h(c) \leq h(a) + 2$. Interestingly, they also construct an *infinite* (countable) poset for which their value of δ is optimal (and so the natural infinite analogue of the $\frac{1}{3}$-$\frac{2}{3}$ conjecture is false). In order to formulate this result, one needs a probability measure on the set of all linear extensions of the considered poset. Their poset is *thin*, meaning that the maximum size of an antichain is bounded by a constant, and the probability measure is obtained by taking a limit over a sequence of finite intervals in the poset.

The proofs of the efficient comparison theorem do not provide an efficient algorithm for actually computing suitable elements a, b. General methods for estimating the volume of convex bodies, mentioned in Section 13.2, yield a polynomial-time randomized algorithm.

Kahn and Kim [KK95] gave a deterministic polynomial-time adaptive sorting procedure that sorts any given n-element poset $(X, \preceq)$ by $O(\log(e(\preceq)))$ comparisons. We at least mention some interesting concepts in their algorithm. Instead of the order polytope, they consider the *chain polytope*; this the convex hull of the characteristic vectors of all antichains in $(X, \preceq)$. Equivalently, it is the stable set polytope STAB(G) (see Section 12.1) of the *comparability graph G* of $(X, \preceq)$, where $G = G(\preceq) = (X, \{\{x, y\}: x \prec y \text{ or } y \prec x\})$. As was shown by Stanley [Sta86], the chain polytope has the same volume as the order polytope. The next key notion is the *entropy* of a graph. For a given graph $G = (V, E)$ and a probability distribution $p: V \to [0, 1]$ on its vertices, the entropy $H(G, p)$ can be defined as $\min_{x \in \text{STAB}(G)} \left(-\sum_{v \in V} p_v \log_2 x_v \right)$ (there are several equivalent definitions). Graph entropy was introduced by Körner [Kör73], and he and his coworkers achieved remarkable results in extremal set theory and related fields using this concept (see, e.g., Gargano, Körner, and Vaccaro [GKV94]). The entropy can be approximated in deterministic polynomial time, and the adaptive sorting algorithm of Kahn and Kim chooses the next comparison as one that increases the entropy of the comparability graph as much as possible (this need not always be an "efficient comparison" in the sense of Theorem 12.3.1).

Exercises

1. Let $(X, \preceq)$ be a finite poset. Prove that if $\preceq$ is not a linear ordering, then there always exist $a, b \in X$ with $|h(a) - h(b)| < 1$. [1]
2. Show that the center of gravity of a simplex with vertices $a_0, a_1, \ldots, a_d$ is the same as the center of gravity of its vertex set. [2]
3. Let K be a bounded convex body in $\mathbf{R}^n$, h a hyperplane passing through the center of gravity of K, and K_1 and K_2 the parts into which K is divided by h.
   (a) Prove that $\text{vol}(K_1), \text{vol}(K_2) \geq (\frac{n}{n+1})^n \text{vol}(K)$. [3]
   (b) Show that the bound in (a) cannot be improved in general. [2]

13

Volumes in High Dimension

We begin with comparing the volume of the n-dimensional cube with the volume of the unit ball inscribed in it, in order to realize that volumes of "familiar" bodies behave quite differently in high dimensions from what the 3-dimensional intuition suggests. Then we calculate that any convex polytope in the unit ball B^n whose number of vertices is at most polynomial in n occupies only a tiny fraction of B^n in terms of volume. This has interesting consequences for deterministic algorithms for approximating the volume of a given convex body: If they look only at polynomially many points of the considered body, then they are unable to distinguish a gigantic ball from a tiny polytope. Finally, we prove a classical result, John's lemma, which states that for every n-dimensional symmetric convex body K there are two similar ellipsoids with ratio $\sqrt{n}$ such that the smaller ellipsoid lies inside K and the larger one contains K. So, in a very crude scale where the ratio $\sqrt{n}$ can be ignored, each symmetric convex body looks like an ellipsoid.

Besides presenting nice and important results, this chapter could help the reader in acquiring proficiency and intuition in geometric computations, which are skills obtainable mainly by practice. Several calculations of non-trivial length are presented in detail, and while some parts do not require any great ideas, they still contain useful small tricks.

13.1 Volumes, Paradoxes of High Dimension, and Nets

In the next section we are going to estimate the volumes of various convex polytopes. Here we start, more modestly, with the volumes of the simplest bodies.

The ball in the cube. Let V_n denote the volume of the n-dimensional ball B^n of unit radius. A neat way of calculating V_n is indicated in Exercise 2;

the result, which can be verified in various other ways and found in many books of formulas, is

$$V_n = \frac{\pi^{n/2}}{\Gamma(\frac{n}{2}+1)} = \frac{\pi^{\lfloor n/2 \rfloor} 2^{\lceil n/2 \rceil}}{\prod_{i:\, 0 \le 2i < n}(n - 2i)}.$$

Here $\Gamma(x) = \int_0^\infty t^{x-1}e^{-t}\,dt$ is the usual gamma function, with $\Gamma(k+1) = k!$ for natural numbers k.

Let us compare the volume of the unit cube $[0,1]^n$ with that of the inscribed ball (of radius $\frac{1}{2}$).

(Using Exercise 1, the reader may want to add the crosspolytope inscribed in both bodies to the comparison.) For dimension $n = 3$, the volume of the ball is about 0.52, but for $n = 11$ it is already less than 10^{-3}. Using Stirling's formula, we find that it behaves roughly like $(\frac{2\pi e}{n})^{n/2}$. For large n, the inscribed ball is thus like a negligible dust particle in the cube, as far as the volume is concerned.

This can be experienced if one tries to generate random points uniformly distributed in the unit ball B^n. A straightforward method is first to generate a random point x in the cube $[-1,1]^n$, by producing n independent random numbers $x_1, x_2, \ldots, x_n \in [-1,1]$. If $\|x\| > 1$, then x is discarded and the experiment is repeated, and if $\|x\| \le 1$, then x is the desired random point in the unit ball. This works reasonably in dimensions below 10, say, but in dimension 20, we expect about 40 million discarded points for each accepted point, and the method is rather useless.

Another way of comparing the ball and the cube is to picture the sizes of the n-dimensional ball having the same volume as the unit cube:

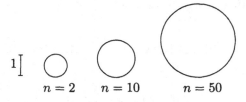

$$1\big[\qquad \underset{n=2}{\bigcirc} \qquad \underset{n=10}{\bigcirc} \qquad \underset{n=50}{\bigcirc}$$

For large n, the radius grows approximately like $0.24\sqrt{n}$. This indicates that the n-dimensional unit cube is actually quite a huge body; for example, its diameter (the length of the longest diagonal) is $\sqrt{n}$. Here is another example illustrating the largeness of the unit cube quite vividly.

Balls enclosing a ball. Place balls of radius $\frac{1}{2}$ into each of the 2^n vertices of the unit cube $[0,1]^n$ so that they touch along the edges of the cube, and consider the ball concentric with the cube and just touching the other balls:

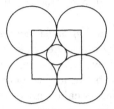

Obviously, this ball is quite small, and it is fully contained in the cube, right? No: Already for $n = 5$ it starts protruding out through the facets.

Proper pictures. If a planar sketch of a high-dimensional convex body should convey at least a partially correct intuition about the distribution of the mass, say for the unit cube, it is perhaps best to give up the convexity in the drawing! According to Milman [Mil98], a "realistic" sketch of a high-dimensional convex body might look like this:

Strange central sections: the Busemann–Petty problem. Let K and L be convex bodies in $\mathbf{R}^n$ symmetric about 0, and suppose that for every hyperplane h passing through 0, we have $\mathrm{vol}_{n-1}(K \cap h) \leq \mathrm{vol}_{n-1}(L \cap h)$. It seems very plausible that this should imply $\mathrm{vol}(K) \leq \mathrm{vol}(L)$; this conjecture of Busemann and Petty used to be widely believed (after all, it was known that if the volumes of the sections are equal for all h, then $K = L$). But as it turned out, it is true only for $n \leq 4$, while in dimensions $n \geq 5$ it can fail! In fact, for large dimensions, one of the counterexamples is the unit cube and the ball of an appropriate radius: It is known that all sections of the unit cube have volume at most $\sqrt{2}$, while in large dimensions, the unit-volume ball has sections of volume about $\sqrt{e}$.

Nets in a sphere. We conclude this section by introducing a generally useful tool. Let $S^{n-1} = \{x \in \mathbf{R}^n : \|x\| = 1\}$ denote the unit sphere in $\mathbf{R}^n$ (note that S^2 is the 2-dimensional sphere living in $\mathbf{R}^3$). We are given a number $\eta > 0$, and we want to place a reasonably small finite set N of points on S^{n-1} in such a way that each $x \in S^{n-1}$ has some point of N at distance no larger than η. Such an N is called η-*dense* in S^{n-1}. For example, the set $N = \{e_1, -e_1, \ldots, e_n, -e_n\}$ of the $2n$ orthonormal unit vectors of the standard basis is $\sqrt{2}$-dense. But it is generally difficult to find good explicit constructions for arbitrary η and n. The following simple but clever existential argument yields an η-dense set whose size has essentially the best possible order of magnitude.

Let us call a subset $N \subseteq S^{n-1}$ η-*separated* if every two distinct points of N have (Euclidean) distance greater than η. In a sense, this is opposite to being η-dense.

In order to construct a small η-dense set, we start with the empty set and keep adding points one by one. The trick is that we do not worry about η-density along the way, but we always keep the current set η-separated. Clearly, if no more points can be added, the current set must be η-dense.

The result of this algorithm is called an η-*net*.[1] That is, $N \subseteq S^{n-1}$ is an η-net if it is an inclusion-maximal η-separated subset of S^{n-1}; i.e., if N is η-separated but $N \cup \{x\}$ is not η-separated for any $x \in S^{n-1} \setminus N$. (These definitions apply to an arbitrary metric space in place of S^{n-1}.) A volume argument bounds the maximum size of an η-net.

13.1.1 Lemma (Size of η-nets in the sphere). *For each $\eta \in (0, 1]$, any η-net $N \subseteq S^{n-1}$ satisfies*

$$|N| \leq \left(\frac{4}{\eta}\right)^n.$$

Later on, we will check that for η small, no η-dense set can be much smaller (Exercise 14.1.3).

Proof. For each $x \in N$, consider the ball of radius $\frac{\eta}{2}$ centered at x. These balls are all disjoint, and they are contained in the ball $B(0, 1+\eta) \subseteq B(0, 2)$. Therefore, $\mathrm{vol}(B(0, 2)) \geq |N| \, \mathrm{vol}(B(0, \frac{\eta}{2}))$, and since $\mathrm{vol}(B(0, r))$ in $\mathbf{R}^n$ is proportional to r^n, the lemma follows. $\square$

Bibliography and remarks. Most of the material of this section is well known and standard. As for the Busemann–Petty problem, which we are not going to pursue any further in this book, information can be found, e.g., in Gardner, Koldobski, and Schlumprecht [GKS99] (recent unified solution for all dimensions), in Ball [Bal], or in the *Handbook of Convex Geometry* [GW93].

Exercises

1. Calculate the volume of the n-dimensional crosspolytope, i.e., the convex hull of $\{e_1, -e_1, \ldots, e_n, -e_n\}$, where e_i is the ith vector in the standard basis of $\mathbf{R}^n$. ②
2. (Ball volume via the Gaussian distribution)
   (a) Let $I_n = \int_{\mathbf{R}^d} e^{-\|x\|^2} \, \mathrm{d}x$, where $\|x\| = (x_1^2 + \cdots + x_n^2)^{1/2}$ is the Euclidean norm. Express I_n using I_1. ②

---

[1] Not to be confused with the notion of ε-net considered in Chapter 10; unfortunately, the same name is customarily used for two rather unrelated concepts.

(b) Express I_n using $V_n = \text{vol}(B^n)$ and a suitable one-dimensional integral, by considering the contribution to I_n of the spherical shell with inner radius r and outer radius $r + dr$. ③

(c) Calculate I_n by using (b) for $n = 2$ and (a). ①

(d) Integrating by parts, set up a recurrence and calculate the integral appearing in (b). Compute V_n. ③

This calculation appears in Pisier [Pis89] (also see Ball [Bal97]).

3. Let $X \subset S^{n-1}$ be such that every two points of X have (Euclidean) distance at least $\sqrt{2}$. Prove that $|X| \leq 2n$. ④

13.2 Hardness of Volume Approximation

The theorem in this section can be regarded as a variation on one of the "paradoxes of high dimension" mentioned in the previous section, namely, that the volume of the ball inscribed in the unit cube becomes negligible as the dimension grows. The theorem addresses a dual situation: the volume of a convex polytope inscribed in the unit ball.

13.2.1 Theorem. *Let B^n denote the unit ball in $\mathbf{R}^n$, and let P be a convex polytope contained in B^n and having at most N vertices. Then*

$$\frac{\text{vol}(P)}{\text{vol}(B^n)} \leq \left(\frac{C \ln(\frac{N}{n}+1)}{n} \right)^{n/2}.$$

with an absolute constant C.

Thus, unless the number of vertices is exponential in n, the polytope is very tiny compared to the ball.

For $N > ne^{n/C}$, the bound in the theorem is greater than 1, and so it makes little sense, since we always have $\text{vol}(P) \leq \text{vol}(B^n)$. Thus, a reasonable range of N is $n+1 \leq N \leq c^N$ for some positive constant $c > 0$. It turns out that the bound is tight in this range, up to the value of C, as discussed in the next section. This may be surprising, since the elementary proof below makes seemingly quite rough estimates.

Let us remark that the weaker bound

$$\frac{\text{vol}(P)}{\text{vol}(B^n)} \leq \left(\frac{C \ln N}{n} \right)^{n/2} \tag{13.1}$$

is somewhat easier to prove than the one in Theorem 13.2.1. The difference between these two bounds is immaterial for $N > n^2$, say. It becomes significant, for example, for comparing the largest possible volume of a polytope in B^n with $n \log n$ vertices with the volume of the largest simplex in B^n.

Application to hardness of volume approximation. Computing or estimating the volume of a given convex body in $\mathbf{R}^n$, with n large, is a fundamental algorithmic problem. Many combinatorial counting problems can be reduced to it, such as counting the number of linear extension of a given poset, as we saw in Section 12.3. Since many of these counting problems are computationally intractable, one cannot expect to compute the volume precisely, and so approximation up to some multiplicative factor is sought.

It turns out that no polynomial-time deterministic algorithm can generally achieve approximation factor better than exponential in the dimension. A concrete lower bound, derived with help of Theorem 13.2.1, is $(cn/\log n)^n$. This can also be almost achieved: An algorithm is known with factor $(c'n)^n$.

In striking contrast to this, there are *randomized* polynomial-time algorithms that can approximate the volume within a factor of $(1+\varepsilon)$ for each fixed $\varepsilon > 0$ with high probability. Here "randomized" means that the algorithm makes random decisions (like coin tosses) during its computation; it does not imply any randomness of the input. These are marvelous developments, but they are not treated in this book. We only briefly explain the relation of Theorem 13.2.1 to the deterministic volume approximation.

To understand this connection, one needs to know how the input convex body is presented to an algorithm. A general convex body cannot be exactly described by finitely many parameters, so caution is certainly necessary. One way of specifying certain convex bodies, namely, convex polytopes, is to give them as convex hulls of finite point sets (V-presentation) or as intersections of finite sets of half-spaces (H-presentation). But there are many other computationally important convex bodies that are not polytopes, or have no polynomial-size V-presentation or H-presentation. We will meet an example in Section 15.5, where the convex body lives in the space of $n \times n$ real matrices and is the intersection of a polytope with the cone consisting of all positive semidefinite matrices.

In order to abstract the considerations from the details of the presentation of the input body, the *oracle model* was introduced for computation with convex bodies. If $K \subset \mathbf{R}^n$ is a convex body, a *membership oracle* for K is, roughly speaking, an algorithm (subroutine, black box) that for any given input point $x \in \mathbf{R}^n$ outputs YES if $x \in K$ and NO if $x \notin K$.

This is simplified, because in order to be able to compute with the body, one needs to assume more. Namely, K should contain a ball $B(0,r)$ and be contained in a ball $B(0,R)$, where R and $r > 0$ are written using at most polynomially many digits. On the other hand, the oracle need not (and often cannot) be exact, so a wrong answer is allowed for points very close to the boundary. These are important but rather technical issues, and we will ignore them. Let us note that a polynomial-time membership oracle can be constructed for both V-presented and H-presented polytopes, as well as for many other bodies.

Let us now assume that a deterministic algorithm approximates the volume of each convex body given by a suitable membership oracle. First we call the algorithm with $K = B^n$, the unit ball. The algorithm asks the oracle about some points $\{x_1, x_2, \ldots, x_N\}$, gets the correct answers, and outputs an estimate for $\mathrm{vol}(B^n)$. Next, we call the algorithm with the body $K = \mathrm{conv}(\{x_1, x_2, \ldots, x_N\} \cap B^n)$. The answers of the oracle are exactly the same, and since the algorithm has no other information about the body K and it is deterministic, it has to output the same volume estimate as it did for B^n. But by Theorem 13.2.1, $\mathrm{vol}(B^n)/\mathrm{vol}(K) \geq (cn/\ln(N/n+1))^{n/2}$, and so the error of the approximation must be at least this factor. If N, the number of oracle calls, is polynomial in n, it follows that the error is at least $(c'n/\log n)^{n/2}$.

By more refined consideration, one can improve the lower bound to approximately the square of the quantity just given. The idea is to input the dual body K^* into the algorithm, too, for which it gets the same answers, and then use a deep result (the inverse Blaschke–Santaló inequality) stating that $\mathrm{vol}(K)\,\mathrm{vol}(K^*) \geq c^n/n!$ for any centrally symmetric n-dimensional convex body K, with an absolute constant $c > 0$ (some technical steps are omitted here). This improvement is interesting because, as was remarked above, for symmetric convex bodies it almost matches the performance of the best known algorithm.

Idea of the proof of Theorem 13.2.1. Let V be the set of vertices of the polytope $P \subset B^n$, $|V| = N$. We choose a suitable parameter $k < n$ and prove that for every $x \in P$, there is a k-tuple J of points of V such that x is close to $\mathrm{conv}(J)$. Then $\mathrm{vol}(P)$ is simply estimated as $\binom{N}{k}$ times the maximum possible volume of the appropriate neighborhood of the convex hull of k points in B^n. Here is the first step towards realizing this program.

13.2.2 Lemma. *Let S in $\mathbf{R}^n$ be an n-dimensional simplex, i.e., the convex hull of $n+1$ affinely independent points, and let $R = R(S)$ and $\rho = \rho(S)$ be the circumradius and inradius of S, respectively, that is, the radius of the smallest enclosing ball and of the largest inscribed ball. Then $\frac{R}{\rho} \geq n$.*

Proof. We first sketch the proof of an auxiliary claim: *Among all simplices contained in B^n, the regular simplex inscribed in B^n has the largest volume.* The volume of a simplex is proportional to the $(n-1)$-dimensional volume of its base times the corresponding height. It follows that in a maximum-volume simplex S inscribed in B^n, the hyperplane passing through a vertex v of S and parallel to the facet of S not containing v is tangent to B^n, for otherwise, v could be moved to increase the height:

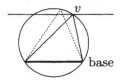

It can be easily shown (Exercise 2) that this property characterizes the regular simplex (so the regular simplex is even the unique maximum).

Another, slightly more difficult, argument shows that if S is a simplex of minimum volume circumscribed about B^n, then each facet of S touches B^n at its center of gravity (Exercise 3), and it follows that the volume is minimized by the regular simplex circumscribed about B^n.

Let $S_0 \subset B^n$ be a simplex. We consider two auxiliary regular simplices S_1 and S_2, where S_1 is inscribed in B^n and S_2 satisfies $\mathrm{vol}(S_2) = \mathrm{vol}(S_0)$. Since $\mathrm{vol}(S_1) \geq \mathrm{vol}(S_0) = \mathrm{vol}(S_2)$, S_1 is at least as big as S_2, and so $\rho(S_0) \leq \rho(S_2) \leq \rho(S_1)$. A calculation shows that $\rho(S_1) = \frac{1}{n}$ (Exercise 1(a)). $\square$

Let F be a j-dimensional simplex in $\mathbf{R}^n$. We define the *orthogonal ρ-neighborhood F_ρ* of F as the set of all $x \in \mathbf{R}^n$ for which there is a $y \in F$ such that the segment xy is orthogonal to F and $\|x - y\| \leq \rho$. The next drawing shows orthogonal neighborhoods in $\mathbf{R}^3$ of a 1-simplex and of a 2-simplex:

 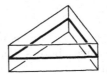

The orthogonal ρ-neighborhood of F can be expressed as the Cartesian product of F with a ρ-ball of dimension $n-j$, and so $\mathrm{vol}_n(F_\rho) = \mathrm{vol}_j(F) \cdot \rho^{n-j} \cdot \mathrm{vol}_{n-j}(B^{n-j})$.

13.2.3 Lemma. *Let S be an n-dimensional simplex contained in B^n, let $x \in S$, and let k be an integer parameter, $1 \leq k \leq n$. Then there is a k-tuple J of affinely independent vertices of S such that x lies in the orthogonal ρ-neighborhood of $\mathrm{conv}(J)$, where*

$$\rho = \rho(n, k) = \left(\sum_{i=k}^{n} \frac{1}{i^2} \right)^{1/2}.$$

Proof. We proceed by induction on $n - k$. For $n = k$, this is Lemma 13.2.2: Consider the largest ball centered at x and contained in S; it has radius at most $\frac{1}{n}$, it touches some facet F of S at a point y, and the segment xy is perpendicular to F, witnessing $x \in F_{1/n}$.

For $k < n$, using the case $k = n$, let S' be a facet of S and $x' \in S'$ a point at distance at most $\frac{1}{n}$ from S' with $xx' \perp S'$. By the inductive assumption, we find a $(k-1)$-face F of S' and a point $y \in F$ with $\|x' - y\| \leq \rho(n-1, k)$ and $x'y \perp F$. Here is an illustration for $n = 3$ and $k = 2$:

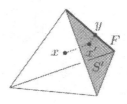

Then $xx' \perp x'y$ (because the whole of S' is perpendicular to xx'), and so $\|x - y\|^2 = \|x - x'\|^2 + \|x' - y\|^2 \leq \rho(n, k)^2$. Finally, $xy \perp F$, since both the vectors $x' - y$ and $x - x'$ lie in the orthogonal complement of the linear subspace generated by $F - y$. $\square$

Proof of Theorem 13.2.1. By Carathéodory's theorem and Lemma 13.2.3, $P = \text{conv}(V)$ is covered by the union of all the orthogonal ρ-neighborhoods $\text{conv}(J)_\rho$, $J \in \binom{V}{k}$, where $\rho = \rho(n, k)$ is as in the lemma. The maximum $(k-1)$-dimensional volume of $\text{conv}(J)$ is no larger than the $(k-1)$-dimensional volume of the regular $(k-1)$-simplex inscribed in B^{k-1}, which is

$$M(k-1) = \left(\frac{k}{k-1} \right)^{(k-1)/2} \frac{\sqrt{k}}{(k-1)!} ;$$

see Exercise 1(b). (If we only want to prove the weaker estimate (13.1) and do not care about the value of C, then $M(k-1)$ can also be trivially estimated by $\text{vol}_{k-1}(B^{k-1})$ or even by 2^{k-1}.)

What remains is calculation. We have

$$\frac{\text{vol}(P)}{\text{vol}(B^n)} \leq \binom{N}{k} \cdot M(k-1) \cdot \rho(n, k)^{n-k+1} \cdot \frac{\text{vol}_{n-k+1}(B^{n-k+1})}{\text{vol}(B^n)} . \qquad (13.2)$$

We first estimate

$$\rho(n, k)^2 = \sum_{i=k}^{n} \frac{1}{i^2} \leq \sum_{i=k}^{n} \frac{1}{i(i-1)} = \sum_{i=k}^{n} \left(\frac{1}{i-1} - \frac{1}{i} \right) = \frac{1}{k-1} - \frac{1}{n} \leq \frac{1}{k-1} .$$

We now set

$$k = \left\lfloor \frac{n}{\ln(\frac{N}{n}+1)} \right\rfloor$$

(for obtaining the weaker estimate (13.1), the simpler value $k = \lfloor \frac{n}{\ln N} \rfloor$ is more convenient). We may assume that $\ln N$ is much smaller than n, for otherwise, the bound in the theorem is trivially valid, and so k is larger than any suitable constant. In particular, we can ignore the integer part in the definition of k.

For estimating the various terms in (13.2), it is convenient to work with the natural logarithm of the quantities. The logarithm of the bound we are heading for is $\frac{n}{2}(\ln\ln(\frac{N}{n}+1) - \ln n + O(1))$, and so terms up to $O(n)$ can be ignored if we do not care about the value of the constant C. Further, we find that $k \ln k = k \ln n - k \ln\ln(\frac{N}{n}+1) = k \ln n + O(n)$. This is useful for estimating $\ln(k!) = k \ln k - O(k) = k \ln n - O(n)$.

Now, we can bound the logarithms of the terms in (13.2) one by one. We have $\ln \binom{N}{k} \leq k \ln N - \ln(k!) = k(\ln(\frac{N}{n}) + \ln n) - \ln(k!) \leq n + k \ln n - k \ln n + O(n) = O(n)$; this term is negligible. Next, $\ln M(k-1)$ contributes about $-\ln(k!) = -k \ln n + O(n)$. The main contribution comes from the term $\ln \rho(n, k)^{n-k+1} \leq -(n-k)\ln\sqrt{k} + O(n) = \frac{n}{2}(-\ln n + \ln\ln(\frac{N}{n}+1)) + \frac{k}{2}\ln n +$

$O(n)$. Finally $\ln(\mathrm{vol}_{n-k+1}(B^{n-k+1})/\mathrm{vol}(B^n)) = \ln(\Gamma(\frac{n}{2}+1)/\Gamma(\frac{n-k+1}{2}+1)) + O(n) \leq \ln n^{k/2} + O(n) = \frac{k}{2}\ln n + O(n)$. The term $-k\ln n$ originating from $M(k-1)$ cancels out nicely with the two terms $\frac{k}{2}\ln n$, and altogether we obtain $\frac{n}{2}(-\ln n + \ln\ln(\frac{N}{n}+1) + O(1))$ as claimed in the theorem.   □

Bibliography and remarks. Our presentation of Theorem 13.2.1 mostly follows Bárány and Füredi [BF87]. They pursued the hardness of deterministic volume approximation, inspired by an earlier result of Elekes [Ele86] (see Exercise 5). They proved the weaker bound (13.1); the stronger bound in Theorem 13.2.1, in a slightly different form, was obtained in their subsequent paper [BF88].

Theorem 13.2.1 was also derived by Carl and Pajor [CP88] from a work of Carl [Car85] (they provide similar near-tight bounds for ℓ_p-balls).

A dual version of Theorem 13.2.1 was independently discovered by Gluskin [Glu89] and by Bourgain, Lindenstrauss, and Milman [BLM89]. The dual setting deals with the minimum volume of the intersection of N symmetric slabs in $\mathbf{R}^n$. Namely, let $u_1, u_2, \ldots, u_N \in \mathbf{R}^n$ be given (nonzero) vectors, and let $K = \bigcap_{i=1}^{N}\{x \in \mathbf{R}^n : |\langle u_i, x\rangle| \leq 1\}$ (the width of the ith slab is $\frac{1}{\|u_i\|}$). The dual analogue of Theorem 13.2.1 is this: Whenever all $\|u_i\| \leq 1$, we have $\mathrm{vol}(B^n)/\mathrm{vol}(K) \leq \left(\frac{C}{n}\ln(\frac{N}{n}+1)\right)^{n/2}$. A short and beautiful proof can be found in Ball's handbook chapter [Bal]. There are also bounds based on the sum of norms of the u_i. Namely, for all $p \in [1, \infty)$, we have $\mathrm{vol}(K)^{1/n} \geq \frac{2}{\sqrt{p/2} \cdot R}$, where

$R = \left(\frac{1}{n}\sum_{i=1}^{N}\|u_i\|^p\right)^{1/p}$ (Euclidean norms!), as was proved by Ball and Pajor [BP90]; it also follows from Gluskin's work [Glu89]. For $p = 2$, this result was established earlier by Vaaler. It has the following nice reformulation: The intersection of the cube $[-1, 1]^N$ with any n-flat through 0 has n-dimensional volume at least 2^n (see [Bal] for more information and related results).

The setting with slabs and that of Theorem 13.2.1 are connected by the *Blaschke–Santaló inequality*[2] and the *inverse Blaschke–Santaló inequality*. The former states that $\mathrm{vol}(K)\,\mathrm{vol}(K^*) \leq \mathrm{vol}(B^n)^2 \leq c_1^n/n!$ for every centrally symmetric convex body in $\mathbf{R}^n$ (or, more generally, for every convex body K having 0 as the center of gravity). It allows one the passage from the setting with slabs to the setting of Theorem 13.2.1: If the intersection of the slabs $\{x: |\langle u_i, x\rangle| \leq 1\}$ has large volume, then $\mathrm{conv}\{u_1, \ldots, u_N\}$ has small volume. The inverse Blaschke–Santaló inequality, as was mentioned in the text, asserts that $\mathrm{vol}(K)\,\mathrm{vol}(K^*) \geq c^n/n!$ for a suitable $c > 0$, and it can thus be used

---

[2] In the literature one often finds it as either Blaschke's inequality or Santaló's inequality. Blaschke proved it for $n \leq 3$ and Santaló for all n; see, e.g., the chapter by Lutwak in the *Handbook of Convex Geometry* [GW93].

for the reverse transition. It is much more difficult than the Blaschke–Santaló inequality and it was proved by Bourgain and Milman; see, e.g., [Mil98] for discussion and references.

Let us remark that the weaker bound $\left(\frac{C}{n}(\ln N)\right)^{n/2}$ is relatively easy to prove in the dual setting with slabs (Exercise 14.1.4), which together with the Blaschke–Santaló inequality gives (13.1).

Theorem 13.2.1 concerns the situation where $\mathrm{vol}(P)$ is small compared to $\mathrm{vol}(B^n)$. The smallest number of vertices of P such that $\mathrm{vol}(P) \geq (1-\varepsilon)\,\mathrm{vol}(B^n)$ for a small $\varepsilon > 0$ was investigated by Gordon, Reisner, and Schütt [GRS97]. In an earlier work they constructed polytopes with N vertices giving $\varepsilon = O(nN^{-2/(n-1)})$, and in the paper mentioned they proved that this is asymptotically optimal for $N \geq (Cn)^{(n-1)/2}$, with a suitable constant C.

The oracle model for computation with convex bodies was introduced by Grötschel, Lovász, and Schrijver [GLS88]. A deterministic polynomial-time algorithm approximating the volume of a convex body given by a suitable oracle (weak separation oracle) achieving the approximation factor $n!(1+\varepsilon)$, for every $\varepsilon > 0$, was given by Betke and Henk [BH93] (the geometric idea goes back at least to Macbeath [Mac50]). The algorithm chooses an arbitrary direction v_1 and finds the supporting hyperplanes h_1^+ and h_1^- of K perpendicular to v_1. Let p_1^+ and p_1^- be contact points of h_1^+ and h_1^- with K. The next direction v_2 is chosen perpendicular to the affine hull of $\{p_1^+, p_1^-\}$, etc.

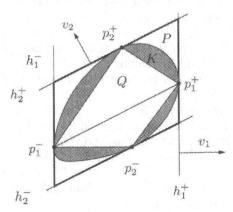

After n steps, the n pairs of hyperplanes determine a parallelotope $P \supseteq K$, while $Q = \mathrm{conv}\{p_1^+, p_1^-, \ldots, p_n^+, p_n^-\} \subseteq K$, and it is not hard to show that $\mathrm{vol}(P)/\mathrm{vol}(Q) \leq n!$ (the extra factor $(1+\varepsilon)^n$ arises because the oracle is not exact).

The first polynomial-time randomized algorithm for approximating the volume with arbitrary precision was discovered by Dyer, Frieze, and Kannan [DFK91]. Its parameters have been improved many times since then; see, e.g., Kannan, Lovász, and Simonovits [KLS97]. A recent success of these methods is a polynomial-time approximation algorithm for the permanent of a nonnegative matrix by Jerrum, Sinclair, and Vigoda [JSV01].

By considerations partially indicated in Exercise 4, Bárány and Füredi [BF87] showed that in deterministic polynomial time one cannot approximate the width of a convex body within a factor better than $\Omega(\sqrt{n/\log n})$. Brieden, Gritzmann, Kannan, Klee, Lovász, and Simonovits [BGK$^+$99] provided a matching upper bound (up to a constant), and they showed that in this case even randomized algorithms are not more powerful. They also considered a variety of other parameters of the convex body, such as diameter, inradius, and circumradius, attaining similar results and improving many previous bounds from [GLS88].

Lemma 13.2.2 appears in Fejes Tóth [Tót65].

Exercises

1. (a) Calculate the inradius and circumradius of a regular n-dimensional simplex. [3]
   (b) Calculate the volume of the regular n-dimensional simplex inscribed in the unit ball B^n. [2]
2. Suppose that the vertices of an n-dimensional simplex S lie on the sphere S^{n-1} and for each vertex v, the hyperplane tangent to S^{n-1} at v is parallel to the facet of S opposite to v. Check that S is regular. [2]
3. Let $S \subset \mathbf{R}^n$ be a simplex circumscribed about B^n and let F be a facet of S touching B^n at a point c. Show that if c is not the center of gravity of F, then there is another simplex S' (arising by slightly moving the hyperplane that determines the facet F) that contains B^n and has volume smaller than $\mathrm{vol}(S)$. [4]
4. The *width* of a convex body K is the minimum distance of two parallel hyperplanes such that K lies between them. Prove that the convex hull of N points in B^n has width at most $O(\sqrt{(\ln N)/n})$. [3]
5. (A weaker but simpler estimate) Let $V \subset \mathbf{R}^n$ be a finite set. Prove that $\mathrm{conv}(V) \subseteq \bigcup_{v \in V} B(\frac{1}{2}v, \frac{1}{2}\|v\|)$, where $B(x, r)$ is the ball of radius r centered at x. Deduce that the convex hull of N points contained in B^n has volume at most $\frac{N}{2^n} \mathrm{vol}(B^n)$. [4]
   This is essentially the argument of Elekes [Ele86].

13.3 Constructing Polytopes of Large Volume

For all N in the range $2n \leq N \leq 4^n$, we construct a polytope $P \subset B^n$ with N vertices containing a ball of radius $r = \Omega(((\ln \frac{N}{n})/n)^{1/2})$. This shows that the bound in Theorem 13.2.1 is tight for $N \geq 2n$, since $\mathrm{vol}(P)/\mathrm{vol}(B^n) \geq r^n$. We begin with two extreme cases.

First we construct a k-dimensional polytope $P_0 \subset B^k$ with 4^k vertices containing the ball $\frac{1}{2}B^k$. There are several possible ways; the simplest is based on η-nets. We choose a 1-net $V \subset S^{k-1}$ and set $P_0 = \mathrm{conv}(V)$. According to Lemma 13.1.1, we have $N = |V| \leq 4^k$. If there were an x with $\|x\| = \frac{1}{2}$ not lying in P_0,

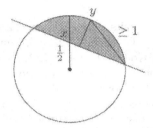

then the separating hyperplane passing through x and avoiding P_0 would define a cap (shaded) whose center y would be at distance at least 1 from V.

Another extreme case is with $N = 2q$ vertices in dimension $n = q$. Then we can take the crosspolytope, i.e., the convex hull of the vectors $e_1, -e_1, \ldots, e_q, -e_q$, where $(e_1, \ldots, e_q)$ is the standard orthonormal basis. The radius of the inscribed ball is $r = \frac{1}{\sqrt{q}}$, which matches the asserted formula.

Next, suppose that $n = qk$ for integers q and k and set $N = q4^k$. From $N = q4^k = \frac{n}{k}4^k$ we have $\frac{N}{n} = 4^k/k \geq e^k$ and $k \leq \ln \frac{N}{n}$, and so $q \geq n/\ln \frac{N}{n}$. Hence it suffices to construct an N-vertex polytope $P \subset B^n$ containing the ball rB^n with $r = \frac{1}{2\sqrt{q}}$.

The construction of P is a combination of the two constructions above. We interpret $\mathbf{R}^n$ as the product $\mathbf{R}^k \times \mathbf{R}^k \times \cdots \times \mathbf{R}^k$ (q factors). In each of the copies of $\mathbf{R}^k$, we choose a polytope P_0 with 4^k vertices as above, and we let P be the convex hull of their union. More formally,

$$P = \mathrm{conv}\Big\{ \underbrace{(0,0,\ldots,0}_{(i-1)k\times}, x_1, x_2, \ldots, x_k, 0, 0, \ldots, 0) \colon (x_1, \ldots, x_k) \in V,$$

$$i = 1, 2, \ldots, q \Big\},$$

where V is the vertex set of P_0.

We want to show that P contains the ball rB^n, $r = \frac{1}{2\sqrt{q}}$. Let x be a point of norm $\|x\| \leq r$ and let $x^{(i)}$ be the vector obtained from x by retaining the coordinates in the ith block, i.e., in positions $(i-1)k+1, \ldots, ik$, and setting all the other coordinates to 0. These $x^{(i)}$ are pairwise orthogonal, and x lies in the q-dimensional subspace spanned by them. Let $y^{(i)} = \frac{x^{(i)}}{2\|x^{(i)}\|}$ be the vector of length $\frac{1}{2}$ in the direction of $x^{(i)}$. Each $y^{(i)}$ is contained in P, since P_0 contains the ball of radius $\frac{1}{2}$. The convex hull of the $y^{(i)}$ is a q-dimensional crosspolytope of circumradius $\frac{1}{2}$, and so it contains all vectors of norm $\frac{1}{2\sqrt{q}}$ in the subspace spanned by the $x^{(i)}$, including x.

This construction assumes that n and N are of a special form, but it is not difficult to extend the bounds to all $n \geq 2$ and all N in the range $2n \leq N \leq 4^n$ by monotonicity considerations; we omit the details. This proves that the bound in Theorem 13.2.1 is tight up to the value of the constant C for $2n \leq N \leq 4^n$. $\qquad\qquad\square$

Bibliography and remarks. Several proofs are known for the lower bound almost matching Theorem 13.2.1 (Bárány and Füredi [BF87], Carl and Pajor [CP88], Kochol [Koc94]). In Bárány and Füredi [BF87], the appropriate polytope is obtained essentially as the convex hull of N random points on S^{n-1} (for technical reasons, d special vertices are added), and the volume estimate is derived from an exact formula for the expected surface measure of the convex hull of N random points on S^{n-1} due to Buchta, Müller, and Tichy [BMT95].

The idea of the beautifully simple construction in the text is due to Kochol [Koc94]. His treatment of the basic case with exponentially large N is different, though: He takes points of a suitably scaled integer lattice contained in B^k for V, which yields an efficient construction (unlike the argument with a 1-net used in the text, which is only existential).

Exercises

1. (Polytopes in B^n with polynomially many facets)

   (a) Show that the cube inscribed in the unit ball B^n, which is a convex polytope with $2n$ facets, has volume of a larger order of magnitude than any convex polytope in B^n with polynomially many vertices (and so, concerning volume, "facets are better than vertices"). [2]

   (b) Prove that the inradius of any convex polytope with N facets contained in B^n is at most $O\left(\sqrt{(\ln(N/n+1))/n}\right)$ (and so, in this respect, facets are not better than vertices). [3]

   These observations are from Brieden and Kochol [BK00].

13.4 Approximating Convex Bodies by Ellipsoids

One of the most important issues in the life of convex bodies is their approximation by ellipsoids, since ellipsoids are in many respects the simplest imaginable compact convex bodies. The following result tells us how well they can generally be approximated (or how badly, depending on the point of view).

13.4.1 Theorem (John's lemma). *Let $K \subset \mathbf{R}^n$ be a bounded closed convex body with nonempty interior. Then there exists an ellipsoid E_{in} such that*

$$E_{\text{in}} \subseteq K \subseteq E_{\text{out}},$$

where E_{out} is E_{in} expanded from its center by the factor n. If K is symmetric about the origin, then we have the improved approximation

$$E_{\text{in}} \subseteq K \subseteq E_{\text{out}} = \sqrt{n} \cdot E_{\text{in}}.$$

Thus, K can be approximated from outside and from inside by similar ellipsoids with ratio $1 : n$, or $1 : \sqrt{n}$ for the centrally symmetric case. Both these ratios are the best possible in general, as is shown by K being the regular simplex in the general case and the cube in the centrally symmetric case.

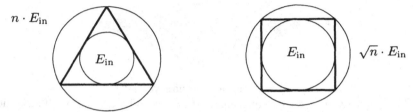

In order to work with ellipsoids, we need a rigorous definition. A suitable one is to consider ellipsoids as *affine images of the unit ball*: If B^n denotes the unit ball in $\mathbf{R}^n$, an ellipsoid E is a set $E = f(B^n)$, where $f \colon \mathbf{R}^n \to \mathbf{R}^n$ is an affine map of the form $f \colon x \mapsto Ax + c$. Here x is regarded as a column vector, $c \in \mathbf{R}^n$ is a translation vector, and A is a nonsingular $n \times n$ matrix. A very simple case is that of $c = 0$ and A a diagonal matrix with positive entries $a_1, a_2, \ldots, a_n$ on the diagonal. Then

$$E = \left\{ x \in \mathbf{R}^n \colon \frac{x_1^2}{a_1^2} + \frac{x_2^2}{a_2^2} + \cdots + \frac{x_n^2}{a_n^2} \leq 1 \right\}, \qquad (13.3)$$

as is easy to check; this is an ellipsoid with center at 0 and with semiaxes $a_1, a_2, \ldots, a_n$. In this case we have $\text{vol}(E) = a_1 a_2 \cdots a_n \cdot \text{vol}(B^n)$. An arbitrary ellipsoid E can be brought to this form by a suitable translation and rotation about the origin. In the language of linear algebra, this corresponds to diagonalizing a positive definite matrix using an orthonormal basis consisting of its eigenvectors; see Exercise 1.

Proof of Theorem 13.4.1. In both cases in the theorem, E_{in} is chosen as an ellipsoid of the largest possible volume contained in K. Easy compactness considerations show that a maximum-volume ellipsoid exists. In fact, it is also unique, but we will not prove this. (Alternatively, the proof can be done starting with the smallest-volume ellipsoid enclosing K, but this has some technical disadvantages. For example, its existence is not so obvious.)

We prove only the centrally symmetric case of John's lemma. The non-symmetric case follows the same idea, but the calculations are different and more complicated, and we leave them to Exercise 2.

So we suppose that K is symmetric about 0, and we fix the ellipsoid E_{in} of maximum volume contained in K. It is easily seen that E_{in} can be assumed to be symmetric, too. We make a linear transformation so that E_{in} becomes the unit ball B^n. Assuming that the enlarged ball $\sqrt{n} \cdot B^n$ does not contain K, we derive a contradiction by exhibiting an ellipsoid $E' \subseteq K$ with $\mathrm{vol}(E') > \mathrm{vol}(B^n)$.

We know that there is a point $x \in K$ with $\|x\| > \sqrt{n}$. For convenience, we may suppose that $x = (s, 0, 0, \ldots, 0)$, $s > \sqrt{n}$. To finish the proof, we check that the region $R = \mathrm{conv}(B^n \cup \{-x, x\})$

contains an ellipsoid E' of volume larger than $\mathrm{vol}(B^n)$.

The calculation is a little unpleasant but not so bad, after all. The region R is a rotational body; all the sections by hyperplanes perpendicular to the x_1-axis are balls. We naturally also choose E' with this property: The semiaxis in the x_1-direction is some $a > 1$, while the slice with the hyperplane $\{x_1 = 0\}$ is a ball of a suitable radius $b < 1$. We have $\mathrm{vol}(E') = ab^{n-1}\,\mathrm{vol}(B^n)$, and so we want to choose a and b such that $ab^{n-1} > 1$ and $E' \subseteq R$. By the rotational symmetry, it suffices to consider the planar situation and make sure that the ellipsis with semiaxes a and b is contained in the planar region depicted above.

In order to avoid direct computation of a tangent to the ellipsis, we multiply the x_1-coordinate of all points by the factor $\frac{b}{a}$. This turns our ellipsis into the dashed ball of radius b:

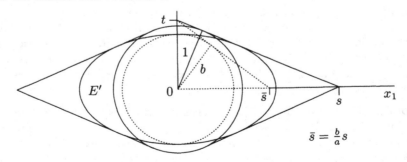

$$\bar{s} = \frac{b}{a}s$$

A bit of trigonometry yields

$$t = \frac{s}{\sqrt{s^2 - 1}}, \quad b = \frac{\bar{s}t}{\sqrt{\bar{s}^2 + t^2}} = \frac{bs}{\sqrt{b^2 s^2 + a^2 - b^2}}.$$

This leads to $a^2 = s^2(1 - b^2) + b^2$. We now choose b just a little smaller than 1; a suitable parameterization is $b^2 = 1 - \varepsilon$ for a small $\varepsilon > 0$. We want to show that $ab^{n-1} > 1$, and for convenience, we work with the square. We have

$$a^2 b^{2(n-1)} = (1 + \varepsilon(s^2 - 1))(1 - \varepsilon)^{n-1}.$$

The Maclaurin series of the right-hand side in the variable ε is $1 + (s^2 - n)\varepsilon + O(\varepsilon^2)$. Since $s^2 > n$, the expression indeed exceeds 1 for all sufficiently small $\varepsilon > 0$. Theorem 13.4.1 is proved. □

Bibliography and remarks. Theorem 13.4.1 was obtained by John [Joh48]. He actually proved a stronger statement, which can be quite useful in many applications. Roughly speaking, it says that the maximum-volume inscribed ellipsoid has many points of contact with K that "fix" it within K. The statement and proof are nicely explained in Ball [Bal97].

As was remarked in the text, the maximum-volume ellipsoid contained in K is unique. The same is true for the minimum-volume enclosing ellipsoid of K; a proof of the latter fact is outlined in Exercise 3. The uniqueness was proved independently by several authors, and the oldest such results seem to be due to Löwner (see Danzer, Grünbaum, and Klee [DGK63] for references). The minimum-volume enclosing ellipsoid is sometimes called the *Löwner–John ellipsoid*, but in other sources the same name refers to the maximum-volume inscribed ellipsoid.

The exact computation of the smallest enclosing ellipsoid for a given convex body K is generally hard. For example, it is NP-hard to compute the smallest enclosing ellipsoid of a given finite set if the dimension is a part of input (there are linear-time algorithms for every fixed dimension; see, e.g., Matoušek, Sharir, and Welzl [MSW96]). But under suitable algorithmic assumptions on the way that a convex body K is given (weak separation oracle), it is possible to compute in polynomial time an enclosing ellipsoid such that its shrinking by a factor of roughly $n^{3/2}$ (roughly n in the centrally symmetric case) is contained in K (if K is given as an H-polytope, then these factors can be improved to the nearly worst-case optimal $n+1$ and $\sqrt{n+1}$, respectively). Finding such approximating ellipsoids is a basic subroutine in other important algorithms; see Grötschel, Lovász, and Schrijver [GLS88] for more information.

There are several other significant ellipsoids associated with a given convex body that approximate it in various ways; see, e.g., Lindenstrauss and Milman [LM93] and Tomczak-Jaegermann [TJ89].

Exercises

1. Let E be the ellipsoid $f(B^n)$, where $f: x \mapsto Ax$ for an $n \times n$ nonsingular matrix A.
   (a) Show that $E = \{x \in \mathbf{R}^n: x^T B x \leq 1\}$. What is the matrix B? [2]
   (b) Recall or look up appropriate theorems in linear algebra showing that there is an orthonormal matrix T such that $B' = TBT^{-1}$ is a diagonal matrix with the eigenvalues of B on the diagonal (check and use the fact that B is positive definite in our case). [2]
   (c) What is the geometric meaning of T, and what is the relation of the entries of TBT^{-1} to the semiaxes of the ellipsoid E? [2]
2. Prove the part of Theorem 13.4.1 dealing with not necessarily symmetric convex bodies. [5]
3. (Uniqueness of the smallest enclosing ellipsoid) Let $X \subset \mathbf{R}^n$ be a bounded set that is not contained in a hyperplane (i.e., it contains $n+1$ affinely independent points). Let $\mathcal{E}(X)$ be the set of all ellipsoids in $\mathbf{R}^n$ containing X.
   (a) Prove that there exists an $E_0 \in \mathcal{E}(X)$ with $\mathrm{vol}(E_0) = \inf\{\mathrm{vol}(E): E \in \mathcal{E}(X)\}$. (Show that the infimum can be taken over a suitable compact subset of $\mathcal{E}(X)$.) [4]
   (b) Let E_1, E_2 be ellipsoids in $\mathbf{R}^n$; check that after a suitable affine transformation of coordinates, we may assume that $E_1 = \{x \in \mathbf{R}^n: \sum_{i=1}^n \frac{x_i^2}{a_i^2} \leq 1\}$ and $E_2 = \{x \in \mathbf{R}^n: \|x - c\|^2 \leq 1\}$. Define $E = \{x \in \mathbf{R}^n: \frac{1}{2}\sum_{i=1}^n \frac{x_i^2}{a_i^2} + \frac{1}{2}\sum_{i=1}^n (x_i - c_i)^2 \leq 1\}$. Verify that $E_1 \cap E_2 \subseteq E$, that E is an ellipsoid, and that $\mathrm{vol}(E) \geq \min(\mathrm{vol}(E_1), \mathrm{vol}(E_2))$, with equality only if $E_1 = E_2$. Conclude that the smallest-volume enclosing ellipsoid of X is unique. [4]
4. (Uniqueness of the smallest enclosing ball)
   (a) In analogy with Exercise 3, prove that for every bounded set $X \subset \mathbf{R}^n$, there exists a unique minimum-volume ball containing X. [3]
   (b) Show that if $X \subset \mathbf{R}^n$ is finite then the smallest enclosing ball is determined by at most $n+1$ points of X; that is, there exists an at most $(n+1)$-point subset of X whose smallest enclosing ball is the same as that of X. [3]
5. (a) Let $P \subset \mathbf{R}^2$ be a convex polygon with n vertices. Prove that there are three consecutive vertices of P such that the area of their convex hull is at most $O(n^{-3})$ times the area of P. [4]
   (b) Using (a) and the fact that every triangle with vertices at integer points has area at least $\frac{1}{2}$ (check!), prove that every convex n-gon with integral vertices has area $\Omega(n^3)$. [3]

   Remark. Rényi and Sulanke [RS64] proved that the worst case in (a) is the regular convex n-gon.

14
Measure Concentration and Almost Spherical Sections

In the first two sections we are going to discuss measure concentration on a high-dimensional unit sphere. Roughly speaking, measure concentration says that if $A \subseteq S^{n-1}$ is a set occupying at least half of the sphere, then almost all points of S^{n-1} are quite close to A, at distance about $O(n^{-1/2})$. Measure concentration is an extremely useful technical tool in high-dimensional geometry. From the point of view of probability theory, it provides tail estimates for random variables defined on S^{n-1}, and in this respect it resembles Chernoff-type tail estimates for the sums of independent random variables. But it is of a more general nature, more like tail estimates for Lipschitz functions on discrete spaces obtained using martingales.

The second main theme of this chapter is almost-spherical sections of convex bodies. Given a convex body $K \subset \mathbf{R}^n$, we want to find a k-dimensional subspace L of $\mathbf{R}^n$ such that $K \cap L$ is almost spherical; i.e., it contains a ball of some radius r and is contained in the concentric ball of radius $(1+\varepsilon)r$. A remarkable Ramsey-type result, Dvoretzky's theorem, shows that with k being about $\varepsilon^{-2} \log n$, such a k-dimensional almost-spherical section exists for every K. We also include an application concerning convex polytopes, showing that a high-dimensional centrally symmetric convex polytope cannot have both a small number of vertices and a small number of facets.

Both measure concentration and the existence of almost-spherical sections are truly high-dimensional phenomena, practically meaningless in the familiar dimensions 2 and 3. The low-dimensional intuition is of little use here, but perhaps by studying many results and examples one can develop intuition on what to expect in high dimensions.

We present only a few selected results from an extensive and well-developed theory of high-dimensional convexity. Most of it was built in the so-called *local theory of Banach spaces*, which deals with the geometry of finite-dimensional subspaces of various Banach spaces. In the literature, the theorems are usually formulated in the language of Banach spaces, so instead of symmetric convex bodies, one speaks about norms, and so on. Here we introduce some rudimentary terminology concerning normed spaces, but we express most of the notions in geometric language, hoping to make it more accessible to nonspecialists in Banach spaces. So, for example, in the formulation of Dvoretzky's theorem, we do not speak about the Banach–Mazur distance to an inner product norm but rather about almost spherical convex bodies. On the other hand, for a more serious study of this theory, the language of normed spaces seems necessary.

14.1 Measure Concentration on the Sphere

Let P denote the usual surface measure on the unit Euclidean sphere S^{n-1}, scaled so that all of S^{n-1} has measure 1 (a rigorous definition will be mentioned later). This P is a probability measure, and we often think of S^{n-1} as a probability space. For a set $A \subseteq S^{n-1}$, $P[A]$ is the P-measure of A and also the probability that a random point of S^{n-1} falls into A. The letter P should suggest "probability of," and the notation $P[A]$ is analogous to $\text{Prob}[A]$ used elsewhere in the book.

Measure concentration on the sphere can be approached in two steps. The first step is the observation, interesting but rather easy to prove, that for large n, most of S^{n-1} lies quite close to the "equator." For example, the following diagram shows the width of the band around the equator that contains 90% of the measure, for various dimensions n:

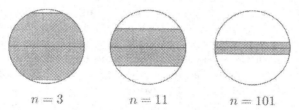

$$n = 3 \qquad\qquad n = 11 \qquad\qquad n = 101$$

That is, if the width of the gray stripe is $2w$, then

$$P\left[\{x \in S^{n-1}: -w \le x_n \le w\}\right] = 0.9.$$

As we will see later, w is of order $n^{-1/2}$ for large n. (Of course, one might ask why the measure is concentrated just around the "equator" $x_n = 0$. But counterintuitive as it may sound, it is concentrated around *any* equator, i.e., near any hyperplane containing the origin.)

The second, considerably deeper, step shows that the measure on S^{n-1} is concentrated not only around the equator, but near the boundary of any (measurable) subset $A \subset S^{n-1}$ covering half of the sphere. Here is a precise quantitative formulation.

14.1.1 Theorem (Measure concentration for the sphere). *Let $A \subseteq S^{n-1}$ be a measurable set with $P[A] \geq \frac{1}{2}$, and let A_t denote the t-neighborhood of A, that is, the set of all $x \in S^{n-1}$ whose Euclidean distance to A is at most t. Then*

$$1 - P[A_t] \leq 2e^{-t^2 n/2}.$$

Thus, if A occupies half of the sphere, almost all points of the sphere lie at distance at most $O(n^{-1/2})$ from A; only extremely small reserves can vegetate undisturbed by the nearness of A. (There is nothing very special about measure $\frac{1}{2}$ here; see Exercise 1 for an analogous result with $P[A] = \alpha \in (0, \frac{1}{2})$.) To recover the concentration around the equator, it suffices to choose A as the northern hemisphere and then as the southern hemisphere.

We present a simple and direct geometric proof of a slightly weaker version of Theorem 14.1.1, with $-t^2 n/4$ in the exponent instead of $-t^2 n/2$. It deals with both the steps mentioned above in one stroke.

It is based on the Brunn–Minkowski inequality: $\mathrm{vol}(A)^{1/n} + \mathrm{vol}(B)^{1/n} \leq \mathrm{vol}(A+B)^{1/n}$ for any nonempty compact sets $A, B \subset \mathbf{R}^n$ (Theorem 12.2.2). We actually use a slightly different version of the inequality, which resembles the well known inequality between the arithmetic and geometric means, at least optically:

$$\mathrm{vol}(\tfrac{1}{2}(A+B)) \geq \sqrt{\mathrm{vol}(A)\,\mathrm{vol}(B)}. \tag{14.1}$$

This is easily derived from the usual version: We have $\mathrm{vol}(\tfrac{1}{2}(A+B))^{1/n} \geq \mathrm{vol}(\tfrac{1}{2}A)^{1/n} + \mathrm{vol}(\tfrac{1}{2}B)^{1/n} = \tfrac{1}{2}(\mathrm{vol}(A)^{1/n} + \mathrm{vol}(B)^{1/n}) \geq (\mathrm{vol}(A)\,\mathrm{vol}(B))^{1/2n}$ by the inequality $\tfrac{1}{2}(a+b) \geq \sqrt{ab}$.

Proof of a weaker version of Theorem 14.1.1. For a set $A \subseteq S^{n-1}$, we define $\tilde{A}$ as the union of all the segments connecting the points of A to 0: $\tilde{A} = \{\alpha x : x \in A, \alpha \in [0,1]\} \subseteq B^n$. Then we have

$$P[A] = \mu(\tilde{A}),$$

where $\mu(\tilde{A}) = \mathrm{vol}(\tilde{A})/\mathrm{vol}(B^n)$ is the normalized volume of $\tilde{A}$; in fact, this can be taken as the definition of $P[A]$.

Let $t \in [0,1]$, let $P[A] \geq \frac{1}{2}$, and let $B = S^{n-1} \setminus A_t$. Then $\|a - b\| \geq t$ for all $a \in A$, $b \in B$.

14.1.2 Lemma. *For any $\tilde{x} \in \tilde{A}$ and $\tilde{y} \in \tilde{B}$, we have $\|\frac{\tilde{x}+\tilde{y}}{2}\| \leq 1 - t^2/8$.*

Proof of the lemma. Let $\tilde{x} = \alpha x$, $\tilde{y} = \beta y$, $x \in A$, $y \in B$:

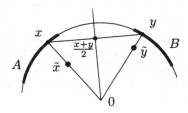

First we calculate, by the Pythagorean theorem and by elementary calculus,

$$\left\| \frac{x+y}{2} \right\| \le \sqrt{1 - \frac{t^2}{4}} \le 1 - \frac{t^2}{8}.$$

For passing to $\tilde{x}$ and $\tilde{y}$, we may assume that $\beta = 1$. Then

$$\left\| \frac{\tilde{x}+\tilde{y}}{2} \right\| = \left\| \frac{\alpha x + y}{2} \right\| \le \alpha \left\| \frac{x+y}{2} \right\| + (1-\alpha) \left\| \frac{y}{2} \right\|$$

$$= \alpha(1 - \tfrac{t^2}{8}) + (1-\alpha)(1 - \tfrac{1}{2}) \le 1 - \tfrac{t^2}{8}.$$

The lemma is proved.

By the lemma, the set $\frac{1}{2}(\tilde{A} + \tilde{B})$ is contained in the ball of radius $1 - t^2/8$ around the origin. Applying Brunn–Minkowski in the form (14.1) to $\tilde{A}$ and $\tilde{B}$, we have

$$\left(1 - \frac{t^2}{8}\right)^n \ge \mu(\tfrac{1}{2}(\tilde{A} + \tilde{B})) \ge \sqrt{\mu(\tilde{A})\mu(\tilde{B})} = \sqrt{\mathrm{P}[A]\,\mathrm{P}[B]} \ge \sqrt{\tfrac{1}{2}\mathrm{P}[B]}.$$

So

$$\mathrm{P}[B] \le 2\left(1 - \frac{t^2}{8}\right)^{2n} \le 2e^{-t^2 n/4}.$$

$\square$

Bibliography and remarks.   The simple proof of the slightly weaker measure concentration result for the sphere shown in this section is due to Arías-de-Reyna, Ball, and Villa [ABV98]. More about the history of measure concentration and related results will be mentioned in the next section.

Exercises

1. Derive the following from Theorem 14.1.1: If $A \subseteq S^{n-1}$ satisfies $\mathrm{P}[A] \ge \alpha$, $0 < \alpha \le \frac{1}{2}$, then $1 - \mathrm{P}[A_t] \le 2e^{-(t-t_0)^2 n/2}$, where t_0 is such that $2e^{-t_0^2 n/2} < \alpha$. ③

2. Let $A, B \subset S^{n-1}$ be measurable sets with distance at least $2t$. Prove that $\min(\mathrm{P}[A], \mathrm{P}[B]) \le 2e^{-t^2 n/2}$. ②

3. Use Theorem 14.1.1 to show that any 1-dense set in the unit sphere S^{n-1} has at least $\frac{1}{2}e^{n/8}$ points. ②

4. Let $K = \bigcap_{i=1}^{N} \{x \in \mathbf{R}^n : |\langle u_i, x \rangle| \le 1\}$ be the intersection of symmetric slabs determined by unit vectors $u_1, \ldots, u_N \in \mathbf{R}^n$. Using Theorem 14.1.1, prove that $\mathrm{vol}(B^n)/\mathrm{vol}(K) \le (\frac{C}{n} \ln N)^{n/2}$ for a suitable constant C. ③ The relation to Theorem 13.2.1 is explained in the notes to Section 13.2.

14.2 Isoperimetric Inequalities and More on Concentration

The usual proof of Theorem 14.1.1 (measure concentration) has two steps. First, $P[A_t]$ is bounded for A the hemisphere (which is elementary calculus), and second, it is shown that among all sets A of measure $\frac{1}{2}$, the hemisphere has the smallest $P[A_t]$. The latter result is an example of an *isoperimetric inequality*.

Before we formulate this inequality, let us begin with the mother of all isoperimetric inequalities, the one for planar geometric figures. It states that among all planar geometric figures with a given perimeter, the circular disk has the largest possible area. (This is well known but not so easy to prove rigorously.) More general isoperimetric inequalities are usually formulated using the volume of a neighborhood instead of "perimeter." They claim that among all sets of a given volume in some metric space under consideration, a ball of that volume has the smallest volume of the t-neighborhood:

(In the picture, assuming that the dark areas are the same, the light gray area is the smallest for the disk.) Letting $t \to 0$, one can get a statement involving the perimeter or surface area. But the formulation with t-neighborhood makes sense even in spaces where "surface area" is not defined; it suffices to have a metric and a measure on the considered space.

Here is this "neighborhood" form of isoperimetric inequality for the Euclidean space $\mathbf{R}^n$ with Lebesgue measure.

14.2.1 Proposition. *For any compact set $A \subset \mathbf{R}^d$ and any $t \geq 0$, we have* $\mathrm{vol}(A_t) \geq \mathrm{vol}(B_t)$, *where B is a ball of the same volume as A.*

Although we do not need this particular result in the further development, let us digress and mention a nice proof using the Brunn–Minkowski inequality (Theorem 12.2.2).

Proof. By rescaling, we may assume that B is a ball of unit radius. Then $A_t = A + tB$, and so

$$\mathrm{vol}(A_t) = \mathrm{vol}(A + tB) \geq \left(\mathrm{vol}(A)^{1/n} + t\,\mathrm{vol}(B)^{1/n}\right)^n$$
$$= (1 + t)^n\,\mathrm{vol}(B) = \mathrm{vol}(B_t).$$

$\square$

For the sphere S^{n-1} with the usual Euclidean metric inherited from $\mathbf{R}^n$, an r-ball is a spherical cap, i.e., an intersection of S^{n-1} with a half-space. The isoperimetric inequality states that for all measurable sets $A \subseteq S^{n-1}$ and all $t \geq 0$, we have $P[A_t] \geq P[C_t]$, where C is a spherical cap with $P[C] = P[A]$. We are not going to prove this; no really simple proof seems to be known.

The measure concentration on the sphere (Theorem 14.1.1) is a rather direct consequence of this isoperimetric inequality, by the argument already indicated above. If $P[A] = \frac{1}{2}$, then $P[A_t] \geq P[C_t]$, where C is a cap with $P[C] = \frac{1}{2}$, i.e., a hemisphere. Thus, it suffices to estimate the measure of the complementary cap $S^{n-1} \setminus C_t$.[1]

Gaussian concentration. There are many other metric probability spaces with measure concentration phenomena analogous to Theorem 14.1.1. Perhaps the most important one is $\mathbf{R}^n$ with the Euclidean metric and with the n-dimensional Gaussian measure γ given by

$$\gamma(A) = (2\pi)^{-n/2} \int_A e^{-\|x\|^2/2}\, dx.$$

This is a probability measure on $\mathbf{R}^n$ corresponding to the n-dimensional normal distribution. Let $Z_1, Z_2, \ldots, Z_n$ be independent real random variables, each of them with the standard normal distribution $N(0,1)$, i.e., such that

$$\text{Prob}[Z_i \leq z] = \tfrac{1}{\sqrt{2\pi}} \int_{-\infty}^{z} e^{-t^2/2}\, dt$$

for all $z \in \mathbf{R}$. Then the vector $(Z_1, Z_2, \ldots, Z_n) \in \mathbf{R}^n$ is distributed according to the measure γ. This γ is spherically symmetric; the density function $(2\pi)^{-n/2} e^{-\|x\|^2/2}$ depends only on the distance of x from the origin. The distance of a point chosen at random according to this distribution is sharply concentrated around $\sqrt{n}$, and in many respects, choosing a random point according to γ is similar to choosing a random point from the uniform distribution on the sphere $\sqrt{n}\, S^{n-1}$.

The isoperimetric inequality for the Gaussian measure claims that among all sets A with given $\gamma(A)$, a half-space has the smallest possible measure of the t-neighborhood. By simple calculation, this yields the corresponding theorem about measure concentration for the Gaussian measure:

14.2.2 Theorem (Gaussian measure concentration). *Let a measurable set $A \subseteq \mathbf{R}^n$ satisfy $\gamma(A) \geq \frac{1}{2}$. Then $\gamma(A_t) \geq 1 - e^{-t^2/2}$.*

---

[1] Theorem 14.1.1 provides a good upper bound for the measure of a spherical cap, but sometimes a lower bound is useful, too. Here are fairly precise estimates; for convenience they are expressed with a different parameterization. Let $C(\tau) = \{x \in S^{n-1}: x_1 \geq \tau\}$ denote the spherical cap of height $1 - \tau$. Then for $0 \leq \tau \leq \sqrt{2/n}$, we have $\frac{1}{12} \leq P[C(\tau)] \leq \frac{1}{2}$, and for $\sqrt{2/n} \leq \tau < 1$, we have

$$\frac{1}{6\tau\sqrt{n}} (1 - \tau^2)^{(n-1)/2} < P[C(\tau)] < \frac{1}{2\tau\sqrt{n}} (1 - \tau^2)^{(n-1)/2}.$$

These formulas are taken from Brieden et al. [BGK+99].

Note that the dimension does not appear in this inequality, and indeed the Gaussian concentration has infinite-dimensional versions as well. Measure concentration on S^{n-1}, with slightly suboptimal constants, can be proved as an easy consequence of the Gaussian concentration; see, for example, Milman and Schechtman [MS86] (Appendix V) or Pisier [Pis89].

Most of the results in the sequel obtained using measure concentration on the sphere can be derived from the Gaussian concentration as well. In more advanced applications the Gaussian concentration is often technically preferable, but here we stick to the perhaps more intuitive measure concentration on the sphere.

Other important "continuous" spaces with concentration results similar to Theorem 14.1.1 include the n-dimensional torus (the n-fold Cartesian product $S^1 \times \cdots \times S^1 \subset \mathbf{R}^{2n}$) and the group $\mathrm{SO}(n)$ of all rotations around the origin in $\mathbf{R}^n$ (see Section 14.4 for a little more about $\mathrm{SO}(n)$).

Discrete metric spaces. Similar concentration inequalities also hold in many discrete metric spaces encountered in combinatorics. One of the simplest examples is the n-dimensional Hamming cube $C_n = \{0,1\}^n$. The points are n-component vectors of 0's and 1's, and their Hamming distance is the number of positions where they differ. The "volume" of a set $A \subseteq \{0,1\}^n$ is defined as $\mathrm{P}[A] = \frac{1}{2^n}|A|$. An r-ball B is the set of all 0/1 vectors that differ from a given vector in at most r coordinates, and so its volume is $\mathrm{P}[B] = 2^{-n}\left(1 + \binom{n}{1} + \binom{n}{2} + \cdots + \binom{n}{r}\right)$. The isoperimetric inequality for the Hamming cube, due to Harper, is exactly of the form announced above:

If $A \subseteq C_n$ is any set with $\mathrm{P}[A] \geq \mathrm{P}[B]$, then $\mathrm{P}[A_t] \geq \mathrm{P}[B_t]$.

Of course, if A is an r-ball, then A_t is an $(r+t)$-ball and we have equality. Suitable estimates (tail estimates for the binomial distribution in probability theory) then give an analogue of Theorem 14.1.1:

14.2.3 Theorem (Measure concentration for the cube). *Let $A \subseteq C_n$ satisfy $\mathrm{P}[A] \geq \frac{1}{2}$. Then $1 - \mathrm{P}[A_t] \leq e^{-t^2/2n}$.*

This is very similar to the situation for S^{n-1}, only the scaling is different: While the Hamming cube C_n has diameter n, and the interesting range of t is from about $\sqrt{n}$ to n, the sphere S^{n-1} has diameter 2, and the interesting t are in the range from about $\frac{1}{\sqrt{n}}$ to 2.

Another significant discrete metric space with similar measure concentration is the space S_n of all permutations of $\{1, 2, \ldots, n\}$ (i.e., bijective mappings $\{1, 2, \ldots, n\} \to \{1, 2, \ldots, n\}$). The distance of two permutations p_1 and p_2 is $|\{i\colon p_1(i) \neq p_2(i)\}|$, and the measure is the usual uniform probability measure on S_n, where every single permutation has measure $\frac{1}{n!}$. Here a measure concentration inequality reads $1 - \mathrm{P}[A_t] \leq e^{-(t-3\sqrt{n})^2/8n}$ for all $A \subseteq S_n$ with $\mathrm{P}[A] \geq \frac{1}{2}$. The expander graphs, to be discussed in Section 15.5, also offer an example of spaces with measure concentration; see Exercise 15.5.7.

Bibliography and remarks. A modern treatment of measure concentration is the book Ledoux [Led01], to which we refer for more material and references. A concise introduction to concentration of Lipschitz functions and discrete isoperimetric inequalities, including some very recent material and combinatorial applications, is contained in the second edition of the book by Alon and Spencer [AS00d]. Older material on measure concentration in discrete metric spaces, with martingale proofs and several combinatorial examples, can be found in Bollobás's survey [Bol87]. For isoperimetric inequalities and measure concentration on manifolds see also Gromov [Gro98] (or Gromov's appendix in [MS86]).

The Euclidean isoperimetric inequality (the ball has the smallest surface for a given volume) has a long and involved history. It has been "known" since antiquity, but full and rigorous proofs were obtained only in the nineteenth century; see, e.g., Talenti [Tal93] for references. The quick proof via Brunn–Minkowski is taken from Pisier [Pis89].

The exact isoperimetric inequality for the sphere was first proved (according to [FLM77]) by Schmidt [Sch48]. Figiel, Lindenstrauss, and Milman [FLM77] have a 3-page proof based on symmetrization.

Measure concentration on the sphere and on other spaces was first recognized as an important general tool in the local theory of Banach spaces, and its use was mainly pioneered by Milman. Several nice surveys with numerous applications, mainly in Banach spaces but also elsewhere, are available, such as Lindenstrauss [Lin92], Lindenstrauss and Milman [LM93], Milman [Mil98], and some chapters of the book Benyamini and Lindenstrauss [BL99].

The Gaussian isoperimetric inequality was obtained by Borell [Bor75] and independently by Sudakov and Tsirel'son [ST74]. A proof can also be found in Pisier [Pis89]. Ball [Bal97] derives a slightly weaker version of the Gaussian concentration directly using the Prékopa–Leindler inequality mentioned in the notes to Section 12.2. The exact isoperimetric inequality for the Hamming cube is due to Harper [Har66]. We will indicate a short proof of measure concentration for product spaces, including the Hamming cube, in the notes to the next section.

More recently, very significant progress was made in the area of measure concentration and similar inequalities, especially on product spaces, mainly associated with the name of Talagrand; see, for instance, [Tal95] or the already mentioned book [Led01]. Talagrand's proof method, which works by establishing suitable one-dimensional inequalities and extending them to product spaces by a clever induction, also gives most of the concentration results previously obtained with the help of martingales.

Many new isoperimetric and concentration inequalities, as well as new proofs of known results, have been obtained by a function theoretic (as opposed to geometric) approach. Here concentration inequalities are usually derived from other types of inequalities, such as logarithmic Sobolev inequalities (estimating the entropy of a random variable). One advantage of this is that while concentration inequalities usually do not behave well under products, entropy estimates extend to products automatically, and so it suffices to prove one-dimensional versions.

Reverse isoperimetric inequality. The smallest possible surface area of a set with given volume is determined by the isoperimetric inequality. In the other direction, the surface area can be arbitrarily large for a given volume, but a meaningful question is obtained if one considers affine-equivalence classes of convex bodies. The following reverse isoperimetric inequality was proved by Ball (see [Bal97] or [Bal]): For every n-dimensional convex body C there exists an affine image $\tilde{C}$ of unit volume whose surface area is no larger than the surface area of the n-dimensional unit-volume regular simplex. Among symmetric convex bodies, the extremal body is the cube.

14.3 Concentration of Lipschitz Functions

Here we derive a form of the measure concentration that is very suitable for applications. It says that any Lipschitz function on a high-dimensional sphere is tightly concentrated around its expectation. (Any measurable real function $f\colon S^{n-1} \to \mathbf{R}$ can be regarded as a random variable, and its expectation is given by $\mathbf{E}[f] = \int_{S^{n-1}} f(x)\, \mathrm{d}\mathrm{P}(x)$.)

We recall that a mapping f between metric spaces is C-*Lipschitz*, where $C > 0$ is a real number, if the distance of $f(x)$ and $f(y)$ is never larger than C times the distance of x and y. We first show that a 1-Lipschitz function $f\colon S^{n-1} \to \mathbf{R}$ is concentrated around its median. The *median* of a real-valued function f is defined as

$$\mathrm{med}(f) = \sup\{t \in \mathbf{R}\colon \mathrm{P}[f \leq t] \leq \tfrac{1}{2}\}.$$

Here P is the considered probability measure on the domain of f; in our case, it is the normalized surface measure on S^{n-1}. The notation $\mathrm{P}[f \leq t]$ is the usual probability-theory shorthand for $\mathrm{P}\big[\{x \in S^{n-1}\colon f(x) \leq t\}\big]$. The following lemma looks obvious, but an actual proof is perhaps not completely obvious:

14.3.1 Lemma. *Let $f\colon \Omega \to \mathbf{R}$ be a measurable function on a space Ω with a probability measure* P. *Then*

$$\mathrm{P}[f < \mathrm{med}(f)] \leq \tfrac{1}{2} \text{ and } \mathrm{P}[f > \mathrm{med}(f)] \leq \tfrac{1}{2}.$$

Proof. The first inequality can be derived from the σ-additivity of the measure P:

$$P[f < \text{med}(f)] = \sum_{k=1}^{\infty} P\left[\text{med}(f) - \tfrac{1}{k-1} < f \le \text{med}(f) - \tfrac{1}{k}\right]$$
$$= \sup_{k \ge 1} P\left[f \le \text{med}(f) - \tfrac{1}{k}\right] \le \tfrac{1}{2}.$$

The second inequality follows similarly. □

We are ready to prove that any 1-Lipschitz function $S^{n-1} \to \mathbf{R}$ is concentrated around its median:

14.3.2 Theorem (Lévy's lemma). *Let $f: S^{n-1} \to \mathbf{R}$ be 1-Lipschitz. Then for all $t \in [0,1]$,*

$$P[f > \text{med}(f) + t] \le 2e^{-t^2 n/2} \quad \text{and} \quad P[f < \text{med}(f) - t] \le 2e^{-t^2 n/2}.$$

For example, on 99% of S^{n-1}, the function f attains values deviating from $\text{med}(f)$ by at most $3.5n^{-1/2}$.

Proof. We prove only the first inequality. Let $A = \{x \in S^{n-1}: f(x) \le \text{med}(f)\}$. By Lemma 14.3.1, $P[A] \ge \tfrac{1}{2}$. Since f is 1-Lipschitz, we have $f(x) \le \text{med}(f) + t$ for all $x \in A_t$. Therefore, by Theorem 14.1.1, we get $P[f > \text{med}(f) + t] \le 1 - P[A_t] \le 2e^{-t^2 n/2}$. □

The median is generally difficult to compute. But for a 1-Lipschitz function, it cannot be too far from the expectation, which is usually easier to estimate:

14.3.3 Proposition. *Let $f: S^{n-1} \to \mathbf{R}$ be 1-Lipschitz. Then*

$$|\text{med}(f) - \mathbf{E}[f]| \le 12n^{-1/2}.$$

Proof.

$$|\text{med}(f) - \mathbf{E}[f]| \le \mathbf{E}[|f - \text{med}(f)|] \le \sum_{k=0}^{\infty} \frac{k+1}{\sqrt{n}} \cdot P\left[|f - \text{med}(f)| \ge \tfrac{k}{\sqrt{n}}\right]$$
$$\le n^{-1/2} \sum_{k=0}^{\infty} (k+1) \cdot 4e^{-k^2/2} \le 12n^{-1/2}$$

(the numerical estimate of the last sum is not important; it is important that it converges to some constant, which is obvious). □

We derive a consequence of Lévy's lemma on finding k-dimensional subspaces where a given Lipschitz function is almost constant. But first we need some notions and results.

Random rotations and random subspaces. We want to speak about a random k-dimensional (linear) subspace of $\mathbf{R}^n$. We thus need to specify a probability measure on the set of all k-dimensional linear subspaces of $\mathbf{R}^n$ (so-called *Grassmann manifold* or *Grassmannian*). An elegant way of doing this is via random rotations.

A rotation ρ is an isometry of $\mathbf{R}^n$ fixing the origin and preserving the orientation. In algebraic terms, ρ is a linear mapping $x \mapsto Ax$ given by an orthonormal matrix A with determinant 1. The result of performing the rotation ρ on the standard orthonormal basis $(e_1, \ldots, e_n)$ in $\mathbf{R}^n$ is an n-tuple of orthonormal vectors, and these vectors are the columns of A.

The group of all rotations in $\mathbf{R}^n$ around the origin with the operation of composition (corresponding to multiplication of the matrices) is denoted by $\mathrm{SO}(n)$, which stands for the *special orthogonal* group. With the natural topology (obtained by regarding the corresponding matrices as points in $\mathbf{R}^{n^2}$), it is a compact group. By a general theorem in the theory of topological groups, there is a unique Borel probability measure on $\mathrm{SO}(n)$ (the *Haar measure*) that is invariant under the action of the elements of $\mathrm{SO}(n)$. Here is a more concrete description of this probability measure. To obtain a random rotation ρ, we first choose a vector $a_1 \in S^{n-1}$ uniformly at random. Then we pick a_2 orthogonal to a_1; this a_2 is drawn from the uniform distribution on the $(n-2)$-dimensional sphere that is the intersection of S^{n-1} with the hyperplane perpendicular to a_1 and passing through 0. Then a_3 is chosen from the unit sphere within the $(n-2)$-dimensional subspace perpendicular to a_1 and a_2, and so on.

In the sequel we need only the following intuitively obvious fact about a random rotation $\rho \in \mathrm{SO}(n)$: For every fixed $u \in S^{n-1}$, $\rho(u)$ is a random vector of S^{n-1}. Therefore, if $u \in S^{n-1}$ is fixed, $A \subseteq S^{n-1}$ is measurable, and $\rho \in \mathrm{SO}(n)$ is random, then the probability of $\rho(u) \in A$ equals $\mathrm{P}[A]$.

Let L_0 be the k-dimensional subspace spanned by the first k coordinate vectors $e_1, e_2, \ldots, e_k$. A random k-dimensional linear subspace $L \subset \mathbf{R}^n$ can be defined as $\rho(L_0)$, where $\rho \in \mathrm{SO}(n)$ is a random rotation.

By Lévy's lemma, a 1-Lipschitz function on S^{n-1} is "almost constant" on a subset A occupying almost all of S^{n-1}. Generally we do not know anything about the shape of such an A. But the next proposition shows that the almost-constant behavior can be guaranteed on the intersection of S^{n-1} with a linear subspace of $\mathbf{R}^n$ of relatively large dimension.

14.3.4 Proposition (Subspace where a Lipschitz function is almost constant). *Let $f\colon S^{n-1} \to \mathbf{R}$ be a 1-Lipschitz function and let $\delta \in (0,1]$. Then there is a linear subspace $L \subseteq \mathbf{R}^n$ such that all values of f restricted to $S^{n-1} \cap L$ are in the interval $[\mathrm{med}(f) - \delta, \mathrm{med}(f) + \delta]$ and*

$$\dim L \geq \frac{\delta^2}{8 \log(8/\delta)} \cdot n - 1.$$

Proof. Let L_0 be the subspace spanned by the first $k = \lceil n\delta^2/8\log\frac{8}{\delta} - 1\rceil$ coordinate vectors. Fix a $\frac{\delta}{2}$-net N (as defined above Lemma 13.1.1) in $S^{n-1}\cap L_0$. Let $\rho \in \mathrm{SO}(n)$ be a random rotation. For $x \in N$, $\rho(x)$ is a random point, and so by Lévy's lemma, the probability that $|f(\rho(x)) - \mathrm{med}(f)| > \frac{\delta}{2}$ for at least one point $x \in N$ is no more than $|N| \cdot 4e^{-\delta^2 n/8}$. Using the bound $|N| \le (\frac{8}{\delta})^k$ from Lemma 13.1.1, we calculate that with a positive probability, $|f(y) - \mathrm{med}(f)| \le \frac{\delta}{2}$ for all $y \in \rho(N)$.

We choose a ρ with this property and let $L = \rho(L_0)$. For each $x \in S^{n-1}\cap L$, there is some $y \in \rho(N)$ with $\|x-y\| \le \frac{\delta}{2}$, and since f is 1-Lipschitz, we obtain $|f(x) - \mathrm{med}(f)| \le |f(x) - f(y)| + |f(y) - \mathrm{med}(f)| \le \delta$. $\qquad\square$

Bibliography and remarks. Lévy's lemma and a measure concentration result similar to Theorem 14.1.1 were found by Lévy [Lév51].

Analogues of Lévy's lemma for other spaces with measure concentration follow by the same argument. On the other hand, a measure concentration inequality for sets follows from concentration of Lipschitz functions (a Lévy's lemma) on the considered space (Exercise 1). For some spaces, concentration of Lipschitz functions can be proved directly. Often this is done using martingales (see [Led01], [AS00d], [MS86], [Bol87]). Here we outline a proof without martingales (following [Led01]) for product spaces.

Let Ω be a space with a probability measure P and a metric ρ. The *Laplace functional* $E = E_{\Omega,\mathrm{P},\rho}$ is a function $(0,\infty) \to \mathbf{R}$ defined by

$$E(\lambda) = \sup\left\{\mathbf{E}\left[e^{\lambda f}\right] : f\colon \Omega \to \mathbf{R} \text{ is 1-Lipschitz and } \mathbf{E}[f] = 0\right\}.$$

First we show that a bound on $E(\lambda)$ implies concentration of Lipschitz functions. Assume that $E(\lambda) \le e^{a\lambda^2/2}$ for some $a > 0$ and all $\lambda > 0$, and let $f\colon\Omega \to \mathbf{R}$ be 1-Lipschitz. We may suppose that $\mathbf{E}[f] = 0$. Using Markov's inequality for the random variable $Y = e^{\lambda f}$, we have $\mathrm{P}[f \ge t] = \mathrm{P}[Y \ge e^{t\lambda}] \le \mathbf{E}[Y]/e^{t\lambda} \le E(\lambda)/e^{t\lambda} \le e^{a\lambda^2/2 - \lambda t}$, and setting $\lambda = \frac{t}{a}$ yields $\mathrm{P}[f \ge t] \le e^{-t^2/2a}$.

Next, for some spaces, $E(\lambda)$ can be bounded directly. Here we show that if (Ω,ρ) has diameter at most 1, then $E(\lambda) \le e^{-\lambda^2/2}$. This can be proved by the following elegant trick. First we note that $e^{\mathbf{E}[f]} \le \mathbf{E}\left[e^f\right]$ for any f, by Jensen's inequality in integral form, and so if $\mathbf{E}[f] = 0$, then $\mathbf{E}\left[e^{-f}\right] \ge 1$. Then, for a 1-Lipschitz f with $\mathbf{E}[f] = 0$, we calculate

$$\mathbf{E}\left[e^{\lambda f}\right] = \int_\Omega e^{\lambda f(x)}\,\mathrm{dP}(x)$$

$$\le \left(\int e^{-\lambda f(y)}\,\mathrm{dP}(y)\right)\left(\int e^{\lambda f(x)}\,\mathrm{dP}(x)\right)$$

$$= \int\int e^{\lambda(f(x)-f(y))}\,\mathrm{dP}(x)\,\mathrm{dP}(y)$$

$$= \sum_{i=0}^{\infty} \int \int \frac{(\lambda(f(x) - f(y)))^i}{i!} \, \mathrm{dP}(x) \, \mathrm{dP}(y).$$

For i even, we can bound the integrand by $\lambda^i/i!$, since $|f(x) - f(y)| \leq 1$. For odd i, the integral vanishes by symmetry. The resulting bound is $\sum_{k=0}^{\infty} \lambda^{2k}/(2k)! \leq e^{\lambda^2/2}$. (If the diameter is D, then we obtain $E(\lambda) \leq e^{D^2\lambda^2/2}$.)

Finally, we prove that the Laplace functional is submultiplicative. Let $(\Omega_1, \mathrm{P}_1, \rho_1)$ and $(\Omega_2, \mathrm{P}_2, \rho_2)$ be spaces, let $\Omega = \Omega_1 \times \Omega_2$, $\mathrm{P} = \mathrm{P}_1 \times \mathrm{P}_2$, and $\rho = \rho_1 + \rho_2$ (that is, $\rho((x,y),(x',y')) = \rho_1(x,x') + \rho_2(y,y')$). We claim that $E_{\Omega,\mathrm{P},\rho}(\lambda) \leq E_{\Omega_1,\mathrm{P}_1,\rho_1}(\lambda) \cdot E_{\Omega_2,\mathrm{P}_2,\rho_2}(\lambda)$. To verify this, let $f: \Omega \to \mathbf{R}$ be 1-Lipschitz with $\mathbf{E}[f] = 0$, and set $g(y) = \mathbf{E}_x[f(x,y)] = \int_{\Omega_1} f(x,y) \, \mathrm{dP}_1(x)$. We observe that g, being a weighted average of 1-Lipschitz functions, is 1-Lipschitz. We have

$$\mathbf{E}[e^{\lambda f}] = \int_{\Omega_2} \int_{\Omega_1} e^{\lambda f(x,y)} \, \mathrm{dP}_1(x) \, \mathrm{dP}_2(y)$$

$$= \int_{\Omega_2} e^{\lambda g(y)} \left(\int_{\Omega_1} e^{\lambda(f(x,y) - g(y))} \, \mathrm{dP}_1(x) \right) \mathrm{dP}_2(y).$$

The function $x \mapsto f(x,y) - g(y)$ is 1-Lipschitz and has zero expectation for each y, and the inner integral is at most $E_{\Omega_1,\mathrm{P}_1,\rho_1}(\lambda)$. Since g is 1-Lipschitz and $\mathbf{E}[g] = 0$, we have $\int_{\Omega_2} e^{\lambda g(y)} \, \mathrm{dP}_2(y) \leq E_{\Omega_2,\mathrm{P}_2,\rho_2}(\lambda)$ and we are done.

By combining the above, we obtain, among others, that if each of n spaces $(\Omega_i, \mathrm{P}_i, \rho_i)$ has diameter at most 1 and $(\Omega, \mathrm{P}, \rho)$ is the product, then $\mathrm{P}[f \geq \mathbf{E}[f] + t] \leq e^{-t^2/2n}$ for all 1-Lipschitz $f: \Omega \to \mathbf{R}$. In particular, this applies to the Hamming cube.

Proposition 14.3.4 is due to Milman [Mil69], [Mil71].

Exercises

1. Derive the measure concentration on the sphere (Theorem 14.1.1) from Lévy's lemma. ☑

14.4 Almost Spherical Sections: The First Steps

For a real number $t \geq 1$, we call a convex body K *t-almost spherical* if it contains a (Euclidean) ball B of some radius r and it is contained in the concentric ball of radius tr.

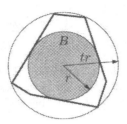

Given a centrally symmetric convex body $K \subset \mathbf{R}^n$ and $\varepsilon > 0$, we are interested in finding a k-dimensional (linear) subspace L, with k as large as possible, such that the "section" $K \cap L$ is $(1+\varepsilon)$-almost spherical.

Ellipsoids. First we deal with ellipsoids, where the existence of large spherical sections is not very surprising. But in the sequel it gives us additional freedom: Instead of looking for a $(1+\varepsilon)$-spherical section of a given convex body, we can as well look for a $(1+\varepsilon)$-ellipsoidal section, while losing only a factor of at most 2 in the dimension. This means that we are free to transform a given body by any (nonsingular) affine map, which is often convenient. Let us remark that in the local theory of Banach spaces, almost-ellipsoidal sections are usually as good as almost-spherical ones, and so the following lemma is often not even mentioned.

14.4.1 Lemma (Ellipsoids have large spherical sections). *For any $(2k-1)$-dimensional ellipsoid E, there is a k-flat L passing through the center of E such that $E \cap L$ is a Euclidean ball.*

Proof. Let $E = \left\{ x \in \mathbf{R}^{2k-1} \colon \sum_{i=1}^{2k-1} \frac{x_i^2}{a_i^2} \leq 1 \right\}$ with $0 < a_1 \leq a_2 \leq \cdots \leq a_{2k-1}$. We define the k-dimensional linear subspace L by a system of $k - 1$ linear equations. The ith equation is

$$ x_i \sqrt{\frac{1}{a_i^2} - \frac{1}{a_k^2}} = x_{2k-i} \sqrt{\frac{1}{a_k^2} - \frac{1}{a_{2k-i}^2}}, $$

$i = 1, 2, \ldots, k-1$. It is chosen so that

$$ \frac{x_i^2}{a_i^2} + \frac{x_{2k-i}^2}{a_{2k-i}^2} = \frac{1}{a_k^2} (x_i^2 + x_{2k-i}^2) $$

for $x \in L$. It follows that for $x \in L$, we have $x \in E$ if and only if $\|x\| \leq a_k$, and so $E \cap L$ is a ball of radius a_k. The reader is invited to find a geometric meaning of this proof and/or express it in the language of eigenvalues.   □

To make formulas simpler, we consider only the case $\varepsilon = 1$ (2-almost spherical sections) in the rest of this section. An arbitrary $\varepsilon > 0$ can always be handled very similarly.

The cube. The cube $[-1, 1]^n$ is a good test case for finding almost-spherical sections; it seems hard to imagine how a cube could have very round slices. In some sense, this intuition is not totally wrong, since the almost-spherical sections of a cube can have only logarithmic dimension, as we verify next. (But the n-dimensional crosspolytope has $(1+\varepsilon)$-spherical sections of dimension as high as $c(\varepsilon)n$, and yet it does not look any rounder than the cube; so much for the intuition.)

The intersection of the cube with a k-dimensional linear subspace of $\mathbf{R}^n$ is a k-dimensional convex polytope with at most $2k$ facets.

14.4.2 Lemma. *Let P be a k-dimensional 2-almost spherical convex polytope. Then P has at least $\frac{1}{2} e^{k/8}$ facets.*

Therefore, any 2-almost spherical section of the cube has dimension at most $O(\log n)$.

Proof of Lemma 14.4.2. After a suitable affine transform, we may assume $\frac{1}{2} B^k \subseteq P \subseteq B^k$. Each point $x \in S^{k-1}$ is separated from P by one of the facet hyperplanes. For each facet F of P, the facet hyperplane h_F cuts off a cap C_F of S^{k-1}, and these caps together cover all of S^{k-1}. The cap C_F is at distance at least $\frac{1}{2}$ from the hemisphere defined by the hyperplane h'_F parallel to h_F and passing through 0.

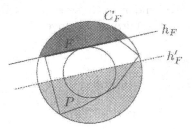

By Theorem 14.1.1 (measure concentration), we have $\mathrm{P}[C_F] \leq 2e^{-k/8}$. $\qquad\square$

Next, we show that the n-dimensional cube actually does have 2-almost spherical sections of dimension $\Omega(\log n)$. First we need a k-dimensional 2-almost spherical polytope with 4^k facets. We note that if P is a convex polytope with $B^k \subset P \subset tB^k$, then the dual polytope P^* satisfies $\frac{1}{t} B^k \subset P^* \subset B^k$ (Exercise 1). So it suffices to construct a k-dimensional 2-almost spherical polytope with 4^k *vertices*, and this was done in Section 13.3: We can take any 1-net in S^{k-1} as the vertex set. (Let us remark that an exponential lower bound for the number of vertices also follows from Theorem 13.2.1.)

By at most doubling the number of facets, we may assume that our k-dimensional 2-almost spherical polytope is centrally symmetric. It remains to observe that *every k-dimensional centrally symmetric convex polytope P with $2n$ facets is an affine image of the section $[-1, 1]^n \cap L$ for a suitable k-dimensional linear subspace $L \subseteq \mathbf{R}^n$.* Indeed, such a P can be expressed as the intersection $\bigcap_{i=1}^{n} \{x \in \mathbf{R}^k : |\langle a_i, x \rangle| \leq 1\}$, where $\pm a_1, \ldots, \pm a_n$ are suitably normalized normal vectors of the facets of P. Let $f: \mathbf{R}^k \to \mathbf{R}^n$ be the linear map given by

$$f(x) = (\langle a_1, x \rangle, \langle a_2, x \rangle, \ldots, \langle a_n, x \rangle).$$

Since P is bounded, the a_i span all of $\mathbf{R}^k$, and so f has rank k. Consequently, its image $L = f(\mathbf{R}^k)$ is a k-dimensional subspace of $\mathbf{R}^n$. We have $P = f^{-1}([-1,1]^n)$, and so the intersection $[-1,1]^n \cap L$ is the affine image of P.

We see that the n-dimensional cube has 2-almost ellipsoidal sections of dimension $\Omega(\log n)$ (as well as 2-almost spherical sections, by Lemma 14.4.1).

Next, we make preparatory steps for finding almost-spherical sections of arbitrary centrally symmetric convex bodies. These considerations are most conveniently formulated in the language of norms.

Reminder on norms. We recall that a *norm* on a real vector space Z is a mapping that assigns a nonnegative real number $\|x\|_Z$ to each $x \in Z$ such that $\|x\|_Z = 0$ implies $x = 0$, $\|\alpha x\|_Z = |\alpha| \cdot \|x\|_Z$ for all $\alpha \in \mathbf{R}$, and the triangle inequality holds: $\|x + y\|_Z \leq \|x\|_Z + \|y\|_Z$. (Since we have reserved $\|\cdot\|$ for the Euclidean norm, we write other norms with various subscripts, or occasionally we use the symbol $|\cdot|$.)

Norms are in one-to-one correspondence with closed bounded convex bodies symmetric about 0 and containing 0 in their interior. Here we need only one direction of this correspondence: Given a convex body K with the listed properties, we assign to it the norm $\|\cdot\|_K$ given by

$$\|x\|_K = \min \left\{ t > 0 : \frac{x}{t} \in K \right\} \quad (x \neq 0).$$

Here is an illustration:

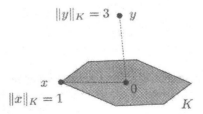

It is easy to verify the axioms of the norm (the convexity of K is needed for the triangle inequality). The body K is the unit ball of the norm $\|\cdot\|_K$. The norm of points *decreases* by blowing up the body K.

General body: the first attempt. Let $K \subset \mathbf{R}^n$ be a convex body defining a norm (i.e., closed, bounded, symmetric, 0 in the interior). Let us define the function $f_K: S^{n-1} \to \mathbf{R}$ as the restriction of the norm $\|\cdot\|_K$ on S^{n-1}; that is, $f_K(x) = \|x\|_K$. We note that K is t-almost spherical if (and only if) there is a number $a > 0$ such that $a \leq f(x) \leq ta$ for all $x \in S^{n-1}$. So for finding a large almost-spherical section of K, we need a linear subspace L such that

f does not vary too much on $S^{n-1} \cap L$, and this is where Proposition 14.3.4, about subspaces where a Lipschitz function is almost constant, comes in.

Of course, that proposition has its assumptions, and one of them is that f_K is 1-Lipschitz. A sufficient condition for that is that K should contain the unit ball:

14.4.3 Observation. *Suppose that the convex body K contains the R-ball $B(0, R)$. Then $\|x\|_K \leq \frac{1}{R}\|x\|$ for all x, and the function $x \mapsto \|x\|_K$ is $\frac{1}{R}$-Lipschitz with respect to the Euclidean metric.* $\square$

Then we can easily prove the following result.

14.4.4 Proposition. *Let $K \subset \mathbf{R}^n$ be a convex body defining a norm and such that $B^n \subseteq K$, and let $m = \mathrm{med}(f_K)$, where f_K is as above. Then there exists a 2-almost-spherical section of K of dimension at least*

$$\Omega\left(\frac{nm^2}{\log(24/m)}\right).$$

Proof. By Observation 14.4.3, f_K is 1-Lipschitz. Let us set $\delta = \frac{m}{3}$ (note that $B^n \subseteq K$ also implies $m \leq 1$). Proposition 14.3.4 shows that there is a subspace L such that $f_K \in \left[\frac{2}{3}m, \frac{4}{3}m\right]$ on $S^{n-1} \cap L$, where

$$\dim L = \Omega\left(\frac{n\delta^2}{\log(8/\delta)}\right) = \Omega\left(\frac{nm^2}{\log(24/m)}\right). \qquad (14.2)$$

The section $K \cap L$ is 2-almost spherical. $\square$

A slight improvement. It turns out that the factor $\log(24/m)$ in the result just proved can be eliminated by a refined argument, which uses the fact that f_K comes from a norm.

14.4.5 Theorem. *With the assumptions as in Proposition 14.4.4, a 2-almost spherical section exists of dimension at least $\beta n m^2$, where $\beta > 0$ is an absolute constant.*

Proof. The main new observation is that for our f_K, we can afford a much less dense net N in the proof of Proposition 14.3.4. Namely, it suffices to let N be a $\frac{1}{5}$-net in S^{k-1}, where $k = \lceil \beta m^2 n \rceil$.

If $\beta > 0$ is sufficiently small, Lévy's lemma gives the existence of a rotation ρ such that $\frac{14}{15}m \leq f_K(y) \leq \frac{16}{15}m$ for all $y \in \rho(N)$; this is exactly as in the proof of Proposition 14.3.4. It remains to verify $\frac{2}{3}m \leq f_K(x) \leq \frac{4}{3}m$ for all $x \in S^{n-1} \cap L$, where $L = \rho(L_0)$. This is implied by the following claim with $a = \frac{16}{15}m$ and $|\cdot| = \|\cdot\|_K$:

Claim. *Let N be a $\frac{1}{5}$-net in S^{k-1} with respect to the Euclidean metric, and let $|\cdot|$ be a norm on $\mathbf{R}^k$ satisfying $\frac{7}{8}a \leq |y| \leq a$ for all $y \in N$ and for some number $a > 0$. Then $\frac{5}{8}a \leq |x| \leq \frac{5}{4}a$ for all $x \in S^{k-1}$.*

To prove the claim, we begin with the upper bound (this is where the new trick lies). Let $M = \max\{|x|: x \in S^{k-1}\}$ and let $x_0 \in S^{k-1}$ be a point where M is attained. Choose a $y_0 \in N$ at distance at most $\frac{1}{5}$ from x_0, and let $z = (x_0 - y_0)/\|x_0 - y_0\|$ be the unit vector in the direction of $x_0 - y_0$. Then $M = |x_0| \leq |y_0| + |x_0 - y_0| \leq a + \|x_0 - y_0\| \cdot |z| \leq a + \frac{1}{5}M$. The resulting inequality $M \leq a + \frac{1}{5}M$ yields $M \leq \frac{5}{4}a$.

The lower bound is now routine: If $x \in S^{k-1}$ and $y \in N$ is at distance at most $\frac{1}{5}$ from it, then $|x| \geq |y| - |x - y| \geq \frac{7}{8}a - \frac{1}{5} \cdot \frac{5}{4}a \geq \frac{5}{8}a$. The claim, as well as Theorem 14.4.5, is proved.   □

Theorem 14.4.5 yields almost-spherical sections of K, *provided that* we can estimate $\mathrm{med}(f_K)$ (after rescaling K so that $B^n \subseteq K$). We must warn that this in itself does not yet give almost spherical sections for every K (Dvoretzky's theorem), and another twist is needed, shown in Section 14.6. But in order to reap some benefits from the hard work done up until now, we first explain an application to convex polytopes.

Bibliography and remarks. As was remarked in the text, almost-spherical and almost-ellipsoidal sections are seldom distinguished in the local theory of Banach spaces, where symmetric convex bodies are considered up to isomorphism, i.e., up to a nonsingular linear transform. If K_1 and K_2 are symmetric convex bodies in $\mathbf{R}^n$, their *Banach–Mazur distance* $d(K_1, K_2)$ is defined as the smallest positive t for which there is a linear transform T such that $T(K_1) \subseteq K_2 \subseteq t \cdot T(K_2)$. So a symmetric convex body K is t-almost ellipsoidal if and only if $d(K, B^n) \leq t$. It turns out that every two symmetric compact convex bodies $K_1, K_2 \subset \mathbf{R}^n$ satisfy $d(K_1, K_2) \leq \sqrt{n}$. The logarithm of the Banach–Mazur distance is a metric on the space of compact symmetric convex bodies in $\mathbf{R}^n$.

Lemma 14.4.1 appears in Dvoretzky [Dvo61]. Theorem 14.4.5 is from Figiel, Lindenstrauss, and Milman [FLM77].

There are several ways of proving that the n-dimensional crosspolytope has almost spherical sections of dimension $\Omega(n)$ (but, perhaps surprisingly, no explicit construction of such a section seems to be known). A method based on Theorem 14.4.5 is indicated in Exercise 14.6.2. A somewhat more direct way, found by Schechtman, is to let the section L be the image of the linear map $f: \mathbf{R}^{cn} \to \mathbf{R}^n$ whose matrix has entries ± 1 chosen uniformly and independently at random ($c > 0$ is a suitable small constant). The proof uses martingales (Azuma's inequality); see, e.g., Milman and Schechtman [MS86]. The existence of a C-almost spherical section of dimension $\frac{n}{2}$, with a suitable constant C, is a consequence of a theorem of Kashin: If B_1^n denotes the crosspolytope and ρ is a random rotation, then $B_1^n \cap \rho(B_1^n)$ is 32-almost spherical with a positive probability; see Ball [Bal97] for an insightful exposition. The previously mentioned methods do not provide a dimension this large, but Kashin's result does not give $(1+\varepsilon)$-almost spherical sections for small ε.

Exercises

1. Let K be a convex body containing 0 in its interior. Check that $K \subseteq B^n$ if and only if $B^n \subseteq K^*$ (recall that $K^* = \{x \in \mathbf{R}^k \colon \langle x, y \rangle \le 1 \text{ for all } y \in K\}$). Derive that if $B^k \subset K \subset tB^k$, then $\frac{1}{t}B^k \subset K^* \subset B^k$. [2]

14.5 Many Faces of Symmetric Polytopes

Can an n-dimensional convex polytope have both few vertices and few facets? Yes, an n-simplex has $n+1$ vertices and $n+1$ facets. What about a *centrally symmetric polytope*? The n-dimensional cube has only $2n$ facets but 2^n vertices. Its dual, the crosspolytope (regular octahedron for $n = 3$), has few vertices but many facets. It turns out that every centrally symmetric polytope has many facets or many vertices.

14.5.1 Theorem. *There is a constant $\alpha > 0$ such that for any centrally symmetric n-dimensional convex polytope P, we have $\log f_0(P) \cdot \log f_{n-1}(P) \ge \alpha n$ (recall that $f_0(P)$ denotes the number of vertices and $f_{n-1}(P)$ the number of facets).*

For the cube, the expression $\log f_0(P) \cdot \log f_{n-1}(P)$ is about $n \log n$, which is even slightly larger than the lower bound in the theorem. However, polytopes can be constructed with both $\log f_0(P)$ and $\log f_{n-1}(P)$ bounded by $O(\sqrt{n})$ (Exercise 1).

Proof of Theorem 14.5.1. We use the dual polytope P^* with $f_0(P) = f_{n-1}(P^*)$, and we prove the theorem in the equivalent form $\log f_{n-1}(P^*) \cdot \log f_{n-1}(P) \ge \alpha n$.

John's lemma (Theorem 13.4.1) claims that for any symmetric convex body K, there exists a (nonsingular) linear map that transforms K into a $\sqrt{n}$-almost spherical body. We can thus assume that the considered n-dimensional polytope P is $\sqrt{n}$-almost spherical (this is crucial for the proof).

After rescaling, we may suppose $B^n \subset P \subset \sqrt{n}\,B^n$. Letting $m = \text{med}(f_P)$, where f_P is the restriction of $\|\cdot\|_P$ on S^{n-1} as usual, Theorem 14.4.5 tells us that there is a linear subspace L of $\mathbf{R}^n$ with $P \cap L$ being 2-almost spherical and with $\dim(L) = \Omega(nm^2)$. Thus, since any k-dimensional 2-almost spherical polytope has $e^{\Omega(k)}$ facets, we have $\log f_{n-1}(P) = \Omega(nm^2)$.

Now, we look at P^*. Since $B^n \subset P \subset \sqrt{n}\,B^n$, by Exercise 14.4.1 we have $n^{-1/2}B^n \subset P^* \subset B^n$. In order to apply Theorem 14.4.5, we set $\tilde{P} = \sqrt{n}\,P^*$, and obtain a 2-almost spherical section $\tilde{L}$ of $\tilde{P}$ of dimension $\Omega(n\tilde{m}^2)$, where $\tilde{m} = \text{med}(f_{\tilde{P}})$. This implies $\log f_{n-1}(P^*) = \Omega(n\tilde{m}^2)$.

It remains to observe the following inequality:

14.5.2 Lemma. *Let P be a polytope in $\mathbf{R}^n$ defining a norm and let P^* be the dual polytope. Then we have* $\mathrm{med}(f_P)\,\mathrm{med}(f_{P^*}) \geq 1$.

We leave the easy proof as Exercise 2. Since $\tilde{m} = \mathrm{med}(f_{P^*})/\sqrt{n}$, we finally obtain

$$\log f_{n-1}(P) \cdot \log f_{n-1}(P^*) = \Omega(n^2 m^2 \tilde{m}^2)$$
$$= \Omega(n\,\mathrm{med}(f_P)^2\,\mathrm{med}(f_{P^*})^2) = \Omega(n).$$

This concludes the proof of Theorem 14.5.1.   □

Bibliography and remarks. Theorem 14.5.1, as well as the example in Exercise 1, is due to Figiel, Lindenstrauss, and Milman [FLM77]. Most of the tools in the proof come from earlier papers of Milman [Mil69], [Mil71].

Exercises

1. Construct an n-dimensional convex polytope P with $\log f_0(P) = \Omega(\sqrt{n})$ and $\log f_{n-1}(P) = \Omega(\sqrt{n})$, thereby demonstrating that Theorem 14.5.1 is asymptotically optimal. Start with the interval $[0,1] \subset \mathbf{R}^1$, and alternate the operations $(\cdot)^*$ (passing to the dual polytope) and $\times$ (Cartesian product) suitably; see Exercise 5.5.1 for some properties of the Cartesian product of polytopes. ③
   The polytopes obtained from $[0,1]$ by a sequence of these operations are called *Hammer polytopes*, and they form an important class of examples.
2. Let K be a bounded centrally symmetric convex body in $\mathbf{R}^n$ containing 0 in its interior, and let K^* be the dual body.
   (a) Show that $\|x\|_K \cdot \|x\|_{K^*} \geq 1$ for all $x \in S^{n-1}$. ①
   (b) Let $f, g: S^{n-1} \to \mathbf{R}$ be (measurable) functions with $f(x)g(x) \geq 1$ for all $x \in S^{n-1}$. Show that $\mathrm{med}(f)\,\mathrm{med}(g) \geq 1$. ②

14.6 Dvoretzky's Theorem

Here is the remarkable Ramsey-type result in high-dimensional convexity promised at the beginning of this chapter.

14.6.1 Theorem (Dvoretzky's theorem). *For any natural number k and any real $\varepsilon > 0$, there exists an integer $n = n(k, \varepsilon)$ with the following property. For any n-dimensional centrally symmetric convex body $K \subseteq \mathbf{R}^n$, there exists a k-dimensional linear subspace $L \subseteq \mathbf{R}^n$ such that the section $K \cap L$ is $(1+\varepsilon)$-almost spherical.*

   The best known estimates give $n(k, \varepsilon) = e^{O(k/\varepsilon^2)}$.

Thus, no matter how "edgy" a high-dimensional K may be, there is always a slice of not too small dimension that is almost a Euclidean ball. Another way of expressing the statement is that any normed space of a sufficiently large dimension contains a large subspace on which the norm is very close to the Euclidean norm (with a suitable choice of a coordinate system in the subspace). Note that the Euclidean norm is the *only* norm with this universal property, since all sections of the Euclidean ball are again Euclidean balls.

As we saw in Section 14.4, the n-dimensional cube shows that the largest dimension of a 2-almost spherical section is only $O(\log n)$ in the worst case.

The assumption that K is symmetric can in fact be omitted; it suffices to require that 0 be an interior point of K. The proof of this more general version is not much more difficult than the one shown below.

We prove Dvoretzky's theorem only for $\varepsilon = 1$, since in Section 14.4 we prepared the tools for this particular setting. But the general case is not very different.

Preliminary considerations. Since affine transforms of K are practically for free in view of Lemma 14.4.1, we may assume that $B^n \subseteq K \subseteq \sqrt{n}\,B^n$ by John's lemma (Theorem 13.4.1). So the norm induced by K satisfies $n^{-1/2}\|x\| \le \|x\|_K \le \|x\|$ for all x. If f_K is the restriction of $\|\cdot\|_K$ to S^{n-1}, we have the obvious bound $\operatorname{med}(f_K) \ge n^{-1/2}$. Immediate application of Theorem 14.4.5 shows the existence of a 2-almost spherical section of K of dimension $\Omega(n\operatorname{med}(f_K)^2) = \Omega(1)$, so this approach gives nothing at all! On the other hand, it *just* fails, and a small improvement in the order of magnitude of the lower bound for $\operatorname{med}(f_K)$ already yields Dvoretzky's theorem.

We will not try to improve the estimate for $\operatorname{med}(f_K)$ directly. Instead, we find a relatively large subspace $Z \subset \mathbf{R}^n$ such that the section $K \cap Z$ can be enclosed in a not too large parallelotope P. Then we estimate, by direct computation, $\operatorname{med}(f_P)$ (over the unit sphere in Z).

The selection of the subspace Z is known as the *Dvoretzky–Rogers lemma*. We present a version with a particularly simple proof, where $\dim Z \approx n/\log n$. (For our purposes, we would be satisfied with even much weaker estimates, say $\dim Z \ge n^\delta$ for some fixed $\delta > 0$, but on the other hand, another proof gives even $\dim Z = \frac{n}{2}$.)

14.6.2 Lemma (A version of the Dvoretzky–Rogers lemma). *Let $K \subset \mathbf{R}^n$ be a centrally symmetric convex body. Then there exist a linear subspace $Z \subset \mathbf{R}^n$ of dimension $k = \lfloor \frac{n}{\log_2 n} \rfloor$, an orthonormal basis $u_1, u_2, \ldots, u_k$ of Z, and a nonsingular linear transform T of $\mathbf{R}^n$ such that if we let $\tilde{K} = T(K) \cap Z$, then $\|x\|_{\tilde{K}} \le \|x\|$ for all $x \in Z$ and $\|u_i\|_{\tilde{K}} \ge \frac{1}{2}$ for all $i = 1, 2, \ldots, k$.*

Geometrically, the lemma asserts that $\tilde{K}$ is sandwiched between the unit ball B^k and a parallelotope P as in the picture:

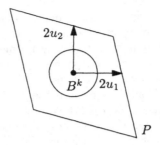

(The lemma claims that the points $2u_i$ are outside of K or on its boundary, and P is obtained by separating these points from K by hyperplanes.)

Proof. By John's lemma, we may assume $B^n \subseteq K \subseteq tB^n$, where $t = \sqrt{n}$. Interestingly, the full power of John's lemma is not needed here; the same proof works with, say, $t = n$ or $t = n^{10}$, only the bound for k would become worse by a constant factor.

Let $X_0 = \mathbf{R}^n$ and $K_0 = K$. Here is the main idea of the proof. The current body K_i is enclosed between an inner ball and an outer ball. Either K_i approaches the inner ball sufficiently closely at "many" places, and in this case we can construct the desired $u_1, \ldots, u_k$, or it stays away from the inner ball on a "large" subspace. In the latter case, we can restrict to that subspace and inflate the inner ball. But since the outer ball remains the same, the inflation of the inner ball cannot continue indefinitely. A precise argument follows; for notational reasons, instead of inflating the inner ball, we will shrink the body and the outer ball.

We consider the following condition:

(∗) Each linear subspace $Y \subseteq X_0$ with $\dim(X_0) - \dim(Y) < k$ contains a vector u with $\|u\| = 1$ and $\|u\|_{K_0} \geq \frac{1}{2}$.

This condition may or may not be satisfied. If it holds, we construct the orthonormal basis $u_1, u_2, \ldots, u_k$ by an obvious induction. If it is not satisfied, we obtain a subspace X_1 of dimension greater than $n - k$ such that $\|x\|_{K_0} \leq \frac{1}{2}\|x\|$ for all $x \in X_1$. Thus, $K_0 \cap X_1$ is twice "more spherical" than K_0. Setting $K_1 = \frac{1}{2}(K_0 \cap X_1)$, we have

$$\tfrac{2}{t} \|\cdot\| \leq \|\cdot\|_{K_1} \leq \|\cdot\|.$$

We again check the condition (∗) with X_1 and K_1 instead of X_0 and K_0. If it holds, we find the u_i within X_1, and if it does not, we obtain a subspace X_2 of dimension greater than $n - 2k$, etc. After the ith step, we have

$$\tfrac{2^i}{t} \|\cdot\| \leq \|\cdot\|_{K_i} \leq \|\cdot\|.$$

This construction cannot proceed all the way to step $i = i_0 = \lfloor \log_2 n \rfloor$, since $2^{i_0} > t = \sqrt{n}$. Thus, the condition (∗) must hold for X_{i_0-1} at the latest. We have $\dim X_{i_0-1} > n - (i_0 - 1)k \geq k$, and so the required basis $u_1, \ldots, u_k$ can be constructed. □

The parallelotope is no worse than the cube. From now on, we work within the subspace Z as in Lemma 14.6.2. For convenient notation, we assume that Z is all of $\mathbf{R}^n$ and K is as $\tilde{K}$ in the above lemma, i.e., $B^n \subseteq K$ and $\|u_i\|_K \geq \frac{1}{2}$, $i = 1, 2, \ldots, n$, where $u_1, \ldots, u_n$ is an orthonormal basis of $\mathbf{R}^n$. (Note that the reduction of the dimension from n to $n/\log n$ is nearly insignificant for the estimate of $n(k, \varepsilon)$ in Dvoretzky's theorem.)

The goal is to show that $\mathrm{med}(f_K) = \Omega\big(\sqrt{(\log n)/n}\,\big)$, where f_K is $\|\cdot\|_K$ restricted to S^{n-1}. Instead of estimating $\mathrm{med}(f_K)$, we bound the expectation $\mathbf{E}[f_K]$. Since f_K is 1-Lipschitz (we have $B^n \subseteq K$), the difference $|\mathrm{med}(f_K) - \mathbf{E}[f_K]|$ is $O(n^{-1/2})$ by Proposition 14.3.3, which is negligible compared to the lower bound we are heading for.

We have $\|\cdot\|_K \geq \|\cdot\|_P$, where P is the parallelotope as in the illustration to Lemma 14.6.2. So we actually bound $\mathbf{E}[f_P]$ from below.

First we show, by an averaging trick, that $\mathbf{E}[f_P] \geq \mathbf{E}[f_C]$, where $f_C(x) = \frac{1}{2}\|x\|_\infty = \frac{1}{2}\max_i |x_i|$ is the norm induced by the cube C of side 4. The idea of the averaging is to consider, together with a point $x = \sum_{i=1}^n \alpha_i u_i \in S^{n-1}$, the 2^n points of the form $\sum_{i=1}^n \sigma_i \alpha_i u_i$, where $\sigma \in \{-1, 1\}^n$ is a vector of signs. For any measurable function $f_P \colon S^{n-1} \to \mathbf{R}$, we have

$$\int_{S^{n-1}} \sum_{\sigma \in \{-1,1\}^n} f_P\Big(\sum_{i=1}^n \sigma_i \alpha_i u_i\Big)\, \mathrm{dP}(\alpha) = \sum_\sigma \int_{S^{n-1}} f_P\Big(\sum_{i=1}^n \sigma_i \alpha_i u_i\Big)\, \mathrm{dP}(\alpha)$$

$$= 2^n \int_{S^{n-1}} f_P(x)\, \mathrm{dP}(x) = 2^n\, \mathbf{E}[f_P].$$

The following lemma with $v_i = \alpha_i u_i$ and $|\cdot| = \|\cdot\|_P$ implies that the integrand on the left-hand side is always at least $2^n \max_i \|\alpha_i u_i\|_P \geq 2^n \cdot \frac{1}{2} \max_i |\alpha_i|$, and so indeed $\mathbf{E}[f_P] \geq \mathbf{E}[f_C]$.

14.6.3 Lemma. *Let $v_1, v_2, \ldots, v_n$ be arbitrary vectors in a normed space with norm $|\cdot|$. Then*

$$\sum_{\sigma \in \{-1,1\}^n} \Big|\sum_{i=1}^n \sigma_i v_i\Big| \geq 2^n \max_i |v_i|.$$

The proof is left as Exercise 1. It remains to estimate $\mathbf{E}[f_C]$ from below.

14.6.4 Lemma. *For a suitable positive constant c and for all n we have*

$$\mathbf{E}[f_C] = \tfrac{1}{2} \int_{S^{n-1}} \|x\|_\infty\, \mathrm{dP}(x) \geq c\sqrt{\frac{\log n}{n}},$$

where $\|x\|_\infty = \max_i |x_i|$ is the ℓ_∞ (or maximum) norm.

Note that once this lemma is proved, Dvoretzky's theorem (with $\varepsilon = 1$) follows from what we have already done and from Theorem 14.4.5.

Proof of Lemma 14.6.4. There are various proofs; a neat way is based on the generally useful fact that the n-dimensional normal distribution is spherically symmetric around the origin. We use probabilistic terminology. Let $Z_1, Z_2, \ldots, Z_n$ be independent random variables, each of them with the standard normal distribution $N(0,1)$. As was mentioned in Section 14.1, the random vector $Z = (Z_1, Z_2, \ldots, Z_n)$ has a spherically symmetric (Gaussian) distribution, and consequently, the random variable $\frac{Z}{\|Z\|}$ is uniformly distributed in S^{n-1}. Thus

$$\mathbf{E}[f_C] = \tfrac{1}{2} \mathbf{E}\left[\frac{\|Z\|_\infty}{\|Z\|}\right].$$

We show first, that we have $\|Z\| \leq \sqrt{3n}$ with probability at least $\frac{2}{3}$, and second, that for a suitable constant $c_1 > 1$, $\|Z\|_\infty \geq c_1 \sqrt{\log n}$ holds with probability at least $\frac{2}{3}$. It follows that both these events occur simultaneously with probability at least $\frac{1}{3}$, and so $\mathbf{E}[f_C] \geq c\sqrt{\log n / n}$ as claimed.

As for the Euclidean norm $\|Z\|$, we obtain $\mathbf{E}[\|Z\|^2] = n\mathbf{E}[Z_1^2] = n$, since an $N(0,1)$ random variable has variance 1. By Markov's inequality, $\mathrm{Prob}[\|Z\| \geq \sqrt{3n}] = \mathrm{Prob}[\|Z\|^2 \geq 3\mathbf{E}[\|Z\|^2]] \leq \frac{1}{3}$.

Further, by the independence of the Z_i we have

$$\mathrm{Prob}[\|Z\|_\infty \leq z] = \mathrm{Prob}[|Z_i| \leq z \text{ for all } i = 1, 2, \ldots, n]$$
$$= \mathrm{Prob}[|Z_1| \leq z]^n = \left(1 - \tfrac{2}{\sqrt{2\pi}} \int_z^\infty e^{-t^2/2}\, \mathrm{d}t\right)^n.$$

We can estimate $\int_z^\infty e^{-t^2/2}\, \mathrm{d}t \geq \int_z^{z+1} e^{-t^2/2}\, \mathrm{d}t \geq e^{-(z+1)^2/2}$. Thus, setting $z = \sqrt{\ln n} - 1$, we have $\mathrm{Prob}[\|Z\|_\infty \leq z] \leq (1 - \tfrac{2}{\sqrt{2\pi}} n^{-1/2})^n$, which is below $\frac{1}{3}$ for sufficiently large n. Lemma 14.6.4 is proved. $\qquad \square$

Bibliography and remarks. Dvoretzky and Rogers [DR50] investigated so-called unconditional convergence in infinite-dimensional Banach spaces, and as an auxiliary result, they proved a statement similar to Lemma 14.6.2, with the dimension of the subspace about $\sqrt{n}$. They used the largest inscribed ellipsoid and a variational argument (somewhat similar to the proof of John's lemma). The lemma actually holds with an $\frac{n}{2}$-dimensional subspace; for a proof due to Johnson, again using the largest inscribed ellipsoid, see Benyamini and Lindenstrauss [BL99]. The proof of Lemma 14.6.2 presented in this section is from Figiel, Lindenstrauss, and Milman [FLM77].

Dvoretzky's theorem was conjectured by Grothendieck [Gro56] and first proved by Dvoretzky [Dvo59], [Dvo61]. His proof was quite complicated, and the estimate for the dimension of the almost spherical section was somewhat worse than that in Theorem 14.6.1. Since then, several other proofs have appeared; see Lindenstrauss [Lin92] for an insightful summary. The proof shown above essentially follows Figiel et al. [FLM77], who improved and streamlined Milman's proof [Mil71] based on measure concentration. A modern proof using measure concentration for the Gaussian measure instead of that for the sphere can be found in Pisier [Pis89]. Gordon [Gor88] has a proof with more probability-theoretic flavor, using certain inequalities for Gaussian random variables (an extension of the so-called Slepian's lemma).

The dependence of the dimension of the almost spherical section on n is of order $\log n$, which is tight, as we have seen. In terms of ε, the proof presented gives a bound proportional to $\varepsilon^2 / \log \frac{1}{\varepsilon}$, and the best known general bound is proportional to ε^2 (Gordon [Gor88]).

A version of Dvoretzky's theorem for not necessarily symmetric convex bodies was established by Larman and Mani [LM75], and Gordon's proof [Gor88] is also formulated in this setting.

For $x \in \mathbf{R}^n$, let $\|x\|_p = (|x_1|^p + \cdots + |x_n|^p)^{1/p}$ denote the ℓ_p-norm of x. Here $p \in [1, \infty)$, and for the limit case $p = \infty$ we have $\|x\|_\infty = \max_i |x_i|$. For not too large p, the unit balls of ℓ_p-norms have much larger almost spherical sections than is guaranteed by Dvoretzky's theorem. For $p \in [1, 2]$, the dimension of a $(1+\varepsilon)$-almost spherical section is $c_\varepsilon n$, and for $p \geq 2$, it is $c_\varepsilon n^{2/p}$. These results are obtained by the probabilistic method, and no explicitly given sections with comparable dimensions seem to be known; see, e.g., [MS86]. There are many other estimates on the dimension of almost spherical sections, for example in terms of the so-called *type* and *cotype* of a Banach space, as well as bounds for the dimension of almost spherical *projections*. For example, by a result of Milman, for any centrally symmetric n-dimensional convex body K there is a section of an orthogonal projection of K that is $(1+\varepsilon)$-almost spherical and has dimension at least $c(\varepsilon)n$ (which is surprising, since both for sections alone and for projections alone the dimension of an almost spherical section can be only logarithmic). Such things and much more information can be found in the books Milman and Schechtman [MS86], Pisier [Pis89], and Tomczak-Jaegermann [TJ89].

Exercises

1. Prove Lemma 14.6.3. ④
2. (Large almost spherical sections of the crosspolytope) Use Theorem 14.4.5 and the method of the proof of Lemma 14.6.4 for proving that the n-dimensional unit ball of the ℓ_1-norm has a 2-almost spherical section of dimension at least cn, for a suitable constant $c > 0$. ④

15

Embedding Finite Metric Spaces into Normed Spaces

15.1 Introduction: Approximate Embeddings

We recall that a *metric space* is a pair (X, ρ), where X is a set and $\rho\colon X \times X \to [0, \infty)$ is a *metric*, satisfying the following axioms: $\rho(x, y) = 0$ if and only if $x = y$, $\rho(x, y) = \rho(y, x)$, and $\rho(x, y) + \rho(y, z) \geq \rho(x, z)$.

A metric ρ on an n-point set X can be specified by an $n \times n$ matrix of real numbers (actually $\binom{n}{2}$ numbers suffice because of the symmetry). Such tables really arise, for example, in microbiology: X is a collection of bacterial strains, and for every two strains, one can obtain their *dissimilarity*, which is some measure of how much they differ. Dissimilarity can be computed by assessing the reaction of the considered strains to various tests, or by comparing their DNA, and so on.[1] It is difficult to see any structure in a large table of numbers, and so we would like to represent a given metric space in a more comprehensible way.

For example, it would be very nice if we could assign to each $x \in X$ a point $f(x)$ in the plane in such a way that $\rho(x, y)$ equals the Euclidean distance of $f(x)$ and $f(y)$. Such representation would allow us to see the structure of the metric space: tight clusters, isolated points, and so on. Another advantage would be that the metric would now be represented by only $2n$ real numbers, the coordinates of the n points in the plane, instead of $\binom{n}{2}$ numbers as before. Moreover, many quantities concerning a point set in the plane can be computed by efficient geometric algorithms, which are not available for an arbitrary metric space.

---

[1] There are various measures of dissimilarity, and not all of them yield a metric, but many do.

This sounds very good, and indeed it is too good to be generally true: It is easy to find examples of small metric spaces that cannot be represented in this way by a planar point set. One example is 4 points, each two of them at distance 1; such points cannot be found in the plane. On the other hand, they exist in 3-dimensional Euclidean space.

Perhaps less obviously, there are 4-point metric spaces that cannot be represented (exactly) in *any* Euclidean space. Here are two examples:

The metrics on these 4-point sets are given by the indicated graphs; that is, the distance of two points is the number of edges of a shortest path connecting them in the graph. For example, in the second picture, the center has distance 1 from the leaves, and the mutual distances of the leaves are 2.

So far we have considered *isometric embeddings*. A mapping $f: X \to Y$, where X is a metric space with a metric ρ and Y is a metric space with a metric σ, is called an isometric embedding if it preserves distances, i.e., if $\sigma(f(x), f(y)) = \rho(x, y)$ for all $x, y \in X$. But in many applications we need not insist on preserving the distance exactly; rather, we can allow some distortion, say by 10%. A notion of an approximate embedding is captured by the following definition.

15.1.1 Definition (*D*-embedding of metric spaces). *A mapping $f: X \to Y$, where X is a metric space with a metric ρ and Y is a metric space with a metric σ, is called a D-embedding, where $D \geq 1$ is a real number, if there exists a number $r > 0$ such that for all $x, y \in X$,*

$$r \cdot \rho(x, y) \leq \sigma(f(x), f(y)) \leq D \cdot r \cdot \rho(x, y).$$

The infimum of the numbers D such that f is a D-embedding is called the distortion of f.

Note that this definition permits scaling of all distances in the same ratio r, in addition to the distortion of the individual distances by factors between 1 and D. If Y is a Euclidean space (or a normed space), we can rescale the image at will, and so we can choose the scaling factor r at our convenience.

Mappings with a bounded distortion are sometimes called *bi-Lipschitz mappings*. This is because the distortion of f can be equivalently defined using the Lipschitz constants of f and of the inverse mapping f^{-1}. Namely, if we define the *Lipschitz norm* of f by $\|f\|_{\text{Lip}} = \sup\{\sigma(f(x), f(y))/\rho(x, y): x, y \in X, x \neq y\}$, then the distortion of f equals $\|f\|_{\text{Lip}} \cdot \|f^{-1}\|_{\text{Lip}}$.

We are going to study the possibility of *D*-embedding of *n*-point metric spaces into Euclidean spaces and into various normed spaces. As usual, we cover only a small sample of results. Many of them are negative, showing that certain metric spaces cannot be embedded too well. But in Section 15.2

we start on an optimistic note: We present a surprising positive result of considerable theoretical and practical importance. Before that, we review a few definitions concerning ℓ_p-spaces.

The spaces ℓ_p and ℓ_p^d. For a point $x \in \mathbf{R}^d$ and $p \in [1, \infty)$, let

$$\|x\|_p = \left(\sum_{i=1}^{d} |x_i|^p \right)^{1/p}$$

denote the ℓ_p-*norm* of x. Most of the time, we will consider the case $p = 2$, i.e., the usual Euclidean norm $\|x\|_2 = \|x\|$. Another particularly important case is $p = 1$, the ℓ_1-norm (sometimes called the Manhattan distance). The ℓ_∞-norm, or *maximum norm*, is given by $\|x\|_\infty = \max_i |x_i|$. It is the limit of the ℓ_p-norms as $p \to \infty$.

Let ℓ_p^d denote the space $\mathbf{R}^d$ equipped with the ℓ_p-norm. In particular, we write ℓ_2^d in order to stress that we mean $\mathbf{R}^d$ with the usual Euclidean norm.

Sometimes we are interested in embeddings into *some* space ℓ_p^d, with p given but without restrictions on the dimension d; for example, we can ask whether there exists some Euclidean space into which a given metric space embeds isometrically. Then it is convenient to speak about ℓ_p, which is the space of all infinite sequences $x = (x_1, x_2, \ldots)$ of real numbers with $\|x\|_p < \infty$, where $\|x\|_p = \left(\sum_{i=1}^{\infty} |x_i|^p \right)^{1/p}$. In particular, ℓ_2 is the *(separable) Hilbert space*. The space ℓ_p contains each ℓ_p^d isometrically, and it can be shown that any finite metric space isometrically embeddable into ℓ_p can be isometrically embedded into ℓ_p^d for some d. (In fact, every n-point subspace of ℓ_p can be isometrically embedded into ℓ_p^d with $d \leq \binom{n}{2}$; see Exercise 15.5.2.)

Although the spaces ℓ_p are interesting mathematical objects, we will not really study them; we only use embeddability into ℓ_p as a convenient shorthand for embeddability into ℓ_p^d for some d.

Bibliography and remarks. This chapter aims at providing an overview of important results concerning low-distortion embeddings of finite metric spaces. The scope is relatively narrow, and we almost do not discuss even closely related areas, such as isometric embeddings. Another recent survey, with fewer proofs and mainly focused on algorithmic aspects, is Indyk [Ind01].

For studying approximate embeddings, it may certainly be helpful to understand isometric embeddings, and here extensive theory is available. For example, several ingenious characterizations of isometric embeddability into ℓ_2 can be found in old papers of Schoenberg (e.g., [Sch38], building on the work of mathematicians like Menger and von Neumann). A recent book concerning isometric embeddings, and embeddings into ℓ_1 in particular, is Deza and Laurent [DL97].

Another closely related area is the investigation of bi-Lipschitz maps, usually $(1+\varepsilon)$-embeddings with $\varepsilon > 0$ small, defined on an open subset of a Euclidean space (or a Banach space) and being local home-omorphisms. These mappings are called *quasi-isometries* (the definition of a quasi-isometry is slightly more general, though), and the main question is how close to an isometry such a mapping has to be, in terms of the dimension and ε; see Benyamini and Lindenstrauss [BL99], Chapters 14 and 15, for an introduction.

Exercises

1. Consider the two 4-point examples presented above (the square and the star); prove that they cannot be isometrically embedded into ℓ_2^2. ☑ Can you determine the minimum necessary distortion for embedding into ℓ_2^2?
2. (a) Prove that a bijective mapping f between metric spaces is a D-embedding if and only if $\|f\|_{\mathrm{Lip}} \cdot \|f^{-1}\|_{\mathrm{Lip}} \leq D$. ☐
   (b) Let (X, ρ) be a metric space, $|X| \geq 3$. Prove that the distortion of an embedding $f \colon X \to Y$, where (Y, σ) is a metric space, equals the supremum of the factors by which f "spoils" the ratios of distances; that is,

$$\sup \left\{ \frac{\sigma(f(x), f(y))/\sigma(f(z), f(t))}{\rho(x,y)/\rho(z,t)} : x, y, z, t \in X,\ x \neq y, z \neq t \right\}.$$

☑

15.2 The Johnson–Lindenstrauss Flattening Lemma

It is easy to show that there is no isometric embedding of the vertex set V of an n-dimensional regular simplex into a Euclidean space of dimension $k < n$. In this sense, the $(n+1)$-point set $V \subset \ell_2^n$ is truly n-dimensional. The situation changes drastically if we do not insist on exact isometry: As we will see, the set V, and any other $(n+1)$-point set in ℓ_2^n, can be almost isometrically embedded into ℓ_2^k with $k = O(\log n)$ only!

15.2.1 Theorem (Johnson–Lindenstrauss flattening lemma). *Let* X *be an* n*-point set in a Euclidean space (i.e.,* $X \subset \ell_2$*), and let* $\varepsilon \in (0, 1]$ *be given. Then there exists a* $(1+\varepsilon)$*-embedding of* X *into* ℓ_2^k*, where* $k = O(\varepsilon^{-2} \log n)$.

This result shows that any metric question about n points in ℓ_2^n can be considered for points in $\ell_2^{O(\log n)}$, if we do not mind a distortion of the distances by at most 10%, say. For example, to represent n points of ℓ_2^n in a computer, we need to store n^2 numbers. To store all of their distances, we need about n^2 numbers as well. But by the flattening lemma, we can store only $O(n \log n)$ numbers and still reconstruct any of the n^2 distances with error at most 10%.

Various proofs of the flattening lemma, including the one below, provide efficient randomized algorithms that find the almost isometric embedding into ℓ_2^k quickly. Numerous algorithmic applications have recently been found: in fast clustering of high-dimensional point sets, in approximate searching for nearest neighbors, in approximate multiplication of matrices, and also in purely graph-theoretic problems, such as approximating the bandwidth of a graph or multicommodity flows.

The proof of Theorem 15.2.1 is based on the following lemma, of independent interest.

15.2.2 Lemma (Concentration of the length of the projection). *For a unit vector $x \in S^{n-1}$, let*

$$f(x) = \sqrt{x_1^2 + x_2^2 + \cdots + x_k^2}$$

be the length of the projection of x on the subspace L_0 spanned by the first k coordinates. Consider $x \in S^{n-1}$ chosen at random. Then $f(x)$ is sharply concentrated around a suitable number $m = m(n,k)$:

$$\mathrm{P}[f(x) \geq m + t] \leq 2e^{-t^2 n/2} \text{ and } \mathrm{P}[f(x) \leq m - t] \leq 2e^{-t^2 n/2},$$

where P is the uniform probability measure on S^{n-1}. For n larger than a suitable constant and $k \geq 10 \ln n$, we have $m \geq \frac{1}{2}\sqrt{\frac{k}{n}}$.

In the lemma, the k-dimensional subspace is fixed and x is random. Equivalently, if x is a fixed unit vector and L is a random k-dimensional subspace of ℓ_2^n (as introduced in Section 14.3), the length of the projection of x on L obeys the bounds in the lemma.

Proof of Lemma 15.2.2. The orthogonal projection $p: \ell_2^n \to \ell_2^k$ given by $(x_1,\ldots,x_n) \mapsto (x_1,\ldots,x_k)$ is 1-Lipschitz, and so f is 1-Lipschitz as well. Lévy's lemma (Theorem 14.3.2) gives the tail estimates as in the lemma with $m = \mathrm{med}(f)$. It remains to establish the lower bound for m. It is not impossibly difficult to do it by elementary calculation (we need to find the measure of a simple region on S^{n-1}). But we can also avoid the calculation by a trick combined with a general measure concentration result.

For random $x \in S^{n-1}$, we have $1 = \mathbf{E}[\|x\|^2] = \sum_{i=1}^n \mathbf{E}[x_i^2]$. By symmetry, $\mathbf{E}[x_i^2] = \frac{1}{n}$, and so $\mathbf{E}[f^2] = \frac{k}{n}$. We now show that, since f is tightly concentrated, $\mathbf{E}[f^2]$ cannot be much larger than m^2, and so m is not too small.

For any $t \geq 0$, we can estimate

$$\frac{k}{n} = \mathbf{E}[f^2] \leq \mathrm{P}[f \leq m+t] \cdot (m+t)^2 + \mathrm{P}[f > m+t] \cdot \max_x (f(x)^2)$$
$$\leq (m+t)^2 + 2e^{-t^2 n/2}.$$

Let us set $t = \sqrt{k/5n}$. Since $k \geq 10 \ln n$, we have $2e^{-t^2 n/2} \leq \frac{2}{n}$, and from the above inequality we calculate $m \geq \sqrt{(k-2)/n} - t \geq \frac{1}{2}\sqrt{k/n}$.

Let us remark that a more careful calculation shows that $m = \sqrt{k/n} + O(\frac{1}{\sqrt{n}})$ for all k.   □

Proof of the flattening lemma (Theorem 15.2.1). We may assume that n is sufficiently large. Let $X \subset \ell_2^n$ be a given n-point set. We set $k = 200\varepsilon^{-2} \ln n$ (the constant can be improved). If $k \geq n$, there is nothing to prove, so we assume $k < n$. Let L be a random k-dimensional linear subspace of ℓ_2^n (obtained by a random rotation of L_0).

The chosen L is a copy of ℓ_2^k. We let $p\colon \ell_2^n \to L$ be the orthogonal projection onto L. Let m be the number around which $\|p(x)\|$ is concentrated, as in Lemma 15.2.2. We prove that for any two distinct points $x, y \in \ell_2^n$, the condition

$$(1 - \tfrac{\varepsilon}{3})m \, \|x - y\| \leq \|p(x) - p(y)\| \leq (1 + \tfrac{\varepsilon}{3})m \, \|x - y\| \qquad (15.1)$$

is violated with probability at most n^{-2}. Since there are fewer than n^2 pairs of distinct $x, y \in X$, there exists some L such that (15.1) holds for all $x, y \in X$. In such a case, the mapping p is a D-embedding of X into ℓ_2^k with $D \leq \frac{1+\varepsilon/3}{1-\varepsilon/3} < 1+\varepsilon$ (for $\varepsilon \leq 1$).

Let x and y be fixed. First we reformulate the condition (15.1). Let $u = x - y$; since p is a linear mapping, we have $p(x) - p(y) = p(u)$, and (15.1) can be rewritten as $(1 - \tfrac{\varepsilon}{3})m \, \|u\| \leq \|p(u)\| \leq (1 + \tfrac{\varepsilon}{3})m \, \|u\|$. This is invariant under scaling, and so we may suppose that $\|u\| = 1$. The condition thus becomes

$$\left| \|p(u)\| - m \right| \leq \tfrac{\varepsilon}{3}m. \qquad (15.2)$$

By Lemma 15.2.2 and the remark following it, the probability of violating (15.2), for u fixed and L random, is at most

$$4e^{-\varepsilon^2 m^2 n/18} \leq 4e^{-\varepsilon^2 k/72} < n^{-2}.$$

This proves the Johnson–Lindenstrauss flattening lemma.   □

Alternative proofs. There are several variations of the proof, which are more suitable from the computational point of view (if we really want to produce the embedding into $\ell_2^{O(\log n)}$).

In the above proof we project the set X on a random k-dimensional subspace L. Such an L can be chosen by selecting an orthonormal basis $(b_1, b_2, \ldots, b_k)$, where $b_1, \ldots, b_k$ is a random k-tuple of unit orthogonal vectors. The coordinates of the projection of x to L are the scalar products $\langle b_1, x \rangle, \ldots, \langle b_k, x \rangle$. It turns out that the condition of orthogonality of the b_i can be dropped. That is, we can pick unit vectors $b_1, \ldots, b_k \in S^{n-1}$ independently at random and define a mapping $p\colon X \to \ell_2^k$ by $x \mapsto$

$(\langle b_1, x \rangle, \ldots, \langle b_k, x \rangle)$. Using suitable concentration results, one can verify that p is a $(1+\varepsilon)$-embedding with probability close to 1. The procedure of picking the b_i is computationally much simpler.

Another way is to choose each component of each b_i from the normal distribution $N(0, 1)$, all the nk choices of the components being independent. The distribution of each b_i in $\mathbf{R}^n$ is rotationally symmetric (as was mentioned in Section 14.1). Therefore, for every fixed $u \in S^{n-1}$, the scalar product $\langle b_i, u \rangle$ also has the normal distribution $N(0, 1)$ and $\|p(u)\|^2$, the squared length of the image, has the distribution of $\sum_{i=1}^k Z_i^2$, where the Z_i are independent $N(0, 1)$. This is the well known Chi-Square distribution with k degrees of freedom, and a strong concentration result analogous to Lemma 15.2.2 can be found in books on probability theory (or derived from general measure-concentration results for the Gaussian measure or from Chernoff-type tail estimates). A still different method, particularly easy to implement but with a more difficult proof, uses independent random vectors $b_i \in \{-1, 1\}^n$.

Bibliography and remarks. The flattening lemma is from Johnson and Lindenstrauss [JL84]. They were interested in the following question: Given a metric space Y, an n-point subspace $X \subset Y$, and a 1-Lipschitz mapping $f \colon X \to \ell_2$, what is the smallest $C = C(n)$ such that there is always a C-Lipschitz mapping $\bar{f} \colon Y \to \ell_2$ extending f? They obtained the upper bound $C = O(\sqrt{\log n})$, together with an almost matching lower bound.

The alternative proof of the flattening lemma using independent normal random variables was given by Indyk and Motwani [IM98]. A streamlined exposition of a similar proof can be found in Dasgupta and Gupta [DG99]. For more general concentration results and techniques using the Gaussian distribution see, e.g., [Pis89], [MS86].

Achlioptas [Ach01] proved that the components of the b_i can also be chosen as independent uniform ± 1 random variables. Here the distribution of $\langle b_i, u \rangle$ does depend on u but the proof shows that for every $u \in S^{n-1}$, the concentration of $\|p(u)\|^2$ is at least as strong as in the case of the normally distributed b_i. This is established by analyzing higher moments of the distribution.

The sharpest known upper bound on the dimension needed for a $(1+\varepsilon)$-embedding of an n-point Euclidean metric is $\frac{16}{\varepsilon^2}(1 + o(1)) \ln n$, where $o(1)$ is with respect to $\varepsilon \to 0$ [IM98], [DG99], [Ach01]. The main term is optimal for the current proof method; see Exercises 3 and 15.3.4.

The Johnson–Lindenstrauss flattening lemma has been applied in many algorithms, both in theory and practice; see the survey [Ind01] or, for example, Kleinberg [Kle97], Indyk and Motwani [IM98], Borodin, Ostrovsky, and Rabani [BOR99].

Exercises

1. Let $x, y \in S^{n-1}$ be two points chosen independently and uniformly at random. Estimate their expected (Euclidean) distance, assuming that n is large. ③
2. Let $L \subseteq \mathbf{R}^n$ be a fixed k-dimensional linear subspace and let x be a random point of S^{n-1}. Estimate the expected distance of x from L, assuming that n is large. ③
3. (Lower bound for the flattening lemma)
   (a) Consider the $n+1$ points $0, e_1, e_2, \ldots, e_n \in \mathbf{R}^n$ (where the e_i are the vectors of the standard orthonormal basis). Check that if these points with their Euclidean distances are $(1+\varepsilon)$-embedded into ℓ_2^k, then there exist unit vectors $v_1, v_2, \ldots, v_n \in \mathbf{R}^k$ with $|\langle v_i, v_j \rangle| \leq 100\varepsilon$ for all $i \neq j$ (the constant can be improved). ②
   (b) Let A be an $n \times n$ symmetric real matrix with $a_{ii} = 1$ for all i and $|a_{ij}| \leq n^{-1/2}$ for all j, j, $i \neq j$. Prove that A has rank at least $\frac{n}{2}$. ④
   (c) Let A be an $n \times n$ real matrix of rank d, let k be a positive integer, and let B be the $n \times n$ matrix with $b_{ij} = a_{ij}^k$. Prove that the rank of B is at most $\binom{k+d}{k}$. ④
   (d) Using (a)–(c), prove that if the set as in (a) is $(1+\varepsilon)$-embedded into ℓ_2^k, where $100n^{-1/2} \leq \varepsilon \leq \frac{1}{2}$, then

$$k = \Omega\left(\frac{1}{\varepsilon^2 \log \frac{1}{\varepsilon}} \log n\right).$$

③

This proof is due to Alon (unpublished manuscript, Tel Aviv University).

15.3 Lower Bounds By Counting

In this section we explain a construction providing many "essentially different" n-point metric spaces, and we derive a general lower bound on the minimum distortion required to embed all these spaces into a d-dimensional normed space. The key ingredient is a construction of graphs without short cycles.

Graphs without short cycles. The *girth* of a graph G is the length of the shortest cycle in G. Let $m(\ell, n)$ denote the maximum possible number of edges of a simple graph on n vertices containing no cycle of length ℓ or shorter, i.e., with girth at least $\ell+1$.

We have $m(2, n) = \binom{n}{2}$, since the complete graph K_n has girth 3. Next, $m(3, n)$ is the maximum number of edges of a triangle-free graph on n vertices, and it equals $\lfloor \frac{n}{2} \rfloor \cdot \lceil \frac{n}{2} \rceil$ by Turán's theorem; the extremal example is the complete bipartite graph $K_{\lfloor n/2 \rfloor, \lceil n/2 \rceil}$. Another simple observation is that for all k, $m(2k+1, n) \geq \frac{1}{2}m(2k, n)$. This is because any graph G has a bipartite subgraph H that contains at least half of the edges of G.[2] So it suffices to care about even cycles and to consider ℓ even, remembering that the bounds for $\ell = 2k$ and $\ell = 2k+1$ are almost the same up to a factor of 2.

Here is a simple general upper bound on $m(\ell, n)$.

15.3.1 Lemma. *For all n and ℓ,*

$$m(\ell, n) \le n^{1+1/\lfloor \ell/2 \rfloor} + n.$$

Proof. It suffices to consider even $\ell = 2k$. Let G be a graph with n vertices and $m = m(2k, n)$ edges. The average degree is $\bar{d} = \frac{2m}{n}$. There is a subgraph $H \subseteq G$ with *minimum* degree at least $\delta = \frac{1}{2}\bar{d}$. Indeed, by deleting a vertex of degree smaller than δ the average degree does not decrease, and so H can be obtained by a repeated deletion of such vertices.

Let v_0 be a vertex of H. The crucial observation is that, since H has no cycle of length $2k$ or shorter, the subgraph of H induced by all vertices at distance at most k from v_0 is a tree:

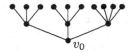

The number of vertices in this tree is at least $1+\delta+\delta(\delta-1)+\cdots+\delta(\delta-1)^{k-1} \ge (\delta-1)^k$, and this is no more than n. So $\delta \le n^{1/k}+1$ and $m = \frac{1}{2}\bar{d}n \le \delta n \le n^{1+1/k} + n$. $\qquad\square$

This simple argument yields essentially the best known upper bound. But it was asymptotically matched only for a few small values of ℓ, namely, for $\ell \in \{4, 5, 6, 7, 10, 11\}$. For $m(4, n)$ and $m(5, n)$, we need bipartite graphs without $K_{2,2}$; these were briefly discussed in Section 4.5, and we recall that they can have up to $n^{3/2}$ edges, as is witnessed by the finite projective plane. The remaining listed cases use clever algebraic constructions.

For the other ℓ, the record is also held by algebraic constructions; they are not difficult to describe, but proving that they work needs quite deep mathematics. For all $\ell \equiv 1 \,(\mathrm{mod}\,4)$ (and not on the list above), they yield $m(\ell, n) = \Omega(n^{1+4/(3\ell-7)})$, while for $\ell \equiv 3 \,(\mathrm{mod}\,4)$, they lead to $m(\ell, n) = \Omega(n^{1+4/(3\ell-9)})$.

Here we prove a weaker but simple lower bound by the probabilistic method.

---

[2] To see this, divide the vertices of G into two classes A and B arbitrarily, and while there is a vertex in one of the classes having more neighbors in its class than in the other class, move such a vertex to the other class; the number of edges between A and B increases in each step. For another proof, assign each vertex randomly to A or B and check that the expected number of edges between A and B is $\frac{1}{2}|E(G)|$.

15.3.2 Lemma. *For all $\ell \geq 3$ and $n \geq 2$, we have*

$$m(\ell, n) \geq \tfrac{1}{9}\, n^{1+1/(\ell-1)}.$$

Of course, for odd ℓ we obtain an $\Omega(n^{1+1/(\ell-2)})$ bound by using the lemma for $\ell-1$.

Proof. First we note that we may assume $n \geq 4^{\ell-1} \geq 16$, for otherwise, the bound in the lemma is verified by a path, say.

We consider the random graph $G(n,p)$ with n vertices, where each of the $\binom{n}{2}$ possible edges is present with probability p, $0 < p < 1$, and these choices are mutually independent. The value of p is going to be chosen later.

Let E be the set of edges of $G(n,p)$ and let $F \subseteq E$ be the edges contained in cycles of length ℓ or shorter. By deleting all edges of F from $G(n,p)$, we obtain a graph with no cycles of length ℓ or shorter. If we manage to show, for some m, that the expectation $\mathbf{E}[|E \setminus F|]$ is at least m, then there is an instance of $G(n,p)$ with $|E \setminus F| \geq m$, and so there exists a graph with n vertices, m edges, and of girth greater than ℓ.

We have $\mathbf{E}[|E|] = \binom{n}{2}p$. What is the probability that a fixed pair $e = \{u,v\}$ of vertices is an edge of F? First, e must be an edge of $G(n,p)$, which has probability p, and second, there must be path of length between 2 and $\ell-1$ connecting u and v. The probability that all the edges of a given potential path of length k are present is p^k, and there are fewer than n^{k-1} possible paths from u to v of length k. Therefore, the probability of $e \in F$ is at most $\sum_{k=2}^{\ell-1} p^{k+1}n^{k-1}$, which can be bounded by $2p^\ell n^{\ell-2}$, provided that $np \geq 2$. Then $\mathbf{E}[|F|] \leq \binom{n}{2} \cdot 2p^\ell n^{\ell-2}$, and by the linearity of expectation, we have

$$\mathbf{E}[|E \setminus F|] = \mathbf{E}[|E|] - \mathbf{E}[|F|] \geq \binom{n}{2}p\left(1 - 2p^{\ell-1}n^{\ell-2}\right).$$

Now, we maximize this expression as a function of p; a somewhat rough but simple choice is $p = \frac{n^{1/(\ell-1)}}{2n}$, which leads to $\mathbf{E}[|E \setminus F|] \geq \frac{1}{9}n^{1+1/(\ell-1)}$ (the constant can be improved somewhat). The assumption $np \geq 2$ follows from $n \geq 4^{\ell-1}$. Lemma 15.3.2 is proved. $\square$

There are several ways of proving a lower bound for $m(\ell, n)$ similar to that in Lemma 15.3.2, i.e., roughly $n^{1+1/\ell}$; one of the alternatives is indicated in Exercise 1 below. But obtaining a significantly better bound in an elementary way and improving on the best known bounds (of roughly $n^{1+4/3\ell}$) remain challenging open problems.

We now use the knowledge about graphs without short cycles in lower bounds for distortion.

15.3.3 Proposition (Distortion versus dimension). *Let Z be a d-dimensional normed space, such as some ℓ_p^d, and suppose that all n-point metric spaces can be D-embedded into Z. Let ℓ be an integer with $D < \ell \leq 5D$ (it is essential that ℓ be strictly larger than D, while the upper bound is only for technical convenience). Then*

$$d \geq \frac{1}{\log_2 \frac{16D\ell}{\ell - D}} \cdot \frac{m(\ell, n)}{n}.$$

Proof. Let G be a graph with vertex set $V = \{v_1, v_2, \ldots, v_n\}$ and with $m = m(\ell, n)$ edges. Let $\mathcal{G}$ denote the set of all subgraphs $H \subseteq G$ obtained from G by deleting some edges (but retaining all vertices). For each $H \in \mathcal{G}$, we define a metric ρ_H on the set V by $\rho_H(u, v) = \min(\ell, d_H(u, v))$, where $d_H(u, v)$ is the length of a shortest path connecting u and v in H.

The idea of the proof is that $\mathcal{G}$ contains many essentially different metric spaces, and if the dimension of Z were small, then there would not be sufficiently many essentially different placements of n points in Z.

Suppose that for every $H \in \mathcal{G}$ there exists a D-embedding $f_H : (V, \rho_H) \to Z$. By rescaling, we make sure that $\frac{1}{D}\rho_H(u, v) \leq \|f_H(u) - f_H(v)\|_Z \leq \rho_H(u, v)$ for all $u, v \in V$. We may also assume that the images of all points are contained in the ℓ-ball $B_Z(0, \ell) = \{x \in Z : \|x\|_Z \leq \ell\}$.

Set $\beta = \frac{1}{4}(\frac{\ell}{D} - 1)$. We have $0 < \beta \leq 1$. Let N be a β-net in $B_Z(0, \ell)$. The notion of β-net was defined above Lemma 13.1.1, and that lemma showed that a β-net in the $(d-1)$-dimensional Euclidean sphere has cardinality at most $(\frac{4}{\beta})^d$. Exactly the same volume argument proves that in our case $|N| \leq (\frac{4\ell}{\beta})^d$.

For every $H \in \mathcal{G}$, we define a new mapping $g_H : V \to N$ by letting $g_H(v)$ be the nearest point to $f_H(v)$ in N (ties resolved arbitrarily). We prove that for distinct $H_1, H_2 \in \mathcal{G}$, the mappings g_{H_1} and g_{H_2} are distinct.

The edge sets of H_1 and H_2 differ, so we can choose a pair u, v of vertices that form an edge in one of them, say in H_1, and not in the other one (H_2). We have $\rho_{H_1}(u, v) = 1$, while $\rho_{H_2}(u, v) = \ell$, for otherwise, a u–v path in H_2 of length smaller than ℓ and the edge $\{u, v\}$ would induce a cycle of length at most ℓ in G. Thus

$$\|g_{H_1}(u) - g_{H_1}(v)\|_Z < \|f_{H_1}(u) - f_{H_1}(v)\|_Z + 2\beta \leq 1 + 2\beta$$

and

$$\|g_{H_2}(u) - g_{H_2}(v)\|_Z > \|f_{H_2}(u) - f_{H_2}(v)\|_Z - 2\beta \geq \frac{\ell}{D} - 2\beta = 1 + 2\beta.$$

Therefore, $g_{H_1}(u) \neq g_{H_2}(u)$ or $g_{H_1}(v) \neq g_{H_2}(v)$.

We have shown that there are at least $|\mathcal{G}|$ distinct mappings $V \to N$. The number of all mappings $V \to N$ is $|N|^n$, and so

$$|\mathcal{G}| = 2^m \leq |N|^n \leq \left(\frac{4\ell}{\beta}\right)^{nd}.$$

The bound in the proposition follows by calculation. $\square$

15.3.4 Corollary ("Incompressibility" of general metric spaces). *If Z is a normed space such that all n-point metric spaces can be D-embedded into Z, where $D > 1$ is considered fixed and $n \to \infty$, then we have*

- $\dim Z = \Omega(n)$ for $D < 3$,
- $\dim Z = \Omega(\sqrt{n})$ for $D < 5$,
- $\dim Z = \Omega(n^{1/3})$ for $D < 7$.

This follows from Proposition 15.3.3 by substituting the asymptotically optimal bounds for $m(3,n)$, $m(5,n)$, and $m(7,n)$. The constant of proportionality in the first bound goes to 0 as $D \to 3$, and similarly for the other bounds.

The corollary shows that there is no normed space of dimension significantly smaller than n in which one could represent all n-point metric spaces with distortion smaller than 3. So, for example, one cannot save much space by representing a general n-point metric space by the coordinates of points in some suitable normed space.

It is very surprising that, as we will see later, it *is* possible to 3-embed all n-point metric spaces into a particular normed space of dimension close to $\sqrt{n}$. So the value 3 for the distortion is a real threshold! Similar thresholds occur at the values 5 and 7. Most likely this continues for all odd integers D, but we cannot prove this because of the lack of tight bounds for the number of edges in graphs without short cycles.

Another consequence of Proposition 15.3.3 concerns embedding into Euclidean spaces, without any restriction on dimension.

15.3.5 Proposition (Lower bound on embedding into Euclidean spaces). *For all n, there exist n-point metric spaces that cannot be embedded into ℓ_2 (i.e., into any Euclidean space) with distortion smaller than $c \log n / \log \log n$, where $c > 0$ is a suitable positive constant.*

Proof. If an n-point metric space is D-embedded into ℓ_2^n, then by the Johnson–Lindenstrauss flattening lemma, it can be $(2D)$-embedded into ℓ_2^d with $d \leq C \log n$ for some specific constant C.

For contradiction, suppose that $D \leq c_1 \log n / \log \log n$ with a sufficiently small $c_1 > 0$. Set $\ell = 4D$ and assume that ℓ is an integer. By Lemma 15.3.2, we have $m(\ell, n) \geq \frac{1}{9} n^{1+1/(\ell-1)} \geq C_1 n \log n$, where C_1 can be made as large as we wish by adjusting c_1. So Proposition 15.3.3 gives $d \geq \frac{C_1}{5} \log n$. If $C_1 > 5C$, we have a contradiction. $\square$

In the subsequent sections the lower bound in Proposition 15.3.5 will be improved to $\Omega(\log n)$ by a completely different method, and then we will see that this latter bound is tight.

Bibliography and remarks. The problem of constructing small graphs with given girth and minimum degree has a rich history; see, e.g., Bollobás [Bol85] for most of the earlier results.

In the proof of Lemma 15.3.1 we have derived that any graph of minimum degree δ and girth $2k+1$ has at least $1 + \delta \sum_{i=0}^{k-1} (\delta-1)^i$ vertices, and a similar lower bound for girth $2k$ is $2 \sum_{i=0}^{k-1} (\delta-1)^i$. Graphs

attaining these bounds (they are called *Moore graphs* for odd girth
and *generalized polygon graphs* for even girth) are known to exist only
in very few cases (see, e.g., Biggs [Big93] for a nice exposition). Alon,
Hoory, and Linial [AHL01] proved by a neat argument using random
walks that the same formulas still bound the number of vertices from
below if δ is the *average* degree (rather than minimum degree) of the
graph. But none of this helps improve the bound on $m(\ell, n)$ by any
substantial amount.

The proof of Lemma 15.3.2 is a variation on well known proofs by
Erdős.

The constructions mentioned in the text attaining the asymptotically optimal value of $m(\ell, n)$ for several small ℓ are due to Benson
[Ben66] (constructions with similar properties appeared earlier in Tits
[Tit59], where they were investigated for different reasons). As for the
other ℓ, graphs with the parameters given in the text were constructed
by Lazebnik, Ustimenko, and Woldar [LUW95], [LUW96] by algebraic
methods, improving on earlier bounds (such as those in Lubotzky,
Phillips, Sarnak [LPS88]; also see the notes to Section 15.5).

Proposition 15.3.5 and the basic idea of Proposition 15.3.3 were
invented by Bourgain [Bou85]. The explicit use of graphs without
short cycles and the detection of the "thresholds" in the behavior
of the dimension as a function of the distortion appeared in Matoušek
[Mat96b].

Proposition 15.3.3 implies that a normed space that should accom-
modate *all* n-point metric spaces with a given small distortion must
have large dimension. But what if we consider just one n-point metric
space M, and we ask for the minimum dimension of a normed space Z
such that M can be D-embedded into Z? Here Z can be "customized"
to M, and the counting argument as in the proof of Proposition 15.3.3
cannot work. By a nice different method, using the rank of certain
matrices, Arias-de-Reyna and Rodríguez-Piazza [AR92] proved that
for each $D < 2$, there are n-point metric spaces that do not D-embed
into any normed space of dimension below $c(D)n$, for some $c(D) > 0$.
In [Mat96b] their technique was extended, and it was shown that for
any $D > 1$, the required dimension is at least $c(\lfloor D \rfloor)n^{1/2\lfloor D \rfloor}$, so for a
fixed D it is at least a fixed power of n. The proof again uses graphs
without short cycles. An interesting open problem is whether the possibility of selecting the norm in dependence on the metric can ever
help substantially. For example, we know that if we want one normed
space for all n-point metric spaces, then a linear dimension is needed
for all distortions below 3. But the lower bounds in [AR92], [Mat96b]
for a customized normed space force linear dimension only for distortion $D < 2$. Can every n-point metric space M be 2.99-embedded, say,
into some normed space $Z = Z(M)$ of dimension $o(n)$?

We have examined the tradeoff between dimension and distortion when the distortion is a fixed number. One may also ask for the minimum distortion if the dimension d is fixed; this was considered in Matoušek [Mat90b]. For fixed d, all ℓ_p-norms on $\mathbf{R}^d$ are equivalent up to a constant, and so it suffices to consider embeddings into ℓ_2^d. Considering the n-point metric space with all distances equal to 1, a simple volume argument shows that an embedding into ℓ_2^d has distortion at least $\Omega(n^{1/d})$. The exponent can be improved by a factor of roughly 2; more precisely, for any $d \geq 1$, there exist n-point metric spaces requiring distortion $\Omega\left(n^{1/\lfloor (d+1)/2 \rfloor}\right)$ for embedding into ℓ_2^d (these spaces are even isometrically embeddable into ℓ_2^{d+1}). They are obtained by taking a q-dimensional simplicial complex that cannot be embedded into $\mathbf{R}^{2q}$ (a Van Kampen–Flores complex; for modern treatment see, e.g., [Sar91] or [Živ97]), considering a geometric realization of such a complex in $\mathbf{R}^{2q+1}$, and filling it with points uniformly (taking an η-net within it for a suitable η, in the metric sense); see Exercise 3 below for the case $q = 1$. For $d = 1$ and $d = 2$, this bound is asymptotically tight, as can be shown by an inductive argument [Mat90b]. It is also almost tight for all even d. An upper bound of $O(n^{2/d} \log^{3/2} n)$ for the distortion is obtained by first embedding the considered metric space into ℓ_2^n (Theorem 15.7.1), and then projecting on a random d-dimensional subspace; the analysis is similar to the proof of the Johnson–Lindenstrauss flattening lemma. It would be interesting to close the gap for odd $d \geq 3$; the case $d = 1$ suggests that perhaps the lower bound might be the truth. It is also rather puzzling that the (suspected) bound for the distortion for fixed dimension, $D \approx n^{1/\lfloor (d+1)/2 \rfloor}$, looks optically similar to the (suspected) bound for dimension given the distortion (Corollary 15.3.4), $d \approx n^{1/\lfloor (D+1)/2 \rfloor}$. Is this a pure coincidence, or is it trying to tell us something?

Exercises

1. (Erdős–Sachs construction) This exercise indicates an elegant proof, by Erdős and Sachs [ES63], of the existence of graphs without short cycles whose number of edges is not much smaller than in Lemma 15.3.2 and that are *regular*. Let $\ell \geq 3$ and $\delta \geq 3$.

   (a) (Starting graph) For all δ and ℓ, construct a finite δ-regular graph $G(\delta, \ell)$ with no cycles of length ℓ or shorter; the number of vertices does not matter. One possibility is by double induction: Construct $G(\delta+1, \ell)$ using $G(\delta, \ell)$ and $G(\delta', \ell-1)$ with a suitable δ'. ▣

   (b) Let G be a δ-regular graph of girth at least $\ell+1$ and let u and v be two vertices of G at distance at least $\ell+2$. Delete them together with their incident edges, and connect their neighbors by a matching:

Check that the resulting graph still does not contain any cycle of length at most ℓ. [2]

(c) Show that starting with a graph as in (a) and reducing it by the operations as in (b), we arrive at a δ-regular graph of girth $\ell+1$ and with at most $1 + \delta + \delta(\delta-1) + \cdots + \delta(\delta-1)^\ell$ vertices. What is the resulting asymptotic lower bound for $m(n,\ell)$, with ℓ fixed and $n \to \infty$? [1]

2. (Sparse spanners) Let G be a graph with n vertices and with positive real weights on edges, which represent the edge lengths. A subgraph H of G is called a *t-spanner* of G if the distance of any two vertices u, v in H is no more than t times their distance in G (both the distances are measured in the shortest-path metric). Using Lemma 15.3.1, prove that for every G and every integer $t \geq 2$, there exists a t-spanner with $O\left(n^{1+1/\lfloor t/2 \rfloor}\right)$ edges. [4]

3. Let G_n denote the graph arising from K_5, the complete graph on 5 vertices, by subdividing each edge $n-1$ times; that is, every two of the original vertices of K_5 are connected by a path of length n. Prove that the vertex set of G_n, considered as a metric space with the graph-theoretic distance, cannot be embedded into the plane with distortion smaller than $const \cdot n$. [3]

4. (Another lower bound for the flattening lemma)
   (a) Given $\varepsilon \in (0, \frac{1}{2})$ and n sufficiently large in terms of ε, construct a collection $\mathcal{V}$ of ordered n-tuples of points of ℓ_2^n such that the distance of every two points in each $V \in \mathcal{V}$ is between two suitable constants, no two $V \neq V' \in \mathcal{V}$ can have the same $(1+\varepsilon)$-embedding (that is, there are i, j such that the distances between the ith point and the jth point in V and in V' differ by a factor of at least $1+\varepsilon$), and $\log |\mathcal{V}| = \Omega(\varepsilon^{-2} n \log n)$. [4]
   (b) Use (a) and the method of this section to prove a lower bound of $\Omega(\frac{1}{\varepsilon^2 \log \frac{1}{\varepsilon}} \log n)$ for the dimension in the Johnson–Lindenstrauss flattening lemma. [2]

15.4 A Lower Bound for the Hamming Cube

We have established the existence of n-point metric spaces requiring the distortion close to $\log n$ for embedding into ℓ_2 (Proposition 15.3.5), but we have not constructed any specific metric space with this property. In this section we prove a weaker lower bound, only $\Omega(\sqrt{\log n})$, but for a specific and very simple space: the Hamming cube. Later on, we extend the proof method and exhibit metric spaces with $\Omega(\log n)$ lower bound, which turns out to be optimal. We recall that C_m denotes the space $\{0,1\}^m$ with the Hamming (or ℓ_1) metric, where the distance of two $0/1$ sequences is the number of places where they differ.

15.4.1 Theorem. *Let $m \geq 2$ and $n = 2^m$. Then there is no D-embedding of the Hamming cube C_m into ℓ_2 with $D < \sqrt{m} = \sqrt{\log_2 n}$. That is, the natural embedding, where we regard $\{0,1\}^m$ as a subspace of ℓ_2^m, is optimal.*

The reader may remember, perhaps with some dissatisfaction, that at the beginning of this chapter we mentioned the 4-cycle as an example of a metric space that cannot be isometrically embedded into any Euclidean space, but we gave no reason. Now, we are obliged to rectify this, because the 4-cycle is just the 2-dimensional Hamming cube.

The intuitive reason why the 4-cycle cannot be embedded isometrically is that if we embed the vertices so that the edges have the right length, then at least one of the diagonals is too short. We make this precise using a slightly more complicated notation than necessary, in anticipation of later developments.

Let V be a finite set, let ρ be a metric on V, and let $E, F \subseteq \binom{V}{2}$ be nonempty sets of pairs of points of V. As our running example, $V = \{v_1, \ldots, v_4\}$ is the set of vertices of the 4-cycle, ρ is the graph metric on it, $E = \{\{v_1, v_2\}, \{v_2, v_3\}, \{v_3, v_4\}, \{v_4, v_1\}\}$ are the edges, and $F = \{\{v_1, v_3\}, \{v_2, v_4\}\}$ are the diagonals.

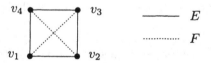

Let us introduce the abbreviated notation

$$\rho^2(E) = \sum_{\{u,v\} \in E} \rho(u,v)^2.$$

We consider the ratio

$$R_{E,F}(\rho) = \sqrt{\frac{\rho^2(F)}{\rho^2(E)}};$$

the subscripts E, F will be omitted unless there is danger of confusion. For our 4-cycle, $R(\rho)$ is a kind of ratio of "diagonals to edges" but with quadratic averages of distances, and it equals $\sqrt{2}$ (right?).

Next, let $f: V \to \ell_2^d$ be a D-embedding of the considered metric space into a Euclidean space. This defines another metric σ on V: $\sigma(u,v) = \|f(u) - f(v)\|$. With the same E and F, let us now look at the ratio $R(\sigma)$.

If f is a D-embedding, then $R(\sigma) \geq R(\rho)/D$. But according to the idea mentioned above, in any embedding of the 4-cycle into a Euclidean space, the diagonals are always too short, and so $R(\sigma)$ can be expected to be smaller than $\sqrt{2}$ in this case. This is confirmed by the following lemma, which (with $x_i = f(v_i)$) shows that $R(\sigma) \leq 1$ and therefore $D \geq \sqrt{2}$.

15.4.2 Lemma (Short diagonals lemma). *Let* x_1, x_2, x_3, x_4 *be arbitrary points in a Euclidean space. Then*

$$\|x_1 - x_3\|^2 + \|x_2 - x_4\|^2 \leq \|x_1 - x_2\|^2 + \|x_2 - x_3\|^2 + \|x_3 - x_4\|^2 + \|x_4 - x_1\|^2.$$

Proof. Four points can be assumed to lie in $\mathbf{R}^3$, so one could start some stereometric calculations. But a better way is to observe that it suffices to prove the lemma for points on the real line! Indeed, for the x_i in some $\mathbf{R}^d$ we can write the 1-dimensional inequality for each coordinate and then add these inequalities together. (This is the reason for using squares in the definition of the ratio $R(\sigma)$: Squares of Euclidean distances split into the contributions of individual coordinates, and so they are easier to handle than the distances themselves.)

If the x_i are real numbers, we calculate

$$(x_1 - x_2)^2 + (x_2 - x_3)^2 + (x_3 - x_4)^2 + (x_4 - x_1)^2 - (x_1 - x_3)^2 - (x_2 - x_4)^2$$

$$= (x_1 - x_2 + x_3 - x_4)^2 \geq 0,$$

and this is the desired inequality. $\qquad\square$

Proof of Theorem 15.4.1. We proceed as in the 2-dimensional case. Let $V = \{0, 1\}^m$ be the vertex set of C_m, let ρ be the Hamming metric, let E be the set of edges of the cube (pairs of points at distance 1), and let F be the set of the long diagonals. The long diagonals are pairs of points at distance m, or in other words, pairs $\{u, \overline{u}\}$, $u \in V$, where $\overline{u}$ is the vector arising from u by changing 0's to 1's and 1's to 0's.

We have $|E| = m2^{m-1}$ and $|F| = 2^{m-1}$, and we calculate $R_{E,F}(\rho) = \sqrt{m}$. If σ is a metric on V induced by some embedding $f: V \to \ell_2^d$, we want to show that $R_{E,F}(\sigma) \leq 1$; this will give the theorem. So we need to prove that $\sigma^2(F) \leq \sigma^2(E)$. This follows from the inequality for the 4-cycle (Lemma 15.4.2) by a convenient induction.

The basis for $m = 2$ is directly Lemma 15.4.2. For larger m, we divide the vertex set V into two parts V_0 and V_1, where V_0 are the vectors with the last component 0, i.e., of the form $u0$, $u \in \{0, 1\}^{m-1}$. The set V_0 induces an $(m-1)$-dimensional subcube. Let E_0 be its edge set and F_0 the set of its long diagonals; that is, $F_0 = \{\{u0, \overline{u}0\}: u \in \{0, 1\}^{m-1}\}$, and similarly for E_1 and F_1. Let $E_{01} = E \setminus (E_0 \cup E_1)$ be the edges of the m-dimensional cube going between the two subcubes. By induction, we have

$$\sigma^2(F_0) \leq \sigma^2(E_0) \text{ and } \sigma^2(E_1) \leq \sigma^2(F_1).$$

For $u \in \{0, 1\}^{m-1}$, we consider the quadrilateral with vertices $u0, \overline{u}0, \overline{u}1, u1$; for $u = 00$, it is indicated in the picture:

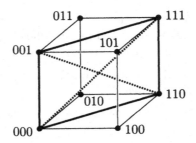

Its sides are two edges of E_{01}, one diagonal from F_0 and one from F_1, and its diagonals are from F. If we write the inequality of Lemma 15.4.2 for this quadrilateral and sum up over all such quadrilaterals (they are 2^{m-2}, since u and $\bar{u}$ yield the same quadrilaterals), we get

$$\sigma^2(F) \le \sigma^2(E_{01}) + \sigma^2(F_0) + \sigma^2(F_1).$$

By the inductive assumption for the two subcubes, the right-hand side is at most $\sigma^2(E_{01}) + \sigma^2(E_0) + \sigma^2(E_1) = \sigma^2(E)$.   □

Bibliography and remarks.   Theorem 15.4.1, found by Enflo [Enf69], is probably the first result showing an unbounded distortion for embeddings into Euclidean spaces. Enflo considered the problem of uniform embeddability among Banach spaces, and the distortion was an auxiliary device in his proof.

Exercises

1. Consider the second graph in the introductory section, the star with 3 leaves, and prove a lower bound of $\frac{2}{\sqrt{3}}$ for the distortion required to embed into a Euclidean space. Follow the method used for the 4-cycle. [3]
2. (Planar graphs badly embeddable into ℓ_2) Let $G_0, G_1, \dots$ be the following graphs:

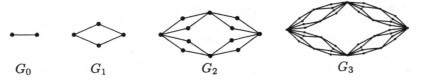

G_{i+1} is obtained from G_i by replacing each edge by a square with two new vertices. Using the short diagonals lemma and the method of this section, prove that any Euclidean embedding of G_m (with the graph metric) requires distortion at least $\sqrt{m+1}$. [4]
This result is due to Newman and Rabinovich [NR01].

3. (Almost Euclidean subspaces) Prove that for every k and $\varepsilon > 0$ there exists $n = n(k, \varepsilon)$ such that every n-point metric space (X, ρ) contains a k-point subspace that is $(1+\varepsilon)$-embeddable into ℓ_2. Use Ramsey's theorem. [5]

This result is due to Bourgain, Figiel, and Milman [BFM86]; it is a kind of analogue of Dvoretzky's theorem for metric spaces.

15.5 A Tight Lower Bound via Expanders

Here we provide an explicit example of an n-point metric space that requires distortion $\Omega(\log n)$ for embedding into any Euclidean space. It is the vertex set of a constant-degree expander G with the graph metric. In the proof we are going to use bounds on the second eigenvalue of G, but for readers not familiar with the important notion of expander graphs, we first include a little wider background.

Roughly speaking, expanders are graphs that are sparse but well connected. If a model of an expander is made with vertices being little balls and edges being thin strings, it is difficult to tear off any subset of vertices, and the more vertices we want to tear off, the larger effort that is needed.

More formally, we define the *edge expansion* (also called the *conductance*) $\Phi(G)$ of a graph $G = (V, E)$ as

$$\min \left\{ \frac{e(A, V \setminus A)}{|A|} : A \subset V, 1 \leq |A| \leq \tfrac{1}{2}|V| \right\},$$

where $e(A, B)$ is the number of edges of G going between A and B. One can say, still somewhat imprecisely, that a graph G is a good expander if $\Phi(G)$ is not very small compared to the average degree of G.

In this section, we consider r-regular graphs for a suitable constant $r \geq 3$, say $r = 3$. We need r-regular graphs with an arbitrary large number n of vertices and with edge expansion bounded below by a positive constant independent of n. Such graphs are usually called *constant-degree expanders*.[3]

It is useful to note that, for example, the edge expansion of the $n \times n$ planar square grid tends to 0 as $n \to \infty$. More generally, it is known that constant-degree expanders cannot be planar; they must be much more tangled than planar graphs.

The existence of constant-degree expanders is not difficult to prove by the probabilistic method; for every fixed $r \geq 3$, random r-regular graphs provide very good expanders. With considerable effort, explicit constructions have been found as well; see the notes to this section.

---

[3] A rigorous definition should be formulated for an infinite *family* of graphs. A family $\{G_1, G_2, \ldots\}$ of r-regular graphs with $|V(G_i)| \to \infty$ as $i \to \infty$ is a family of constant-degree expanders if the edge expansion of all G_i is bounded below by a positive constant independent of i.

Let us remark that several notions similar to edge expansion appear in the literature, and each of them can be used for quantifying how good an expander a given graph is (but they usually lead to an equivalent notion of a family of constant-degree expanders). Often it is also useful to consider nonregular expanders or expanders with larger than constant degree, but regular constant-degree expanders are probably used most frequently.

Now, we pass to the second eigenvalue. For our purposes it is most convenient to talk about eigenvalues of the Laplacian of the considered graph. Let $G = (V, E)$ be an r-regular graph. The *Laplacian matrix L_G* of G is an $n \times n$ matrix, $n = |V|$, with both rows and columns indexed by the vertices of G, defined by

$$(L_G)_{uv} = \begin{cases} r & \text{for } u = v, \\ -1 & \text{if } u \neq v \text{ and } \{u, v\} \in E(G), \\ 0 & \text{otherwise.} \end{cases}$$

It is a symmetric positive semidefinite real matrix, and it has n real eigenvalues $\mu_1 = 0 \leq \mu_2 \leq \cdots \leq \mu_n$. The *second eigenvalue $\mu_2 = \mu_2(G)$* is a fundamental parameter of the graph G.[4]

Somewhat similar to edge expansion, $\mu_2(G)$ describes how much G "holds together," but in a different way. The edge expansion and $\mu_2(G)$ are related but they do *not* determine each other. For every r-regular graph G, we have $\mu_2(G) \geq \frac{\Phi(G)^2}{4r}$ (see, e.g., Lovász [Lov93], Exercise 11.31 for a proof) and $\mu_2(G) \leq 2\Phi(G)$ (Exercise 6). Both the lower and the upper bound can almost be attained for some graphs.

For our application below, we need the following fact: There are constants r and $\beta > 0$ such that for sufficiently many values of n (say for at least one n between 10^k and 10^{k+1}), there exists an n-vertex r-regular graph G with $\mu_2(G) \geq \beta$. This follows from the existence results for constant-degree expanders mentioned above (random 3-regular graphs will do, for example), and actually most of the known explicit constructions of expanders bound the second eigenvalue directly.

We are going to use the lower bound on $\mu_2(G)$ via the following fact:

For all real vectors $(x_v)_{v \in V}$ with $\sum_{v \in V} x_v = 0$, we have $x^T L_G x \geq \mu_2 \|x\|$. (15.3)

To understand what is going on here, we recall that every symmetric real $n \times n$ matrix has n real eigenvalues (not necessarily distinct), and the corresponding n unit eigenvectors $b_1, b_2, \ldots, b_n$ form an orthonormal basis of $\mathbf{R}^n$. For the

---

[4] The notation μ_i for the eigenvalues of L_G is not standard. We use it in order to distinguish these eigenvalues from the eigenvalues $\lambda_1 \geq \lambda_2 \geq \cdots \geq \lambda_n$ of the *adjacency matrix A_G* usually considered in the literature, where $(A_G)_{uv} = 1$ if $\{u, v\} \in E(G)$ and $(A_G)_{uv} = 0$ otherwise. Here we deal exclusively with regular graphs, for which the eigenvalues of A_G are related to those of L_G in a very simple way: $\lambda_i = r - \mu_i$, $i = 1, 2 \ldots, n$, for any r-regular graph.

matrix L_G, the unit eigenvector b_1 belonging to the eigenvalue $\mu_1 = 0$ is $n^{-1/2}(1, 1, \ldots, 1)$. So the condition $\sum_{v \in V} x_v = 0$ means the orthogonality of x to b_1, and we have $x = \sum_{i=1}^{n} \alpha_i b_i$ for suitable real α_i with $\alpha_1 = 0$. We calculate, using $x^T b_i = \alpha_i$,

$$x^T L_G x = \sum_{i=2}^{n} x^T (\alpha_i L_G b_i) = \sum_{i=2}^{n} \alpha_i \mu_i x^T b_i = \sum_{i=2}^{n} \alpha_i^2 \mu_i \geq \mu_2 \sum_{i=2}^{n} \alpha_i^2 = \mu_2 \|x\|^2.$$

This proves (15.3), and we can also see that $x = b_2$ yields equality in (15.3). So we can write $\mu_2 = \min\{x^T L_G x \colon \|x\| = 1, \sum_{v \in V} x_v = 0\}$ (this is a special case of the variational definition of eigenvalues discussed in many textbooks of linear algebra).

Now, we are ready to prove the main result of this section.

15.5.1 Theorem (Expanders are badly embeddable into ℓ_2). *Let G be an r-regular graph on an n-element vertex set V with $\mu_2(G) \geq \beta$, where $r \geq 3$ and $\beta > 0$ are constants, and let ρ be the shortest-path metric on V. Then the metric space (V, ρ) cannot be D-embedded into a Euclidean space for $D \leq c \log n$, where $c = c(r, \beta) > 0$ is independent of n.*

Proof. We again consider the ratios $R_{E,F}(\rho)$ and $R_{E,F}(\sigma)$ as in the proof for the cube (Theorem 15.4.1). This time we let E be the edge set of G, and $F = \binom{V}{2}$ are all pairs of distinct vertices. In the graph metric all pairs in E have distance 1, while most pairs in F have distance about $\log n$, as we will check below. On the other hand, it turns out that in any embedding into ℓ_2 such that all the distances in E are at most 1, a typical distance in F is only $O(1)$. The calculations follow.

We have $\rho^2(E) = |E| = \frac{nr}{2}$. To bound $\rho^2(F)$ from below, we observe that for each vertex v_0, there are at most $1 + r + r(r-1) + \cdots + r(r-1)^{k-1} \leq r^k + 1$ vertices at distance at most k from v_0. So for $k = \log_r \frac{n-1}{2}$, at least half of the pairs in F have distance more than k, and we obtain $\rho^2(F) = \Omega(n^2 k^2) = \Omega(n^2 \log^2 n)$. Thus

$$R_{E,F}(\rho) = \Omega\left(\sqrt{n} \cdot \log n\right).$$

Let $f \colon V \to \ell_2^d$ be an embedding into a Euclidean space, and let σ be the metric induced by it on V. To prove the theorem, it suffices to show that $R_{E,F}(\sigma) = O(\sqrt{n})$; that is,

$$\sigma^2(F) = O(n\sigma^2(E)).$$

By the observation in the proof of Lemma 15.4.2 about splitting into coordinates, it is enough to prove this inequality for a one-dimensional embedding. So for every choice of real numbers $(x_v)_{v \in V}$, we want to show that

$$\sum_{\{u,v\} \in F} (x_u - x_v)^2 = O(n) \sum_{\{u,v\} \in E} (x_u - x_v)^2. \tag{15.4}$$

By adding a suitable number to all the x_v, we may assume that $\sum_{v \in V} x_v = 0$. This does not change anything in (15.4), but it allows us to relate both sides to the Euclidean norm of the vector x.

We calculate, using $\sum_{v \in V} x_v = 0$,

$$\sum_{\{u,v\} \in F} (x_u - x_v)^2 = (n-1) \sum_{v \in V} x_v^2 - \sum_{u \neq v} x_u x_v = n \sum_{v \in V} x_v^2 - \left(\sum_{v \in V} x_v\right)^2 = n \|x\|^2.$$

For the right-hand side of (15.4), the Laplace matrix enters:

$$\sum_{\{u,v\} \in E} (x_u - x_v)^2 = r \sum_{v \in V} x_v^2 - 2 \sum_{\{u,v\} \in E} x_u x_v = x^T L_G x \geq \mu_2 \|x\|^2,$$

the last inequality being (15.3). This establishes (15.4) and concludes the proof of Theorem 15.5.1.   □

The proof actually shows that the maximum of $R_{E,F}(\sigma)$ over all Euclidean metrics σ equals $\sqrt{\mu_2/n}$ (which is an interesting geometric interpretation of μ_2). The maximum is attained for the σ induced by the mapping $V \to \mathbf{R}$ specified by b_2, the eigenvector belonging to μ_2.

The cone of squared ℓ_2-metrics and universality of the lower-bound method. For the Hamming cubes, we obtained the exact minimum distortion required for a Euclidean embedding. This was due to the lucky choice of the sets E and F of point pairs. As we will see below, a "lucky" choice, leading to an exact bound, exists for every finite metric space if we allow for sets of *weighted* pairs. Let (V, ρ) be a finite metric space and let $\eta, \varphi \colon \binom{V}{2} \to [0, \infty)$ be weight functions. We define

$$\rho^2(\eta) = \sum_{\{u,v\} \in \binom{V}{2}} \eta(u,v) \rho(u,v)^2$$

and similarly for $\rho^2(\varphi)$, and we let

$$R_{\eta,\varphi}(\rho) = \sqrt{\frac{\rho^2(\varphi)}{\rho^2(\eta)}}.$$

15.5.2 Proposition. *Let (V, ρ) be a finite metric space and let $D \geq 1$ be the smallest number such that (V, ρ) can be D-embedded into ℓ_2. Then there are weight functions $\eta, \varphi \colon \binom{V}{2} \to [0, \infty)$ such that $R_{\eta,\varphi}(\rho) \geq D$ and $R_{\eta,\varphi}(\sigma) \leq 1$ for any metric σ induced on V by an embedding into ℓ_2.*

Thus, the exact lower bound for the embeddability into Euclidean spaces always has an "easy" proof, provided that we can guess the right weight functions η and φ. (As we will see below, there is even an efficient algorithm for deciding D-embeddability into ℓ_2.)

Proposition 15.5.2 is included mainly because of generally useful concepts appearing in its proof.

Let V be a fixed n-point set. An arbitrary function $\varphi\colon \binom{V}{2} \to \mathbf{R}$, assigning a real number to each unordered pair of points of V, can be represented by a point in $\mathbf{R}^N$, where $N = \binom{n}{2}$; the coordinates of such a point are indexed by pairs $\{u,v\} \in \binom{V}{2}$. For example, the set of all metrics on V corresponds to a subset of $\mathbf{R}^N$ called the *metric cone* (also see the notes to Section 5.5). As is not difficult to verify, it is an N-dimensional convex polyhedron in $\mathbf{R}^N$. Its combinatorial structure has been studied intensively.

In the proof of Proposition 15.5.2 we will not work with the metric cone but rather with the *cone of squared Euclidean metrics*, denoted by $\mathcal{L}_2$. We define
$$\mathcal{L}_2 = \left\{ \left(\|f(u) - f(v)\|^2 \right)_{\{u,v\}\in\binom{V}{2}} \colon f\colon V \to \ell_2 \right\} \subset \mathbf{R}^N.$$

15.5.3 Observation. *The set $\mathcal{L}_2$ is a convex cone.*

Proof. Clearly, if $x \in \mathcal{L}_2$, then $\lambda x \in \mathcal{L}_2$ for all $\lambda \geq 0$, and so it suffices to verify that if $x, y \in \mathcal{L}_2$, then $x + y \in \mathcal{L}_2$. Let $x, y \in \mathcal{L}_2$ correspond to embeddings $f\colon V \to \ell_2^k$ and $g\colon V \to \ell_2^m$, respectively. We define a new embedding $h\colon V \to \ell_2^{k+m}$ by concatenating the coordinates of f and g; that is,
$$h(v) = (f(v)_1, \dots, f(v)_k, g(v)_1, \dots, g(v)_m) \in \ell_2^{k+m}.$$
The point of $\mathcal{L}_2$ corresponding to h is $x + y$. $\qquad\square$

Proof of Proposition 15.5.2. Suppose that (V, ρ) cannot be D-embedded into any Euclidean space. We are going to exhibit η and φ with $R_{\eta,\varphi}(\rho) \geq D$ and $R_{\eta,\varphi}(\sigma) \leq 1$ for every Euclidean σ. The claim of the proposition is easily derived from this by a compactness argument.

Let $\mathcal{L}_2 \subset \mathbf{R}^N$ be the cone of squared Euclidean metrics on V as above and let
$$\mathcal{K} = \Big\{ (x_{uv})_{\{u,v\}\in\binom{V}{2}} \in \mathbf{R}^N\colon \text{ there exists an } r > 0 \text{ with}$$
$$r^2 \rho(u,v)^2 \leq x_{uv} \leq D^2 r^2 \rho(u,v)^2 \text{ for all } u, v \Big\}.$$

This $\mathcal{K}$ includes all squares of metrics arising by D-embeddings of (V, ρ). But not all elements of $\mathcal{K}$ are necessarily squares of metrics, since the triangle inequality may be violated. Since there is no Euclidean D-embedding of (V, ρ), we have $\mathcal{K} \cap \mathcal{L}_2 = \emptyset$. Both $\mathcal{K}$ and $\mathcal{L}_2$ are convex sets in $\mathbf{R}^N$, and so they can be separated by a hyperplane, by the separation theorem (Theorem 1.2.4). Moreover, since $\mathcal{L}_2$ is a cone and $\mathcal{K}$ is a cone minus the origin 0, the separating hyperplane has to pass through 0. So there is an $a \in \mathbf{R}^N$ such that
$$\langle a, x \rangle \geq 0 \text{ for all } x \in \mathcal{K} \text{ and } \langle a, x \rangle \leq 0 \text{ for all } x \in \mathcal{L}_2. \tag{15.5}$$

Using this a, we define the desired η and φ, as follows:

$$\eta(u, v) = \begin{cases} a_{uv} & \text{if } a_{uv} \geq 0, \\ 0 & \text{otherwise;} \end{cases}$$

$$\varphi(u, v) = \begin{cases} -a_{uv} & \text{if } a_{uv} < 0, \\ 0 & \text{otherwise.} \end{cases}$$

First we show that $R_{\eta,\varphi}(\rho) \geq D$. To this end, we employ the property (15.5) for the following $x \in \mathcal{K}$:

$$x_{uv} = \begin{cases} D^2 \rho(u, v)^2 & \text{if } a_{uv} \geq 0, \\ \rho(u, v)^2 & \text{if } a_{uv} < 0. \end{cases}$$

Then $\langle a, x \rangle \geq 0$ boils down to $D^2 \rho^2(\eta) - \rho^2(\varphi) \geq 0$, which means that $R_{\eta,\varphi}(\rho) \geq D$.

Next, let σ be a metric induced by a Euclidean embedding of V. This time we apply $\langle a, x \rangle \leq 0$ with the $x \in \mathcal{L}_2$ corresponding to σ, i.e., $x_{uv} = \sigma(u, v)^2$. This yields $\sigma^2(\eta) - \sigma^2(\varphi) \leq 0$, and so $R_{\eta,\varphi}(\sigma) \leq 1$. This proves Proposition 15.5.2. □

Algorithmic remark: Euclidean embeddings and semidefinite programming. The problem of deciding whether a given n-point metric space (V, ρ) admits a D-embedding into ℓ_2 (i.e., into a Euclidean space without restriction on the dimension), for a given $D \geq 1$, can be solved by a polynomial-time algorithm. Let us stress that the dimension of the target Euclidean space cannot be prescribed in this method. If we insist that the embedding be into ℓ_2^d, for some given d, we obtain a different algorithmic problem, and it is not known how hard it is. Many other similar-looking embedding problems are known to be NP-hard, such as the problem of D-embedding into ℓ_1.

The algorithm for D-embedding into ℓ_2 is based on a powerful technique called *semidefinite programming*, where the problem is expressed as the existence of a positive semidefinite matrix in a suitable convex set of matrices.

Let (V, ρ) be an n-point metric space, let $f: V \to \mathbf{R}^n$ be an embedding, and let X be the $n \times n$ matrix whose columns are indexed by the elements of V and such that the vth column is the vector $f(v) \in \mathbf{R}^n$. The matrix $Q = X^T X$ has both rows and columns indexed by the points of V, and the entry q_{uv} is the scalar product $\langle f(u), f(v) \rangle$.

The matrix Q is positive semidefinite, since for any $x \in \mathbf{R}^n$, we have $x^T Q x = (x^T X^T)(X x) = \|X x\|^2 \geq 0$. (In fact, as is not too difficult to check, a real symmetric $n \times n$ matrix P is positive semidefinite if *and only if* it can be written as $X^T X$ for some real $n \times n$ matrix X.)

Let $\sigma(u, v) = \|f(u) - f(v)\| = \langle f(u) - f(v), f(u) - f(v) \rangle^{1/2}$. We can express

$$\sigma(u, v)^2 = \langle f(u), f(u) \rangle + \langle f(v), f(v) \rangle - 2\langle f(u), f(v) \rangle = q_{uu} + q_{vv} - 2q_{uv}.$$

Therefore, the space (V, ρ) can be D-embedded into ℓ_2 if and only if there exists a symmetric real positive semidefinite matrix Q whose entries satisfy the following constraints:

$$\rho(u,v)^2 \le q_{uu} + q_{vv} - 2q_{uv} \le D^2\rho(u,v)^2$$

for all $u,v \in V$. These are linear inequalities for the unknown entries of Q.

The problem of finding a positive semidefinite matrix whose entries satisfy a given system of linear inequalities can be solved efficiently, in time polynomial in the size of the unknown matrix Q and in the number of the linear inequalities. The algorithm is not simple; we say a little more about it in the remarks below.

Bibliography and remarks. Theorem 15.5.1 was proved by Linial, London, and Rabinovich [LLR95]. This influential paper introduced methods and results concerning low-distortion embeddings, developed in local theory of Banach spaces, into theoretical computer science, and it gave several new results and algorithmic applications. It is very interesting that using low-distortion Euclidean embeddings, one obtains algorithmic results for certain graph problems that until then could not be attained by other methods, although the considered problems look purely graph-theoretic without any geometric structure. A simple but important example is presented at the end of Section 15.7.

The bad embeddability of expanders was formulated and proved in [LLR95] in connection with the problem of multicommodity flows in graphs. The proof was similar to the one shown above, but it established an $\Omega(\log n)$ bound for embedding into ℓ_1. The result for Euclidean spaces is a corollary, since every finite Euclidean metric space can be isometrically embedded into ℓ_1 (Exercise 5). An inequality similar to (15.4) was used, but with squares of differences replaced by absolute values of differences. Such an inequality was well known for expanders. The method of [LLR95] was generalized for embeddings to ℓ_p-spaces with arbitrary p in [Mat97]; it was shown that the minimum distortion required to embed all n-point metric spaces into ℓ_p is of order $\frac{\log n}{p}$, and a matching upper bound was proved by the method shown in Section 15.7.

The proof of Theorem 15.5.1 given in the text can easily be extended to prove a lower bound for ℓ_1-embeddability as well. It actually shows that distortion $\Omega(\log n)$ is needed for approximating the expander metric by a squared Euclidean metric, and every ℓ_1-metric is a squared Euclidean metric. Squared Euclidean metrics do not generally satisfy the triangle inequality, but that is not needed in the proof. Those squared Euclidean metrics that do satisfy the triangle inequality are sometimes called the *metrics of negative type*. Not all of these metrics are ℓ_1-metrics, but a challenging conjecture (made by Linial and independently by Goemans) states that perhaps they are not very far from ℓ_1-metrics: Each metric of negative type might be embeddable into ℓ_1 with distortion bounded by a universal constant. If true, this would have significant algorithmic consequences: Many problems can be formulated as optimization over the cone of all ℓ_1-metrics, which is computationally intractable, and the metrics of negative type would provide a good and algorithmically manageable approximation.

The formulation of the minimum distortion problem for Euclidean embeddings as semidefinite programming is also due to [LLR95], as well as Proposition 15.5.2. These ideas were further elaborated and applied in examples by Linial and Magen [LM00]. The proof of Proposition 15.5.2 given in the text is simpler than that in [LLR95], and it extends to ℓ_p-embeddability (Exercise 4), unlike the formulation of the D-embedding problem as a semidefinite program. It was communicated to me by Yuri Rabinovich.

A further significant progress in lower bounds for ℓ_2-embeddings of graphs was made by Linial, Magen, and Naor [LMN01]. They proved that the metric of every r-regular graph, $r > 2$, of girth g requires distortion at least $\Omega(\sqrt{g})$ for embedding into ℓ_2 (an $\Omega(g)$ lower bound was conjectured in [LLR95]). They give two proofs, one based on the concept of Markov type of a metric space due to Ball [Bal92] and another that we now outline (adapted to the notation of this section). Let $G = (V, E)$ be an r-regular graph of girth $2t+1$ or $2t+2$ for some integer $t \geq 1$, and let ρ be the metric of G. We set $F = \{\{u, v\} \in \binom{V}{2}: \rho(u, v) = t\}$; note that the graph $H = (V, F)$ is s-regular for $s = r(r-1)^{t-1}$. Calculating $R_{E,F}(\rho)$ is trivial, and it remains to bound $R_{E,F}(\sigma)$ for all Euclidean metrics σ on V, which amounts to finding the largest $\beta > 0$ such that $\sigma^2(E) - \beta \cdot \sigma^2(F) \geq 0$ for all σ. Here it suffices to consider line metrics σ; so let $x_v \in \mathbf{R}$ be the image of v in the embedding $V \to \mathbf{R}$ inducing σ. We may assume $\sum_{v \in V} x_v = 0$ and, as in the proof in the text, $\sigma^2(E) = \sum_{\{u,v\} \in E}(x_u - x_v)^2 = x^T L_G x = x^T(rI - A_G)x^T$, where I is the identity matrix and A_G is the adjacency matrix of G, and similarly for $\sigma^2(F)$. So we require $x^T C x \geq 0$ for all x with $\sum_{v \in V} x_v = 0$, where $C = (r - \beta s)I - A_G + \beta A_H$. It turns out that there is a degree-t polynomial $P_t(x)$ such that $A_H = P_t(A_G)$ (here we need that the girth of G exceeds $2t$). This $P_t(x)$ is called the *Geronimus polynomial*, and it is not hard to derive a recurrence for it: $P_0(x) = 1$, $P_1(x) = x$, $P_2(x) = x^2 - r$, and $P_t(x) = xP_{t-1}(x) - (r-1)P_{t-2}(x)$ for $t > 2$. So $C = Q(A)$ for $Q(x) = r - \beta s - x + P_t(x)$. As is well known, all the eigenvalues of A lie in the interval $[-r, r]$, and so if we make sure that $Q(x) \geq 0$ for all $x \in [-r, r]$, all eigenvalues of C are nonnegative, and our condition holds. This leaves us with a nontrivial but doable calculus problem whose discussion we omit.

Semidefinite programming. The general problem of semidefinite programming is to optimize a linear function over a set of positive definite $n \times n$ matrices defined by a system of linear inequalities. This is a convex set in the space of all real $n \times n$ matrices, and in principle it is not difficult to construct a polynomial-time membership oracle for it (see the explanation following Theorem 13.2.1). Then the *ellipsoid method* can solve the optimization problem in polynomial time; see Grötschel, Lovász and Schrijver [GLS88]. More practical algorithms are based on interior point methods. Semidefinite programming is an extremely powerful tool in combinatorial optimization and other areas. For example, it provides the only known polynomial-time algorithms for computing the chromatic number of perfect graphs and the best known approximation algorithms for several fundamental NP-hard graph-theoretic problems. Lovász's recent lecture notes [Lov] are a beautiful concise introduction. Here we outline at least one lovely application, concerning the approximation of the maximum cut in a graph, in Exercise 8 below.

The second eigenvalue. The investigation of graph eigenvalues constitutes a well established part of graph theory; see, e.g., Biggs [Big93] for a nice introduction. The second eigenvalue of the Laplace matrix as an important graph parameter was first considered by Fiedler [Fie73] (who called it the *algebraic connectivity*). Tanner [Tan84] and Alon and Milman [AM85] gave a lower bound for the so-called vertex expansion of a regular graph (a notion similar to edge expansion) in terms of $\mu_2(G)$, and a reverse relation was proved by Alon [Alo86a].

There are many useful analogies of graph eigenvalues with the eigenvalues of the Laplace operator Δ on manifolds, whose theory is classical and well developed; this is pursued to a considerable depth in Chung [Chu97]. This point of view prefers the eigenvalues of the Laplacian matrix of a graph, as considered in this section, to the eigenvalues of the adjacency matrix. In fact, for nonregular graphs, a still closer correspondence with the setting of manifolds is obtained with a differently normalized Laplacian matrix $\mathcal{L}_G$: $(\mathcal{L}_G)_{v,v} = 1$ for all $v \in V(G)$, $(\mathcal{L}_G)_{uv} = -(\deg_G(u) \deg_G(v))^{-1/2}$ for $\{u, v\} \in E(G)$, and $(\mathcal{L}_G)_{uv} = 0$ otherwise.

Expanders have been used to address many fundamental problems of computer science in areas such as network design, theory of computational complexity, coding theory, on-line computation, and cryptography; see, e.g., [RVW00] for references.

For random graphs, parameters such as edge expansion or vertex expansion are usually not too hard to estimate (the technical difficulty of the arguments depends on the chosen model of a random graph). On the other hand, estimating the second eigenvalue of a random r-regular graph is quite challenging, and a satisfactory answer is known only for r large (and even); see Friedman, Komlós, and Szemerédi [FKS89] or Friedman [Fri91]. Namely, with high probability, a random r-regular graph with r even has $\lambda_2 \leq 2\sqrt{r-1} + O(\log r)$. Here the number of vertices n is assumed to be sufficiently large in terms of r and the $O(\cdot)$ notation is with respect to $r \to \infty$. At the same time, for every fixed $r \geq 3$ and any r-regular graph on n vertices, $\lambda_2 \geq 2\sqrt{r-1} - o(1)$, where this time $o(\cdot)$ refers to $n \to \infty$. So random graphs are almost optimal for large r.

For many of the applications of expanders, random graphs are not sufficient, and explicit constructions are required. In fact, explicitly constructed expanders often serve as substitutes for truly random graphs; for example, they allow one to convert some probabilistic algorithms into deterministic ones (derandomization) or reduce the number of random bits required by a probabilistic algorithm.

Explicit construction of expanders was a big challenge, and it has led to excellent research employing surprisingly deep results from classical areas of mathematics (group theory, number theory, harmonic analysis, etc.). In the analysis of such constructions, one usually bounds the second eigenvalue (rather than edge expansion or vertex expansion). After the initial breakthrough by Margulis in 1973 and several other works in this direction (see, e.g., [Mor94] or [RVW00] for references), explicit families of constant-degree expanders matching the quality of random graphs in several parameters (and even superseding them in some respects) were constructed by Lubotzky, Phillips, and Sarnak [LPS88] and independently by Margulis [Mar88]. Later Morgenstern [Mor94] obtained similar results for many more values of the parameters (degree and number of vertices). In particular, these constructions achieve $\lambda_2 \leq 2\sqrt{r-1}$, which is asymptotically optimal, as was mentioned earlier.

For illustration, here is one of the constructions (from [LPS88]). Let $p \neq q$ be primes with $p, q \equiv 1 \pmod{4}$ and such that p is a quadratic nonresidue modulo q, let i be an integer with $i^2 \equiv -1 \pmod{q}$, and let F denote the field of residue classes modulo q. The vertex set $V(G)$ consists of all 2×2 nonsingular matrices over F. Two matrices $A, B \in V(G)$ are connected by an edge iff AB^{-1} is a matrix of the form $\begin{pmatrix} a_0+ia_1 & a_2+ia_3 \\ -a_2+ia_3 & a_0-ia_1 \end{pmatrix}$, where a_0, a_1, a_2, a_3 are integers with $a_0^2 + a_1^2 + a_2^2 + a_3^2 = p$, $a_0 > 0$, a_0 odd, and a_1, a_2, a_3 even. By a theorem of Jacobi, there are exactly $p+1$ such vectors (a_0, a_1, a_2, a_3), and it follows that the graph is $(p+1)$-regular with $q(q^2-1)$ vertices. A family of constant-degree expanders is obtained by fixing p, say $p = 5$, and letting $q \to \infty$.

Reingold, Vadhan, and Wigderson [RVW00] discovered an explicit construction of a different type. Expanders are obtained from a constant-size initial graph by iterating certain sophisticated product operations. Their parameters are somewhat inferior to those from [Mar88], [LPS88], [Mor94], but the proof is relatively short, and it uses only elementary linear algebra.

Exercises

1. Show that every real symmetric positive semidefinite $n \times n$ matrix can be written as $X^T X$ for a real $n \times n$ matrix X. ③

2. (Dimension for isometric ℓ_p-embeddings)
   (a) Let V be an n-point set and let $N = \binom{n}{2}$. Analogous to the set $\mathcal{L}_2$ defined in the text, let $\mathcal{L}_1^{(\text{fin})} \subset \mathbf{R}^N$ be the set of all metrics on V induced by embeddings $f : V \to \ell_1^k$, $k = 1, 2, \ldots$. Show that $\mathcal{L}_1^{(\text{fin})}$ is the convex hull of *line pseudometrics*,[5] i.e., pseudometrics induced by mappings $f : V \to \ell_1^1$. ②
   (b) Prove that any metric from $\mathcal{L}_1^{(\text{fin})}$ can be isometrically embedded into ℓ_1^N. That is, any n-point set in some ℓ_1^k can be realized in ℓ_1^N. ④ (Examples show that one cannot do much better and that dimension $\Omega(n^2)$ is necessary, in contrast to Euclidean embeddings, where dimension $n-1$ always suffices.)
   (c) Let $\mathcal{L}_1 \subset \mathbf{R}^N$ be all metrics induced by embeddings of V into ℓ_1 (the space of infinite sequences with finite ℓ_1-norm). Show that $\mathcal{L}_1 = \mathcal{L}_1^{(\text{fin})}$, and thus that any n-point subset of ℓ_1, can be realized in ℓ_1^N. ③
   (d) Extend the considerations in (a)–(c) to ℓ_p-metrics with arbitrary $p \in [1, \infty)$. ③
   See Ball [Bal90] for more on the dimension of isometric ℓ_p-embeddings.

3. With the notation as in Exercise 2, show that every line pseudometric ν on an n-point set V is a nonnegative linear combination of at most $n-1$ *cut pseudometrics*: $\nu = \sum_{i=1}^{n-1} \alpha_i \tau_i$, $\alpha_1, \ldots, \alpha_{n-1} \geq 0$, where each τ_i is a cut pseudometric, i.e., a line pseudometric induced by a mapping $\psi_i : V \to \{0, 1\}$. (Consequently, by Exercise 2(a), every finite metric isometrically embeddable into ℓ_1 is a nonnegative linear combination of cut pseudometrics.) ③

4. (An ℓ_p-analogue of Proposition 15.5.2) Let $p \in [1, \infty)$ be fixed. Using Exercise 2, formulate and prove an appropriate ℓ_p-analogue of Proposition 15.5.2. ③

5. (Finite ℓ_2-metrics embed isometrically into ℓ_p)
   (a) Let p be fixed. Check that if for all $\varepsilon > 0$, a finite metric space (V, ρ) can be $(1+\varepsilon)$-embedded into some ℓ_p^k, $k = k(\varepsilon)$, then (V, ρ) can be isometrically embedded into ℓ_p^N, where $N = \binom{|V|}{2}$. Use Exercise 2. ②
   (b) Prove that every n-point set in ℓ_2 can be isometrically embedded into ℓ_p^N. ②

6. (The second eigenvalue and edge expansion) Let G be an r-regular graph with n vertices, and let $A, B \subseteq V$ be disjoint. Prove that the number of edges connecting A to B is at least $e(A, B) \geq \mu_2(G) \cdot \frac{|A| \cdot |B|}{n}$ (use (15.3) with a suitable vector x), and deduce that $\Phi(G) \geq \frac{1}{2} \mu_2(G)$. ④

---

[5] A pseudometric ν satisfies all the axioms of a metric except that we may have $\nu(x, y) = 0$ even for two distinct points x and y.

7. (Expansion and measure concentration) Let us consider the vertex set of a graph G as a metric probability space, with the usual graph metric and with the uniform probability measure P (each vertex has measure $\frac{1}{n}$, $n = |V(G)|$). Suppose that $\Phi = \Phi(G) > 0$ and that the maximum degree of G is Δ. Prove the following measure concentration inequality: If $A \subseteq V(G)$ satisfies $P[A] \geq \frac{1}{2}$, then $1 - P[A_t] \leq \frac{1}{2}e^{-t\Phi/\Delta}$, where A_t denotes the t-neighborhood of A. ▣3

8. (The Goemans–Williamson approximation to MAXCUT) Let $G = (V, E)$ be a given graph and let $n = |V|$. The MAXCUT problem for G is to find the maximum possible number of "crossing" edges for a partition $V = A \dot\cup B$ of the vertex set into two disjoint subsets, i.e., $\max_{A \subseteq V} e(A, V \setminus A)$. This is an NP-complete problem. The exercise outlines a geometric randomized algorithm that finds an approximate solution using semidefinite programming.

(a) Check that the MAXCUT problem is equivalent to computing

$$M_{\text{opt}} = \max\left\{ \frac{1}{2} \sum_{\{u,v\} \in E} (1 - x_u x_v) : x_v \in \{-1, 1\}, v \in V \right\}.$$

▣2

(b) Let

$$M_{\text{relax}} = \max\left\{ \frac{1}{2} \sum_{\{u,v\} \in E} (1 - \langle y_u, y_v \rangle) : y_v \in \mathbf{R}^n, \|y_v\| = 1, v \in V \right\}.$$

Clearly, $M_{\text{relax}} \geq M_{\text{opt}}$. Verify that this relaxed version of the problem is an instance of a semidefinite program, that is, the maximum of a linear function over the intersection of a polytope with the cone of all symmetric positive semidefinite real matrices. ▣2

(c) Let $(y_v : v \in V)$ be some system of unit vectors in $\mathbf{R}^n$ for which M_{relax} is attained. Let $r \in \mathbf{R}^n$ be a random unit vector, and set $x_v = \text{sgn}\langle y_v, r \rangle$, $v \in V$. Let $M_{\text{approx}} = \frac{1}{2}\sum_{\{u,v\} \in E}(1 - x_u x_v)$ for these x_v. Show that the expectation, with respect to the random choice of r, of M_{approx} is at least $0.878 \cdot M_{\text{relax}}$ (consider the expected contribution of each edge separately). So we obtain a polynomial-time randomized algorithm producing a solution to MAXCUT whose expected value is at least about 88% of the optimal solution. ▣4

Remark. This algorithm is due to Goemans and Williamson [GW95]. Later, Håstad [Hås97] proved that no polynomial-time algorithm can produce better approximation in the worst case than about 94% unless P=NP (also see Feige and Schechtman [FS01] for nice mathematics showing that the Goemans–Williamson value 0.878... is, in a certain sense, optimal for approaches based on semidefinite programming).

15.6 Upper Bounds for ℓ_∞-Embeddings

In this section we explain a technique for producing low-distortion embeddings of finite metric spaces. Although we are mainly interested in Euclidean embeddings, here we begin with embeddings into the space ℓ_∞, which are somewhat simpler. We derive almost tight upper bounds.

Let (V, ρ) be an arbitrary metric space. To specify an embedding

$$f \colon (V, \rho) \to \ell_\infty^d$$

means to define d functions $f_1, \ldots, f_d \colon V \to \mathbf{R}$, the coordinates of the embedded points. If we aim at a D-embedding, without loss of generality we may require it to be nonexpanding, which means that $|f_i(u) - f_i(v)| \le \rho(u, v)$ for all $u, v \in V$ and all $i = 1, 2, \ldots, d$. The D-embedding condition then means that for every pair $\{u, v\}$ of points of V, there is a coordinate $i = i(u, v)$ that "takes care" of the pair: $|f_i(u) - f_i(v)| \ge \frac{1}{D}\rho(u, v)$.

One of the key tricks in constructions of such embeddings is to take each f_i as the distance to some suitable subset $A_i \subseteq V$; that is, $f_i(u) = \rho(u, A_i) = \max_{a \in A_i} \rho(u, a)$. By the triangle inequality, we have $|\rho(u, A_i) - \rho(v, A_i)| \le \rho(u, v)$ for any $u, v \in V$, and so such an embedding is automatically nonexpanding. We "only" have to choose a suitable collection of the A_i that take care of all pairs $\{u, v\}$.

We begin with a simple case: an old observation showing that every finite metric space embeds isometrically into ℓ_∞.

15.6.1 Proposition (Fréchet's embedding). *Let (V, ρ) be an arbitrary n-point metric space. Then there is an isometric embedding $f \colon V \to \ell_\infty^n$.*

Proof. Here the coordinates in ℓ_∞^n are indexed by the points of V, and the vth coordinate is given by $f_v(u) = \rho(u, v)$. In the notation above, we thus put $A_v = \{v\}$. As we have seen, the embedding is nonexpanding by the triangle inequality. On the other hand, the coordinate v takes care of the pairs $\{u, v\}$ for all $u \in V$:

$$\|f(u) - f(v)\|_\infty \ge |f_v(u) - f_v(v)| = \rho(u, v).$$

□

The dimension of the image in this embedding can be reduced a little; for example, we can choose some $v_0 \in V$ and remove the coordinate corresponding to v_0, and the above proof still works. To reduce the dimension significantly, though, we have to pay the price of distortion. For example, from Corollary 15.3.4 we know that for distortions below 3, the dimension must generally remain at least a fixed fraction of n. We prove an upper bound on the dimension needed for embeddings with a given distortion, which nearly matches the lower bounds in Corollary 15.3.4:

15.6.2 Theorem. *Let $D = 2q-1 \geq 3$ be an odd integer and let (V, ρ) be an n-point metric space. Then there is a D-embedding of V into ℓ_∞^d with*

$$d = O(qn^{1/q} \ln n).$$

Proof. The basic scheme of the construction is as explained above: Each coordinate is given by the distance to a suitable subset of V. This time the subsets are chosen at random with suitable densities.

Let us consider two points $u, v \in V$. What are the sets A such that $|\rho(u, A) - \rho(v, A)| \geq \Delta$, for a given real $\Delta > 0$? For some $r \geq 0$, they must intersect the closed r-ball around u and avoid the open $(r+\Delta)$-ball around v; schematically,

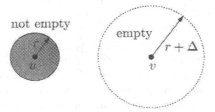

or conversely (with the roles of u and v interchanged).

In the favorable situation where the closed r-ball around u does not contain many fewer points of V than the open $(r+\Delta)$-ball around v, a random A with a suitable density has a reasonable chance to work. Generally we have no control over the distribution of points around u and around v, but by considering several suitable balls simultaneously, we can find a good pair of balls. We also do not know the right density needed for the sample to work, but since we have many coordinates, we can take samples of essentially all possible densities.

Now we begin with the formal proof. We define an auxiliary parameter $p = n^{-1/q}$, and for $j = 1, 2, \ldots, q$, we introduce the probabilities $p_j = \min(\frac{1}{2}, p^j)$. Further, let $m = \lceil 24n^{1/q} \ln n \rceil$. For $i = 1, 2, \ldots, m$ and $j = 1, 2, \ldots, q$, we choose a random subset $A_{ij} \subseteq V$. The sets (and the corresponding coordinates in ℓ_∞^{mq}) now have double indices, and the index j influences the "density" of A_{ij}. Namely, each point $v \in V$ has probability p_j of being included into A_{ij}, and these events are mutually independent. The choices of the A_{ij}, too, are independent for distinct indices i and j. Here is a schematic illustration of the sampling:

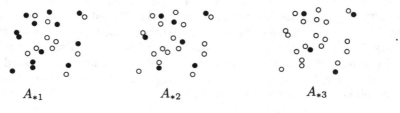

A_{*1} $\qquad\qquad\qquad$ A_{*2} $\qquad\qquad\qquad$ A_{*3} $\qquad$ $\cdots$

We divide the coordinates in ℓ_∞^d into q blocks by m coordinates. For $v \in V$, we let

$$f(v)_{ij} = \rho(v, A_{ij}), \quad i = 1, 2, \ldots, m, \ j = 1, 2, \ldots, q.$$

We claim that with a positive probability, this $f \colon V \to \ell_\infty^{mq}$ is a D-embedding. We have already noted that f is nonexpanding, and the following lemma serves for showing that with a positive probability, every pair $\{u, v\}$ is taken care of.

15.6.3 Lemma. *Let u, v be two distinct points of V. Then there exists an index $j \in \{1, 2, \ldots, q\}$ such that if the set A_{ij} is chosen randomly as above, then the probability of the event*

$$|\rho(u, A_{ij}) - \rho(v, A_{ij})| \geq \tfrac{1}{D}\,\rho(u, v) \tag{15.6}$$

is at least $\frac{p}{12}$.

First, assuming this lemma, we finish the proof of the theorem. To show that f is a D-embedding, it suffices to show that with a nonzero probability, for every pair $\{u, v\}$ there are i, j such that the event (15.6) in the lemma occurs for the set A_{ij}. Consider a fixed pair $\{u, v\}$ and select the appropriate index j as in the lemma. The probability that the event (15.6) does not occur for any of the m indices i is at most $(1 - \frac{p}{12})^m \leq e^{-pm/12} \leq n^{-2}$. Since there are $\binom{n}{2} < n^2$ pairs $\{u, v\}$, the probability that we fail to choose a good set for any of the pairs is smaller than 1. $\qquad\square$

Proof of Lemma 15.6.3. Set $\Delta = \frac{1}{D}\rho(u, v)$. Let $B_0 = \{u\}$, let B_1 be the (closed) Δ-ball around v, let B_2 be the (closed) 2Δ-ball around $u, \ldots$, finishing with B_q, which is a $q\Delta$-ball around u (if q is even) or around v (if q is odd). The parameters are chosen so that the radii of B_{q-1} and B_q add up to $\rho(u, v)$; that is, the last two balls just touch (recall that $D = 2q-1$):

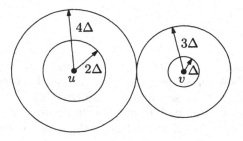

Let n_t denote number of points of V in B_t.

We want to select an index j such that

$$n_t \geq n^{(j-1)/q} \quad \text{and} \quad n_{t+1} \leq n^{j/q}. \tag{15.7}$$

To this end, we divide the interval $[1, n]$ into q intervals $I_1, I_2, \ldots, I_q$, where

$$I_j = \left[n^{(j-1)/q}, n^{j/q} \right].$$

If the sequence $(n_1, n_2, \ldots, n_q)$ is not monotone increasing, i.e., if $n_{t+1} < n_t$ for some t, then (15.7) holds for the j such that I_j contains n_t. On the other hand, if $1 = n_0 \leq n_1 \leq \ldots \leq n_q \leq n$, then by the pigeonhole principle, there exist t and j such that the interval I_j contains both n_t and n_{t+1}. Then (15.7) holds for this j as well.

In this way, we have selected the index j whose existence is claimed in the lemma. We will show that with probability at least $\frac{p}{12}$, the set A_{ij}, randomly selected with point probability p_j, includes a point of B_t (event E_1) and is disjoint from the interior of B_{t+1} (event E_2); such an A_{ij} satisfies (15.6). Since B_t and the interior of B_{t+1} are disjoint, the events E_1 and E_2 are independent.

We calculate

$$\text{Prob}[E_1] = 1 - \text{Prob}[A_{ij} \cap B_t = \emptyset] = 1 - (1 - p_j)^{n_t} \geq 1 - e^{-p_j n_t}.$$

Using (15.7), we have $p_j n_t \geq p_j n^{(j-1)/q} = p_j p^{-j+1} = \min(\frac{1}{2}, p^j) p^{-j+1} \geq \min(\frac{1}{2}, p)$. For $p \geq \frac{1}{2}$, we get $\text{Prob}[E_1] \geq 1 - e^{-1/2} > \frac{1}{3} \geq \frac{p}{3}$, while for $p < \frac{1}{2}$, we have $\text{Prob}[E_1] \geq 1 - e^{-p}$, and a bit of calculus verifies that the last expression is well above $\frac{p}{3}$ for all $p \in [0, \frac{1}{2})$.

Further,

$$\text{Prob}[E_2] \geq (1 - p_j)^{n_{t+1}} \geq (1 - p_j)^{n^{j/q}} \geq (1 - p_j)^{1/p_j} \geq \tfrac{1}{4}$$

(since $p_j \leq \frac{1}{2}$). Thus $\text{Prob}[E_1 \cap E_2] \geq \frac{p}{12}$, which proves the lemma. $\square$

Bibliography and remarks. The embedding method discussed in this section was found by Bourgain [Bou85], who used it to prove Theorem 15.7.1 explained in the subsequent section. Theorem 15.6.2 is from [Mat96b].

Exercises

1. (a) Find an isometric embedding of ℓ_1^d into $\ell_\infty^{2^d}$. ③
   (b) Explain how an embedding as in (a) can be used to compute the diameter of an n-point set in ℓ_1^d in time $O(d2^d n)$. ③
2. Show that if the unit ball K of some finite-dimensional normed space is a convex polytope with $2m$ facets, then that normed space embeds isometrically into ℓ_∞^m. ②
   (Using results on approximation of convex bodies by polytopes, this yields useful approximate embeddings of arbitrary norms into ℓ_∞^k.)
3. Deduce from Theorem 15.6.2 that every n-point metric space can be D-embedded into ℓ_2^k with $D = O(\log^2 n)$ and $k = O(\log^2 n)$. ②

15.7 Upper Bounds for Euclidean Embeddings

By a method similar to the one shown in the previous section, one can also prove a tight upper bound on Euclidean embeddings; the method was actually invented for this problem.

15.7.1 Theorem (Bourgain's embedding into ℓ_2). *Every n-point metric space (V, ρ) can be embedded into a Euclidean space with distortion at most $O(\log n)$.*

The overall strategy of the embedding is similar to the embedding into ℓ_∞^d in the proof of Theorem 15.6.2. The coordinates in ℓ_2^d are given by distances to suitable subsets. The situation is slightly more complicated than before: For embedding into ℓ_∞^d, it was enough to exhibit one coordinate "taking care" of each pair, whereas for the Euclidean embedding, many of the coordinates will contribute significantly to every pair. Here is the appropriate analogue of Lemma 15.6.3.

15.7.2 Lemma. *Let $u, v \in V$ be two distinct points. Then there exist real numbers $\Delta_1, \Delta_2, \ldots, \Delta_q \geq 0$ with $\Delta_1 + \cdots + \Delta_q = \frac{1}{4}\rho(u,v)$, where $q = \lfloor \log_2 n \rfloor + 1$, and such that the following holds for each $j = 1, 2, \ldots, q$: If $A_j \subseteq V$ is a randomly chosen subset of V, with each point of V included in A_j independently with probability 2^{-j}, then the probability P_j of the event*

$$|\rho(u, A_j) - \rho(v, A_j)| \geq \Delta_j$$

satisfies $P_j \geq \frac{1}{12}$.

Proof. We fix u and v. We define $r_q = \frac{1}{4}\rho(u,v)$ and for $j = 0, 1, \ldots, q-1$, we let r_j be the smallest radius such that both $|B(u, r_j)| \geq 2^j$ and $|B(v, r_j)| \geq 2^j$ where, as usual, $B(x, r) = \{y \in V : \rho(x, y) \leq r\}$. We are going to show that the claim of the lemma holds with $\Delta_j = r_j - r_{j-1}$.

Fix $j \in \{1, 2, \ldots, q\}$ and let $A_j \subseteq V$ be a random sample with point probability 2^{-j}. By the definition of r_j, $|B^\circ(u, r_j)| < 2^j$ or $|B^\circ(v, r_j)| < 2^j$, where $B^\circ(x, r) = \{y \in V : \rho(x, y) < r\}$ denotes the open ball (this holds for $j = q$, too, because $|V| \leq 2^q$). We choose the notation u, v so that $|B^\circ(u, r_j)| < 2^j$. A random set A_j is good if it intersects $B(v, r_{j-1})$ and misses $B^\circ(u, r_j)$. The former set has cardinality at least 2^{j-1} and the latter at most 2^j. The calculation of the probability that A_j has these properties is identical to the calculation in the proof of Lemma 15.6.3 with $p = \frac{1}{2}$.   □

In the subsequent proof of Theorem 15.7.1 we will construct the embedding in a slightly roundabout way, which sheds some light on what is really going on. Define a *line pseudometric* on V to be any pseudometric ν induced by a mapping $\varphi: V \to \mathbf{R}$, that is, given by $\nu(u, v) = |\varphi(u) - \varphi(v)|$. For each $A \subseteq V$, let ν_A be the line pseudometric corresponding to the mapping $v \mapsto \rho(v, A)$. As we have noted, each ν_A is *dominated* by ρ, i.e., $\nu_A \leq \rho$

(the inequality between two (pseudo)metrics on the same point set means inequality for each pair of points).

The following easy lemma shows that if a metric ρ on V can be approximated by a convex combination of line pseudometrics, each of them dominated by ρ, then a good embedding of (V, ρ) into ℓ_2 exists.

15.7.3 Lemma. *Let (V, ρ) be a finite metric space, and let $\nu_1, \ldots, \nu_N$ be line pseudometrics on V with $\nu_i \leq \rho$ for all i and such that*

$$\sum_{i=1}^{N} \alpha_i \nu_i \geq \frac{1}{D} \rho$$

for some nonnegative $\alpha_1, \ldots, \alpha_N$ summing up to 1. Then (V, ρ) can be D-embedded into ℓ_2^N.

Proof. Let $\varphi_i \colon V \to \mathbf{R}$ be a mapping inducing the line pseudometric ν_i. We define the embedding $f \colon V \to \ell_2^N$ by

$$f(v)_i = \sqrt{\alpha_i} \cdot \varphi_i(v).$$

Then, on the one hand,

$$\|f(u) - f(v)\|^2 = \sum_{i=1}^{N} \alpha_i \nu_i(u, v)^2 \leq \rho(u, v)^2,$$

because all ν_i are dominated by ρ and $\sum \alpha_i = 1$. On the other hand,

$$\|f(u) - f(v)\| = \left(\sum_{i=1}^{N} \alpha_i \nu_i(u, v)^2 \right)^{1/2} = \left(\sum_{i=1}^{N} \alpha_i \right)^{1/2} \left(\sum_{i=1}^{N} \alpha_i \nu_i(u, v)^2 \right)^{1/2}$$

$$\geq \sum_{i=1}^{N} \alpha_i \nu_i(u, v)$$

by Cauchy–Schwarz, and the latter expression is at least $\frac{1}{D} \rho(u, v)$ by the assumption. $\qquad \square$

Proof of Theorem 15.7.1. As was remarked above, each of the line pseudometrics ν_A corresponding to the mapping $v \mapsto \rho(v, A)$ is dominated by ρ. It remains to observe that Lemma 15.7.2 provides a convex combination of these line pseudometrics that is bounded from below by $\frac{1}{48q} \cdot \rho$. The coefficient of each ν_A in this convex combination is given by the probability of A appearing as one of the sets A_j in Lemma 15.7.2. More precisely, write $\pi_j(A)$ for the probability that a random subset of V, with points picked independently with probability 2^{-j}, equals A. Then the claim of Lemma 15.7.2 implies, for every pair $\{u, v\}$,

$$\sum_{A \subseteq V} \pi_j(A) \cdot \nu_A(u,v) \geq \tfrac{1}{12} \Delta_j.$$

Summing over $j = 1, 2, \ldots, q$, we have

$$\sum_{A \subseteq V} \left(\sum_{j=1}^{q} \pi_j(A) \right) \cdot \nu_A(u,v) \geq \tfrac{1}{12} \cdot \sum_{j=1}^{q} \Delta_j = \tfrac{1}{48} \, \rho(u,v).$$

Dividing by q and using $\sum_{A \subseteq V} \pi_j(A) = 1$, we arrive at

$$\sum_{A \subseteq V} \alpha_A \nu_A \geq \frac{1}{48q} \, \rho, \qquad \sum_{A \subseteq V} \alpha_A = 1,$$

with $\alpha_A = \frac{1}{q} \sum_{j=1}^{q} \pi_j(A)$. Lemma 15.7.3 now gives embeddability into ℓ_2 with distortion at most $48q$. Theorem 15.7.1 is proved. □

Remarks. Almost the same proof with a slight modification of Lemma 15.7.3 shows that for each $p \in [1, \infty)$, every n-point metric space can be embedded into ℓ_p with distortion $O(\log n)$; see Exercise 1.

The proof as stated produces an embedding into space of dimension 2^n, since there are 2^n subsets $A \subseteq V$, each of them yielding one coordinate. To reduce the dimension, one can argue that not all the sets A are needed: by suitable Chernoff-type estimates, it follows that it is sufficient to choose $O(\log n)$ random sets with point probability 2^{-j}, i.e., $O(\log^2 n)$ sets altogether (Exercise 2). Of course, for Euclidean embeddings, an even better dimension $O(\log n)$ is obtained using the Johnson–Lindenstrauss flattening lemma, but for other ℓ_p, no flattening lemma is available.

An algorithmic application: approximating the sparsest cut. We know that every n-point metric space can be $O(\log n)$-embedded into ℓ_1^d with $d = O(\log^2 n)$. By inspecting the proof, it is not difficult to give a randomized algorithm that computes such an embedding in polynomial expected time. We show a neat algorithmic application to a graph-theoretic problem.

Let $G = (V, E)$ be a graph. A *cut* in G is a partition of V into two nonempty subsets A and $B = V \setminus A$. The *density* of the cut (A, B) is $\frac{e(A,B)}{|A| \cdot |B|}$, where $e(A, B)$ is the number of edges connecting A and B. Given G, we would like to find a cut of the smallest possible density. This problem is NP-hard, and here we discuss an efficient algorithm for finding an approximate answer: a cut whose density is at most $O(\log n)$ times larger than the density of the sparsest cut, where $n = |V|$ (this is the best known approximation guarantee for any polynomial-time algorithm). Note that this also allows us to approximate the edge expansion of G (discussed in Section 15.5) within a multiplicative factor of $O(\log n)$.

First we reformulate the problem equivalently using cut pseudometrics. A *cut pseudometric* on V is a pseudometric τ corresponding to some cut (A, B), with $\tau(u, v) = \tau(v, u) = 1$ for $u \in A$ and $v \in B$ and $\tau(u, v) = 0$ for $u, v \in A$ or

$u, v \in B$. In other words, a cut pseudometric is a line pseudometric induced by a mapping $\psi: V \to \{0, 1\}$ (excluding the trivial case where all of V gets mapped to the same point). Letting $F = \binom{V}{2}$, the density of the cut (A, B) can be written as $\tau(E)/\tau(F)$, where τ is the corresponding cut pseudometric and $\tau(E) = \sum_{\{u,v\} \in E} \tau(u, v)$. Therefore, we would like to minimize the ratio $R_1(\tau) = \tau(E)/\tau(F)$ over all cut pseudometrics τ.

In the first step of the algorithm we relax the problem, and we find a pseudometric, not necessarily a cut one, minimizing the ratio $R_1(\rho) = \rho(E)/\rho(F)$. This can be done efficiently by linear programming. The minimized function looks nonlinear, but we can get around this by a simple trick: We postulate the additional condition $\rho(F) = 1$ and minimize the linear function $\rho(E)$. The variables in the linear program are the $\binom{n}{2}$ numbers $\rho(u, v)$ for $\{u, v\} \in F$, and the constraints are $\rho(u, v) \geq 0$ (for all u, v), $\rho(F) = 1$, and those expressing the triangle inequalities for all triples $u, v, w \in V$.

Having computed a ρ_0 minimizing $R_1(\rho)$, we find a D-embedding f of (V, ρ_0) into some ℓ_1^d with $D = O(\log n)$. If σ_0 is the pseudometric induced on V by this f, we clearly have $R_1(\sigma_0) \leq D \cdot R_1(\rho_0)$. Now since σ_0 is an ℓ_1-pseudometric, it can be expressed as a nonnegative linear combination of suitable cut pseudometrics (Exercise 15.5.3): $\sigma_0 = \sum_{i=1}^{N} \alpha_i \tau_i$, $\alpha_1, \ldots, \alpha_N > 0$, $N \leq d(n-1)$. It is not difficult to check that $R_1(\sigma_0) \geq \min\{R_1(\tau_i): i = 1, 2, \ldots, N\}$ (Exercise 3). Therefore, at least one of the τ_i is a cut pseudometric satisfying $R_1(\tau_i) \leq R_1(\sigma_0) \leq D \cdot R_1(\rho_0) \leq D \cdot R_1(\tau_0)$, where τ_0 is a cut pseudometric with the smallest possible $R_1(\tau_0)$. Therefore, the cut corresponding to this τ_i has density at most $O(\log n)$ times larger than the sparsest possible cut.

Bibliography and remarks.   Theorem 15.7.1 is due to Bourgain [Bou85]. The algorithmic application to approximating the sparsest cut uses the idea of an algorithm for a somewhat more complicated problem (multicommodity flow) found by Linial et al. [LLR95] and independently by Aumann and Rabani [AR98].

We will briefly discuss further results proved by variations of Bourgain's embedding technique. Many of them have been obtained in the study of approximation algorithms and imply strong algorithmic results.

Tree metrics.   Let $\mathcal{G}$ be a class of graphs and consider a graph $G \in \mathcal{G}$. Each positive weight function $w: E(G) \to (0, \infty)$ defines a metric on $V(G)$, namely the shortest-path metric, where the length of a path is the sum of the weights of its edges. All subspaces of the resulting metric spaces are referred to as $\mathcal{G}$-*metrics*. A *tree metric* is a $\mathcal{T}$-metric for $\mathcal{T}$ the class of all trees. Tree metrics generally behave much better than arbitrary metrics, but for embedding problems they are far from trivial.

Bourgain [Bou86] proved, using martingales, a surprising lower bound for embedding tree metrics into ℓ_2: A tree metric on n points requires distortion $\Omega(\sqrt{\log \log n})$ in the worst case. His example is the complete binary tree with unit edge lengths, and for that example, he also constructed an embedding with $O(\sqrt{\log\log n}\,)$ distortion. For embedding the complete binary tree into ℓ_p, $p > 1$, the distortion is $\Omega((\log\log n)^{\min(1/2,1/p)})$, with the constant of proportionality depending on p and tending to 0 as $p \to 1$. (For Banach-space specialists, we also remark that all tree metrics can be embedded into a given Banach space Z with bounded distortion if and only if Z is not superreflexive.) In Matoušek [Mat99b] it was shown that the complete binary tree is essentially the worst example; that is, every n-point tree metric can be embedded into ℓ_p with distortion $O((\log\log n)^{\min(1/2,1/p)})$. An alternative, elementary proof was given for the matching lower bound (see Exercise 5 for a weaker version). Another proof of the lower bound, very short but applying only for embeddings into ℓ_2, was found by Linial and Saks [LS02] (Exercise 6).

In the notes to Section 15.3 we mentioned that general n-point metric spaces require worst-case distortion $\Omega(n^{1/\lfloor(d+1)/2\rfloor})$ for embedding into ℓ_2^d, $d \geq 2$ fixed. Gupta [Gup00] proved that for n-point tree metrics, $O(n^{1/(d-1)})$-embeddings into ℓ_2^d are possible. The best known lower bound is $\Omega(n^{1/d})$, from a straightforward volume argument. Babilon, Matoušek, Maxová, and Valtr [BMMV02] showed that every n-vertex tree with unit-length edges can be $O(\sqrt{n}\,)$-embedded into ℓ_2^2.

Planar-graph metrics and metrics with excluded minor. A *planar-graph metric* is a $\mathcal{P}$-metric with $\mathcal{P}$ standing for the class of all planar graphs (the shorter but potentially confusing term *planar metric* is used in the literature). Rao [Rao99] proved that every n-point planar-graph metric can be embedded into ℓ_2 with distortion only $O(\sqrt{\log n}\,)$, as opposed to $\log n$ for general metrics. More generally, the same method shows that whenever H is a fixed graph and $Excl(H)$ is the class of all graphs not containing H as a minor, then $Excl(H)$-metrics can be $O(\sqrt{\log n}\,)$-embedded into ℓ_2. For a matching lower bound, valid already for the class $Excl(K_4)$ (series-parallel graphs), and consequently for planar-graph metrics; see Exercise 15.4.2.

We outline Rao's method of embedding. We begin with graphs where all edges have unit weight (this is the setting in [Rao99], but our presentation differs in some details), and then we indicate how graphs with arbitrary edge weights can be treated. The main new ingredient in Rao's method, compared to Bourgain's approach, is a result of Klein, Plotkin, and Rao [KPR93] about a decomposition of graphs with an excluded minor into pieces of low diameter. Here is the decomposition procedure.

Let G be a graph, let ρ be the corresponding graph metric (with all edges having unit length), and let Δ be an integer parameter. We fix a vertex $v_0 \in V(G)$ arbitrarily, we choose an integer $r \in \{0, 1, \ldots, \Delta-1\}$ uniformly at random, and we let $B_1 = \{v \in V(G) : \rho(v, v_0) \equiv$

$r \pmod{\Delta}$}. By deleting the vertices of B_1 from G, the remaining vertices are partitioned into connected components; this is the first level of the decomposition. For each of these components of $G \setminus B_1$, we repeat the same procedure; Δ remains unchanged and r is chosen anew at random (but we can use the same r for all the components). Let B_2 be the set of vertices deleted from G in this second round, taken together for all the components. The second level of the decomposition consists of the connected components of $G \setminus (B_1 \cup B_2)$, and decompositions of levels 3, 4, ... can be produced similarly. The following schematic drawing illustrates the two-level decomposition; the graph is marked as the gray area, and the vertices of B_1 and B_2 are indicated by the solid and dashed arcs, respectively.

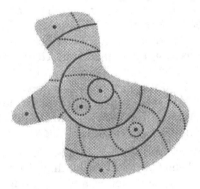

For planar graphs, it suffices to use a 3-level decomposition, and for every fixed graph H, there is a suitable $k = k(H)$ such that a k-level decomposition is appropriate for all graphs $G \in Excl(H)$.

Let $B = B_1 \cup \cdots \cup B_k$; this can be viewed as the boundary of the components in the k-level decomposition. Here are the key properties of the decomposition:

(i) For each vertex $v \in V(G)$, we have $\rho(v, B) \geq c_1 \Delta$ with probability at least c_2, for suitable constants $c_1, c_2 > 0$. The probability is with respect to the random choices of the parameters r at each level of the decomposition. (This is not hard to see; for example, in the first level of the decomposition, for every fixed v, $\rho(v, v_0)$ is some fixed number and it has a good chance to be at least $c_1 \Delta$ away, modulo Δ, from a random r.)

(ii) Each component in the resulting decomposition has diameter at most $O(\Delta)$. (This is not so easy to prove, and it is where one needs $k = k(H)$ sufficiently large. For $H = K_{3,3}$, which includes the case of planar graphs, the proof is a relatively simple case analysis.)

Next, we describe the embedding of $V(G)$ into ℓ_2 in several steps. First we consider Δ and the decomposition as above fixed, and we let

$C_1, \ldots, C_m$ be the components of $G \setminus B$. For all the C_i, we choose random signs $\sigma(C_i) \in \{-1, +1\}$ uniformly and independently. For a vertex $x \in V(G)$, we define $\sigma(x) = 0$ if $x \in B$ and $\sigma(x) = \sigma(C_i)$ if $x \in V(C_i)$. Then we define the mapping $\varphi_{B,\sigma}: V(G) \to \mathbf{R}$ by $\varphi_{B,\sigma}(v) = \sigma(x) \cdot \rho(x, B)$ (the distance of x to the boundary signed by the component's sign). This $\varphi_{B,\sigma}$ induces a line pseudometric $\nu_{B,\sigma}$, and it is easy to see that $\nu_{B,\sigma}$ is dominated by ρ.

Let C be a constant such that all the C_i have diameter at most $C\Delta$, and let $x, y \in V(G)$ be such that $C\Delta < \rho(x,y) \leq 2C\Delta$. Such x and y certainly lie in distinct components, and $\sigma(x) \neq \sigma(y)$ with probability $\frac{1}{2}$. With probability at least c_2, we have $\rho(x, B) \geq c_1\Delta$, and so with a fixed positive probability, $\nu_{B,\sigma}$ places x and y at distance at least $c_1\Delta$.

Now, we still keep Δ fixed and consider $\nu_{B,\sigma}$ for all possible B and σ. Letting $\alpha_{B,\sigma}$ be the probability that a particular pair (B, σ) results from the decomposition procedure, we have

$$\sum_{B,\sigma} \alpha_{B,\sigma} \nu_{B,\sigma}(x,y) = \Omega(\rho(x,y))$$

whenever $C\Delta < \rho(x,y) \leq 2C\Delta$. As in the proof of Lemma 15.7.3, this yields a 1-Lipschitz embedding $f_\Delta: V(G) \to \ell_2^N$ (for some N) that shortens distances for pairs x, y as above by at most a constant factor. (It is not really necessary to use all the possible pairs (B, σ) in the embedding; it is easy to show that $const \cdot \log n$ independent random B and σ will do.)

To construct the final embedding $f: V(G) \to \ell_2$, we let $f(v)$ be the concatenation of the vectors f_Δ for $\Delta \in \{2^j: 1 \leq 2^j \leq \mathrm{diam}(G)\}$. No distance is expanded by more than $O(\sqrt{\log \mathrm{diam}(G)}) = O(\sqrt{\log n})$, and the contraction is at most by a constant factor, and so we have an embedding into ℓ_2 with distortion $O(\sqrt{\log n})$.

Why do we get a better bound than for Bourgain's embedding? In both cases we have about $\log n$ groups of coordinates in the embedding. In Rao's embedding we know that for every pair (x, y), one of the groups contributes at least a fixed fraction of $\rho(x,y)$ (and no group contributes more than $\rho(x,y)$). Thus, the sum of squares of the contributions is between $\rho(x,y)^2$ and $\rho(x,y)^2 \log n$. In Bourgain's embedding (with a comparable scaling) no group contributes more than $\rho(x,y)$, and the sum of the contributions of all groups is at least a fixed fraction of $\rho(x,y)$. But since we do not know how the contributions are distributed among the groups, we can conclude only that the sum of squares of the contributions is between $\rho(x,y)^2/\log n$ and $\rho(x,y)^2 \log n$.

It remains to sketch the modifications of Rao's embedding for a graph G with arbitrary nonnegative weights on edges. For the unweighted case, we defined B_1 as the vertices lying exactly at the given distances from v_0. In the weighted case, there need not be vertices exactly at these distances, but we can add artificial vertices by subdividing the appropriate edges; this is a minor technical issue. A more serious problem is that the distances $\rho(x,y)$ can be in a very wide range, not just from 1 to n. We let Δ run through all the relevant powers of 2 (that is, such that $C\Delta < \rho(x,y) \leq 2C\Delta$ for some $x \neq y$), but for producing the decomposition for a particular Δ, we use a modified graph G_Δ obtained from G by contracting all edges shorter than $\frac{\Delta}{2n}$. In this way, we can have many more than $\log n$ values of Δ, but only $O(\log n)$ of them are relevant for each pair (x,y), and the analysis works as before.

Gupta, Newman, Rabinovich, and Sinclair [GNRS99] proved that any $Excl(K_4)$-metric, as well as any $Excl(K_{2,3})$-metric, can be $O(1)$-embedded into ℓ_1, and they conjectured that for any H, $Excl(H)$-metrics might be $O(1)$-embeddable into ℓ_1 (the constant depending on H).

Volume-respecting embeddings. Feige [Fei00] introduced an interesting strengthening of the notion of the distortion of an embedding, concerning embeddings into Euclidean spaces. Let $f : (V, \rho) \to \ell_2$ be an embedding that for simplicity we require to be 1-Lipschitz (nonexpanding). The usual distortion of f is determined by looking at pairs of points, while Feige's notion takes into account all k-tuples for some $k \geq 2$. For example, if V has 3 points, every two with distance 1, then the following two embeddings into ℓ_2^2 have about the same distortion:

But while the left embedding is good in Feige's sense for $k = 3$, the right one is completely unsatisfactory. For a k-point set $P \subset \ell_2$, define $\mathrm{Evol}(P)$ as the $(k-1)$-dimensional volume of the simplex spanned by P (so $\mathrm{Evol}(P) = 0$ if P is affinely dependent). For a k-point metric space (S, ρ), the *volume* $\mathrm{Vol}(S)$ is defined as $\sup_f \mathrm{Evol}(f(S))$, where the supremum is over all 1-Lipschitz $f : S \to \ell_2$. An embedding $f : (V, \rho) \to \ell_2$ is (k, D) *volume-respecting* if for every k-point subset $S \subseteq V$, we have $D \cdot \mathrm{Evol}(f(S))^{1/(k-1)} \geq \mathrm{Vol}(S)^{1/(k-1)}$. For D small, this means that the image of any k-tuple spans nearly as large a volume as it possibly can for a 1-Lipschitz map. (Note, for example, that an isometric embedding of a path into ℓ_2 is *not* volume-respecting.)

Feige showed that $\mathrm{Vol}(S)$ can be approximated quite well by an intrinsic parameter of the metric space (not referring to embeddings), namely, by the *tree volume* $\mathrm{Tvol}(S)$, which equals the products of the edge lengths in a minimum spanning tree on S (with respect to the metric on S). Namely, $\mathrm{Vol}(S) \leq \frac{1}{(k-1)!} \mathrm{Tvol}(S) \leq 2^{(k-2)/2} \mathrm{Vol}(S)$. He proved that for any n-point metric space and all $k \geq 2$, the embedding as in the proof of Theorem 15.7.1 is $(k, O(\log n + \sqrt{k \log n \log k}))$ volume-respecting (the result in the conference version of his paper is slightly weaker).

The notion of volume-respecting embeddings currently still looks somewhat mysterious. In an attempt to convey some feeling about it, we outline Feige's application and indicate the use of the volume-respecting condition in it. He considered the problem of approximating the *bandwidth* of a given n-vertex graph G. The bandwidth is the minimum, over all bijective maps $\varphi: V(G) \rightarrow \{1, 2, \ldots, n\}$, of $\max\{|\varphi(u) - \varphi(v)|: \{u, v\} \in E(G)\}$ (so it has the flavor of an approximate embedding problem). Computing the bandwidth is NP-hard, but Feige's ingenious algorithm approximates it within a factor of $O((\log n)^{\text{const}})$. The algorithm has two main steps: First, embed the graph (as a metric space) into ℓ_2^m, with m being some suitable power of $\log n$, by a (k, D) volume-respecting embedding f, where $k = \log n$ and D is as small as one can get. Second, let λ be a random line in ℓ_2^m and let $\psi(v)$ denote the orthogonal projection of $f(v)$ on λ. This $\psi: V(G) \rightarrow \lambda$ is almost surely injective, and so it provides a linear ordering of the vertices, that is, a bijective map $\varphi: V(G) \rightarrow \{1, 2, \ldots, n\}$, and this is used for estimating the bandwidth.

To indicate the analysis, we need the notion of *local density* of the graph G: $\mathrm{ld}(G) = \max\{|B(v, r)|/r: v \in V(G), r = 1, 2, \ldots, n\}$, where $B(v, r)$ are all vertices at distance at most r from v. It is not hard to see that $\mathrm{ld}(G)$ is a lower bound for the bandwidth, and Feige's analysis shows that $O(\mathrm{ld}(G)(\log n)^{\text{const}})$ is an upper bound.

One first verifies that with high probability, if $\{u, v\} \in E(G)$, then the images $\psi(u)$ and $\psi(v)$ on λ are close; concretely, $|\psi(u) - \psi(v)| \leq \Delta = O(\sqrt{(\log n)/m})$. For proving this, it suffices to know that f is 1-Lipschitz, and it is an immediate consequence of measure concentration on the sphere. If b is the bandwidth obtained from the ordering given by ψ, then some interval of length Δ on λ contains the images of b vertices. Call a k-tuple $S \subset V(G)$ *squeezed* if $\psi(S)$ lies in an interval of length Δ. If b is large, then there are many squeezed S. On the other hand, one proves that, not surprisingly, if $\mathrm{ld}(G)$ is small, then $\mathrm{Vol}(S)$ is large for all but a few k-tuples $S \subset V(G)$. Now, the volume-respecting condition enters: If $\mathrm{Vol}(S)$ is large, then $\mathrm{conv}(f(S))$ has large $(k-1)$-dimensional volume. It turns out that the projection of a convex set in ℓ_2^m with large $(k-1)$-dimensional volume on a random line is unlikely to be short, and so S with large $\mathrm{Vol}(S)$ is unlikely to be squeezed. Thus, by estimating the number of squeezed k-tuples in two ways, one gets an inequality bounding b from above in terms of $\mathrm{ld}(G)$.

Vempala [Vem98] applied volume-respecting embeddings in another algorithmic problem, this time concerning arrangement of graph vertices in the plane. Moreover, he also gave alternative proof of some of Feige's lemmas. Rao in the already mentioned paper [Rao99] also obtained improved volume-respecting embeddings for planar metrics.

Bartal's trees. As we have seen, in Bourgain's method, for a given metric ρ one constructs a convex combination $\sum \alpha_i \nu_i \geq \frac{1}{D} \rho$, where ν_i are line pseudometrics dominated by ρ. An interesting "dual" result was found by Bartal [Bar96], following earlier work in this direction by Alon, Karp, Peleg, and West [AKPW95]. He approximated a given ρ by a convex combination $\sum_{i=1}^{N} \alpha_i \tau_i$, where this time the inequalities go in the opposite direction: $\tau_i \geq \rho$ and $\sum \alpha_i \tau_i \leq D\rho$, with $D = O(\log^2 n)$ (later he improved this to $O(\log n \log \log n)$ in [Bar98]). The τ_i are not line metrics (and in general they cannot be), but they are tree metrics, and even of a special form, the so-called *hierarchically well-separated trees.* This means that τ_i is given as the shortest-path metric of a rooted tree with weighted edges such that the distances from each vertex to all of its sons are the same, and if v is a son of u, and w a son of v, then $\tau_i(u, v) \geq K \cdot \tau_i(v, w)$, where $K \geq 1$ is a parameter that can be set at will (and the constant in the bound on D depends on it).

This result has been used in approximation algorithms for problems involving metric spaces, according to the following scheme: Choose $i \in \{1, 2, \ldots, N\}$ at random, with each i having probability α_i, solve the problem in question for the tree metric τ_i, and show that the expected value of the solution is not very far from the optimal solution for the original metric ρ.

Since tree metrics embed isometrically into ℓ_1, Bartal's result also implies $O(\log n \log \log n)$-embeddability of all n-point metric spaces into ℓ_1, which is just a little weaker than Bourgain's approach (and it also implies that $\Omega(\log n)$ is a lower bound in Bartal's setting). For a simpler proof of a weaker version Bartal's result see Indyk [Ind01].

Exercises

1. (Embedding into ℓ_p) Prove that under the assumptions of Lemma 15.7.3, the metric space (V, ρ) can be D-embedded into ℓ_p^N, $1 \leq p \leq \infty$, with distortion at most D. (You may want to start with the rather easy cases $p = 1$ and $p = \infty$, and use Hölder's inequality for an arbitrary p.) ③

2. (Dimension reduction for the embedding)
   (a) Let $E_1, \ldots, E_m$ be independent events, each of them having probability at least $\frac{1}{12}$. Prove that the probability of no more than $\frac{m}{24}$ of the E_i occurring is at most e^{-cm}, for a sufficiently small positive constant c. Use suitable Chernoff-type estimates or direct estimates of binomial coefficients. ③

(b) Modify the proof of Theorem 15.7.1 as follows: For each $j = 1, 2, \ldots, q$, pick sets A_{ij} independently at random, $i = 1, 2, \ldots, m$, where the points are included in A_{ij} with probability 2^{-j} and where $m = C \log n$ for a sufficiently large constant C. Using (a) and Lemmas 15.7.2 and 15.7.3, prove that with a positive probability, the embedding $f \colon V \to \ell_2^{qm}$ given by $f(v)_{ij} = \rho(v, A_{ij})$ has distortion $O(\log n)$. ③

3. Let $a_1, a_2, \ldots, a_n,\ b_1, b_2, \ldots, b_n,\ \alpha_1, \alpha_2, \ldots, \alpha_n$ be positive real numbers. Show that

$$\frac{\alpha_1 a_1 + \alpha_2 a_2 + \cdots + \alpha_n a_n}{\alpha_1 b_1 + \alpha_2 b_2 + \cdots + \alpha_n b_n} \geq \min\{\frac{a_1}{b_1}, \frac{a_2}{b_2}, \ldots, \frac{a_n}{b_n}\}.$$

②

4. Let P_n be the metric space $\{0, 1, \ldots, n\}$ with the metric inherited from $\mathbf{R}$ (or a path of length n with the graph metric). Prove the following Ramsey-type result: For every $D > 1$ and every $\varepsilon > 0$ there exists an $n = n(D, \varepsilon)$ such that whenever $f \colon P_n \to (Z, \sigma)$ is a D-embedding of P_n into some metric space, then there are $a < b < c$, $b = \frac{a+c}{2}$, such that f restricted to the subspace $\{a, b, c\}$ of P_n is a $(1+\varepsilon)$-embedding. That is, if a sufficiently long path is D-embedded, then it contains a scaled copy of a path of length 2 embedded with distortion close to 1. ④

Can you extend the proof so that it provides a scaled copy of a path of length k?

5. (Lower bound for embedding trees into ℓ_2)

(a) Show that for every $\varepsilon > 0$ there exists $\delta > 0$ with the following property. Let $x_0, x_1, x_2, x_2' \in \ell_2$ be points such that $\|x_0 - x_1\|, \|x_1 - x_2\|, \|x_1 - x_2'\| \in [1, 1 + \delta]$ and $\|x_0 - x_2\|, \|x_0 - x_2'\| \in [2, 2 + \delta]$ (so all the distances are almost like the graph distances in the following tree, except possibly for the one marked by a dotted line).

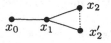

Then $\|x_2 - x_2'\| \leq \varepsilon$; that is, the remaining distance must be very short. ③

(b) Let $T_{k,m}$ denote the complete k-ary tree of height m; the following picture shows $T_{3,2}$:

Show that for every r and m there exists k such that whenever the leaves of $T_{k,m}$ are colored by r colors, there is a subtree of $T_{k,m}$ isomorphic to $T_{2,m}$ with all leaves having the same color. ②

(c) Use (a), (b), and Exercise 4 to prove that for any $D > 1$ there exist m and k such that the tree $T_{k,m}$ considered as a metric space with the shortest-path metric cannot be D-embedded into ℓ_2. [4]

6. (Another lower bound for embedding trees into ℓ_2)
   (a) Let $x_0, x_1, \ldots, x_n$ be arbitrary points in a Euclidean space (we think of them as images of the vertices of a path of length n under some embedding). Let $\Gamma = \{(a, a+2^k, a+2^{k+1}): a = 0, 1, 2, \ldots, a+2^{k+1} \leq n, k = 0, 1, 2 \ldots\}$. Prove that

$$\sum_{(a,b,c) \in \Gamma} \frac{\|x_a - 2x_b + x_c\|^2}{(c-a)^2} \leq \sum_{a=0}^{n-1} \|x_a - x_{a+1}\|^2;$$

   this shows that an average triple (x_a, x_b, x_c) is "straight" (and provides an alternative solution to Exercise 4 for $Z = \ell_2$). [3]
   (b) Prove that the complete binary tree $T_{2,m}$ requires $\Omega(\sqrt{\log m})$ distortion for embedding into ℓ_2. Consider a nonexpanding embedding $f: V(T_{2,m}) \to \ell_2$ and sum the inequalities as in (a) over all images of the root-to-leaf paths. [4]

7. (Bourgain's embedding of complete binary trees into ℓ_2) Let $B_m = T_{2,m}$ be the complete binary tree of height m (notation as in Exercise 5). We identify the vertices of B_m with words of length at most m over the alphabet $\{0, 1\}$: The root of B_m is the empty word, and the sons of a vertex w are the vertices $w0$ and $w1$. We define the embedding $f: V(B_m) \to \ell_2^{|V(B_m)|-1}$, where the coordinates in the range of f are indexed by the vertices of B_m distinct from the root, i.e., by nonempty words. For a word $w \in V(B_m)$ of length a, let $f(w)_u = \sqrt{a-b+1}$ if u is a nonempty initial segment of w of length b, and $f(w)_u = 0$ otherwise. Prove that this embedding has distortion $O(\sqrt{\log m})$. [4]

8. Prove that any finite tree metric can be isometrically embedded into ℓ_1. [3]

9. (Low-dimensional embedding of trees)
   (a) Let T be a tree (in the graph-theoretic sense) on $n \geq 3$ vertices. Prove that there exist subtrees T_1 and T_2 of T that share a single vertex and no edge and together cover T, such that $\min(|V(T_1)|, |V(T_2)|) \leq 1+\frac{2}{3}n$. [3]

   (b) Using (a), prove that every tree metric space with n points can be isometrically embedded into ℓ_∞^d with $d = O(\log n)$. [4]
   This result is from [LLR95].

What Was It About? An Informal Summary

Chapter 1

- Linear and affine notions (dependence, hull, subspace, mapping); hyperplane, k-flat.
- General position: Degenerate configurations have measure zero in the space of all configurations, provided that degeneracy can be described by countably many polynomial equations.
- Convex set, hull, combination.
- Separation theorem: Disjoint convex sets can be separated by a hyperplane; strictly so if one of them is compact and the other closed.
- Theorems involving the dimension: Helly (if $\mathcal{F}$ is a finite family of convex sets with empty intersection, then there is a subfamily of at most $d+1$ sets with empty intersection), Radon ($d+2$ points can be partitioned into two subsets with intersecting convex hulls), Carathéodory (if $x \in \text{conv}(X)$, then $x \in \text{conv}(Y)$ for some at most $(d+1)$-point $Y \subseteq X$).
- Centerpoint of X: Every half-space containing it contains at least $\frac{1}{d+1}$ of X. It always exists by Helly. Ham-sandwich: Any d mass distributions in $\mathbf{R}^d$ can be simultaneously bisected by a hyperplane.

Chapter 2

- Minkowski's theorem: A 0-symmetric convex body of volume larger than 2^d contains a nonzero integer point.
- General lattice: a discrete subgroup of $(\mathbf{R}^d, +)$. It can be written as the set of all integer linear combinations of at most d linearly independent vectors (basis). Determinant = volume of the parallelotope spanned by a basis.
- Minkowski for general lattices: Map the lattice onto $\mathbf{Z}^d$ by a linear mapping.

Chapter 3

- Erdős–Szekeres theorem: Every sufficiently large set in the plane in general position contains k points in convex position. How large? Exponential in k.
- What about k-holes (vertex sets of empty convex k-gons)? For $k = 5$ yes (in sufficiently large sets), for $k \geq 7$ no (Horton sets), $k = 6$ is a challenging open problem.

Chapter 4

- Szemerédi–Trotter theorem: m distinct points and n distinct lines in the plane have at most $O(m^{2/3}n^{2/3} + m + n)$ incidences.
- This is tight in the worst case. Example for $m = n$: Use the $k \times 4k^2$ grid and lines $y = ax + b$ with $a = 0, 1, \ldots, 2k-1$ and $b = 0, 1, \ldots, 2k^2-1$.
- Crossing number theorem: A simple graph with n vertices and $m \geq 4n$ edges needs $\Omega(m^3/n^2)$ crossings. Proof: At least $m-3n$ crossings, since planar graphs have fewer than $3n$ edges, then random sampling.
- Forbidden bipartite subgraphs: A graph on n vertices without $K_{r,s}$ has $O(n^{2-1/r})$ edges.
- Cutting lemma: Given n lines and r, the plane can be subdivided into $O(r^2)$ generalized triangles such that the interior of each triangle is intersected by at most $\frac{n}{r}$ lines. Proof of a weaker version: Triangulate the arrangement of a random sample and show that triangles intersected by many lines won't survive. Application: geometric divide-and-conquer.
- For unit distances and distinct distances in the plane, bounds can be proved, but a final answer seems to be far away.

Chapter 5

- Geometric duality: Sends a point a to the hyperplane $\langle a, x \rangle = 1$ and vice versa; preserves incidences and sidedness.
- Convex polytope: the convex hull of a finite set and also the intersection of finitely many half-spaces.
- Face, vertex, edge, facet, ridge. A polytope is the convex hull of its vertices. A face of a face is a face. Face lattice. Duality turns it upside down. Simplex. Simple and simplicial polytopes.
- The convex hull of n points in $\mathbf{R}^d$ can have as many as $\Omega(n^{\lfloor d/2 \rfloor})$ facets; cyclic polytopes.
- This is as bad as it can get: Given the number of vertices, cyclic polytopes maximize the number of faces in each dimension (upper bound theorem).
- Gale transform: An n-point sequence in $\mathbf{R}^d$ (affinely spanning $\mathbf{R}^d$) is mapped to a sequence of n vectors in $\mathbf{R}^{n-d-1}$. Properties: a simple linear algebra. Faces of the convex hull go to subsets whose complement contains 0 in the convex hull.

- 3-dimensional polytopes are nice: Their graphs correspond to vertex 3-connected planar graphs (Steinitz theorem), and they can be realized with rational coordinates. From dimension 4 on, bad things can happen (irrational or doubly exponential coordinates may be required, recognition is difficult).
- Voronoi diagram. It is the projection of a convex polyhedron in dimension one higher (lifting using the paraboloid). Delaunay triangulation (defined using empty balls; dual to the Voronoi diagram).

Chapter 6

- Arrangement of hyperplanes (faces, vertices, edges, facets, cells). For d fixed, there are $O(n^d)$ faces.
- Clarkson's theorem on levels: At most $O(n^{\lfloor d/2 \rfloor} k^{\lceil d/2 \rceil})$ vertices are at level at most k. Proof: Express the expected number of level-0 vertices of a random sample in two ways!
- Zone theorem: The zone of a hyperplane has $O(n^{d-1})$ vertices. Proof: Delete a random hyperplane, and look at how many zone faces are sliced into two by adding it back.
- Proof of the cutting lemma by a finer sampling argument: Vertically decompose the arrangement of a sample taken with probability p, show that the number of trapezoids intersected by at least tnp lines decreases exponentially with t, take $\frac{1}{t}$-cuttings within the trapezoids.
- Canonical triangulation, cutting lemma in $\mathbf{R}^d$ ($O(r^d)$ simplices).
- Milnor–Thom theorem: The arrangement of the zero sets of n polynomials of degree at most D in d real variables has at most $O(Dn/d)^d$ faces.
- Most arrangements of pseudolines are nonstretchable (by Milnor–Thom). Similarly for many other combinatorial descriptions of geometric configurations; usually most of them cannot be realized.

Chapter 7

- Davenport–Schinzel sequences of order s (no $abab\ldots$ with $s+2$ letters); maximum length $\lambda_s(n)$. Correspond to lower envelopes of curves: The curves are graphs of functions defined everywhere, every two intersecting at most s times. Lower envelopes of segments yield DS sequences of order 3.
- $\lambda_3 = \Theta(n\alpha(n))$; $\lambda_s(n)$ is almost linear for every fixed s.
- The lower envelope of n algebraic surface patches in $\mathbf{R}^d$, as well as a single cell in their arrangement, have complexity $O(n^{d-1+\epsilon})$. Charging schemes and more random sampling.

Chapter 8

- Fractional Helly theorem: If a family of n convex sets has $\alpha\binom{n}{d+1}$ intersecting $(d+1)$-tuples, then there is a point common to at least $\frac{\alpha}{d+1}n$ of the sets.
- Colored Carathéodory theorem: If each of $d+1$ sets contains 0 in the convex hull, then we can pick one point from each set so that the convex hull of the picked points contains 0.
- Tverberg's theorem: $(d+1)(r-1)+1$ points can be partitioned into r subsets with intersecting convex hulls (the number is the smallest conceivable one: $r-1$ simplices plus one extra point).
- Colored Tverberg theorem: Given points partitioned into $d+1$ color classes by t points each, we can choose r disjoint rainbow subsets with intersecting convex hulls, $t = t(d,r)$. Only topological proofs are known.

Chapter 9

- The dimension is considered fixed in this chapter. First selection lemma: Given n points, there exists a point contained in a fixed fraction of all simplices with vertices in the given points.
- Second selection lemma: If $\alpha\binom{n}{d+1}$ of the simplices are marked, we can find a point in many of the marked simplices (at least $\Omega(\alpha^{s_d}\binom{n}{d+1})$). Needs colored Tverberg and Erdős–Simonovits.
- Order type. Same-type lemma: Given n points in general position and k fixed, one can find k disjoint subsets of size $\Omega(n)$, all of whose transversals have the same order type.
- A hypergraph regularity lemma: For an $\varepsilon > 0$ and a k-partite hypergraph of density bounded below by a constant $\beta > 0$ and with color classes $X_1, \ldots, X_n$ of size n, we can choose subsets $Y_1 \subseteq X_1, \ldots, Y_k \subseteq X_k$, $|Y_1| = \cdots = |Y_k| \geq cn$, $c = (k, \beta, \varepsilon) > 0$, such that any $Z_1 \subseteq Y_1, \ldots, Z_k \subseteq Y_k$ with $|Z_i| \geq \varepsilon|Y_i|$ induce some edge.
- Positive-fraction selection lemma: Given n red, n white, and n blue points in the plane, we can choose $\frac{n}{12}$ points of each color so that all red–white–blue triangles have a common point; similarly in $\mathbf{R}^d$.

Chapter 10

- Set systems; transversal number τ, packing number ν. Fractional transversal and fractional packing; $\nu^* = \tau^*$ by LP duality.
- Epsilon net, shattered set, VC-dimension. Shatter function lemma: A set system on n points with VC-dimension d has at most $\sum_{k=0}^{d}\binom{n}{k}$ sets.
- Epsilon net theorem: A random sample of $C\frac{d}{\varepsilon}\log\frac{1}{\varepsilon}$ points in a set system of VC-dimension d is an ε-net with high probability. In particular, ε-nets exist of size depending only on d and ε.
- Corollary: $\tau = O(\tau^* \log \tau^*)$ for bounded VC-dimension.

- Half-spaces in $\mathbf{R}^d$ have VC-dimension $d+1$. Lifting (Veronese map) and the shatter function lemma show that systems of sets in $\mathbf{R}^d$ definable by Boolean combinations of a bounded number of bounded-degree polynomial inequalities have bounded VC-dimension.
- Weak epsilon nets for convex sets: Convex sets have infinite VC-dimension, but given a finite set X and $\varepsilon > 0$, we can choose a weak ε-net of size at most $f(d,\varepsilon)$, that is, a set (generally not a subset of X) that intersects every convex C with $|C \cap X| \geq \varepsilon|X|$.
- Consequently, τ is bounded by a function of τ^* for any finite system of convex sets in $\mathbf{R}^d$.
- Alon–Kleitman (p,q)-theorem: Let $\mathcal{F}$ be a system of convex sets such that among every p sets, some q intersect ($p \geq q \geq d+1$). Then $\tau(\mathcal{F})$ is bounded by a function of d, p, q. Proof: First bound ν^* using fractional Helly; then τ is bounded in terms of $\tau^* = \nu^*$ as above.
- A similar (p,q)-theorem for hyperplane transversals of convex sets (even though no Helly theorem!).

Chapter 11

- k-sets, k-facets (only for sets in general position!), halving facets. Dual: cells of level k, vertices of level k. The k-set problem is still unsolved. Straightforward bounds from Clarkson's theorem on levels.
- Bounds for halving facets yield bounds for k-facets sensitive to k.
- A recursive planar construction with a superlinear number of halving edges.
- Lovász lemma: No line intersects more than $O(n^{d-1})$ halving facets. Proof: When a moving line crosses the convex hull of $d-1$ points of X, the number of halving facets intersected changes by 1 (halving-facet interleaving lemma).
- Implies an upper bound of $O(n^{d-\delta(d)})$ for halving facets by the second selection lemma.
- In the plane a continuous motion argument proves that the crossing number of the halving-edge graph is $O(n^2)$, and consequently, it has $O(n^{4/3})$ edges by the crossing number theorem. This is the best we can do in the plane, although $O(n^{1+\varepsilon})$ for every fixed $\varepsilon > 0$ is suspected.

Chapter 12

- Perfect graph ($\chi = \omega$ hereditarily). weak perfect graph conjecture (now theorem): A graph is perfect iff its complement is.
- Proof via the polytope $\{x \in \mathbf{R}^V : x \geq 0, x(K) \leq 1 \text{ for every clique } K\}$.
- Brunn's slice volume inequality: For a compact convex $C \subset \mathbf{R}^{n+1}$, $\mathrm{vol}_n(\{x \in C : x_1 = t\})^{1/n}$ is a concave function of t (as long as the slices do not miss the body).

- Brunn–Minkowski inequality: $\mathrm{vol}(A)^{1/n} + \mathrm{vol}(B)^{1/n} \leq \mathrm{vol}(A+B)^{1/n}$ for nonempty compact $A, B \subset \mathbf{R}^n$.
- A partially ordered set with N linear extensions can be sorted by $O(\log N)$ comparisons. There always exists a comparison that reduces the number of linear extensions by a fixed fraction: Compare elements whose average heights differ by less than 1.
- Order polytope: $0 \leq x \leq 1$, $x_a \leq x_b$ whenever $a \preceq b$. Linear extensions correspond to congruent simplices and good comparison to dividing the volume evenly by a hyperplane $x_a = x_b$. The best ratio is not known (conjectured to be $\frac{1}{3} : \frac{2}{3}$).

Chapter 13

- Volumes and other things in high dimensions behave differently from what we know in $\mathbf{R}^2$ and $\mathbf{R}^3$. For example, the ball inscribed in the unit cube has a tiny volume.
- An η-net is an inclusion-maximal η-separated set. It is mainly useful because it is η-dense. In S^{n-1}, a simple volume argument yields η-nets of size at most $(4/\eta)^n$.
- An N-vertex convex polytope inscribed in the unit ball B^n occupies at most $O(\ln(\frac{N}{n}+1)/n)^{n/2}$ of the volume of B^n. Thus, with polynomially many vertices, the error of deterministic volume approximation is exponential in the worst case.
- Polytopes with such volume can be constructed: For $N = 2n$ use the crosspolytope, for $N = 4^n$ a 1-net in the dual S^{n-1}, and interpolate using a product.
- Ellipsoid: an affine image of B^n. John's lemma: Every n-dimensional convex body has inner and outer ellipsoids with ratio at most n, and a symmetric convex body admits the better ratio $\sqrt{n}$. The maximum-volume inscribed ellipsoid (which is unique) will do as the inner ellipsoid.

Chapter 14

- Measure concentration on S^{n-1}: For any set A occupying half of the sphere, almost all of S^{n-1} is at most $O(n^{-1/2})$ away from A. Quantitatively, $1 - \mathrm{P}[A_t] \leq 2e^{-t^2 n/2}$.
- Similar concentration phenomena in many other high-dimensional spaces: Gaussian measure on $\mathbf{R}^n$, cube $\{0,1\}^n$, permutations, etc.
- Many concentration inequalities can be proved via isoperimetric inequalities. Isoperimetric inequality: Among all sets of given volume, the ball has the smallest volume of a t-neighborhood.
- Lévy's lemma: A 1-Lipschitz function f on S^{n-1} is within $O(n^{-1/2})$ of its median on most of S^{n-1}.
- Consequently (using η-nets), there is a high-dimensional subspace on which f is almost constant (use a random subspace).

- Normed spaces, norm induced by a symmetric convex body.
- For any n-dimensional symmetric convex polytope, $\log(f_0)\log(f_{n-1}) = \Omega(n)$ (many vertices or many facets).
- Dvoretsky's theorem: For every k and $\varepsilon > 0$ there exists n ($n = e^{O(k/\varepsilon^2)}$ suffices) such that any n-dimensional convex body has a k-dimensional $(1+\varepsilon)$-spherical section. In other words, any high-dimensional normed space has an almost Euclidean subspace.

Chapter 15

- Metric space; the distortion of a mapping between two metric spaces, D-embedding. Spaces ℓ_p^d and ℓ_p.
- Flattening lemma: Any n-point Euclidean metric space can be $(1+\varepsilon)$-embedded into ℓ_2^k, $k = O(\varepsilon^{-2}\log n)$ (project on a random k-dimensional subspace).
- Lower bound for D-embedding into a d-dimensional normed space: counting; take all subgraphs of a graph without short cycles and with many edges.
- The m-dimensional Hamming cube needs $\sqrt{m}$ distortion for embedding into ℓ_2 (short diagonals and induction).
- Edge expansion (conductance), second eigenvalue of the Laplacian matrix. Constant-degree expanders need $\Omega(\log n)$ distortion for embedding into ℓ_2 (tight). Method: Compare sums of squared distances over the edges and over all pairs, in the graph and in the target space.
- D-embeddability into ℓ_2 is polynomial-time decidable by semidefinite programming.
- All n-point spaces embed isometrically into ℓ_∞^n. For embeddings with smaller dimension, use distances to random subsets of suitable density as coordinates. A similar method yields $O(\log n)$-embedding into ℓ_2 (or any other ℓ_p).
- Example of algorithmic application: approximating the sparsest cut. Embed the graph metric into ℓ_1 with low distortion; this yields a cut pseudometric defining a sparse cut.

Hints to Selected Exercises

1.2.7(a). The existence of an $x \geq 0$ with $Ax = b$ means that b lies in the convex cone generated by the columns of A. If b is not in the cone, then it can be separated from it as in Exercise 6(b).

1.2.7(b). Apply (a) with the $d \times (n+d)$ matrix $(A \mid I_d)$, where I_d is the identity matrix.

1.3.5(c). $\sqrt{2d/(d+1)}$.

1.3.8(b). By Helly's theorem, $K = \bigcap_{x \in X} \mathrm{conv}(V(x)) \neq \emptyset$. Prove that K is the kernel.

1.3.10(b). Assign the set $H_x = \{(a,b) \in \mathbf{R}^d \times \mathbf{R} \colon \langle a, x \rangle < b\}$ to each $x \in X$ and the set $G_y = \{(a,b) \in \mathbf{R}^d \times \mathbf{R} \colon \langle a, x \rangle \geq b\}$ to each $y \in Y$. Use Helly's theorem.

1.4.1(a). Express γ as $\bigcup_{i=1}^{\infty} C_i$, where $C_1 \subseteq C_2 \subseteq \cdots$ are compact. Then $\mu(\gamma) = \sum_{i=1}^{\infty} \mu(C_{i+1} \setminus C_i)$ by the σ-additivity of μ. (More generally, every Borel probability measure on a separable metric space is regular: The measure of any set can be approximated with arbitrary precision by the measure of a compact set contained in it.)

2.1.4(c). Let $p(x)$ be a polynomial with integer coefficients having α as a root. If $\deg(p) = d$ and $|\alpha - m/n| < n^{d+1}$, say, then $n^d p(m/n)$ is integral, but $|n^d p(m/n)| < 1$ for large n.

2.1.5(a). Seek a nonzero vector in $\mathbf{Z}^3$ close to the line $y = \alpha_1 x$, $y = \alpha_2 x$.

2.2.1. Show that elementary row operations on the matrix, which do not change the determinant, also preserve the volume. Diagonalize the matrix.

3.1.4. Project orthogonally on a suitable plane and apply Erdős–Szekeres.

3.2.4. It suffices to deal with the case $k = 4m$. First prove by induction that a $2m$-point cup contained in a Horton set has at least $2^m - 2m$ points of the set above it.

4.1.2. Place points on two circles lying in orthogonal planes in $\mathbf{R}^4$.

4.3.3. Choose a point set P, one point in each of the m cells. From each top edge, cut off a little segment ab and replace it by the segments ap and pb, where $p \in P$ lies below the edge. Each line is replaced by a polygonal curve.

Consider a graph drawing with P as the vertices and the polygonal curves defining edges.

4.3.4(c). Consider a drawing of G witnessing pair-cr$(G) = k$. At most $2k$ edges are involved in any crossings, and the remaining ones (the *good edges*) form a planar graph. Redraw the edges with crossings so that they do not intersect any of the good edges and, subject to this, have the minimum possible number of crossings.

4.4.1(a). $O(n^{10/7}) = O(n^{1.43})$.

4.4.1(b). Let C_i be the points of C that are the centers of at least 2^i and at most 2^{i+1} circles. We have $|C_i| = q_i \leq n/2^i$. One incidence of a line of the form ℓ_{uv} with a $c \in C_i$ contributes at most 2^{i+2} edges.

4.4.2(b). Look at u, v with $\mu(u, v) \geq 4\sqrt{d_i}$, and suppose that at least half of the uv edges have their partner edges adjacent to u, say. These partner edges connect u to at least $2\sqrt{d_i}$ distinct neighbor vertices. By (a), at most $\sqrt{d_i}/2$ of these partner edges may belong to E_h.

4.4.2(c). We get $|E| = O(|E \setminus E_h|) = O(n^{4/3}d_i^{1/6})$; at the same time, $|E| \geq nd_i/2$. This gives $d_i = O(n^{2/5})$ and $I_{\text{circ}}(n, n) = O(n^{7/5}) = O(n^{1.4})$.

4.7.1. Consider a trapezoid $ABB'A'$; AB is the bottom side and $A'B'$ the top side. Suppose AB is contained in an edge CD of P_j and $A'B'$ is an edge of P_{j+1} (the few other possible cases are discussed similarly). Let A_1 be the intersection of the level $qj + i$ with the vertical line AA', and similarly for B_1. The segments $A'B'$, $A'A_1$, and $B'B_1$ each have at most $q+1$ intersections. Observe that if AA_1 has some a intersections, then CA also has at least a intersections, and similarly for BB_1 and BD. At the same time CD has at most $q+1$ intersections altogether. Therefore, AA_1, AB, and BB_1 have no more than $q+1$ intersections in total.

5.1.9(b). Geometric duality and Helly's theorem.

5.1.9(c). The first segment s_1 is a chord of the unit circle passing near the center. Each s_{i+1} has one endpoint on the unit circle, and the other endpoint almost touches s_i near the center.

5.3.2. Ask in this way: Given a normal vector $a \in \mathbf{R}^d$ of a hyperplane, which vertices maximize the linear function $x \mapsto \langle a, x \rangle$? For example, for the cube, if $a_i > 0$, then x_i has to be $+1$; if $a_i < 0$, then $x_i = -1$; and for $a_i = 0$ both $x_i = \pm 1$ are possible.

5.3.8. If the removed vertices u, v lie in a common 2-face f, let h be the plane defining f; from each vertex there is an edge going "away from h," except for the vertices of a single face $g \neq f$ "opposite" to f. The graph of the face g is connected and can be reached from any other vertex. If u, v do not share a 2-face, pass a plane h through them and one more vertex w. The subgraph on the vertices below h is connected, and so is the subgraph on the vertices above h; they are connected via the vertex w.

5.4.2. Do not forget to check that β is not *contained* in any hyperplane.

5.5.1(c). The simplest example seems to be the product of an n-vertex 4-dimensional cyclic polytope with its dual.

5.7.11(c). Assume $n \geq 2$. If x, y are points on the surface of such an intersection P, coming from the surface of the same ball κ, show that the shorter of the great circle arcs on κ connecting x and y lies entirely on the surface of P (this is a kind of "convexity" of the facets). Infer that each ball contributes at most one facet, and use Euler's formula.

6.1.5. $n! \cdot C_n$, where $C_n = \frac{1}{n+1}\binom{2n}{n}$ is the nth Catalan number.

6.1.6(a). One possibility is a perturbation argument. Another one is a proof by induction, adding one line at a time.

6.1.7(b). Warning: The $\binom{n}{2}$ lines determined by n points in general position are not in general position!

6.2.2(b). Assuming that no s_i is vertical, write $s_i = \{(x, y) \in \mathbf{R}^2 \colon c_i \leq x \leq d_i, y = a_i x + b_i\}$. Whether s_i and s_j intersect can be determined from the signs of the $O(n^2)$ polynomials $a_i - a_j$, $c_i - c_j$, $d_i - d_j$, $c_i(a_i - a_j) + b_i - b_j$, $d_i(a_i - a_j) + b_i - b_j$, $i, j = 1, 2, \ldots, n$.

6.2.2(c). Use the lower bound for the quantity $K(n, n)$ in Chapter 4.

6.3.4(a). First derive $X_W \geq |W| - n$, and then use it for a random sample of the lines.

6.4.3(a). Define an incidence graph between lines and the considered m cells (incidence = the line contributes an edge to the cell). This graphs contains no $K_{2,5}$, since two cells have at most 4 "common tangents."

6.4.3. Each of the given n cells either lies completely within a single triangle Δ_i, or it is in the zone of an edge of some triangle. Use the zone theorem for bounding the total number of edges of cells of the latter type.

6.5.2(a). $\mathbf{E}[X^2] = \sum_{i,j} \mathbf{E}[X_i X_j]$. $\mathbf{E}[X_i X_j] = p^2$ for $i \neq j$ and $\mathbf{E}[X_i^2] = p$. The result is $p^2 n(n-1) + pn$.

7.1.1. Construct the curves from left to right: Start with n horizontal lines on the left and always "bring down" the curve required by the sequence.

7.1.4. Warning: The *abab* subsequence *can* appear!

7.1.8(b). For simplicity assume that all the s_i and t_i are all distinct and let $E = \{s_1, t_1, \ldots, s_n, t_n\}$. Call a vertex v *active* for an interval $I \subseteq \mathbf{R}$ if v appears on the lower envelope of L_t for some $t \in I$ and $I \cap \{s_i, s_j, t_i, t_j\} \neq \emptyset$, where ℓ_i, ℓ_j are the lines defining v. Let $g(I)$ be the number of active vertices for I and let $g(m) = \max\{g(I) \colon |I \cap E| \leq m\}$. Split I in the middle of $E \cap I$ and derive $g(m) \leq O(m) + g(\lfloor m/2 \rfloor) + g(\lceil m/2 \rceil)$.

7.3.2(b). Zero out the first and last 1 in each row. Go through the matrix column by column and write down the row indices of 1's. Deleting contiguous repetitions produces a Davenport–Schinzel sequence with no *ababa*.

7.4.1(b). Given a sequence w witnessing $\psi_s^t(m, n)$, replace each of the m segments in the decomposition of w by the list of its symbols (and erase contiguous repetitions if needed).

8.1.2. Make the sets compact as in the proof of the fractional Helly theorem. Consider all d-element collections $\mathcal{K}$ containing one set from each $\mathcal{C}_i$ but one, and let $v_{\mathcal{K}}$ be the lexicographic minimum of the intersection of $\bigcap \mathcal{K}$. Let $\mathcal{K}_0$ be such that $v = v_{\mathcal{K}_0}$ is the lexicographically largest among all $v_{\mathcal{K}}$, and let i_0 be the index such that $\mathcal{K}_0$ contains no set from $\mathcal{C}_{i_0}$. Show that for each $C \in \mathcal{C}_{i_0}$, v is the minimum of $C \cap \bigcap \mathcal{K}_0$, and in particular, $v \in C$.

8.2.1. Regard $S \cup T$ as a Gale transform of a point sequence and reformulate the problem using that sequence. Or lift $S \cup T$ into $\mathbf{R}^{d+1}$ suitably.

9.2.2(b). For $d = 3$: Choose k points on the moment curve, say, and replace each by a cluster of n/k points. Use all tetrahedra having two vertices in one cluster and the other two vertices in another cluster. There are about n^4/k^2 such tetrahedra, and no point is contained in more than n^4/k^4 of them if the clusters are small and k is not too large compared to n.

9.3.1(b). Be careful with degenerate cases; first determine the dimension of the affine hull of $p_1, \ldots, p_{d+1}$ and test whether p_{d+2} lies in it. Then you may need to use some number of other affinely independent points among the p_i.

9.3.3(a). Let $x_i, x_i' \in X_i$ be such that $(x_1, \ldots, x_{d+1})$ and $(x_1', \ldots, x_{d+1}')$ have different orientations. Let y_i be a point moving along the segment $x_i x_i'$ at constant speed, starting at x_i at time 0 and reaching x_i' at time 1. By continuity of the determinant, all the y_i lie in a common hyperplane at some moment, and this hyperplane intersects the convex hulls of all the X_i.

9.3.3(b). Let the hyperplane h intersect all the C_i, and let $a_i \in h \cap C_i$. Use Radon's lemma.

9.3.3(c). Suppose that $0 \in \mathrm{conv}\big(\bigcup_{i \in I} C_i\big) \cap \mathrm{conv}\big(\bigcup_{j \notin I} C_j\big)$. Then there are points $x_i \in C_i$, $i = 1, 2, \ldots, d+1$, such that $0 \in \mathrm{conv}\{x_i : i \in I\}$ and $0 \in \mathrm{conv}\{x_j : j \notin I\}$. Hence the vectors $\{x_i : i \in I\}$ are linearly dependent, as well as those of $\{x_j : j \notin I\}$. Thus, the linear subspace generated by all the x_i has dimension at most $d-1$.

9.3.5(a). Partition P into 3 sets and apply the same-type lemma. If Y_1, Y_2, Y_3 are the resulting sets, then each line misses at least one $\mathrm{conv}(Y_i)$. Let P' be the Y_i whose convex hull is missed by the largest number of lines of L.

9.3.5(b). First apply (a) with P consisting of the left endpoints of the segments of S. Then apply (a) again with the right endpoints of the remaining segments and the remaining lines. Finally, discard either the lines intersected by all segments or those intersected by no segment.

9.3.5(c). Use (b) twice.

9.4.4. Consider the complete bipartite graphs with classes V_i and V_j, $1 \leq i < j \leq 4$, and color each of their edges randomly either red or blue with equal probability. A triple $\{u, v, w\}$ with $u \in V_i$, $v \in V_j$, $w \in V_k$, $i < j < k$, is present if and only if the edges $\{u, v\}$ and $\{u, w\}$ have distinct colors.

10.1.3. Choose the appropriate number of points independently at random according to the distribution given by an optimal fractional transversal.

10.1.4(a). Let m_k be the number of yet uncovered sets after the last step i such that x_i covered more than k previously uncovered sets ($m_d = |\mathcal{F}|$, $m_0 = 0$). Derive $t \leq \sum_{k=1}^{d} \frac{m_k - m_{k-1}}{k}$ and note that $m_k \leq \nu_k(\mathcal{F})$.

10.1.6(b). By the Farkas lemma, it suffices to check the following: For all $u \in \mathbf{R}^m$, $v \in \mathbf{R}^n$, and $z \in \mathbf{R}$ such that $u \geq 0$, $v \geq 0$, $z \geq 0$, $u^T A \leq zc$, and $Av \geq zb$, we have $u^T b \leq c^T v$. For $z \neq 0$ this is (a), and for $z = 0$ choose $x_0 \in P$ and $y_0 \in D$ and use $u^T b \leq u^T A x_0 \leq 0$ and $c^T v \geq y_0^T A v \geq 0$.

10.2.2. All subsets of size at most d.

10.3.1. 7.

10.3.3. Such a p would have to be 0 on the boundary, but if a polynomial is 0 on a segment, then it is 0 on the whole line containing that segment.

10.3.4(b). Choose a $\frac{1}{r}$-net $S \subseteq L$ for the set system $(L, \mathcal{T})$ and triangulate the arrangement of S. No dangerous triangle appears in this triangulation.

10.3.6(c). The shattering graph SG_d considered in Exercise 5 contains a subdivision of K_d where each edge is subdivided once. Some care is needed, since some vertices might be both shattering and shattered in G.

10.4.1(b). This method gives size $O\left(\varepsilon^{-2^{d-1}}\right)$.

10.4.2(b). (a) yields $f(\varepsilon) \leq \binom{\ell}{2} + \ell f(\ell \varepsilon/3)$; set $\ell = 3/\sqrt{\varepsilon}$. The exponent of $\log \frac{1}{\varepsilon}$ is $\log_2 3$.

10.4.3. We may assume that ε is sufficiently small. Let C be convex with $|C \cap X| \geq \varepsilon n$. Then $C \cap X$ contains points a, b, c such that the shortest of the 3 arcs determined by them, call it α, is at least $\Omega(\varepsilon)$. Show that the triangle abc contains a point of N_i, where i is the smallest with $\varepsilon(1.01)^i/10 > \alpha$.

10.5.2. If x is the last among the lexicographic minima of d-wise intersections of $\mathcal{F}$, the family $\{F \in \mathcal{F}: x \notin F\}$ satisfies the $(p-d, q-d+1)$-condition.

10.5.3(b). By ham-sandwich, choose lines ℓ, ℓ' with $|R_i \cap X| \leq k+1$, where $R_1, \ldots, R_4$ are the "quadrants" determined by ℓ and ℓ'. The point $\ell \cap \ell'$ and centerpoints of $R_i \cap X$ form a transversal.

10.6.1(a). No need to invoke the Alon–Kleitman machinery here.

10.6.1(b). Use Ramsey's theorem.

10.6.2(a). Count the incidences of endpoints with intervals (it can be assumed that all the intervals have distinct endpoints). To get a better β, apply Turán's theorem.

10.6.3. For $\mathcal{F} \subset \mathcal{K}_d^k$ finite, let $\mathcal{G} = \bigcup_{S \in \mathcal{F}} \{S_1, S_2, \ldots, S_k\}$, where $S = S_1 \cup \cdots \cup S_k$ with the S_i convex. If $\mathcal{F}$ has many intersecting $(d+1)$-tuples, then $\mathcal{G}$ has many intersecting $(d+1)$-tuples and so fractional Helly for $\mathcal{F}$, with worse parameters, follows from that for $\mathcal{G}$.

10.6.4. Let $C = f(d+1, d, k)$, where $f(p, d, k)$ is as in Exercise 3, and $h = (d+1)C$. Let $\mathcal{F}'$ be the family of all intersections of C-tuples of sets of $\mathcal{F}$.

This $\mathcal{F}'$ has the $(d+1, d+1)$-property, and so it has a C-point transversal T. Show that some point of T is contained in all members of $\mathcal{F}$.

11.1.4. In $\mathbf{R}^3$: Place the planar construction on $\frac{n}{3}$ points into the xz plane so that all of its points lie very near 0 and all the halving edges are almost parallel to the x-axis. A set A of $\frac{n}{3}$ points is placed on the line $x = 0$, $y = 1$, and the remaining $\frac{n}{3}$ points are the reflected set $-A$.

11.1.5(a). Use the lower bound for $K(n, n)$ in Chapter 4.

11.1.6(a). All the 12 lenses corresponding to such a $K_{3,4}$ are contained in $L \cup U$, and so L intersects U at least 24 times. This is impossible, since U has at most 5 edges and L at most 7 edges (using $\lambda_2(n) \leq 2n-1$).

11.1.6(d). To bound $\nu_k(\mathcal{L})$, fix a k-packing $\mathcal{M} \subseteq \mathcal{L}$, take a random sample $R \subseteq \Gamma$, and consider the family Λ of all lenses ℓ in the arrangement of R "inherited" from $\mathcal{M}$ and such that none of the extremal edges of ℓ are contained in any other lens in the arrangement of R. Extremal edges of a lens are those contained in the lens and adjacent to one of its two end-vertices.

11.3.2. By Exercise 1(a), a vertical line intersects the interior of at most $\sum_{k \in K}(k+1)$ k-edges with $k \in K$. Argue as in the proof of the planar case of Theorem 11.3.3.

11.3.4(b). These halving triangles are not influenced by projecting the other points of X centrally from p_{k+1} on a sphere around p_{k+1}.

11.3.5(a). Let V be the vertex set of a j-facet F entered by ℓ. Among the j points below the hyperplane defined by V we can choose any k points and add them to V, obtaining an S with F being the facet of $\mathrm{conv}(S)$ through which ℓ leaves $\mathrm{conv}(S)$.

11.3.5(b). See the end of Section 5.5 for a similar trick.

11.3.5(c). For $\overline{h}_j = \overline{h}_{n-d-j}$, let X' be the mirror reflection of X by a horizontal hyperplane.

11.3.5(d). Move x far up.

11.3.6(a). Corollary 5.6.3(iii).

11.3.6(b). Use (a) and the formulas expressing the f_k using the h_j and the s_k using the $\overline{h}_j$, respectively.

11.3.8(a). Draw a tiny sphere σ around a vertex incident to at least $3n$ triangles. The intersections of the triangles with σ define a graph drawn on σ. With n vertices and at least $3n$ edges, the graph is nonplanar.

12.1.5. Let v be a vertex of P. First check that there is an $a \in \mathbf{Z}^n$ such that v is the unique vertex minimizing $\langle a, v \rangle$. Moreover, we may assume that $a' = a + (1, 0, \ldots, 0)$, too, has this property. Then $v_1 = \langle a', v \rangle - \langle a, v \rangle \in \mathbf{Z}$.

12.1.6(b). We need that each integral $b \in A\mathbf{R}^n$ is the image of an integer point. Let $\overline{A}$ be a regular $k \times k$ submatrix of A with $k = \mathrm{rank}(A)$; we may assume that $\overline{A}$ is contained in the first k rows and in the first k columns of

A. Let $\bar{b}$ consist of the first k components of b; then $\bar{x} = \bar{A}^{-1}\bar{b}$ is integral by (a). Append $n - k$ zero components to $\bar{x}$.

12.1.6(c). A vertex is determined by some n of the inequalities holding with equality; use (b).

12.1.7(b). It suffices to consider $n = 2^d + 1$. For contradiction, suppose that $\mathbf{Z}^d \cap \bigcap_{i=1}^n \gamma_i = \emptyset$. For $i = 1, 2, \dots, n$, let γ_i' be γ_i translated as far outwards as possible so that $\mathbf{Z}^d \cap \text{int}\left((\bigcap_{j=1}^i \gamma_j') \cap (\bigcap_{j=i+1}^n \gamma_j) \right) = \emptyset$. Show that each γ_i' contributes a facet of $P' = \bigcap_{i=1}^n \gamma_i'$ and there is a $z_i \in \mathbf{Z}^d$ in the relative interior of this facet. Applying (a) to $\{z_1, \dots, z_n\}$ yields a lattice point interior to P'.

12.2.5(b). Suppose $\text{vol}(A), \text{vol}(B) > 0$, fix t with $\text{vol}(A)/(1-t)^n = \text{vol}(B)/t^n$, and set $C = \frac{1}{1-t}A$ and $D = \frac{1}{t}B$.

12.2.7(a). Consider the horizontal slice $F_y = \{x \in \mathbf{R}: f(x) = y\}$, and G_y, H_y defined analogously. We have $\int f = \int_0^1 \text{vol}(F_y)\, dy$. The assumption implies $(1-t)F_y + tG_y \subseteq H_y$. Apply the one-dimensional Brunn–Minkowski to $(1-t)F_y$ and tG_y and integrate over y.

12.2.7(b). Let $f(u)$ be the $(n-1)$-dimensional volume of the slice of C by the hyperplane $x_1 = u$; similarly for $g(u)$ and D and for $h(u)$ and $C+D$.

13.1.1. $2^n/n!$.

13.1.2(b). $I_n = nV_n \int_0^\infty e^{-r^2} r^{n-1}\, dr$.

13.2.3. Fix the coordinate system so that $c = 0$ and F lies in the coordinate hyperplane $h = \{x_n = 0\}$. Since 0 is not the center of gravity, for some i we have $I = \int_F x_i\, dx \neq 0$. Without loss of generality, $i = 1$ and $I > 0$. Let h_1 be h slightly rotated around the flat $\{x_1 = x_n = 0\}$; i.e., $h_1 = \{x \in \mathbf{R}^n: \langle a, x \rangle = 0\}$ with $a = (\varepsilon, 0, \dots, 0, 1)$. Let S_1 be the simplex determined by the same facet hyperplanes as S except that h is replaced by h_1. The difference $\text{vol}(S) - \text{vol}(S_1)$ is proportional to $\varepsilon I + O(\varepsilon^2)$ as $\varepsilon \to 0$. Let h' be a parallel translation of h_1 that touches B^n (near 0), and let S' be the corresponding simplex. Calculation shows that $|\text{vol}(S_1) - \text{vol}(S')| = O(\varepsilon^2)$.

13.2.5. The Thales theorem implies that if $x \notin B(\frac{1}{2}v, \frac{1}{2}\|v\|)$, then v lies in the open half-space γ_x containing 0 and bounded by the hyperplane passing through x and perpendicular to $0x$.

13.3.1(b). Geometric duality and Theorem 13.2.1.

13.4.4(b). Helly's theorem for suitable sets in $\mathbf{R}^{n+1}$.

13.4.5(a). Since the ratio of areas is invariant under affine transforms, we may assume that P contains $B(0, 1)$ and is contained in $B(0, 2)$. Infer that 99% of the edges of P have length $O(\frac{1}{n})$ and 99% of the angles are $\pi - O(\frac{1}{n})$. Then there are two consecutive short edges with angle close to π.

14.1.4. Choose a radius r such that the caps cut off from rB^n by the considered slabs together cover at most half of the surface of rB^n. Then $\text{vol}(K) \geq \text{vol}(K \cap rB^n) \geq \frac{1}{2}r^n$.

14.6.1. Suppose that $\max_i |v_i| = |v_1|$. For any fixed choice of $\sigma_2, \ldots, \sigma_n$, use $\frac{1}{2}(|x+y| + |x-y|) \geq |y|$ with $y = v_1$ and $x = \sum_{i=2}^n \sigma_i v_i$.

14.6.2. We need to bound $n^{-1/2}\mathbf{E}[\|Z\|_1/\|Z\|]$ from below for Z as in Lemma 14.6.4. Each $|Z_i|$ is at least a small constant $\beta > 0$ with probability at least $\frac{1}{2}$; derive that $\|Z\|_1 = \Omega(n)$ with probability at least $\frac{2}{3}$.

15.2.3(b). Let $\lambda_1, \ldots, \lambda_n$ be the eigenvalues of A. The rank is the number of nonzero λ_i. Estimate $\sum \lambda_i^2$ in two ways: First use the trace of $A^T A$, and then the trace of A and Cauchy–Schwarz.

15.2.3(d). If $v_1, \ldots, v_n \in \mathbf{R}^k$, then the matrix A with $a_{ij} = \langle v_i, v_j \rangle$ has rank at most k.

15.3.4(a). Let $n = 2m+1$ and let each n-tuple in $\mathcal{V}$ have the form $(0, e_1, e_2, \ldots, e_m, e_{m+1} + 10\varepsilon w_1, e_{m+1} + 10\varepsilon w_2, \ldots, e_{2m} + 10\varepsilon w_m)$, where each w_i is an $0/1$ vector with $\lfloor \frac{1}{400\varepsilon^2} \rfloor$ ones among the first m positions and zeros elsewhere.

15.4.2. Let $G_i = (V_i, E_i)$, where $V_0 \subset V_1 \subset \cdots \subset V_m$. For each $e \in E_{i-1}$, we have a pair $\{u_e, v_e\}$ of new vertices in G_i in the square that replaces e; let $F_i = \{\{u_e, v_e\}: e \in E_{i-1}\}$. With notation as in the proof of Theorem 15.4.1, put $E = E_m$ and $F = E_0 \cup \bigcup_{i=1}^m F_i$ and show that $R_{E,F}(\rho) = \sqrt{m+1}$, while $R_{E,F}(\sigma) \leq 1$. For the latter, sum up the inequalities $\sigma^2(F_i) + \sigma^2(E_{i-1}) \leq \sigma^2(E_i)$, $i = 1, 2, \ldots, m$, obtained from the short diagonals lemma.

15.4.3. Color the pairs of points; the color of $\{x, y\}$ is the remainder of $\lceil \log_{1+\varepsilon/2} \rho(x,y) \rceil$ modulo r, where r is a sufficiently large integer. Show by induction that a homogeneous set can be embedded satisfactorily.

15.5.2(b). By (a) and Carathéodory's theorem, every metric in $\mathcal{L}_1^{(\text{fin})}$ is a convex combination of at most $N+1$ line metrics. To get rid of the extra $+1$, use the fact that $\mathcal{L}_1^{(\text{fin})}$ is a convex cone.

15.5.8(c). The expectation of $\frac{1}{2}(1 - x_u x_v)$ is the probability that the hyperplane through 0 perpendicular to r separates y_u and y_v, and this equals $\frac{\vartheta}{\pi}$, where $\vartheta \in [0, \pi)$ is the angle of y_u and y_v. On the other hand, the contribution of the edge $\{u, v\}$ to M_{relax} is $\frac{1}{2}(1 - \langle y_u, y_v \rangle) = (1 - \cos \vartheta)/2$. The constant $0.878\ldots$ is the minimum of $\frac{2}{\pi} \cdot \frac{\vartheta}{1 - \cos \vartheta}$, $0 \leq \vartheta \leq \pi$.

15.7.5(c). Suppose that there is a D-embedding f of $T_{k,m}$. For every leaf ℓ, consider f restricted to the path $P^{(\ell)}$ from the root to ℓ, fix a triple $\{a_\ell, b_\ell, c_\ell\}$ of vertices as in Exercise 4 (a scaled copy of P_2), and label the corresponding leaf by the distances of a_ℓ, b_ℓ, c_ℓ from the root. Using (b), choose a $T_{2,m}$ subtree where all leaves have the same labels, consider leaves ℓ and ℓ' of this subtree such that $P^{(\ell)}$ and $P^{(\ell')}$ first meet at $b_\ell = b_{\ell'}$, and use (a) with $x_0 = f(a_\ell)$, $x_1 = f(b_\ell)$, $x_2 = f(c_\ell)$, $x_2' = f(c_{\ell'})$.

15.7.6(a). Sum the parallelogram identities $\frac{1}{(c-a)^2}(\|(x_a - x_b) - (x_b - x_c)\|^2 + \|(x_a - x_b) + (x_b - x_c)\|^2) = \frac{2}{(c-a)^2}(\|x_a - x_b\|^2 + \|x_b - x_c\|^2)$ over $(a, b, c) \in \Gamma$.

Bibliography

The references are sorted alphabetically by the abbreviations (rather than by the authors' names).

[AA92]   P. K. Agarwal and B. Aronov. Counting facets and incidences. *Discrete Comput. Geom.*, 7:359–369, 1992.   (refs: pp. 46, 47)

[AACS98]   P. K. Agarwal, B. Aronov, T. M. Chan, and M. Sharir. On levels in arrangements of lines, segments, planes, and triangles. *Discrete Comput. Geom.*, 19(3):315–331, 1998.   (refs: pp. 269, 270, 271, 286, 287)

[AAHP+98]   A. Andrzejak, B. Aronov, S. Har-Peled, R. Seidel, and E. Welzl. Results on k-sets and j-facets via continuous motion arguments. In *Proc. 14th Annu. ACM Sympos. Comput. Geom.*, pages 192–199, 1998.   (refs: pp. 269, 270, 286)

[AAP+97]   P. K. Agarwal, B. Aronov, J. Pach, R. Pollack, and M. Sharir. Quasi-planar graphs have a linear number of edges. *Combinatorica*, 17:1–9, 1997.   (ref: p. 177)

[AAS01]   P. K. Agarwal, B. Aronov, and M. Sharir. On the complexity of many faces in arrangements of circles. In *Proc. 42nd IEEE Symposium on Foundations of Computer Science*, 2001.   (refs: pp. 47, 70)

[ABFK92]   N. Alon, I. Bárány, Z. Füredi, and D. Kleitman. Point selections and weak ε-nets for convex hulls. *Combin., Probab. Comput.*, 1(3):189–200, 1992.   (refs: pp. 215, 254, 270)

[ABS97]   D. Avis, D. Bremner, and R. Seidel. How good are convex hull algorithms? *Comput. Geom. Theory Appl.*, 7:265–302, 1997.   (ref: p. 106)

[ABV98]   J. Arias-de-Reyna, K. Ball, and R. Villa. Concentration of the distance in finite dimensional normed spaces. *Mathematika*, 45:245–252, 1998.   (ref: p. 332)

[ACE+91]   B. Aronov, B. Chazelle, H. Edelsbrunner, L. J. Guibas,
            M. Sharir, and R. Wenger. Points and triangles in the plane
            and halving planes in space. *Discrete Comput. Geom.*, 6:435–
            442, 1991. (refs: pp. 215, 270)

[Ach01]   D. Achlioptas. Database-friendly random projections. In *Proc.
            20th ACM SIGACT-SIGMOD-SIGART Symposium on Princi-
            ples of Database Systems*, pages 274–281, 2001. (ref: p. 361)

[ACNS82]   M. Ajtai, V. Chvátal, M. Newborn, and E. Szemerédi. Crossing-
            free subgraphs. *Ann. Discrete Math.*, 12:9–12, 1982. (ref: p. 56)

[AEG+94]   B. Aronov, P. Erdős, W. Goddard, D. J. Kleitman, M. Kluger-
            man, J. Pach, and L. J. Schulman. Crossing families. *Combi-
            natorica*, 14:127–134, 1994. (ref: p. 177)

[AEGS92]   B. Aronov, H. Edelsbrunner, L. Guibas, and M. Sharir. The
            number of edges of many faces in a line segment arrangement.
            Combinatorica, 12(3):261–274, 1992. (ref: p. 46)

[AF92]   D. Avis and K. Fukuda. A pivoting algorithm for convex hulls
            and vertex enumeration of arrangements and polyhedra. *Dis-
            crete Comput. Geom.*, 8:295–313, 1992. (ref: p. 106)

[AF00]   N. Alon and E. Friedgut. On the number of permutations avoid-
            ing a given pattern. *J. Combin. Theory, Ser. A*, 81:133–140,
            2000. (ref: p. 177)

[AFH+00]   H. Alt, S. Felsner, F. Hurtado, M. Noy, and E. Welzl. A class
            of point-sets with few *k*-sets. *Comput. Geom. Theor. Appl.*,
            16:95–101, 2000. (ref: p. 270)

[AFR85]   N. Alon, P. Frankl, and V. Rödl. Geometrical realization of set
            systems and probabilistic communication complexity. In *Proc.
            26th IEEE Symposium on Foundations of Computer Science*,
            pages 277–280, 1985. (ref: p. 140)

[AG86]   N. Alon and E. Győri. The number of small semispaces of a
            finite set of points in the plane. *J. Combin. Theory Ser. A*,
            41:154–157, 1986. (ref: p. 145)

[AGHV01]   P. K. Agarwal, L. J. Guibas, J. Hershberger, and E. Veach.
            Maintaining the extent of a moving point set. *Discrete Comput.
            Geom.*, 26:353–374, 2001. (ref: p. 194)

[AH00]   R. Aharoni and P. E. Haxell. Hall's theorem for hypergraphs.
            J. Graph Theory, 35:83–88, 2000. (ref: p. 235)

[Aha01]   R. Aharoni. Ryser's conjecture for tri-partite hypergraphs.
            Combinatorica, 21:1–4, 2001. (ref: p. 235)

[AHL01]   N. Alon, S. Hoory, and N. Linial. The Moore bound for irregular graphs. *Graphs and Combinatorics*, 2001. In press. (ref: p. 367)

[AI88]   F. Aurenhammer and H. Imai. Geometric relations among Voronoi diagrams. *Geom. Dedicata*, 27:65–75, 1988. (ref: p. 121)

[Ajt98]   M. Ajtai. Worst-case complexity, average-case complexity and lattice problems. *Documenta Math. J. DMV*, Extra volume ICM 1998, vol. III:421–428, 1998. (ref: p. 26)

[AK85]   N. Alon and G. Kalai. A simple proof of the upper bound theorem. *European J. Combin.*, 6:211–214, 1985. (ref: p. 103)

[AK92]   N. Alon and D. Kleitman. Piercing convex sets and the Hadwiger Debrunner (p, q)-problem. *Adv. Math.*, 96(1):103–112, 1992. (ref: p. 258)

[AK95]   N. Alon and G. Kalai. Bounding the piercing number. *Discrete Comput. Geom.*, 13:245–256, 1995. (ref: p. 261)

[AK00]   F. Aurenhammer and R. Klein. Voronoi diagrams. In J.-R. Sack and J. Urrutia, editors, *Handbook of Computational Geometry*, pages 201–290. Elsevier Science Publishers B.V. North-Holland, Amsterdam, 2000. (refs: pp. 120, 121)

[AKMM01]   N. Alon, G. Kalai, J. Matoušek, and R. Meshulam. Transversal numbers for hypergraphs arising in geometry. *Adv. Appl. Math.*, 2001. In press. (ref: p. 262)

[AKP89]   N. Alon, M. Katchalski, and W. R. Pulleyblank. The maximum size of a convex polygon in a restricted set of points in the plane. *Discrete Comput. Geom.*, 4:245–251, 1989. (ref: p. 33)

[AKPW95]   N. Alon, R. M. Karp, D. Peleg, and D. West. A graph-theoretic game and its application to the k-server problem. *SIAM J. Computing*, 24(1):78–100, 1995. (ref: p. 398)

[AKV92]   R. Adamec, M. Klazar, and P. Valtr. Generalized Davenport–Schinzel sequences with linear upper bound. *Discrete Math.*, 108:219–229, 1992. (ref: p. 176)

[Alo]   N. Alon. Covering a hypergraph of subgraphs. *Discrete Math.* In press. (ref: p. 262)

[Alo86a]   N. Alon. Eigenvalues and expanders. *Combinatorica*, 6:83–96, 1986. (ref: p. 381)

[Alo86b]   N. Alon. The number of polytopes, configurations, and real matroids. *Mathematika*, 33:62–71, 1986. (ref: p. 140)

[Alo98]   N. Alon. Piercing d-intervals. *Discrete Comput. Geom.*, 19:333–334, 1998. (ref: p. 262)

[ALPS01]   N. Alon, H. Last, R. Pinchasi, and M. Sharir. On the complexity of arrangements of circles in the plane. *Discrete Comput. Geom.*, 26:465–492, 2001. (ref: p. 271)

[AM85]   N. Alon and V. D. Milman. λ_1, isoperimetric inequalities for graphs, and superconcentrators. *J. Combinatorial Theory, Ser. B*, 38(1):73–88, 1985. (ref: p. 381)

[Ame96]   N. Amenta. A short proof of an interesting Helly-type theorem. *Discrete Comput. Geom.*, 15:423–427, 1996. (ref: p. 261)

[AMS94]   B. Aronov, J. Matoušek, and M. Sharir. On the sum of squares of cell complexities in hyperplane arrangements. *J. Combin. Theory Ser. A*, 65:311–321, 1994. (refs: pp. 47, 152)

[AMS98]   P. K. Agarwal, J. Matoušek, and O. Schwarzkopf. Computing many faces in arrangements of lines and segments. *SIAM J. Comput.*, 27(2):491–505, 1998. (ref: p. 162)

[APS93]   B. Aronov, M. Pellegrini, and M. Sharir. On the zone of a surface in a hyperplane arrangement. *Discrete Comput. Geom.*, 9(2):177–186, 1993. (ref: p. 151)

[AR92]   J. Arias-de-Reyna and L. Rodríguez-Piazza. Finite metric spaces needing high dimension for Lipschitz embeddings in Banach spaces. *Israel J. Math.*, 79:103–113, 1992. (ref: p. 367)

[AR98]   Y. Aumann and Y. Rabani. An $O(\log k)$ approximate min-cut max-flow theorem and approximation algorithm. *SIAM J. Comput.*, 27(1):291–301, 1998. (ref: p. 392)

[Aro00]   B. Aronov. A lower bound for Voronoi diagram complexity. Manuscript, Polytechnic University, Brooklyn, New York, 2000. (refs: pp. 123, 192)

[ARS99]   N. Alon, L. Rónyai, and T. Szabó. Norm-graphs: variations and applications. *J. Combin. Theory Ser. B*, 76:280–290, 1999. (ref: p. 68)

[AS94]   B. Aronov and M. Sharir. Castles in the air revisited. *Discrete Comput. Geom.*, 12:119–150, 1994. (ref: p. 193)

[AS00a]   P. K. Agarwal and M. Sharir. Arrangements and their applications. In J.-R. Sack and J. Urrutia, editors, *Handbook of Computational Geometry*, pages 49–119. North-Holland, Amsterdam, 2000. (refs: pp. 47, 128, 145, 168, 191)

[AS00b]   P. K. Agarwal and M. Sharir. Davenport–Schinzel sequences and their geometric applications. In J.-R. Sack and J. Urrutia, editors, *Handbook of Computational Geometry*, pages 1–47. North-Holland, Amsterdam, 2000. (ref: p. 168)

[AS00c]   P. K. Agarwal and M. Sharir. Pipes, cigars, and kreplach: The union of Minkowski sums in three dimensions. *Discrete Comput. Geom.*, pages 645–685, 2000. (ref: p. 194)

[AS00d]   N. Alon and J. Spencer. *The Probabilistic Method (2nd edition)*. J. Wiley and Sons, New York, NY, 2000. First edition 1993. (refs: pp. 336, 340)

[AS01a]   B. Aronov and M. Sharir. Cutting circles into pseudo-segments and improved bounds for incidences. *Discrete Comput. Geom.*, 2001. To appear. (refs: pp. 44, 46, 69, 70, 271)

[AS01b]   B. Aronov and M. Sharir. Distinct distances in three dimensions. Manuscript, School of Computer Science, Tel Aviv University, 2001. (ref: p. 45)

[Ass83]   P. Assouad. Density and dimension (in French). *Ann. Inst. Fourier (Grenoble)*, 33:233–282, 1983. (ref: p. 250)

[ASS89]   P. K. Agarwal, M. Sharir, and P. Shor. Sharp upper and lower bounds on the length of general Davenport–Schinzel sequences. *J. Combin. Theory Ser. A*, 52(2):228–274, 1989. (ref: p. 176)

[ASS96]   P. K. Agarwal, M. Sharir, and O. Schwarzkopf. The overlay of lower envelopes and its applications. *Discrete Comput. Geom.*, 15:1–13, 1996. (ref: p. 192)

[AST97]   B. Aronov, M. Sharir, and B. Tagansky. The union of convex polyhedra in three dimensions. *SIAM J. Comput.*, 26:1670–1688, 1997. (ref: p. 194)

[Aur91]   F. Aurenhammer. Voronoi diagrams: A survey of a fundamental geometric data structure. *ACM Comput. Surv.*, 23(3):345–405, September 1991. (ref: p. 120)

[Avi93]   D. Avis. The m-core properly contains the m-divisible points in space. *Pattern Recognit. Lett.*, 14(9):703–705, 1993. (ref: p. 205)

[Bal]   K. Ball. Convex geometry and functional analysis. In W. B. Johnson and J. Lindenstrauss, editors, *Handbook of Banach Spaces*. North-Holland, Amsterdam. In press. (refs: pp. 314, 320, 337)

[Bal90]   K. Ball. Isometric embedding in ℓ_p-spaces. *European J. Combin.*, 11(4):305–311, 1990. (ref: p. 383)

[Bal92]   K. Ball. Markov chains, Riesz transforms and Lipschitz maps.
            Geom. Funct. Anal., 2(2):137–172, 1992.   (ref: p. 380)

[Bal97]   K. Ball. An elementary introduction to modern convex geome-
            try. In S. Levi, editor, *Flavors of Geometry (MSRI Publications
            vol. 31)*, pages 1–58. Cambridge University Press, Cambridge,
            1997.   (refs: pp. viii, 300, 315, 327, 336, 337, 346)

[Bár82]   I. Bárány. A generalization of Carathéodory's theorem. *Discrete
            Math.*, 40:141–152, 1982.   (refs: pp. 198, 199, 210)

[Bár89]   I. Bárány. Intrinsic volumes and f-vectors of random polytopes.
            Math. Ann., 285(4):671–699, 1989.   (ref: p. 99)

[Bar93]   A. I. Barvinok. A polynomial time algorithm for counting inte-
            gral points in polyhedra when the dimension is fixed. In *Proc.
            34th IEEE Symposium on Foundations of Computer Science*,
            pages 566–572, 1993.   (ref: p. 24)

[Bar96]   Y. Bartal. Probabilistic approximation of metric spaces and its
            algorithmic applications. In *Proc. 37th IEEE Symposium on
            Foundations of Computer Science*, pages 184–193, 1996.   (ref:
            p. 398)

[Bar97]   A. I. Barvinok. Lattice points and lattice polytopes. In J. E.
            Goodman and J. O'Rourke, editors, *Handbook of Discrete and
            Computational Geometry*, chapter 7, pages 133–152. CRC Press
            LLC, Boca Raton, FL, 1997.   (refs: pp. 24, 294)

[Bar98]   Y. Bartal. On approximating arbitrary metrics by tree metrics.
            In *Proc. 30th Annu. ACM Sympos. on Theory of Computing*,
            pages 161–168, 1998.   (ref: p. 398)

[Bas98]   S. Basu. On the combinatorial and topological complexity of a
            single cell. In *Proc. 39th IEEE Symposium on Foundations of
            Computer Science*, pages 606–616, 1998.   (ref: p. 193)

[BCM99]   H. Brönnimann, B. Chazelle, and J. Matoušek. Product range
            spaces, sensitive sampling, and derandomization. *SIAM J.
            Comput.*, 28:1552–1575, 1999.   (ref: p. 106)

[BCR98]   J. Bochnak, M. Coste, and M.-F. Roy. *Real Algebraic Geometry*.
            Springer, Berlin etc., 1998. Transl. from the French, revised and
            updated edition.   (refs: pp. 135, 191)

[BD93]   D. Bienstock and N. Dean. Bounds for rectilinear crossing num-
            bers. *J. Graph Theory*, 17(3):333–348, 1993.   (ref: p. 58)

[BDV91]   A. Bialostocki, P. Dierker, and B. Voxman. Some notes on the
            Erdős–Szekeres theorem. *Discrete Math.*, 91(3):231–238, 1991.
            (ref: p. 38)

[Bec83] J. Beck. On the lattice property of the plane and some problems of Dirac, Motzkin and Erdős in combinatorial geometry. *Combinatorica*, 3(3–4):281–297, 1983. (refs: pp. 45, 50)

[Ben66] C. T. Benson. Minimal regular graphs of girth eight and twelve. *Canad. J. Math.*, 18:1091–1094, 1966. (ref: p. 367)

[BEPY91] M. Bern, D. Eppstein, P. Plassman, and F. Yao. Horizon theorems for lines and polygons. In J. Goodman, R. Pollack, and W. Steiger, editors, *Discrete and Computational Geometry: Papers from the DIMACS Special Year*, volume 6 of *DIMACS Series in Discrete Mathematics and Theoretical Computer Science*, pages 45–66. American Mathematical Society, Association for Computing Machinery, Providence, RI, 1991. (ref: p. 151)

[Ber61] C. Berge. Färbungen von Graphen, deren sämtliche bzw. deren ungerade Kreise starr sind (Zusammenfassung). *Wissentschaftliche Zeitschrift, Martin Luther Universität Halle-Wittenberg, Math.-Naturwiss. Reihe*, pages 114–115, 1961. (ref: p. 293)

[Ber62] C. Berge. Sur une conjecture relative au problème des codes optimaux. Communication, 13ème assemblée générale de l'URSI, Tokyo, 1962. (ref: p. 293)

[BF84] E. Boros and Z. Füredi. The number of triangles covering the center of an n-set. *Geom. Dedicata*, 17:69–77, 1984. (ref: p. 210)

[BF87] I. Bárány and Z. Füredi. Computing the volume is difficult. *Discrete Comput. Geom.*, 2:319–326, 1987. (refs: pp. 320, 322, 324)

[BF88] I. Bárány and Z. Füredi. Approximation of the sphere by polytopes having few vertices. *Proc. Amer. Math. Soc.*, 102(3):651–659, 1988. (ref: p. 320)

[BFL90] I. Bárány, Z. Füredi, and L. Lovász. On the number of halving planes. *Combinatorica*, 10:175–183, 1990. (refs: pp. 205, 215, 229, 269, 270, 280)

[BFM86] J. Bourgain, T. Figiel, and V. Milman. On Hilbertian subsets of finite metric spaces. *Israel J. Math.*, 55:147–152, 1986. (ref: p. 373)

[BFT95] G. R. Brightwell, S. Felsner, and W. T. Trotter. Balancing pairs and the cross product conjecture. *Order*, 12(4):327–349, 1995. (ref: p. 308)

[BGK⁺99] A. Brieden, P. Gritzmann, R. Kannan, V. Klee, L. Lovász, and M. Simonovits. Deterministic and randomized polynomial-time approximation of radii. 1999. To appear in *Mathematika*. Preliminary version in *Proc. 39th IEEE Symposium on Foundations of Computer Science*, 1998, pages 244–251. (refs: pp. 322, 334)

[BH93] U. Betke and M. Henk. Approximating the volume of convex bodies. *Discrete Comput. Geom.*, 10:15–21, 1993. (ref: p. 321)

[Big93] N. Biggs. *Algebraic Graph Theory*. Cambridge Univ. Press, Cambridge, 1993. 2nd edition. (refs: pp. 367, 381)

[BK63] W. Bonnice and V. L. Klee. The generation of convex hulls. *Math. Ann.*, 152:1–29, 1963. (ref: p. 8)

[BK00] A. Brieden and M. Kochol. A note on cutting planes, volume approximation and Mahler's conjecture. Manuscript, TU München, 2000. (ref: p. 324)

[BL81] L. J. Billera and C. W. Lee. A proof of the suffiency of McMullen's conditions for f-vectors of simplicial polytopes. *J. Combin. Theory Ser. A*, 31(3):237–255, 1981. (ref: p. 105)

[BL92] I. Bárány and D. Larman. A colored version of Tverberg's theorem. *J. London Math. Soc. II. Ser.*, 45:314–320, 1992. (ref: p. 205)

[BL99] Y. Benyamini and J. Lindenstrauss. *Nonlinear Functional Analysis, Vol. I, Colloquium Publications 48*. American Mathematical Society (AMS), Providence, RI, 1999. (refs: pp. 336, 352, 358)

[BLM89] J. Bourgain, J. Lindenstrauss, and V. Milman. Approximation of zonoids by zonotopes. *Acta Math.*, 162:73–141, 1989. (ref: p. 320)

[BLPS99] W. Banaszczyk, A. E. Litvak, A. Pajor, and S. J. Szarek. The flatness theorem for nonsymmetric convex bodies via the local theory of Banach spaces. *Math. Oper. Res.*, 24(3):728–750, 1999. (ref: p. 24)

[BLŽV94] A. Björner, L. Lovász, R. Živaljević, and S. Vrećica. Chessboard complexes and matching complexes. *J. London Math. Soc.*, 49:25–39, 1994. (ref: p. 205)

[BMMV02] R. Babilon, J. Matoušek, J. Maxová, and P. Valtr. Low-distortion embeddings of trees. In *Proc. Graph Drawing 2001*. Springer, Berlin etc., 2002. In press. (ref: p. 393)

[BMT95]   C. Buchta, J. Müller, and R. F. Tichy. Stochastical approxima-
           tion of convex bodies. *Math. Ann.*, 271:225–235, 1895.   (ref:
           p. 324)

[BO97]   I. Bárány and S. Onn. Colourful linear programming and its
           relatives. *Math. Oper. Res.*, 22:550–567, 1997.   (refs: pp. 199,
           204)

[Bol85]   B. Bollobás. *Random Graphs*. Academic Press (Harcourt Brace
           Jovanovich, Publishers), London-Orlando etc., 1985.   (ref:
           p. 366)

[Bol87]   B. Bollobás. Martingales, isoperimetric inequalities and random
           graphs. In *52. Combinatorics, Eger (Hungary), Colloq. Math.
           Soc. J. Bolyai*, pages 113–139. Math. Soc. J. Bolyai, Budapest,
           1987.   (refs: pp. 336, 340)

[Bor75]   C. Borell. The Brunn–Minkowski inequality in Gauss space.
           Invent. Math., 30(2):207–216, 1975.   (ref: p. 336)

[BOR99]   A. Borodin, R. Ostrovsky, and Y. Rabani. Subquadratic ap-
           proximation algorithms for clustering problems in high dimen-
           sional spaces. In *Proc. 31st Annual ACM Symposium on Theory
           of Computing*, pages 435–444, 1999.   (ref: p. 361)

[Bou85]   J. Bourgain. On Lipschitz embedding of finite metric spaces in
           Hilbert space. *Israel J. Math.*, 52:46–52, 1985.   (refs: pp. 367,
           388, 392)

[Bou86]   J. Bourgain. The metrical interpretation of superreflexivity in
           Banach spaces. *Israel J. Math.*, 56:222–230, 1986.   (ref: p. 392)

[BP90]   K. Ball and A. Pajor. Convex bodies with few faces. *Proc.
           Amer. Math. Soc.*, 110(1):225–231, 1990.   (ref: p. 320)

[BPR96]   S. Basu, R. Pollack, and M.-F. Roy. On the number of cells
           defined by a family of polynomials on a variety. *Mathematika*,
           43:120–126, 1996.   (ref: p. 135)

[Bre93]   G. Bredon. *Topology and Geometry (Graduate Texts in Math-
           ematics 139)*. Springer-Verlag, Berlin etc., 1993.   (ref: p. 4)

[Bro66]   W. G. Brown. On graphs that do not contain a Thomsen graph.
           Canad. Math. Bull., 9:281–285, 1966.   (ref: p. 68)

[Brø83]   A. Brønsted. *An Introduction to Convex Polytopes*. Springer-
           Verlag, New York, NY, 1983.   (ref: p. 85)

[BS89]   J. Bokowski and B. Sturmfels. *Computational Synthetic Geom-
           etry*. Lect. Notes in Math. 1355. Springer-Verlag, Heidelberg,
           1989.   (ref: p. 138)

[BSTY98]   J.-D. Boissonnat, M. Sharir, B. Tagansky, and M. Yvinec. Voronoi diagrams in higher dimensions under certain polyhedral distance functions. *Discrete Comput. Geom.*, 19(4):473–484, 1998. (ref: p. 194)

[BT89]   T. Bisztriczky and G. Fejes Tóth. A generalization of the Erdős–Szekeres convex n-gon theorem. *J. Reine Angew. Math.*, 395:167–170, 1989. (ref: p. 33)

[BV82]   E. O. Buchman and F. A. Valentine. Any new Helly numbers? *Amer. Math. Mon.*, 89:370–375, 1982. (ref: p. 13)

[BV98]   I. Bárány and P. Valtr. A positive fraction Erdős–Szekeres theorem. *Discrete Comput. Geom*, 19:335–342, 1998. (ref: p. 220)

[BVS+99]   A. Björner, M. Las Vergnas, B. Sturmfels, N. White, and G. M. Ziegler. *Oriented Matroids (2nd edition)*. Encyclopedia of Mathematics 46. Cambridge University Press, Cambridge, 1999. (refs: pp. 100, 137, 139, 222)

[Can69]   R. Canham. A theorem on arrangements of lines in the plane. *Israel J. Math.*, 7:393–397, 1969. (ref: p. 46)

[Car07]   C. Carathéodory. Über den Variabilitätsbereich der Koeffizienten von Potenzreihen, die gegebene Werte nicht annehmen. *Math. Ann.*, 64:95–115, 1907. (refs: pp. 8, 98)

[Car85]   B. Carl. Inequalities of Bernstein–Jackson-type and the degree of compactness of operators in Banach spaces. *Ann. Inst. Fourier*, 35(3):79–118, 1985. (ref: p. 320)

[Cas59]   J. Cassels. *An Introduction to the Geometry of Numbers.* Springer-Verlag, Heidelberg, 1959. (ref: p. 20)

[CCPS98]   W. J. Cook, W. H. Cunningham, W. R. Pulleyblank, and A. Schrijver. *Combinatorial Optimization.* Wiley, New York, NY, 1998. (ref: p. 294)

[CEG+90]   K. Clarkson, H. Edelsbrunner, L. Guibas, M. Sharir, and E. Welzl. Combinatorial complexity bounds for arrangements of curves and spheres. *Discrete Comput. Geom.*, 5:99–160, 1990. (refs: pp. 44, 45, 46, 47, 68, 152)

[CEG+93]   B. Chazelle, H. Edelsbrunner, L. Guibas, M. Sharir, and J. Snoeyink. Computing a face in an arrangement of line segments and related problems. *SIAM J. Comput.*, 22:1286–1302, 1993. (ref: p. 162)

[CEG+94]   B. Chazelle, H. Edelsbrunner, L. Guibas, J. Hershberger, R. Seidel, and M. Sharir. Selecting heavily covered points. *SIAM J. Comput.*, 23:1138–1151, 1994. (ref: p. 215)

[CEG⁺95] B. Chazelle, H. Edelsbrunner, M. Grigni, L. Guibas, M. Sharir, and E. Welzl. Improved bounds on weak ϵ-nets for convex sets. *Discrete Comput. Geom.*, 13:1–15, 1995. (ref: p. 254)

[CEGS89] B. Chazelle, H. Edelsbrunner, L. Guibas, and M. Sharir. A singly-exponential stratification scheme for real semi-algebraic varieties and its applications. In *Proc. 16th Internat. Colloq. Automata Lang. Program.*, volume 372 of *Lecture Notes Comput. Sci.*, pages 179–192. Springer-Verlag, Berlin etc., 1989. (ref: p. 162)

[CEM⁺96] K. L. Clarkson, D. Eppstein, G. L. Miller, C. Sturtivant, and S.-H. Teng. Approximating center points with iterative Radon points. *Internat. J. Comput. Geom. Appl.*, 6:357–377, 1996. (ref: p. 16)

[CF90]   B. Chazelle and J. Friedman. A deterministic view of random sampling and its use in geometry. *Combinatorica*, 10(3):229–249, 1990. (refs: pp. 68, 161)

[CGL85]   B. Chazelle, L. J. Guibas, and D. T. Lee. The power of geometric duality. *BIT*, 25:76–90, 1985. (ref: p. 151)

[Cha93a] B. Chazelle. Cutting hyperplanes for divide-and-conquer. *Discrete Comput. Geom.*, 9(2):145–158, 1993. (refs: pp. 69, 162)

[Cha93b] B. Chazelle. An optimal convex hull algorithm in any fixed dimension. *Discrete Comput. Geom.*, 10:377–409, 1993. (ref: p. 106)

[Cha00a] T. M. Chan. On levels in arrangements of curves. In *Proc. 41st IEEE Symposium on Foundations of Computer Science*, pages 219–227, 2000. (refs: pp. 140, 271)

[Cha00b] T. M. Chan. Random sampling, halfspace range reporting, and construction of ($\leq k$)-levels in three dimensions. *SIAM J. Comput.*, 30(2):561–575, 2000. (ref: p. 106)

[Cha00c] B. Chazelle. *The Discrepancy Method*. Cambridge University Press, Cambridge, 2000. (ref: p. 162)

[Chu84]   F. R. K. Chung. The number of different distances determined by n points in the plane. *J. Combin. Theory Ser. A*, 36:342–354, 1984. (ref: p. 45)

[Chu97]   F. Chung. *Spectral Graph Theory*. Regional Conference Series in Mathematics 92. Amer. Math. Soc., Providence, 1997. (ref: p. 381)

[CKS+98] L. P. Chew, K. Kedem, M. Sharir, B. Tagansky, and E. Welzl.
          Voronoi diagrams of lines in 3-space under polyhedral convex
          distance functions. *J. Algorithms*, 29(2):238–255, 1998. (ref:
          p. 192)

[Cla87]   K. L. Clarkson. New applications of random sampling in compu-
          tational geometry. *Discrete Comput. Geom.*, 2:195–222, 1987.
          (refs: pp. 68, 72)

[Cla88a] K. L. Clarkson. Applications of random sampling in computa-
          tional geometry, II. In *Proc. 4th Annu. ACM Sympos. Comput.
          Geom.*, pages 1–11, 1988. (refs: pp. 145, 161)

[Cla88b] K. L. Clarkson.   A randomized algorithm for closest-point
          queries. *SIAM J. Comput.*, 17:830–847, 1988. (ref: p. 161)

[Cla93]   K. L. Clarkson. A bound on local minima of arrangements that
          implies the upper bound theorem. *Discrete Comput. Geom.*,
          10:427–233, 1993. (refs: pp. 103, 280)

[CLO92]   D. Cox, J. Little, and D. O'Shea. *Ideals, Varieties, and Algo-
          rithms.* Springer-Verlag, New York, NY, 1992. (ref: p. 135)

[CP88]   B. Carl and A. Pajor. Gelfand numbers of operators with values
          in a Hilbert space. *Invent. Math.*, 94:479–504, 1988. (refs:
          pp. 320, 324)

[CS89]   K. L. Clarkson and P. W. Shor. Applications of random sam-
          pling in computational geometry, II. *Discrete Comput. Geom.*,
          4:387–421, 1989. (refs: pp. 68, 105, 145, 161)

[CS99]   J. H. Conway and N. J. A. Sloane. *Sphere Packings, Lattices and
          Groups (3rd edition).* Grundlehren der Mathematischen Wis-
          senschaften 290. Springer-Verlag, New York etc., 1999. (ref:
          p. 24)

[CST92]   F. R. K. Chung, E. Szemerédi, and W. T. Trotter. The number
          of different distances determined by a set of points in the Eu-
          clidean plane. *Discrete Comput. Geom.*, 7:1–11, 1992. (ref:
          p. 45)

[Dan63]   G. B. Dantzig. *Linear Programming and Extensions.* Princeton
          University Press, Princeton, NJ, 1963. (ref: p. 93)

[Dan86]   L. Danzer. On the solution of the problem of Gallai about
          circular discs in the Euclidean plane (in German). *Stud. Sci.
          Math. Hung.*, 21:111–134, 1986. (ref: p. 235)

[dBvKOS97] M. de Berg, M. van Kreveld, M. Overmars, and O. Schwarzkopf.
          Computational Geometry: Algorithms and Applications.
          Springer-Verlag, Berlin, 1997. (refs: pp. 116, 122, 162)

[DE94]   T. K. Dey and H. Edelsbrunner. Counting triangle crossings and halving planes. *Discrete Comput. Geom.*, 12:281–289, 1994. (ref: p. 270)

[Del34]   B. Delaunay. Sur la sphère vide. A la memoire de Georges Voronoi. *Izv. Akad. Nauk SSSR, Otdelenie Matematicheskih i Estestvennyh Nauk*, 7:793–800, 1934. (ref: p. 120)

[Dey98]   T. K. Dey. Improved bounds on planar k-sets and related problems. *Discrete Comput. Geom.*, 19:373–382, 1998. (refs: pp. 269, 270, 285)

[DFK91]   M. E. Dyer, A. Frieze, and R. Kannan. A random polynomial time algorithm for approximating the volume of convex bodies. *J. ACM*, 38:1–17, 1991. (ref: p. 321)

[dFPP90]   H. de Fraysseix, J. Pach, and R. Pollack. How to draw a planar graph on a grid. *Combinatorica*, 10(1):41–51, 1990. (ref: p. 94)

[DFPS00]   A. Deza, K. Fukuda, D. Pasechnik, and M. Sato. Generating vertices with symmetries. In *Proc. of the 5th Workshop on Algorithms and Computation, Tokyo University*, pages 1–8, 2000. (ref: p. 106)

[DG99]   S. Dasgupta and A. Gupta. An elementary proof of the Johnson–Lindenstrauss lemma. Technical Report TR-99-06, Intl. Comput. Sci. Inst., Berkeley, CA, 1999. (ref: p. 361)

[DGK63]   L. Danzer, B. Grünbaum, and V. Klee. Helly's theorem and its relatives. In *Convexity*, volume 7 of *Proc. Symp. Pure Math.*, pages 101–180. American Mathematical Society, Providence, 1963. (refs: pp. 8, 12, 13, 327)

[Dil50]   R. P. Dilworth. A decomposition theorem for partially ordered sets. *Annals of Math.*, 51:161–166, 1950. (ref: p. 295)

[Dir42]   G. L. Dirichlet. Verallgemeinerung eines Satzes aus der Lehre von Kettenbrüchen nebst einigen Anwendungen auf die Theorie der Zahlen. In *Bericht über die zur Bekantmachung geeigneten Verhandlungen der Königlich Preussischen Akademie der Wissenschaften zu Berlin*, pages 93–95. 1842. Reprinted in L. Kronecker (editor): *G. L. Dirichlet's Werke* Vol. I, G. Reimer, Berlin 1889, reprinted Chelsea, New York 1969. (ref: p. 21)

[Dir50]   G. L. Dirichlet. Über die Reduktion der positiven quadratischen Formen mit drei unbestimmten ganzen Zahlen. *J. Reine Angew. Math.*, 40:209–227, 1850. (ref: p. 120)

[DL97]   M. M. Deza and M. Laurent. *Geometry of Cuts and Metrics*. Algorithms and Combinatorics 15. Springer-Verlag, Berlin etc., 1997. (refs: pp. 107, 357)

[Dol'92]   V. L. Dol'nikov. A generalization of the ham sandwich theorem. *Mat. Zametki*, 52(2):27–37, 1992. In Russian; English translation in *Math. Notes* 52,2:771–779, 1992. (ref: p. 16)

[Doi73]   J.-P. Doignon. Convexity in cristallographical lattices. *J. Geometry*, 3:71–85, 1973. (ref: p. 295)

[DR50]   A. Dvoretzky and C. A. Rogers. Absolute and unconditional convergence in normed linear spaces. *Proc. Natl. Acad. Sci. USA*, 36:192–197, 1950. (ref: p. 352)

[DS65]   H. Davenport and A. Schinzel. A combinatorial problem connected with differential equations. *Amer. J. Math.*, 87:684–689, 1965. (ref: p. 175)

[Dud78]   R. M. Dudley. Central limit theorems for empirical measures. *Ann. Probab.*, 6:899–929, 1978. (ref: p. 250)

[DV02]   H. Djidjev and I. Vrťo. An improved lower bound for crossing numbers. In *Proc. Graph Drawing 2001*. Springer, Berlin etc., 2002. In press. (ref: p. 57)

[Dvo59]   A. Dvoretzky. A theorem on convex bodies and applications to Banach spaces. *Proc. Natl. Acad. Sci. USA*, 45:223–226, 1959. Errata. Ibid. 1554. (ref: p. 352)

[Dvo61]   A. Dvoretzky. Some results on convex bodies and Banach spaces. In *Proc. Int. Symp. Linear Spaces 1960*, pages 123–160. Jerusalem Academic Press, Jerusalem; Pergamon, Oxford, 1961. (refs: pp. 346, 352)

[Dwo97]   C. Dwork. Positive applications of lattices to cryptography. In *Proc. 22dn International Symposium on Mathematical Foundations of Computer Science (Lect. Notes Comput. Sci. 1295)*, pages 44–51. Springer, Berlin, 1997. (ref: p. 26)

[Eck85]   J. Eckhoff. An upper-bound theorem for families of convex sets. *Geom. Dedicata*, 19:217–227, 1985. (ref: p. 197)

[Eck93]   J. Eckhoff. Helly, Radon and Carathéodory type theorems. In P. M. Gruber and J. M. Wills, editors, *Handbook of Convex Geometry*. North-Holland, Amsterdam, 1993. (refs: pp. 8, 12, 13)

[Ede89]   H. Edelsbrunner. The upper envelope of piecewise linear functions: Tight complexity bounds in higher dimensions. *Discrete Comput. Geom.*, 4:337–343, 1989. (ref: p. 186)

[Ede98]   H. Edelsbrunner. Geometry of modeling biomolecules. In P. K. Agarwal, L. E. Kavraki, and M. Mason, editors, *Proc. Workshop Algorithmic Found. Robot*. A. K. Peters, Natick, MA, 1998. (ref: p. 122)

[Edm65]   J. Edmonds. Maximum matching and a polyhedron with 0,1-vertices. *J. Res. National Bureau of Standards (B)*, 69:125–130, 1965. (ref: p. 294)

[EE94]   Gy. Elekes and P. Erdős. Similar configurations and pseudo grids. In K. Böröczky et al., editors, *Intuitive Geometry. Proceedings of the 3rd International Conference Held in Szeged, Hungary, From 2 To 7 September, 1991*, Colloq. Math. Soc. Janos Bolyai. 63, pages 85–104. North-Holland, Amsterdam, 1994. (refs: pp. 47, 51)

[EFPR93] P. Erdős, Z. Füredi, J. Pach, and I. Ruzsa. The grid revisited. *Discrete Math.*, 111:189–196, 1993. (ref: p. 47)

[EGS90]   H. Edelsbrunner, L. Guibas, and M. Sharir. The complexity of many cells in arrangements of planes and related problems. *Discrete Comput. Geom.*, 5:197–216, 1990. (ref: p. 46)

[EHP89]   P. Erdős, D. Hickerson, and J. Pach. A problem of Leo Moser about repeated distances on the sphere. *Amer. Math. Mon.*, 96:569–575, 1989. (ref: p. 45)

[EKZ01]   D. Eppstein, G. Kuperberg, and G. M. Ziegler. Fat 4-polytopes and fatter 3-spheres. Manuscript, TU Berlin, 2001. (ref: p. 107)

[Ele86]   Gy. Elekes. A geometric inequality and the complexity of computing the volume. *Discrete Comput. Geom.*, 1:289–292, 1986. (refs: pp. 320, 322)

[Ele97]   Gy. Elekes. On the number of sums and products. *Acta Arith.*, 81(4):365–367, 1997. (ref: p. 50)

[Ele99]   Gy. Elekes. On the number of distinct distances and certain algebraic curves. *Period. Math. Hung.*, 38(3):173–177, 1999. (ref: p. 48)

[Ele01]   Gy. Elekes. Sums versus products in number theory, algebra and Erdős geometry. In G. Halász et al., editors, *Paul Erdős and His Mathematics*. J. Bolyai Math. Soc., Budapest, 2001. In press. (refs: pp. 47, 48, 49, 54)

[ELSS73] P. Erdős, L. Lovász, A. Simmons, and E. Straus. Dissection graphs of planar point sets. In J. N. Srivastava, editor, *A Survey of Combinatorial Theory*, pages 139–154. North-Holland, Amsterdam, Netherlands, 1973. (refs: pp. 269, 276)

[Enf69]   P. Enflo. On the nonexistence of uniform homeomorphisms between L_p-spaces. *Ark. Mat.*, 8:103–105, 1969. (ref: p. 372)

[EOS86]   H. Edelsbrunner, J. O'Rourke, and R. Seidel. Constructing
            arrangements of lines and hyperplanes with applications. *SIAM
            J. Comput.*, 15:341–363, 1986. (ref: p. 151)

[EP71]   P. Erdős and G. Purdy. Some extremal problems in geometry.
            J. Combin. Theory, 10(3):246–252, 1971. (ref: p. 50)

[Epp95]   D. Eppstein. Dynamic Euclidean minimum spanning trees and
            extrema of binary functions. *Discrete Comput. Geom.*, 13:111–
            122, 1995. (ref: p. 124)

[Epp98]   D. Eppstein. Geometric lower bounds for parametric matroid
            optimization. *Discrete Comput. Geom.*, 20:463–476, 1998. (ref:
            p. 271)

[ER00]   Gy. Elekes and L. Rónyai. A combinatorial problem on poly-
            nomials and rational functions. *J. Combin. Thoery Ser. B*,
            89(1):1–20, 2000. (ref: p. 48)

[Erd46]   P. Erdős. On a set of distances of *n* points. *Amer. Math.
            Monthly*, 53:248–250, 1946. (refs: pp. 44, 45, 53, 54, 68)

[Erd60]   P. Erdős. On sets of distances of *n* points in Euclidean space.
            Publ. Math. Inst. Hungar. Acad. Sci., 5:165–169, 1960. (ref:
            p. 45)

[ES35]   P. Erdős and G. Szekeres. A combinatorial problem in geometry.
            Compositio Math., 2:463–470, 1935. (refs: pp. 32, 33)

[ES63]   P. Erdős and H. Sachs. Regular graphs with given girth and
            minimal number of knots (in German). *Wiss. Z. Martin-Luther-
            Univ. Halle-Wittenberg, Math.-Naturwiss. Reihe*, 12:251–258,
            1963. (ref: p. 368)

[ES83]   P. Erdős and M. Simonovits. Supersaturated graphs and hy-
            pergraphs. *Combinatorica*, 3:181–192, 1983. (ref: p. 215)

[ES96]   H. Edelsbrunner and N. R. Shah. Incremental topological flip-
            ping works for regular triangulations. *Algorithmica*, 15:223–241,
            1996. (ref: p. 121)

[ES00]   A. Efrat and M. Sharir. On the complexity of the union of
            fat objects in the plane. *Discrete Comput. Geom.*, 23:171–189,
            2000. (ref: p. 194)

[ESS93]   H. Edelsbrunner, R. Seidel, and M. Sharir. On the zone theorem
            for hyperplane arrangements. *SIAM J. Comput.*, 22(2):418–429,
            1993. (ref: p. 151)

[EVW97]   H. Edelsbrunner, P. Valtr, and E. Welzl. Cutting dense point
            sets in half. *Discrete Comput. Geom.*, 17(3):243–255, 1997.
            (refs: pp. 270, 273)

[EW85]   H. Edelsbrunner and E. Welzl. On the number of line sepa-
          rations of a finite set in the plane. *Journal of Combinatorial
          Theory Ser. A*, 38:15–29, 1985. (ref: p. 269)

[EW86]   H. Edelsbrunner and E. Welzl. Constructing belts in two-
          dimensional arrangements with applications. *SIAM J. Com-
          put.*, 15:271–284, 1986. (ref: p. 75)

[Far94]   G. Farkas. Applications of Fourier's mechanical principle (in
          Hungarian). *Math. Termés. Értesítő*, 12:457–472, 1893/94. Ger-
          man translation in *Math. Nachr. Ungarn* 12:1–27, 1895. (ref:
          p. 8)

[Fei00]   U. Feige. Approximating the bandwidth via volume respecting
          embeddings. *J. Comput. Syst. Sci*, 60:510–539, 2000. (ref:
          p. 396)

[Fel97]   S. Felsner. On the number of arrangements of pseudolines.
          Discrete Comput. Geom., 18:257–267, 1997. (ref: p. 139)

[FH92]   Z. Füredi and P. Hajnal. Davenport–Schinzel theory of matri-
          ces. *Discrete Math.*, 103:233–251, 1992. (ref: p. 177)

[Fie73]   M. Fiedler. Algebraic connectivity of graphs. *Czechosl. Math.
          J.*, 23(98):298–305, 1973. (ref: p. 381)

[FKS89]   J. Friedman, J. Kahn, and E. Szemerédi. On the second eigen-
          value of random regular graphs. In *Proceedings of the Twenty
          First Annual ACM Symposium on Theory of Computing*, pages
          587–598, 1989. (ref: p. 381)

[FLM77]   T. Figiel, J. Lindenstrauss, and V. D. Milman. The dimension of
          almost spherical sections of convex bodies. *Acta Math.*, 139:53–
          94, 1977. (refs: pp. 336, 346, 348, 352, 353)

[FR01]   P. Frankl and V. Rödl. Extremal problems on set systems.
          Random Structures and Algorithms, 2001. In press. (refs:
          pp. 226, 227)

[Fre73]   G. A. Freiman. *Foundations of a Structural Theory of Set Addi-
          tion*. Translations of Mathematical Monographs. Vol. 37. Amer-
          ican Mathematical Society, Providence, RI, 1973. (ref: p. 47)

[Fre76]   M. L. Fredman. How good is the information theory bound in
          sorting? *Theor. Comput. Sci.*, 1:355–361, 1976. (ref: p. 308)

[Fri91]   J. Friedman. On the second eigenvalue and random walks in
          random d-regular graphs. *Combinatorica*, 11:331–362, 1991.
          (ref: p. 381)

[FS01]   U. Feige and G. Schechtman. On the optimality of the random hyperplane rounding technique for MAXCUT. In *Proc. 33rd Annual ACM Symposium on Theory of Computing*, 2001. (ref: p. 384)

[Ful70]   D. R. Fulkerson. The perfect graph conjecture and pluperfect graph theorem. In R. C. Bose et al., editors, *Proc. of the Second Chapel Hill Conference on Combinatorial Mathematics and Its Applications*, pages 171–175. Univ. of North Carolina, Chapel Hill, North Carolina, 1970. (ref: p. 293)

[Für96]   Z. Füredi. New asymptotics for bipartite Turán numbers. *J. Combin. Theory Ser. A*, 75:141–144, 1996. (ref: p. 68)

[Gal56]   D. Gale. Neighboring vertices on a convex polyhedron. In H. W. Kuhn and A. W. Tucker, editors, *Linear Inequalities and Related Systems*, Annals of Math. Studies 38, pages 255–263. Princeton University Press, Princeton, 1956. (ref: p. 114)

[Gal63]   D. Gale. Neighborly and cyclic polytopes. In V. Klee, editor, *Convexity*, volume 7 of *Proc. Symp. Pure Math.*, pages 225–232. American Mathematical Society, 1963. (ref: p. 98)

[GGL95]   R. L. Graham, M. Grötschel, and L. Lovász, editors. *Handbook of Combinatorics*. North-Holland, Amsterdam, 1995. (refs: pp. viii, 85)

[GJ00]   E. Gawrilow and M. Joswig. polymake: a framework for analyzing convex polytopes. In G. Kalai and G. M. Ziegler, editors, *Polytopes—Combinatorics and Computation*, pages 43–74. Birkhäuser, Basel, 2000. Software available at http://www.math.tu-berlin.de/diskregeom/polymake/. (ref: p. 85)

[GKS99]   R. J. Gardner, A. Koldobsky, and T. Schlumprecht. An analytic solution to the Busemann–Petty problem on sections of convex bodies. *Annals of Math.*, 149:691–703, 1999. (ref: p. 314)

[GKV94]   L. Gargano, J. Körner, and U. Vaccaro. Capacities: From information theory to extremal set theory. *J. Combin. Theory, Ser. A*, 68(2):296–316, 1994. (ref: p. 309)

[GL87]   P. M. Gruber and C. G. Lekkerkerker. *Geometry of Numbers*. North-Holland, Amsterdam, 2nd edition, 1987. (ref: p. 20)

[GLS88]   M. Grötschel, L. Lovász, and A. Schrijver. *Geometric Algorithms and Combinatorial Optimization*, volume 2 of *Algorithms and Combinatorics*. Springer-Verlag, Berlin etc., 1988. 2nd edition 1993. (refs: pp. 24, 26, 293, 321, 322, 327, 381)

[Glu89]   E. D. Gluskin. Extremal properties of orthogonal paral-
            lelepipeds and their applications to the geometry of Banach
            spaces. *Math. USSR Sbornik*, 64(1):85–96, 1989. (ref: p. 320)

[GM90]   H. Gazit and G. L. Miller. Planar separators and the Euclidean
            norm. In *Proc. 1st Annu. SIGAL Internat. Sympos. Algorithms*.
            Information Processing Society of Japan, Springer-Verlag, Au-
            gust 1990. (ref: p. 57)

[GNRS99]   A. Gupta, I. Newman, Yu. Rabinovich, and A. Sinclair. Cuts,
            trees and ℓ_1-embeddings of graphs. In *Proc. 40th IEEE Sym-
            posium on Foundations of Computer Science*, pages 399–409,
            1999. Also sumbitted to *Combinatorica*. (ref: p. 396)

[GO97]   J. E. Goodman and J. O'Rourke, editors. *Handbook of Discrete
            and Computational Geometry*. CRC Press LLC, Boca Raton,
            FL, 1997. (refs: pp. viii, 85)

[Goo97]   J. E. Goodman. Pseudoline arrangements. In J. E. Goodman
            and J. O'Rourke, editors, *Handbook of Discrete and Computa-
            tional Geometry*, pages 83–110. CRC Press LLC, Boca Raton,
            FL, 1997. (refs: pp. 136, 139)

[Gor88]   Y. Gordon. Gaussian processes and almost spherical sections of
            convex bodies. *Ann. Probab.*, 16:180–188, 1988. (ref: p. 353)

[Gow98]   W. T. Gowers. A new proof of Szemeredi's theorem for
            arithmetic progressions of length four. *Geom. Funct. Anal.*,
            8(3):529–551, 1998. (refs: pp. 48, 227)

[GP84]   J. E. Goodman and R. Pollack. On the number of k-subsets
            of a set of n points in the plane. *J. Combin. Theory Ser. A*,
            36:101–104, 1984. (ref: p. 145)

[GP86]   J. E. Goodman and R. Pollack. Upper bounds for configurations
            and polytopes in $\Re^d$. *Discrete Comput. Geom.*, 1:219–227, 1986.
            (ref: p. 140)

[GP93]   J. E. Goodman and R. Pollack. Allowable sequences and or-
            der types in discrete and computational geometry. In J. Pach,
            editor, *New Trends in Discrete and Computational Geometry*,
            volume 10 of *Algorithms and Combinatorics*, pages 103–134.
            Springer, Berlin etc., 1993. (ref: p. 220)

[GPS90]   J. E. Goodman, R. Pollack, and B. Sturmfels. The intrinsic
            spread of a configuration in $\Re^d$. *J. Amer. Math. Soc.*, 3:639–
            651, 1990. (ref: p. 138)

[GPW93]   J. E. Goodman, R. Pollack, and R. Wenger. Geometric transver-
            sal theory. In J. Pach, editor, *New Trends in Discrete and*

Computational Geometry, volume 10 of *Algorithms and Combinatorics*, pages 163–198. Springer, Berlin etc., 1993. (ref: p. 262)

[GPW96]   J. E. Goodman, R. Pollack, and R. Wenger. Bounding the number of geometric permutations induced by k-transversals. *J. Combin. Theory Ser. A*, 75:187–197, 1996. (ref: p. 220)

[GPWZ94]   J. E. Goodman, R. Pollack, R. Wenger, and T. Zamfirescu. Arrangements and topological planes. *Amer. Math. Monthly*, 101(10):866–878, 1994. (ref: p. 136)

[Gro56]   A. Grothendieck. Sur certaines classes de suites dans les espaces de Banach et le theorème de Dvoretzky Rogers. *Bol. Soc. Math. Sao Paulo*, 8:81–110, 1956. (ref: p. 352)

[Gro98]   M. Gromov. *Metric Structures for Riemmanian and non-Riemmanian spaces*. Birkhäuser, Basel, 1998. (ref: p. 336)

[GRS97]   Y. Gordon, S. Reisner, and C. Schütt. Umbrellas and polytopal approximation of the Euclidean ball. *J. Approximation Theory*, 90(1):9–22, 1997. Erratum ibid. 95:331, 1998. (ref: p. 321)

[Grü60]   B. Grünbaum. Partitions of mass-distributions and of convex bodies by hyperplanes. *Pac. J. Math.*, 10:1257–1267, 1960. (ref: p. 308)

[Grü67]   B. Grünbaum. *Convex Polytopes*. John Wiley & Sons, New York, NY, 1967. (refs: pp. 85, 114)

[Grü72]   B. Grünbaum. *Arrangements and Spreads*. Regional Conf. Ser. Math. American Mathematical Society, Providence, RI, 1972. (ref: p. 128)

[Gru93]   P. M. Gruber. Geometry of numbers. In P. M. Gruber and J. M. Wills, editors, *Handbook of Convex Geometry (Vol. B)*, pages 739–763. North-Holland, Amsterdam, 1993. (ref: p. 20)

[Gup00]   A. Gupta. Embedding tree metrics into low dimensional Euclidean spaces. *Discrete Comput. Geom.*, 24:105–116, 2000. (ref: p. 393)

[GW93]   P. M. Gruber and J. M. Wills, editors. *Handbook of Convex Geometry (volumes A and B)*. North-Holland, Amsterdam, 1993. (refs: pp. viii, 8, 85, 314, 320)

[GW95]   M. X. Goemans and D. P. Williamson. Improved approximation algorithms for maximum cut and satisfiability problems using semidefinite programming. *J. ACM*, 42:1115–1145, 1995. (ref: p. 384)

[Har66]   L. H. Harper. Optimal numberings and isoperimetric problems on graphs. *J. Combin. Theory*, 1:385–393, 1966. (ref: p. 336)

[Har79]   H. Harborth. Konvexe Fünfecke in ebenen Punktmengen. *Elem. Math.*, 33:116–118, 1979. (ref: p. 37)

[Hås97]   J. Håstad. Some optimal inapproximability results. In *Proc. 29th Annual ACM Symposium on Theory of Computing*, pages 1–10, 1997. (ref: p. 384)

[Hoc96]   D. Hochbaum, editor. *Approximation Algorithms for NP-hard Problems*. PWS Publ. Co., Florence, Kentucky, 1996. (ref: p. 236)

[Hor83]   J. D. Horton. Sets with no empty convex 7-gons. *Canad. Math. Bull.*, 26:482–484, 1983. (ref: p. 37)

[HS86]   S. Hart and M. Sharir. Nonlinearity of Davenport–Schinzel sequences and of generalized path compression schemes. *Combinatorica*, 6:151–177, 1986. (refs: pp. 173, 175)

[HS94]   D. Halperin and M. Sharir. New bounds for lower envelopes in three dimensions, with applications to visibility in terrains. *Discrete Comput. Geom.*, 12:313–326, 1994. (refs: pp. 189, 192)

[HS95]   D. Halperin and M. Sharir. Almost tight upper bounds for the single cell and zone problems in three dimensions. *Discrete Comput. Geom.*, 14:385–410, 1995. (ref: p. 193)

[HW87]   D. Haussler and E. Welzl. Epsilon-nets and simplex range queries. *Discrete Comput. Geom.*, 2:127–151, 1987. (refs: pp. 68, 242, 254)

[IM98]   P. Indyk and R. Motwani. Approximate nearest neighbors: Towards removing the curse of dimensionality. In *Proc. 30th Annual ACM Symposium on Theory of Computing*, pages 604–613, 1998. (ref: p. 361)

[Ind01]   P. Indyk. Algorithmic applications of low-distortion embeddings. In *Proc. 42nd IEEE Symposium on Foundations of Computer Science*, 2001. (refs: pp. 357, 361, 398)

[JL84]   W. B. Johnson and J. Lindenstrauss. Extensions of Lipschitz mappings into a Hilbert space. *Contemp. Math.*, 26:189–206, 1984. (ref: p. 361)

[JM94]   S. Jadhav and A. Mukhopadhyay. Computing a centerpoint of a finite planar set of points in linear time. *Discrete Comput. Geom.*, 12:291–312, 1994. (ref: p. 16)

438   Bibliography

[Joh48]   F. John. Extremum problems with inequalities as subsidiary
                conditions. In *Studies and Essays, presented to R. Courant on
                his 60th birthday, January 8, 1948*, pages 187–204. Interscience
                Publishers, Inc., New York, N. Y., 1948. Reprinted in: J. Moser
                (editor): *Fritz John, Collected papers*, Volume 2, Birkhäuser,
                Boston, Massachusetts, 1985, pages 543–560.   (ref: p. 327)

[JSV01]   M. Jerrum, A. Sinclair, and E. Vigoda.   A polynomial-time
                approximation algorithm for the permanent of a matrix with
                non-negative entries. In *Proc. 33rd Annu. ACM Symposium on
                Theory of Computing*, pages 712–721, 2001. Also available in
                Electronic Colloquium on Computational Complexity, Report
                TR00-079, http://eccc.uni-trier.de/eccc/.   (ref: p. 322)

[Kai97]   T. Kaiser. Transversals of d-intervals. *Discrete Comput. Geom.*,
                18:195–203, 1997.   (ref: p. 262)

[Kal84]   G. Kalai. Intersection patterns of convex sets. *Israel J. Math.*,
                48:161–174, 1984.   (ref: p. 197)

[Kal86]   G. Kalai. Characterization of f-vectors of families of convex
                sets in R^d. II: Sufficiency of Eckhoff's conditions. *J. Combin.
                Theory, Ser. A*, 41:167–188, 1986.   (ref: p. 197)

[Kal88]   G. Kalai. A simple way to tell a simple polytope from its graph.
                J. Combin. Theory, Ser. A, 49(2):381–383, 1988.   (ref: p. 93)

[Kal91]   G. Kalai. The diameter of graphs of convex polytopes and f-
                vector theory. In *Applied Geometry and Discrete Mathematics
                (The Victor Klee Festschrift), DIMACS Series in Discr. Math.
                and Theoret. Comput. Sci. Vol. 4*, pages 387–411. Amer. Math.
                Soc., Providence, RI, 1991.   (ref: p. 104)

[Kal92]   G. Kalai. A subexponential randomized simplex algorithm. In
                Proc. 24th Annu. ACM Sympos. Theory Comput., pages 475–
                482, 1992.   (ref: p. 93)

[Kal97]   G. Kalai. Linear programming, the simplex algorithm and sim-
                ple polytopes. *Math. Program.*, 79B:217–233, 1997.   (ref: p. 93)

[Kal01]   G. Kalai. Combinatorics with a geometric flavor: Some exam-
                ples. In *Visions in Mathematics Towards 2000 (GAFA, special
                volume), part II*, pages 742–792. Birkhäuser, Basel, 2001.   (ref:
                p. 204)

[Kan96]   G. Kant. Drawing planar graphs using the canonical ordering.
                Algorithmica, 16:4–32, 1996.   (ref: p. 94)

[Kár01]   Gy. Károlyi. Ramsey-remainder for convex sets and the Erdős–
                Szekeres theorem. *Discrete Applied Math.*, 109:163–175, 2001.
                (ref: p. 33)

[Kat78] M. Katchalski. A Helly type theorem for convex sets. *Can. Math. Bull.*, 21:121–123, 1978. (ref: p. 13)

[KGT01] D. J. Kleitman, A. Gyárfás, and G. Tóth. Convex sets in the plane with three of every four meeting. *Combinatorica*, 21(2):221–232, 2001. (ref: p. 258)

[Kha89] L. G. Khachiyan. Problems of optimal algorithms in convex programming, decomposition and sorting (in Russian). In Yu. I. Zhuravlev, editor, *The Computer and Choice Problems*, pages 161–205. Nauka, Moscow, 1989. (ref: p. 308)

[Kir03] P. Kirchberger. Über Tschebyschefsche Annäherungsmethoden. *Math. Ann.*, 57:509–540, 1903. (ref: p. 13)

[Kis68] S. S. Kislitsyn. Finite partially ordered sets and their corresponding permutation sets (in Russian). *Mat.Zametki*, 4:511–518, 1968. English translation in *Math. Notes* 4:798-801, 1968. (ref: p. 308)

[KK95] J. Kahn and J.-H. Kim. Entropy and sorting. *J. Assoc. Comput. Machin.*, 51:390–399, 1995. (ref: p. 309)

[KL79] M. Katchalski and A. Liu. A problem of geometry in R^n. *Proc. Amer. Math. Soc.*, 75:284–288, 1979. (ref: p. 197)

[KL91] J. Kahn and N. Linial. Balancing extensions via Brunn–Minkowski. *Combinatorica*, 11(4):363–368, 1991. (ref: p. 308)

[Kla92] M. Klazar. A general upper bound in extremal theory of sequences. *Comment. Math. Univ. Carol.*, 33:737–746, 1992. (ref: p. 176)

[Kla99] M. Klazar. On the maximum length of Davenport–Schinzel sequences. In R. Graham et al., editors, *Contemporary Trends in Discrete Mathematics (DIMACS Series in Discrete Mathematics and Theoretical Computer Science, Vol. 49)*, pages 169–178. Amer. Math. Soc., Providence, RI, 1999. (ref: p. 176)

[Kla00] M. Klazar. The Füredi–Hajnal conjecture implies the Stanley–Wilf conjecture. In D. Krob et al., editors, *Formal Power Series and Algebraic Combinatorics (Proceedings of the 12th FPSAC conference, Moscow, June 25-30, 2000)*, pages 250–255. Springer, Berlin etc., 2000. (ref: p. 177)

[Kle53] V. Klee. The critical set of a convex body. *Amer. J. Math.*, 75:178–188, 1953. (ref: p. 12)

[Kle64] V. Klee. On the number of vertices of a convex polytope. *Canadian J. Math*, 16:701–720, 1964. (refs: pp. 103, 105)

[Kle89]   R. Klein. *Concrete and Abstract Voronoi Diagrams*, volume
            400 of *Lecture Notes Comput. Sci.* Springer-Verlag, Berlin etc.,
            1989. (ref: p. 121)

[Kle97]   J. Kleinberg. Two algorithms for nearest-neighbor search in
            high dimension. In *Proc. 29th Annu. ACM Sympos. Theory
            Comput.*, pages 599–608, 1997. (ref: p. 361)

[KLL88]   R. Kannan, A. K. Lenstra, and L. Lovász. Polynomial factor-
            ization and nonrandomness of bits of algebraic and some tran-
            scendental numbers. *Math. Comput.*, 50(181):235–250, 1988.
            (ref: p. 26)

[KLMR98]   L. E. Kavraki, J.-C. Latombe, R. Motwani, and P. Raghavan.
            Randomized query processing in robot path planning. *J. Com-
            put. Syst. Sci.*, 57:50–60, 1998. (ref: p. 250)

[KLPS86]   K. Kedem, R. Livne, J. Pach, and M. Sharir. On the union of
            Jordan regions and collision-free translational motion amidst
            polygonal obstacles. *Discrete Comput. Geom.*, 1:59–71, 1986.
            (ref: p. 194)

[KLS97]   R. Kannan, L. Lovász, and M. Simonovits. Random walks and
            an $O^*(n^5)$ volume algorithm for convex bodies. *Random Struc.
            Algo.*, 11:1–50, 1997. (ref: p. 322)

[KM97a]   G. Kalai and J. Matoušek. Guarding galleries where every point
            sees a large area. *Israel J. Math*, 101:125–140, 1997. (refs:
            pp. 235, 250)

[KM97b]   M. Karpinski and A. Macintyre. Polynomial bounds for VC
            dimension of sigmoidal and general Pfaffian neural networks.
            J. Syst. Comput. Sci., 54(1):169–176, 1997. (ref: p. 250)

[Koc94]   M. Kochol. Constructive approximation of a ball by polytopes.
            Math. Slovaca, 44(1):99–105, 1994. (ref: p. 324)

[Kol01]   V. Koltun. Almost tight upper bounds for vertical decomposi-
            tions in four dimensions. In *Proc. 42nd IEEE Symposium on
            Foundations of Computer Science*, 2001. (ref: p. 162)

[Kör73]   J. Körner. Coding of an information source having ambigu-
            ous alphabet and the entropy of graphs. In *Inform. The-
            ory, statist. Decision Funct., Random Processes; Transact. 6th
            Prague Conf. 1971*, pages 411–425, 1973. (ref: p. 309)

[KP01]   V. Kaibel and M. E. Pfetsch. Computing the face lattice of
            a polytope from its vertex-facet incidences. Technical Report,
            Inst. für Mathematik, TU Berlin, 2001. (ref: p. 105)

[KPR93]   P. Klein, S. Plotkin, and S. Rao. Excluded minors, network decomposition, and multicommodity flow. In *Proc. 25th Annual ACM Symposium on the Theory of Computing*, pages 682–690, 1993. (ref: p. 393)

[KPT01]   Gy. Károlyi, J. Pach, and G. Tóth. A modular version of the Erdős–Szekeres theorem. *Studia Mathematica Hungarica*, 2001. In press. (ref: p. 38)

[KPW92]   J. Komlós, J. Pach, and G. Woeginger. Almost tight bounds for ε-nets. *Discrete Comput. Geom.*, 7:163–173, 1992. (ref: p. 243)

[KRS96]   J. Kollár, L. Rónyai, and T. Szabó. Norm-graphs and bipartite Turán numbers. *Combinatorica*, 16(3):399–406, 1996. (ref: p. 68)

[KS84]   J. Kahn and M. Saks. Balancing poset extensions. *Order*, 1:113–126, 1984. (ref: p. 308)

[KS96]   J. Komlós and M. Simonovits. Szemerédi's regularity lemma and its applications in graph theory. In D. Miklos et al. editors, *Combinatorics, Paul Erdős Is Eighty.*, Vol. 2, pages 295–352. János Bólyai Mathematical Society, Budapest, 1996. (ref: p. 226)

[KST54]   T. Kővári, V. Sós, and P. Turán. On a problem of k. zarankiewicz. *Coll. Math.*, 3:50–57, 1954. (ref: p. 68)

[KT99]   N. Katoh and T. Tokuyama. Lovász's lemma for the three-dimensional K-level of concave surfaces and its applications. In *Proc. 40th IEEE Symposium on Foundations of Computer Science*, pages 389–398, 1999. (ref: p. 271)

[KV94]   M. Klazar and P. Valtr. Generalized Davenport–Schinzel sequences. *Combinatorica*, 14:463–476, 1994. (ref: p. 176)

[KV01]   Gy. Károlyi and P. Valtr. Point configurations in d-space without large subsets in convex position. *Discrete Comput. Geom.*, 2001. To appear. (ref: p. 33)

[KZ00]   G. Kalai and G. M. Ziegler, editors. *Polytopes—Combinatorics and Computation. DMV-seminar Oberwolfach, Germany, November 1997*. Birkhäuser, Basel, 2000. (ref: p. 85)

[Lar72]   D. G. Larman. On sets projectively equivalent to the vertices of a convex polytope. *Bull. Lond. Math. Soc.*, 4:6–12, 1972. (ref: p. 206)

[Lat91]   J.-C. Latombe. *Robot Motion Planning*. Kluwer Academic Publishers, Boston, 1991. (ref: p. 122)

[Led01]   M. Ledoux. *The Concentration of Measure Phenomenon*, volume 89 of *Mathematical Surveys and Monographs*. Amer. Math. Soc., Providence, RI, 2001. (refs: pp. 336, 340)

[Lee82]   D. T. Lee. On k-nearest neighbor Voronoi diagrams in the plane. *IEEE Trans. Comput.*, C-31:478–487, 1982. (ref: p. 122)

[Lee91]   C. W. Lee. Winding numbers and the generalized lower-bound conjecture. In J.E. Goodman, R. Pollack, and W. Steiger, editors, *Computational Geometry: Papers from the DIMACS special year*, DIMACS Series in Discrete Mathematics and Theoretical Computer Science 6, pages 209–219. Amer. Math. Soc., 1991. (ref: p. 280)

[Lei83]   F. T. Leighton. *Complexity issues in VLSI*. MIT Press, Cambridge, MA, 1983. (ref: p. 57)

[Lei84]   F. T. Leighton. New lower bound techniques for VLSI. *Math. Systems Theory*, 17:47–70, 1984. (ref: p. 56)

[Len83]   H. W. Lenstra. Integer programming with a fixed number of variables. *Math. Oper. Res.*, 8:538–548, 1983. (ref: p. 24)

[Lev26]   F. Levi. Die Teilung der projektiven Ebene durch Gerade oder Pseudogerade. *Ber. Math.-Phys. Kl. sächs. Akad. Wiss. Leipzig*, 78:256–267, 1926. (ref: p. 136)

[Lév51]   P. Lévy. *Problèmes concrets d'analyse fonctionelle*. Gauthier Villars, Paris, 1951. (ref: p. 340)

[Lin84]   N. Linial. The information-theoretic bound is good for merging. *SIAM J. Comput.*, 13:795–801, 1984. (ref: p. 308)

[Lin92]   J. Lindenstrauss. Almost spherical sections; their existence and their applications. In *Jahresbericht der DMV, Jubilaeumstag., 100 Jahre DMV, Bremen/Dtschl. 1990*, pages 39–61, 1992. (refs: pp. 336, 352)

[LLL82]   A. K. Lenstra, H. W. Lenstra, Jr., and L. Lovász. Factoring polynomials with rational coefficients. *Math. Ann.*, 261:514–534, 1982. (ref: p. 25)

[LLR95]   N. Linial, E. London, and Yu. Rabinovich. The geometry of graphs and some its algorithmic applications. *Combinatorica*, 15:215–245, 1995. (refs: pp. 379, 380, 392, 400)

[LM75]   D. Larman and P. Mani. Almost ellipsoidal sections and projections of convex bodies. *Math. Proc. Camb. Philos. Soc.*, 77:529–546, 1975. (ref: p. 353)

[LM93]   J. Lindenstrauss and V. D. Milman. The local theory of normed
            spaces and its applications to convexity. In P. M. Gruber and
            J. M. Wills, editors, *Handbook of Convex Geometry*, pages 1149–
            1220. North-Holland, Amsterdam, 1993. (refs: pp. 327, 336)

[LM00]   N. Linial and A. Magen. Least-distortion Euclidean embeddings
            of graphs: Products of cycles and expanders. *J. Combin. Theory
            Ser. B*, 79:157–171, 2000. (ref: p. 380)

[LMN01]   N. Linial, A. Magen, and N. Naor. Euclidean embeddings of
            regular graphs—the girth lower bound. *Geometric and Func-
            tional Analysis*, 2001. In press. (ref: p. 380)

[LMS94]   C.-Y. Lo, J. Matoušek, and W. L. Steiger. Algorithms for ham-
            sandwich cuts. *Discrete Comput. Geom.*, 11:433, 1994. (ref:
            p. 16)

[LO96]   J.-P. Laumond and M. H. Overmars, editors. *Algorithms for
            Robotic Motion and Manipulation*. A. K. Peters, Wellesley, MA,
            1996. (ref: p. 122)

[Lov]   L. Lovász. Semidefinite programs and combinatorial optimiza-
            tion. In C. Linhares-Sales and B. Reed, editors, *Recent Ad-
            vances in Algorithmic Discrete Mathematics*. Springer, Berlin
            etc. To appear. (refs: pp. 293, 381)

[Lov71]   L. Lovász. On the number of halving lines. *Annal. Univ. Scie.
            Budapest. de Rolando Eötvös Nominatae, Sectio Math.*, 14:107–
            108, 1971. (refs: pp. 269, 280)

[Lov72]   L. Lovász. Normal hypergraphs and the perfect graph conjec-
            ture. *Discrete Math.*, 2:253–267, 1972. (ref: p. 293)

[Lov74]   L. Lovász. Problem 206. *Matematikai Lapok*, 25:181, 1974.
            (ref: p. 198)

[Lov86]   L. Lovász. *An Algorithmic Theory of Numbers, Graphs and
            Convexity*. SIAM Regional Series in Applied Mathematics.
            SIAM, Philadelphia, 1986. (refs: pp. 24, 25)

[Lov93]   L. Lovász. *Combinatorial Problems and Exercises (2nd edition)*.
            Akadémiai Kiadó, Budapest, 1993. (refs: pp. 235, 374)

[LP86]   L. Lovasz and M. D. Plummer. *Matching Theory*, volume 29 of
            Ann. Discrete Math. North-Holland, 1986. (ref: p. 235)

[LPS88]   A. Lubotzky, R. Phillips, and P. Sarnak. Ramanujan graphs.
            Combinatorica, 8:261–277, 1988. (refs: pp. 367, 382)

[LR97] M. Laczkovich and I. Ruzsa. The number of homothetic subsets. In R. L. Graham and J. Nešetřil, editors, *The Mathematics of Paul Erdős, Vol. II*, volume 14 of *Algorithms and Combinatorics*, pages 294–302. Springer, Berlin etc., 1997. (ref: p. 47)

[LS02] N. Linial and M. Saks. The Euclidean distortion of complete binary trees—An elementary proof. *Discr. Comput. Geom.*, 2002. To appear. (ref: p. 393)

[LUW95] F. Lazebnik, V. A. Ustimenko, and A. J. Woldar. A new series of dense graphs of high girth. *Bull. Amer. Math. Soc., New Ser.*, 32(1):73–79, 1995. (ref: p. 367)

[LUW96] F. Lazebnik, V. A. Ustimenko, and A. J. Woldar. A characterization of the components of the graphs $D(k,q)$. *Discrete Math.*, 157(1-3):271–283, 1996. (ref: p. 367)

[LW88] N. Linial L. Lovász and A. Wigderson. Rubber bands, convex embeddings and graph connectivity. *Combinatorica*, 8:91–102, 1988. (ref: p. 92)

[Mac50] A.M. Macbeath. A compactness theorem for affine equivalence-classes of convex regions. *Canad. J. Math*, 3:54–61, 1950. (ref: p. 321)

[Mar88] G. A. Margulis. Explicit group-theoretic constructions of combinatorial schemes and their application to the design of expanders and concentrators (in Russian). *Probl. Peredachi Inf.*, 24(1):51–60, 1988. English translation: *Probl. Inf. Transm.* 24, No.1, 39–46 (1988). (ref: p. 382)

[Mat90a] J. Matoušek. Construction of ϵ-nets. *Discrete Comput. Geom.*, 5:427–448, 1990. (refs: pp. 68, 75)

[Mat90b] J. Matoušek. Bi-Lipschitz embeddings into low-dimensional Euclidean spaces. *Comment. Math. Univ. Carolinae*, 31:589–600, 1990. (ref: p. 368)

[Mat92] J. Matoušek. Efficient partition trees. *Discrete Comput. Geom.*, 8:315–334, 1992. (ref: p. 69)

[Mat96a] J. Matoušek. Note on the colored Tverberg theorem. *J. Combin. Theory Ser. B*, 66:146–151, 1996. (ref: p. 205)

[Mat96b] J. Matoušek. On the distortion required for embedding finite metric spaces into normed spaces. *Israel J. Math.*, 93:333–344, 1996. (refs: pp. 140, 367, 388)

[Mat97] J. Matoušek. On embedding expanders into ℓ_p spaces. *Israel J. Math.*, 102:189–197, 1997. (ref: p. 379)

[Mat98]   J. Matoušek. On constants for cuttings in the plane. *Discrete Comput. Geom.*, 20:427–448, 1998. (ref: p. 75)

[Mat99a]   J. Matoušek. *Geometric Discrepancy (An Illustrated Guide).* Springer-Verlag, Berlin, 1999. (ref: p. 243)

[Mat99b]   J. Matoušek. On embedding trees into uniformly convex Banach spaces. *Israel J. Math,* 114:221–237, 1999. (ref: p. 393)

[Mat01]   J. Matoušek. A lower bound for weak epsilon-nets in high dimension. *Discrete Comput. Geom.*, 2001. In press. (ref: p. 254)

[McM70]   P. McMullen. The maximal number of faces of a convex polytope. *Mathematika,* 17:179–184, 1970. (ref: p. 103)

[McM93]   P. McMullen. On simple polytopes. *Invent. Math.*, 113:419–444, 1993. (ref: p. 105)

[McM96]   P. McMullen. Weights on polytopes. *Discrete Comput. Geom.*, 15:363–388, 1996. (ref: p. 105)

[Mic98]   D. Micciancio. The shortest vector in a lattice is hard to approximate within some constants. In *Proc. 39th IEEE Symposium on Foundations of Computer Science*, pages 92–98, 1998. (ref: p. 25)

[Mil64]   J. W. Milnor. On the Betti numbers of real algebraic varieties. *Proc. Amer. Math. Soc.*, 15:275–280, 1964. (ref: p. 135)

[Mil69]   V. D. Milman. Spectrum of continuous bounded functions on the unit sphere of a Banach space. *Funct. Anal. Appl.*, 3:67–79, 1969. (refs: pp. 341, 348)

[Mil71]   V. D. Milman. New proof of the theorem of Dvoretzky on sections of convex bodies. *Funct. Anal. Appl.*, 5:28–37, 1971. (refs: pp. 341, 348, 353)

[Mil98]   V. D. Milman. Surprising geometric phenomena in high-dimensional convexity theory. In A. Balog et al., editors, *European Congress of Mathematics (ECM), Budapest, Hungary, July 22–26, 1996. Volume II*, pages 73–91. Birkhäuser, Basel, 1998. (refs: pp. 313, 321, 336)

[Min96]   H. Minkowski. *Geometrie der Zahlen.* Teubner, Leipzig, 1896. Reprinted by Johnson, New York, NY 1968. (refs: pp. 20, 300)

[Mne89]   M. E. Mnev. The universality theorems on the classification problem of configuration varieties and convex polytopes varieties. In O. Y. Viro, editor, *Topology and Geometry—Rohlin Seminar*, volume 1346 of *Lecture Notes Math.*, pages 527–544. Springer, Berlin etc., 1989. (ref: p. 138)

[Mor94]   M. Morgenstern. Existence and explicit constructions of $q + 1$ regular Ramanujan graphs for every prime power q. *J. Combin. Theory, Ser. B*, 62(1):44–62, 1994. (ref: p. 382)

[Mos52]   L. Moser. On the different distances determined by n points. *Amer. Math. Monthly*, 59:85–91, 1952. (ref: p. 45)

[MPS+94]   J. Matoušek, J. Pach, M. Sharir, S. Sifrony, and E. Welzl. Fat triangles determine linearly many holes. *SIAM J. Comput.*, 23:154–169, 1994. (ref: p. 194)

[MS71]   P. McMullen and G. C. Shephard. *Convex Polytopes and the Upper Bound Conjecture*, volume 3 of *Lecture Notes*. Cambridge University Press, Cambridge, England, 1971. (refs: pp. 85, 114)

[MS86]   V. D. Milman and G. Schechtman. *Asymptotic Theory of Finite Dimensional Normed Spaces*. Lecture Notes in Math. 1200. Springer-Verlag, Berlin etc., 1986. (refs: pp. 300, 335, 336, 340, 346, 353, 361)

[MS00]   W. Morris and V. Soltan. The Erdős–Szekeres problem on points in convex position—a survey. *Bull. Amer. Math. Soc., New Ser.*, 37(4):437–458, 2000. (ref: p. 32)

[MSW96]   J. Matoušek, M. Sharir, and E. Welzl. A subexponential bound for linear programming. *Algoritmica*, 16:498–516, 1996. (refs: pp. 94, 327)

[Mul93a]   K. Mulmuley. *Computational Geometry: An Introduction Through Randomized Algorithms*. Prentice Hall, Englewood Cliffs, NJ, 1993. (refs: pp. 161, 162)

[Mul93b]   K. Mulmuley. Dehn–Sommerville relations, upper bound theorem, and levels in arrangements. In *Proc. 9th Annu. ACM Sympos. Comput. Geom.*, pages 240–246, 1993. (ref: p. 280)

[Nar00]   W. Narkiewicz. *The Development of Prime Number Theory*. Springer, Berlin etc., 2000. (ref: p. 54)

[NPPS01]   E. Nevo, J. Pach, R. Pinchasi, and M. Sharir. Lenses in arrangements of pseudocircles and their applications. *Discrete Comput. Geom.*, 2001. In press. (ref: p. 271)

[NR01]   I. Newman and Yu. Rabinovich. A lower bound on the distortion of embedding planar metrics into Euclidean space. Manuscript, Computer Science Department, Univ. of Haifa; submitted to *Discrete Comput. Geom.*, 2001. (ref: p. 372)

[Nyk00]   H. Nyklová. Almost empty convex polygons. KAM-DIMATIA Series 498-2000 (technical report), Charles University, Prague, 2000. (ref: p. 39)

[OBS92]   A. Okabe, B. Boots, and K. Sugihara. *Spatial Tessellations: Concepts and Applications of Voronoi Diagrams*. John Wiley & Sons, Chichester, UK, 1992. (ref: p. 120)

[OP49]   O. A. Oleinik and I. B. Petrovskiĭ. On the topology of of real algebraic surfaces (in Russian). *Izv. Akad. Nauk SSSR*, 13:389–402, 1949. (ref: p. 135)

[OS94]   S. Onn and B. Sturmfels. A quantitative Steinitz' theorem. *Beiträge zur Algebra und Geometrie / Contributions to Algebra and Geometry*, 35:125–129, 1994. (ref: p. 94)

[OT91]   P. Orlik and H. Terao. *Arrangements of Hyperplanes*. Springer-Verlag, Berlin etc., 1991. (ref: p. 129)

[ÓY85]   C. Ó'Dúnlaing and C. K. Yap. A "retraction" method for planning the motion of a disk. *J. Algorithms*, 6:104–111, 1985. (ref: p. 122)

[PA95]   J. Pach and P. K. Agarwal. *Combinatorial Geometry*. John Wiley & Sons, New York, NY, 1995. (refs: pp. viii, 20, 24, 44, 45, 50, 53, 56, 57, 92, 243)

[Pac98]   J. Pach. A Tverberg-type result on multicolored simplices. *Comput. Geom.: Theor. Appl.*, 10:71–76, 1998. (refs: pp. 220, 226, 229)

[Pac99]   J. Pach. Geometric graph theory. In J. D. Lamb et al., editors, *Surveys in Combinatorics. Proceedings of the 17th British combinatorial conference, University of Kent at Canterbury, UK, 1999*, Lond. Math. Soc. Lect. Note Ser. 267, pages 167–200. Cambridge University Press, 1999. (ref: p. 56)

[Pin02]   R. Pinchasi. Gallai–Sylvester theorem for pairwise intersecting unit circles. *Discrete Comput. Geom.*, 2002. To appear. (ref: p. 44)

[Pis89]   G. Pisier. *The Volume of Convex Bodies and Banach Space Geometry*. Cambridge University Press, Cambridge, 1989. (refs: pp. 315, 335, 336, 353, 361)

[Pór02]   A. Pór. A partitioned version of the Erdős–Szekeres theorem. *Discrete Comput. Geom.*, 2002. To appear. (ref: p. 220)

[PP01]   J. Pach and R. Pinchasi. On the number of balanced lines. *Discrete Comput. Geom.*, 25:611–628, 2001. (ref: p. 280)

[PR93]   R. Pollack and M.-F. Roy. On the number of cells defined by a set of polynomials. *C. R. Acad. Sci. Paris*, 316:573–577, 1993. (ref: p. 135)

[PS89]   J. Pach and M. Sharir. The upper envelope of piecewise lin-
            ear functions and the boundary of a region enclosed by convex
            plates: combinatorial analysis. *Discrete Comput. Geom.*, 4:291–
            309, 1989. (ref: p. 186)

[PS92]   J. Pach and M. Sharir. Repeated angles in the plane and related
            problems. *J. Combin. Theory Ser. A*, 59:12–22, 1992. (refs:
            pp. 46, 49, 50)

[PS98a]   J. Pach and M. Sharir. On the number of incidences between
            points and curves. *Combinatorics, Probability, and Computing*,
            7:121–127, 1998. (refs: pp. 46, 49, 64)

[PS98b]   J. Pach and J. Solymosi. Canonical theorems for convex sets.
            Discrete Comput. Geom., 19:427–435, 1998. (ref: p. 220)

[PS01]   J. Pach and J. Solymosi. Crossing patterns of segments. *J.
            Combin. Theory Ser. A*, 96:316–325, 2001. (refs: pp. 223, 227)

[PSS88]   R. Pollack, M. Sharir, and S. Sifrony. Separating two sim-
            ple polygons by a sequence of translations. *Discrete Comput.
            Geom.*, 3:123–136, 1988. (ref: p. 176)

[PSS92]   J. Pach, W. Steiger, and E. Szemerédi. An upper bound on the
            number of planar k-sets. *Discrete Comput. Geom.*, 7:109–123,
            1992. (ref: p. 269)

[PSS96]   J. Pach, F. Shahrokhi, and M. Szegedy. Applications of the
            crossing number. *Algorithmica*, 16:111–117, 1996. (ref: p. 57)

[PSS01]   J. Pach, I. Safruti, and M. Sharir. The union of cubes in
            three dimensions. In *Proc. 17th Annu. ACM Sympos. Com-
            put. Geom.*, pages 19–28, 2001. (ref: p. 194)

[PST00]   J. Pach, J. Spencer, and G. Tóth. New bounds for crossing
            numbers,. *Discrete Comput. Geom.*, 24:623–644, 2000. (refs:
            pp. 57, 58)

[PT97]   J. Pach and G. Tóth. Graphs drawn with few crossings per
            edge. *Combinatorica*, 17:427–439, 1997. (ref: p. 56)

[PT98]   J. Pach and G. Tóth. A generalization of the Erdős–Szekeres
            theorem to disjoint convex sets. *Discrete Comput. Geom.*,
            19(3):437–445, 1998. (ref: p. 33)

[PT00]   J. Pach and G. Tóth. Which crossing number is it anyway? *J.
            Combin. Theory Ser. B*, 80:225–246, 2000. (ref: p. 58)

[Rad21]   J. Radon. Mengen konvexer Körper, die einen gemeinsamen
            Punkt enthalten. *Math. Ann.*, 83:113–115, 1921. (ref: p. 12)

[Rad47]   R. Rado. A theorem on general measure. *J. London Math. Soc.*,
            21:291–300, 1947.   (ref: p. 16)

[Rao99]   S. Rao.   Small distortion and volume respecting embeddings
            for planar and Euclidean metrics. In *Proc. 15th Annual ACM
            Symposium on Comput. Geometry*, pages 300–306, 1999.   (refs:
            pp. 393, 398)

[RBG01]   L. Rónyai, L. Babai, and M. K. Ganapathy.   On the number
            of zero-patterns of a sequence of polynomials. *J. Amer. Math.
            Soc.*, 14(3):717–735 (electronic), 2001.   (ref: p. 136)

[Rea68]   J. R. Reay. An extension of Radon's theorem. *Illinois J. Math*,
            12:184–189, 1968.   (ref: p. 204)

[RG97]   J. Richter-Gebert. *Realization Spaces of Polytopes*.   Lecture
            Notes in Mathematics 1643. Springer, Berlin, 1997.   (refs:
            pp. 92, 94, 139)

[RG99]   J. Richter-Gebert. The universality theorems for oriented ma-
            troids and polytopes. In B. Chazelle et al., editors, *Advances
            in Discrete and Computational Geometry*, Contemp. Math. 223,
            pages 269–292. Amer. Math. Soc., Providence, RI, 1999.   (refs:
            pp. 94, 138, 139)

[Rou01a]   J.-P. Roudneff.   Partitions of points into simplices with k-
            dimensional intersection. Part I: The conic Tverberg's theorem.
            European J. Combinatorics, 22:733–743, 2001.   (ref: p. 204)

[Rou01b]   J.-P. Roudneff.   Partitions of points into simplices with k-
            dimensional intersection. Part II: Proof of Reay's conjecture in
            dimensions 4 and 5. *European J. Combinatorics*, 22:745–765,
            2001.   (ref: p. 204)

[RS64]   A. Rényi and R. Sulanke. Über die konvexe Hülle von n zufällig
            gewählten Punkten II. *Z. Wahrsch. Verw. Gebiete*, 3:138–147,
            1964.   (ref: p. 328)

[Rud91]   W. Rudin. *Functional Analysis (2nd edition)*. McGraw-Hill,
            New York, 1991.   (ref: p. 8)

[Ruz94]   I. Z. Ruzsa. Generalized arithmetical progressions and sumsets.
            Acta Math. Hung., 65(4):379–388, 1994.   (ref: p. 47)

[RVW00]   O. Reingold, S. P. Vadhan, and A. Wigderson. Entropy waves,
            the zig-zag graph product, and new constant-degree expanders
            and extractors. In *Proc. 41st IEEE Symposium on Foundations
            of Computer Science*, pages 3–13, 2000.   (refs: pp. 381, 382)

[SA95]   M. Sharir and P. K. Agarwal. *Davenport-Schinzel Sequences and Their Geometric Applications*. Cambridge University Press, Cambridge, 1995. (refs: pp. 168, 172, 173, 176, 181, 191)

[Sal75]   G.T. Sallee. A Helly-type theorem for widths. In *Geom. Metric Lin. Spaces, Proc. Conf. East Lansing 1974*, Lect. Notes Math. 490, pages 227–232. Springer, Berlin etc., 1975. (ref: p. 13)

[Sar91]   K. Sarkaria. A generalized van Kampen–Flores theorem. *Proc. Amer. Math. Soc.*, 11:559–565, 1991. (ref: p. 368)

[Sar92]   K. Sarkaria. Tverberg's theorem via number fields. *Israel J. Math.*, 79:317, 1992. (ref: p. 204)

[Sau72]   N. Sauer. On the density of families of sets. *Journal of Combinatorial Theory Ser. A*, 13:145–147, 1972. (ref: p. 242)

[Sch01]   L. Schläfli. Theorie der vielfachen Kontinuität. *Denkschriften der Schweizerichen naturforschender Gesellschaft*, 38:1–237, 1901. Written in 1850–51. Reprinted in *Ludwig Schläfli, 1814–1895, Gesammelte mathematische Abhandlungen*, Birkhäuser, Basel 1950. (ref: p. 85)

[Sch11]   P. H. Schoute. Analytic treatment of the polytopes regularly derived from the regular polytopes. *Verhandelingen der Koninglijke Akademie van Wetenschappen te Amsterdam*, 11(3), 1911. (ref: p. 85)

[Sch38]   I. J. Schoenberg. Metric spaces and positive definite functions. *Trans. Amer. Math. Soc.*, 44:522–53, 1938. (ref: p. 357)

[Sch48]   E. Schmidt. Die Brunn–Minkowski Ungleichung. *Math. Nachrichten*, 1:81–157, 1948. (ref: p. 336)

[Sch86]   A. Schrijver. *Theory of Linear and Integer Programming*. Wiley-Interscience, New York, NY, 1986. (refs: pp. 8, 24, 25, 85)

[Sch87]   C. P. Schnorr. A hierarchy of polynomial time lattice basis reduction algorithms. *Theor. Comput. Sci.*, 53:201–224, 1987. (ref: p. 25)

[Sch90]   W. Schnyder. Embedding planar graphs on the grid. In *Proc. 1st ACM-SIAM Sympos. Discrete Algorithms*, pages 138–148, 1990. (ref: p. 94)

[Sch93]   R. Schneider. *Convex Bodies: The Brunn-Minkowski Theory*, volume 44 of *Encyclopedia of Mathematics and Its Applications*. Cambridge University Press, Cambridge, 1993. (ref: p. 301)

[Sei91]   R. Seidel. Small-dimensional linear programming and convex
            hulls made easy. *Discrete Comput. Geom.*, 6:423–434, 1991.
            (ref: p. 105)

[Sei95]   R. Seidel. The upper bound theorem for polytopes: an easy
            proof of its asymptotic version. *Comput. Geom. Theory Appl.*,
            5:115–116, 1995. (ref: p. 104)

[Sei97]   R. Seidel. Convex hull computations. In J. E. Goodman and
            J. O'Rourke, editors, *Handbook of Discrete and Computational
            Geometry*, chapter 19, pages 361–376. CRC Press LLC, Boca
            Raton, FL, 1997. (ref: p. 105)

[Sha94]   M. Sharir. Almost tight upper bounds for lower envelopes in
            higher dimensions. *Discrete Comput. Geom.*, 12:327–345, 1994.
            (ref: p. 192)

[Sha01]   M. Sharir.   The Clarkson–Shor technique revisited and ex-
            tended. In *Proc. 17th Annu. ACM Sympos. Comput. Geom.*,
            pages 252–256, 2001. (refs: pp. 145, 146)

[She72]   S. Shelah. A combinatorial problem, stability and order for
            models and theories in infinitary languages. *Pacific J. Math.*,
            41:247–261, 1972. (ref: p. 242)

[Sho91]   P. W. Shor.   Stretchability of pseudolines is NP-hard.   In
            P. Gritzman and B. Sturmfels, editors, *Applied Geometry and
            Discrete Mathematics: The Victor Klee Festschrift*, volume 4 of
            *DIMACS Series in Discrete Mathematics and Theoretical Com-
            puter Science*, pages 531–554. AMS Press, 1991. (ref: p. 138)

[Sib81]   R. Sibson. A brief description of natural neighbour interpola-
            tion. In V. Barnet, editor, *Interpreting Multivariate Data*, pages
            21–36. John Wiley & Sons, Chichester, 1981. (ref: p. 122)

[Sie89]   C. L. Siegel. *Lectures on the Geometry of Numbers. Notes by B.
            Friedman. Rewritten by Komaravolu Chandrasekharan with the
            assistance of Rudolf Suter*. Springer-Verlag, Berlin etc., 1989.
            (ref: p. 20)

[SST84]   J. Spencer, E. Szemerédi, and W. T. Trotter. Unit distances
            in the Euclidean plane. In B. Bollobás, editor, *Graph Theory
            and Combinatorics*, pages 293–303. Academic Press, New York,
            NY, 1984. (ref: p. 45)

[SST01]   M. Sharir, S. Smorodinsky, and G. Tardos.   An improved
            bound for k-sets in three dimensions. *Discrete Comput. Geom.*,
            26:195–204, 2001. (refs: pp. 270, 286)

[ST74]   V. N. Sudakov and B. S. Tsirel'son. Extremal properties of half-spaces for spherically invariant measures (in Russian). *Zap. Naučn. Sem. Leningrad. Otdel. Mat. Inst. Steklov. (LOMI)*, 41:14–24, 1974. Translation in *J. Soviet. Math.* 9:9–18, 1978. (ref: p. 336)

[ST83]   E. Szemerédi and W. Trotter, Jr. A combinatorial distinction between Euclidean and projective planes. *European J. Combin.*, 4:385–394, 1983.   (ref: p. 44)

[ST01]   J. Solymosi and Cs. Tóth. Distinct distances in the plane. *Discrete Comput. Geom.*, 25:629–634, 2001.   (refs: pp. 45, 61)

[Sta75]   R. Stanley. The upper-bound conjecture and Cohen–Macaulay rings. *Stud. Appl. Math.*, 54:135–142, 1975.   (ref: p. 104)

[Sta80]   R. Stanley. The number of faces of a simplical convex polytope. *Adv. Math.*, 35:236–238, 1980.   (ref: p. 105)

[Sta86]   R. P. Stanley. Two poset polytopes. *Discrete Comput. Geom.*, 1:9–23, 1986.   (ref: p. 309)

[Ste26]   J. Steiner. Einige Gesetze über die Theilung der Ebene und des Raumes. *J. Reine Angew. Math.*, 1:349–364, 1826.   (ref: p. 128)

[Ste16]   E. Steinitz. Bedingt konvergente Reihen und konvexe Systeme I; II; III. *J. Reine Angew. Math*, 143; 144; 146:128–175; 1–40; 1–52, 1913; 1914; 1916.   (ref: p. 8)

[Ste22]   E. Steinitz. Polyeder und Raumeinteilungen. *Enzykl. Math. Wiss.*, 3:1–139, 1922. Part 3AB12.   (ref: p. 92)

[Ste85]   H. Steinlein. Borsuk's antipodal theorem and its generalizations and applications: a survey. In A. Granas, editor, *Méthodes topologiques en analyse nonlinéaire*, pages 166–235. Colloq. Sémin. Math. Super., Semin. Sci. OTAN (NATO Advanced Study Institute) 95, Univ. de Montréal Press, Montréal, 1985. (ref: p. 16)

[SU00]   J.-R. Sack and J. Urrutia, editors. *Handbook of Computational Geometry*. North-Holland, Amsterdam, 2000.   (refs: pp. viii, 162)

[SV94]   O. Sýkora and I. Vrťo. On VLSI layouts of the star graph and related networks. *Integration, The VLSI Journal*, 17(1):83–93, 1994.   (ref: p. 57)

[SW01]   M. Sharir and E. Welzl. Balanced lines, halving triangles, and the generalized lower bound theorem. In *Proc. 17th Annu. ACM Sympos. Comput. Geom.*, pages 315–318, 2001.   (refs: pp. 280, 281)

[SY93]   J. R. Sangwine-Yager. Mixed volumes. In P. M. Gruber and
           J. M. Wills, editors, *Handbook of Convex Geometry (Vol. A)*,
           pages 43–71. North-Holland, Amsterdam, 1993. (refs: pp. 300,
           301)

[Syl93]   J. J. Sylvester. Mathematical question 11851. *Educational
           Times*, 59:98, 1893. (ref: p. 44)

[Sze74]   E. Szemerédi. On a problem of Davenport and Schinzel. *Acta
           Arithmetica*, 25:213–224, 1974. (ref: p. 175)

[Sze78]   E. Szemerédi. Regular partitions of graphs. In *Problèmes combi-
           natoires et théorie des graphes, Orsay 1976, Colloq. int. CNRS
           No.260*, pages 399–401. CNRS, Paris, 1978. (ref: p. 226)

[Szé97]   L. Székely. Crossing numbers and hard Erdős problems in dis-
           crete geometry. *Combinatorics, Probability, and Computing*,
           6:353–358, 1997. (refs: pp. 44, 45, 56, 61)

[Tag96]   B. Tagansky. A new technique for analyzing substructures in
           arrangements of piecewise linear surfaces. *Discrete Comput.
           Geom.*, 16:455–479, 1996. (ref: p. 186)

[Tal93]   G. Talenti. The standard isoperimetric theorem. In P. M. Gru-
           ber and J. M. Wills, editors, *Handbook of Convex Geometry
           (Vol. A)*, pages 73–123. North-Holland, Amsterdam, 1993. (ref:
           p. 336)

[Tal95]   M. Talagrand. Concentration of measure and isoperimetric in-
           equalities in product spaces. *Publ. Math. I.H.E.S.*, 81:73–205,
           1995. (ref: p. 336)

[Tam88]   A. Tamir. Improved complexity bounds for center location
           problems on networks by using dynamic data structures. *SIAM
           J. Discr. Math.*, 1:377–396, 1988. (ref: p. 169)

[Tan84]   M. R. Tanner. Explicit concentrators from generalized n-gons.
           SIAM J. Alg. Discr. Methods, 5(3):287–293, 1984. (ref: p. 381)

[Tar75]   R. E. Tarjan. Efficiency of a good but not linear set union
           algorithm. *J. ACM*, 22:215–225, 1975. (ref: p. 175)

[Tar95]   G. Tardos. Transversals of 2-intervals, a topological approach.
           Combinatorica, 15:123–134, 1995. (ref: p. 262)

[Tar01]   G. Tardos. On distinct sums and distinct distances. Manuscript,
           Rényi Institute, Budapest, 2001. (refs: pp. 45, 61, 63)

[Tho65]   R. Thom. On the homology of real algebraic varieties (in
           French). In S.S. Cairns, editor, *Differential and Combinato-
           rial Topology*. Princeton Univ. Press, 1965. (ref: p. 135)

[Tit59]   J. Tits. Sur la trialité et certains groupes qui s'en déduisent. *Publ. Math. I. H. E. S.*, 2:13–60, 1959. (ref: p. 367)

[TJ89]   N. Tomczak-Jaegermann. *Banach-Mazur Distances and Finite-Dimensional Operator Ideals*. Pitman Monographs and Surveys in Pure and Applied Mathematics 38. J. Wiley, New York, 1989. (refs: pp. 327, 353)

[Tót65]   L. Fejes Tóth. *Regular Figures (in German)*. Akadémiai Kiadó Budapest, 1965. (ref: p. 322)

[Tót01a]   Cs. Tóth. The Szemerédi–Trotter theorem in the complex plane. *Combinatorica*, 2001. To appear. (ref: p. 44)

[Tót01b]   G. Tóth. Point sets with many k-sets. *Discrete Comput. Geom.*, 26:187–194, 2001. (refs: pp. 269, 276)

[Tro92]   W. T. Trotter. *Combinatorics and Partially Ordered Sets: Dimension Theory*. Johns Hopkins Series in the Mathematical Sciences. The Johns Hopkins University Press, 1992. (ref: p. 308)

[Tro95]   W. T. Trotter. Partially ordered sets. In R. L. Graham, M. Grötschel, and L. Lovász, editors, *Handbook of Combinatorics*, pages 433–480. North-Holland, Amsterdam, 1995. (ref: p. 308)

[TT98]   H. Tamaki and T. Tokuyama. How to cut pseudo-parabolas into segments. *Discrete Comput. Geom.*, 19:265–290, 1998. (refs: pp. 70, 270)

[Tut60]   W. T. Tutte. Convex representations of graphs. *Proc. London Math. Soc.*, 10(38):304–320, 1960. (ref: p. 92)

[TV93]   H. Tverberg and S. Vrećica. On generalizations of Radon's theorem and the ham sandwich theorem. *European J. Combin.*, 14:259–264, 1993. (ref: p. 204)

[TV98]   G. Tóth and P. Valtr. Note on the Erdős–Szekeres theorem. *Discrete Comput. Geom.*, 19(3):457–459, 1998. (ref: p. 33)

[Tve66]   H. Tverberg. A generalization of Radon's theorem. *J. London Math. Soc.*, 41:123–128, 1966. (ref: p. 203)

[Tve81]   H. Tverberg. A generalization of Radon's theorem II. *Bull. Aust. Math. Soc.*, 24:321–325, 1981. (ref: p. 204)

[Urr00]   J. Urrutia. Art gallery and illumination problems. In J.-R. Sack and J. Urrutia, editors, *Handbook of Computational Geometry*, pages 973–1027. North-Holland, 2000. (ref: p. 250)

[Val92a]   P. Valtr. Convex independent sets and 7-holes in restricted
            planar point sets. *Discrete Comput. Geom.*, 7:135–152, 1992.
            (refs: pp. 33, 37)

[Val92b]   P. Valtr. Sets in R^d with no large empty convex subsets. *Dis-
            crete Appl. Math.*, 108:115–124, 1992. (ref: p. 37)

[Val94]   P. Valtr. Planar point sets with bounded ratios of distances.
            Doctoral Thesis, Mathematik, FU Berlin, 1994. (ref: p. 34)

[Val98]   P. Valtr. Guarding galleries where no point sees a small area.
            Israel J. Math, 104:1–16, 1998. (ref: p. 250)

[Val99a]   P. Valtr. Generalizations of Davenport–Schinzel sequences.
            In R. Graham et al., editors, *Contemporary Trends in Dis-
            crete Mathematics*, volume 49 of *DIMACS Series in Discrete
            Mathematics and Theoretical Computer Science*, pages 349–
            389. Amer. Math. Soc., Providence, RI, 1999. (refs: pp. 176,
            177)

[Val99b]   P. Valtr. On galleries with no bad points. *Discrete and Com-
            putational Geometry*, 21:193–200, 1999. (ref: p. 250)

[Val01]   P. Valtr. A sufficient condition for the existence of large empty
            convex polygons. *Discrete Comput. Geom.*, 2001. To appear.
            (ref: p. 38)

[VC71]   V. N. Vapnik and A. Ya. Chervonenkis. On the uniform con-
            vergence of relative frequencies of events to their probabilities.
            Theory Probab. Appl., 16:264–280, 1971. (refs: pp. 242, 243)

[Vem98]   S. Vempala. Random projection: a new approach to VLSI lay-
            out. In *Proc. 39th IEEE Symposium on Foundations of Com-
            puter Science*, pages 389–395, 1998. (ref: p. 397)

[Vin39]   P. Vincensini. Sur une extension d'un théorème de M. J. Radon
            sur les ensembles de corps convexes. *Bull. Soc. Math. France*,
            67:115–119, 1939. (ref: p. 12)

[Vor08]   G. M. Voronoi. Nouvelles applications des paramètres conti-
            nus à la théorie des formes quadratiques. deuxième Mémoire:
            Recherches sur les parallélloèdres primitifs. *J. Reine Angew.
            Math.*, 134:198–287, 1908. (ref: p. 120)

[VŽ93]   A. Vućić and R. Živaljević. Note on a conjecture of Sierksma.
            Discrete Comput. Geom, 9:339–349, 1993. (ref: p. 205)

[Wag01]   U. Wagner. On the number of corner cuts. *Adv. Appl. Math.*,
            2001. In press. (ref: p. 271)

[War68]   H. E. Warren. Lower bound for approximation by nonlinear
            manifolds. *Trans. Amer. Math. Soc.*, 133:167–178, 1968. (ref:
            p. 135)

[Weg75]   G. Wegner. *d*-collapsing and nerves of families of convex sets.
            Arch. Math., 26:317–321, 1975. (ref: p. 197)

[Wel86]   E. Welzl. More on *k*-sets of finite sets in the plane. *Discrete
            Comput. Geom.*, 1:95–100, 1986. (ref: p. 270)

[Wel88]   E. Welzl. Partition trees for triangle counting and other range
            searching problems. In *Proc. 4th Annu. ACM Sympos. Comput.
            Geom.*, pages 23–33, 1988. (ref: p. 242)

[Wel01]   E. Welzl. Entering and leaving *j*-facets. *Discrete Comput.
            Geom.*, 25:351–364, 2001. (refs: pp. 104, 145, 280, 282)

[Wil99]   A. J. Wilkie. A theorem of the complement and some new o-
            minimal structures. *Sel. Math., New Ser.*, 5(4):397–421, 1999.
            (ref: p. 250)

[Wol97]   T. Wolff. A Kakeya-type problem for circles. *Amer. J. Math.*,
            119(5):985–1026, 1997. (ref: p. 44)

[WS88]   A. Wiernik and M. Sharir. Planar realizations of nonlinear
            Davenport-Schinzel sequences by segments. *Discrete Comput.
            Geom.*, 3:15–47, 1988. (refs: pp. 173, 176)

[WW93]   W. Weil and J. A. Wieacker. Stochastic geometry. In P. M.
            Gruber and J. M. Wills, editors, *Handbook of Convex Geometry
            (Vol. B)*, pages 391–1438. North-Holland, Amsterdam, 1993.
            (ref: p. 99)

[WW01]   U. Wagner and E. Welzl. A continuous analogue of the upper
            bound theorem. *Discrete Comput. Geom.*, 26:205–219, 2001.
            (ref: p. 114)

[Zas75]   T. Zaslavsky. *Facing up to Arrangements: Face-Count Formulas
            for Partitions of Space by Hyperplanes*, volume 154 of *Memoirs
            Amer. Math. Soc.* American Mathematical Society, Providence,
            RI, 1975. (ref: p. 128)

[Zie94]   G. M. Ziegler. *Lectures on Polytopes*, volume 152 of *Graduate
            Texts in Mathematics*. Springer-Verlag, Heidelberg, 1994. Cor-
            rected and revised printing 1998. (refs: pp. viii, 78, 85, 86, 89,
            90, 92, 93, 103, 105, 114, 129, 137)

[Živ97]   R. T. Živaljević. Topological methods. In J. E. Goodman and
            J. O'Rourke, editors, *Handbook of Discrete and Computational
            Geometry*, chapter 11, pages 209–224. CRC Press LLC, Boca
            Raton, FL, 1997. (ref: p. 368)

[Živ98]   R. T. Živaljević. User's guide to equivariant methods in combinatorics. II. *Publ. Inst. Math. (Beograd) (N.S.)*, 64(78):107–132, 1998. (ref: p. 205)

[ŽV90]   R.T. Živaljević and S.T. Vrećica. An extension of the ham sandwich theorem. *Bull. London Math. Soc.*, 22:183–186, 1990. (ref: p. 16)

[ŽV92]   Živaljević and S. Vrećica. The colored Tverberg's problem and complexes of injective functions. *J. Combin. Theory Ser. A*, 61:309–318, 1992. (ref: p. 205)

Index

The index starts with notation composed of special symbols, and Greek letters are listed next. Terms consisting of more than one word mostly appear in several variants, for example, both "convex set" and "set, convex." An entry like "armadillo, 19(8.4.1), 22(Ex. 4)" means that the term is located in theorem (or definition, etc.) 8.4.1 on page 19 and in Exercise 4 on page 22. For many terms, only the page with the term's definition is shown. Names or notation used only within a single proof or remark are usually not indexed at all. For important theorems, the index also points to the pages where they are applied.

Graduate Texts in Mathematics

(continued from page ii)